# Micro and Smart Systems

# Micro and Smart Systems

## Technology and Modeling

**G.K. ANANTHASURESH**
**K.J. VINOY**
**S. GOPALAKRISHNAN**
**K.N. BHAT**
**V.K. AATRE**

*Indian Institute of Science*
*Bangalore INDIA*

John Wiley & Sons, Inc.

| | |
|---|---|
| VP AND EXECUTIVE PUBLISHER | Don Fowley |
| ASSISTANT PUBLISHER | Daniel Sayre |
| SENIOR EDITORIAL ASSISTANT | Katie Singleton |
| EXECUTIVE MARKETING MANAGER | Christopher Ruel |
| MARKETING ASSISTANT | Ashley Tomeck |
| SENIOR PRODUCTION MANAGER | Janis Soo |
| ASSISTANT PRODUCTION EDITOR | Elaine S. Chew |
| EXECUTIVE MEDIA EDITOR | Tom Kulesa |
| MEDIA EDITOR | Wendy Ashenberg |
| MEDIA SPECIALIST | Jennifer Mullin |
| COVER DESIGNER | Wendy Lai |
| COVER IMAGE | Sambuddha Khan |

This book was set in Times by Thomson Digital, Noida, India and printed and bound by Courier Westford, Inc. The cover was printed by Courier Westford, Inc.

This book is printed on acid free paper. ∞

Founded in 1807, John Wiley & Sons, Inc. has been a valued source of knowledge and understanding for more than 200 years, helping people around the world meet their needs and fulfill their aspirations. Our company is built on a foundation of principles that include responsibility to the communities we serve and where we live and work. In 2008, we launched a Corporate Citizenship Initiative, a global effort to address the environmental, social, economic, and ethical challenges we face in our business. Among the issues we are addressing are carbon impact, paper specifications and procurement, ethical conduct within our business and among our vendors, and community and charitable support. For more information, please visit our website: www.wiley.com/go/citizenship.

Evaluation copies are provided to qualified academics and professionals for review purposes only, for use in their courses during the next academic year. These copies are licensed and may not be sold or transferred to a third party. Upon completion of the review period, please return the evaluation copy to Wiley. Return instructions and a free of charge return mailing label are available at www.wiley.com/go/returnlabel. If you have chosen to adopt this textbook for use in your course, please accept this book as your complimentary desk copy. Outside of the United States, please contact your local sales representative.

*Library of Congress Cataloging-in-Publication Data*

Micro and smart systems / G.K. Ananthasuresh . . . [et al.].
    p. cm.
  Includes bibliographical references and index.
  ISBN 978-0-470-91939-2 (acid-free paper)
  1.  Microelectromechanical systems—Design and construction. 2.  Intelligent control systems—Design and construction.  I. Ananthasuresh, G. K.
TK7875.M524 2012
621.381—dc23
                                                      2011029301

Printed in the United States of America
10 9 8 7 6 5 4 3 2 1

*Dedicated to*
*The Institute of Smart Structures and Systems (ISSS)*
*without whose initiative and support this book would not have materialized*

# Preface

If we trace the history of electronics technology over the last six decades, we see that the discovery of the transistor and the development of the integrated circuit (IC) are the key milestones. However, it is miniaturization and the ensuing very-large-scale-integration (VLSI) technologies that really created the electronics and computer revolutions. It is only more recently, within the last couple of decades, that the technology of miniaturization has been extended to mechanical devices and systems; we now have the microelectromechanical system (MEMS) revolution. Complemented by the advances in smart materials, this has led to highly application-oriented microsystems and smart systems.

A microsystem is a system that integrates, on a chip or in a package, one or more of many microdevices: sensors, actuators, electronics, computation, communication, control, power generation, chemical processing, biological reactions, etc. It is now clear that the functionality of such an integrated system will not only be far superior to any other engineered system that we know at the macroscale but will also be able to achieve things well beyond what macroscale integrated systems can do. Smart microelectromechanical systems are collections of microsensors and actuators that can sense their environment and can respond intelligently to changes in that environment by using microcircuit controls. Such microsystems include, in addition to the conventional microelectronics, packaging, integrated antenna structures for command signals, and microelectromechanical structures for desired sensing and actuating functions.

However, micromachined actuators may not be powerful enough to respond to the environment. Using macroscale actuators would defeat the purpose of miniaturization, cost-effective batch-processing, etc. Hence, there is a need to integrate smart material-based actuators with microsystems. This trend is currently being witnessed as this field moves beyond microsensors, which have been the main emphasis in microsystems so far.

Microsystems and smart system technologies have immense application potential in many fields, and in the coming decades, scientists and engineers will be required to design and develop such systems for a variety of applications. It is essential, then, that graduating engineers be exposed to the underlying science and technology of microsystems and smart systems. There are numerous books that cover both microsystems and smart systems separately and a few that cover both. Many of them are suitable for practicing professionals or for advanced-level courses. However, they assume certain fundamentals in various topics of this multidisciplinary field and thus serve the function of a reference book rather than a primary textbook. Many do not emphasize modeling at the fundamental level necessary to be useful at the undergraduate level or for self-study by a reader with background in other disciplines.

This book essentially deals with the basics of microsystem technology and is intended principally as a textbook at the undergraduate level; it can also be used as a background book at the postgraduate level. The book provides an introduction to smart materials and systems. We have tried to present the material without assuming much prior disciplinary background. The aim of this book is to present adequate modeling details so that readers can appreciate the analysis involved in microsystems (and to some extent, smart systems), thereby giving them an in-depth understanding about simulation and design. Therefore, the book will also be useful to practicing researchers in all branches of science and engineering who are interested in applications where they can use this technology. The book presents adequate details on modeling of microsystems and also addresses their fabrication and integration. The engineering of practical applications of microsystems provides areas for multidisciplinary research, already laden with myriad technological issues, and books presently available do not address many of these aspects in sufficient depth. We believe that this book gives a unified treatment of the necessary concepts under a single title.

Anticipating the need for such a technology, the Institute of Smart Structures and Systems (ISSS), an organization dedicated to promoting smart materials and microsystems, was established. This Institute was instrumental not only in mounting a national program and triggering R&D activities in this field in India, but also in creating the required human resources through training courses and workshops. Furthermore, ISSS also initiated a dialogue with Visvesveraya Technological University (VTU), Belgaum, Karnataka, a conglomerate of over 170 Karnataka engineering colleges, to introduce an undergraduate-level course in microsystems and MEMS and to set in motion the creation of a potential syllabus for this course. The culmination of this dialogue is the present book. The material for this book has been taken from several advanced workshops and short courses conducted by the authors over last few years for faculty and students of VTU. A preliminary version of book was used at VTU colleges, where a course on microsystems was first introduced in 2009, and very helpful feedback was received from teachers of this course, who patiently used the draft to teach about 500 students at various colleges. In a sense this book has been class- and student-tested and is a substantially enhanced version of the original draft.

This book has ten chapters covering various topics in microsystems and smart systems including sensors and actuators, microfabrication, modeling, finite-element analysis, modeling and analysis of coupled systems (of great importance in microsystems), electronics and control for microsystems, integration and packaging, and scaling effects in microsystems. The book also includes case studies on a few microsensor systems to illustrate the applications aspects.

In the authors' opinion, the material of the book can be covered in a standard undergraduate one-semester course. The content of Chapters 5 and 6 may be considered optional. The entire book can be covered in a single-semester postgraduate course or a two-semester undergraduate course supplemented by a design and case-study oriented laboratory.

# Acknowledgments

Any project like writing a book depends on the help, advice, and consent of a large number of people. The authors have received such help from many people and it is a pleasure to acknowledge all of them. The trigger for the book was provided by the initiative taken by the Institute of Smart Structures and Systems. The authors acknowledge their indebtedness to ISSS and its presidents, and indeed dedicate the book to ISSS. The authors would like to specially thank Prof. S. Mohan of Indian Institute of Science and Dr. A.R. Upadhya, Director, National Aerospace Laboratories, for their support and encouragement.

While ISSS initiated the writing of the book, it was the support and enthusiasm of the Visvesvaraya Technological University (VTU) that sustained its writing. The authors gratefully acknowledge this support, especially from the former vice-chancellors, Prof. K. Balaveer Reddy and Prof. H.P. Khincha.

A number of VTU faculty and students who attended the workshops based on the preliminary versions of the book provided the all-important feedback necessary to finalize the book. While thanking them all, the authors would like to mention in particular Prof. Premila Manohar (MSR Institute of Technology, Bangalore) and Prof. K. Venkatesh (presently at Jain University, Bangalore). In addition, the significant contributions of Prof. N.G. Kurahatti (presently at East Point College of Engineering and Technology, Bangalore), who compiled part of the contents of Chapter 3 during the initial stages of manuscript preparation, are gratefully acknowledged. The writing of the book would not have been possible without the work put in by several of our post-graduate students. Contributions of P.V, Aman, Santosh Bhargav, A.V. Harikrishnan, Shyamsananth Madhavan, Ipe Mathew, Rizuwana Parveen, Pakeeruraju Podugu, and Jayaprakash Reddy, who collected much of the information presented in Chapter 2, are gratefully appreciated. Also acknowledged for their help are: Subhajit Banerjee, Varun Bollapragada, Vivek Jayabalan, Shymasananth Madhavan, Fatih Mert Ozkeskin, Krishna Pavan, Sudhanshu Shekhar, and Puneet Singh who ran simulations and provided material for Chapter 10. Assistance given by M. S. Deepika and R. Manoj Kumar in creating some of the illustrations is also gratefully acknowledged. The authors thank all their students who read the early manuscripts of this book and provided useful feedback. The credit for the cover image goes to Sambuddha Khan. The image shows a part of the bulk-micromachined accelerometer with a mechanical amplifier developed by him as part of his PhD dissertation.

# A Note to the Reader

Most chapters include worked-out examples, problems given within the text, and end-of-the-chapter exercises.

Some chapters also include exploratory questions marked as ''Your Turn''. They urge the reader to think beyond the scope of the book. They are intended to stimulate the interest of the reader.

Acronyms and notation used in the book are included in separate lists at the beginning of the book. Additionally, a glossary of important terms appears at the end of the book.

An Appendix that appears at the end of the book provides supplementary material for the convenience of the reader.

Typographical oversights, technical mistakes, or any other discrepancies may please be brought to the attention of the authors by sending e-mail to: suresh@mecheng.iisc.ernet.in.

# Acronyms

**μBGA**  Microball-grid array

**μCP**  Microcontact printing

**μTM**  Microtransfer molding

**μTAS**  Micro-total analysis system

**1D**  One-dimensional

**2D**  Two-dimensional

**3D**  Three-dimensional

**ADC or A/D**  Analog-to-digital converter

**AFM**  Atomic force microscopy

**APCVD**  Atmospheric pressure chemical vapor deposition

**ASIC**  Application-specific integrated circuits

**ASIC**  Application-specific-integrated circuit

**BEM**  Boundary element method

**BGA**  Ball-grid array

**BiCMOS**  Bipolar CMOS

**bio-MEMS**  bio-microelectromechanical systems

**BJT**  Bipolar junction transistor

**BSG**  Borosilicate glass

**BST**  Barium strontium titanate

**BW**  Bandwidth

**CMOS**  Complementary metal-oxide-semiconductor

**CMP**  Chemical–mechanical planarization

**CMRR**  Common-mode rejection ratio

**COC**  Cyclic olefin copolymer

**COF**  Chip-on-flex

**CPD**  Critical point drying

**CRT**  Cathode ray tube

**CTE**  Coefficient of thermal expansion

**CVD**  Chemical vapor deposition

**DAC or D/A**  Digital-to-analog converter

**DFT**  Discrete Fourier transform

**DLC**  Diamond-like carbon

**DLP**  Digital light processor

**DMD**  Digital Mirror Device

**DoD**  Drop-on-demand

**DOF**  Degree of freedom

**DRIE**  Deep reactive-ion etching

**DSP**  Digital signal processing

**EDP**  Ethylene diamine pyrocatechol

**EDM**  Electrical discharge machining

**EEPROM**  Erasable programmable read-only memory

**EGS**  Electronic grade silicon

**EMI**  Electromagnetic interference

**ER**  Electrorheological

**ETC**  Electro-thermal-compliant

**ER**  Electro rheological

**FBG**  Fiber Bragg grating

**FCC**  Face-centered cubic

**FCP**  Few-chip package

**FDM**  Finite difference method

**FE**  Finite element

**FEA**  Finite element analysis

**FEM**  Finite element method

**FFT**  Fast Fourier transform

**FPI**  Fabry Perot interferometer

**HDTV**  High-definition television

**HF**  Hydrofluoric

**HVAC**  Heat, ventilation, and air-conditioning

**I/O**  Input/output

**IBE**  Ion beam etching

**IC**  Integrated circuit

**ICP**  Intracranial pressure

**IF**  Intermediate frequency

**IR**  Infrared

**ISR**  Interrupt service routine

**LCD**  Liquid crystal display

**LED**  Light-emitting diode

**LIGA** Lithographie Galvanoformung & Abformung

**LPCVD** Low-pressure chemical vapor deposition

**LPF** Low-pass filter

**LSB** Least significant bit

**LTCC** Low-temperature cofired ceramics

**LVDT** Linear variable differential transformer

**MAP** Manifold-absolute pressure

**MBE** Molecular beam epitaxy

**MCM** Multichip module

**MCM-D** Multichip module – deposited

**MEMS** Microelectromechanical systems

**MGS** Metallurgical grade silicon

**Micro-EDM** Microelectrical discharge machining

**MMF** Magneto motive force

**MMIC** Monolithic microwave integrated circuits

**MOCVD** Metal-organic chemical vapor deposition

**MOEMS** Micro-opto-electromechanical systems

**MOS** Metal-oxide-semiconductor

**MOSFET** Metal oxide semiconductor field-effect transistor

**MR** Magnetorheological

**MS** Metal semiconductor

**nMOS** n-channel MOSFETs

**pMOS** p-channel MOSFET

**ODE** Ordinary differential equation

**Op-amp** Operational amplifier

**PCB** Printed circuit board

**PCR** polymerase chain reaction

**PDE** Partial differential equation

**PBGA** Plastic-ball-grid array

**PDMS** Polydimethylsiloxane

**PECVD** Plasma-enhanced chemical vapor deposition

**PFC** Piezofiber composite

**PHET** Photovoltaic electrochemical etch-stop technique

**PID** Proportional-integral-derivative

**PLC** Programmable logic controller

**PLL** Phase-locked loop

**PMMA** Polymethyl methacrylate

**pMOS** p-channel MOSFETs

**PMPE** Principle of minimum potential energy

**PSG** Phosphosilicate glass

**PTFE** Polytetra-fluoroethylene

**PVC** Polyvinyl chloride

**PVD** Physical vapor deposition

**PVDF** Polyvinylidene fluoride

**PVW** Principle of virtual work

**PZT** Lead zirconate titanate

**R&D** Research and development

**RF** Radio frequency

**RIE** Reactive-ion etching

**RTA** Rapid thermal annealing

**SAC** Successive-approximation converter

**SAW** Surface acoustic wave

**SCS** Single-crystal silicon

**SEM** Scanning electron microscope

**SFB** Silicon fusion bonding

**SFEM** Spectral FEM

**SI unit** International standard unit

**SIP** System-on-a-chip

**SISO** Single-input–single-output

**SMA** Shape-memory-alloy

**SNR** Signal-to-noise ratio

**SOI** Silicon-on-insulator

**SOP** System-on-a-package

**sPROMs** Structurally programmable microfluidic system

**SuMMiT** Santia ultra multi-layer microfabrication technology

**TCR** Temperature coefficient of resistivity

**UV** Ultraviolet

**VCO** Voltage-controlled oscillator

**VED** Vacuum electron devices

**VLSI** Very large-scale integration

**VPE** Vapor phase epitaxy

**WRT** Weighted residual technique

**XFEM** Extended FEM

# Notation

There are 26 letters in the English alphabet and 24 in the Greek alphabet. Using both lower and upper case, we have 100 symbols to denote various quantities. Traditionally, every discipline reserves certain symbols for certain quantities. When we mix disciplines, as happens in interdisciplinary subjects, there are bound to be clashes: the same symbol is used for different quantities in different disciplines (e.g., $R$ for reaction force in mechanics and resistance in electronics). We have made an effort to minimize the overlap of such quantities when they are used in the same chapter. As a result, we use nontraditional symbols for certain quantities. For example, $Y$ is used for Young's modulus instead of $E$ since $E$ is used for the magnitude of the electric field, because they both appear in the same chapter.

Occasionally, we also use subscripts to relate a certain symbol to a discipline (e.g., $k_{th}$ for thermal conductivity). Boldface symbols are used for vectors (e.g., $\mathbf{E}$ for the electric field vector).

The symbols in the list below are arranged in this order: upper-case English, lower-case English, upper-case Greek, and lower-case Greek, all in alphabetical order. For each symbol, boldface symbols appear first and symbols with subscripts or superscripts appear afterwards. If the same symbol is used in two different disciplines, the descriptions are separated by ''OR;'' if the same symbol is used within the same discipline, ''or'' is used.

| | | | |
|---|---|---|---|
| $A$ | Cross-sectional area of a bar or beam OR area of a parallel-plate capacitor or a proof mass | $G$ | Gauge factor |
| $\mathbf{B}$ | Magnetic flux density vector | $\mathbf{H}$ | Magnetic field |
| $A_0$ | Difference mode gain | $I$ | Inertia in general or area of moment of inertia of a beam OR electric current |
| $A_C$ | Common mode gain | $I_{c0}$ | Reverse saturation current |
| $Bo$ | Bond number | $K_d$ | Derivative controller gain |
| $C$ | Capacitance OR a constant | $K_P$ | Proportional controller gain |
| $C_{ox}$ | Gate oxide capacitance per unit area | $K_I$ | Integral controller gain |
| $D$ | Coil diameter of a helical spring OR magnitude of the electric displacement vector OR diffusion constant | $J$ | Polar moment of inertia OR the magnitude of the electric current density |
| | | $K$ | Bulk modulus of a material |
| $D_n$ | Normal component of the electric displacement vector | $K_B$ | Boltzmann constant |
| | | $Kn$ | Knudsen number |
| $DE$ | Dissipated energy | $KE$ | Kinetic energy |
| $\mathbf{E}$ | Electric field vector | $L$ | Length or size OR inductance OR Lagrangian |
| $E_n$ | Normal component of the electric field | | |
| $ESE$ | Electrostatic energy | $\mathbf{M}$ | Magnetization |
| $ESE_c$ | Electrostatic complementary energy | $M$ | Bending moment in a beam or a column OR mass of the proof mass |
| $F$ | Force; occasionally, also the transverse force on a beam or a point force on a body | $MSE_c$ | Magnetostatic coenergy |
| | | $n$ | Number of turns |
| $F_e$ | Electrostatic force | $N_A$ | Acceptor dopant concentration |
| $F_d$ | Damping force | $N_D$ | Donor dopant concentration |

| | | | |
|---|---|---|---|
| $N_o$ | Intrinsic concentration | **t** | Surface force on an elastic body (also called *traction*) |
| $P$ | Axial force in a bar or a beam or a point force on a body OR magnitude of an electric polarization vector | $\mathbf{t}_e$ | Electrostatic force on a conductor |
| PE | Potential energy | $t$ | Time OR thickness OR force (traction) on a surface |
| $Q$ | Electric charge | $u$ | Displacement in general or only $x$-displacement in 2D or 3D objects |
| $R$ | Reaction force OR electrical resistance | | |
| $Re$ | Reynolds number | $v$ | $y$-displacement in 2D or 3D objects |
| $S$ | Sensitivity | $w$ | Width OR transverse displacement of a beam or $z$-displacement in a 3D object |
| $SE$ | Strain energy | | |
| $SE_c$ | Complementary strain energy | $\Delta$ | Deflection OR an increment in a quantity (if followed by another symbol) |
| $T$ | Torque OR temperature | | |
| **U** | Velocity vector | $\alpha$ | Coefficient of thermal expansion OR a constant of proportionality |
| $V$ | Volume OR vertical shear force OR voltage | | |
| $W$ | Work | $\chi_e$ | Electrical susceptibility |
| $V_{\text{th}}$ | Threshold voltage | $\delta$ | Deflection OR an increment in a quantity (if followed by another symbol) |
| $V_{\text{bi}}$ | Built in potential | | |
| $Y$ | Young's modulus, a material property usually denoted by $a$ in mechanics; we use $a$ because $a$ is used for the magnitude of the electric field and for acceleration | $\in$ | Normal strain |
| | | $\varepsilon$ | Permittivity |
| | | $\varepsilon_0$ | Permittivity of free space |
| | | $\varepsilon_r$ | Relative permittivity |
| $\hat{\mathbf{n}}$ | Unit vector usually normal to a surface and directed outward | $\eta$ | Viscosity |
| | | $\phi$ | Twist OR electric potential |
| $a$ | Acceleration | $\gamma$ | Shear strain OR surface tension |
| $b$ | Width of a beam or damping coefficient | $\kappa$ | Torsional spring constant |
| $d\mathbf{l}$ | A differential vector tangential to a path at a point | $\lambda$ | Wave length |
| | | $\mu$ | Permeability |
| $d\mathbf{s}$ | A differential vector normal to a surface | $\mu_n$ | electron mobility |
| $dV$ | Differential volume | $\mu_0$ | Permeability of free space |
| $g$ | Acceleration due to gravity OR gap in parallel-plate capacitor | $v$ | Poisson ratio |
| | | $\theta$ | Slope of a bent beam |
| $g_0$ | Initial gap in parallel-plate capacitor | $\rho$ | Radius of curvature of a straight beam that is bent |
| $h$ | Convective heat transfer coefficient | | |
| $i$ | Electric current | $\rho_e$ | Electrical resistivity |
| $k$ | Spring constant or stiffness in general | $\rho_m$ | Mass density |
| $k_{\text{B}}$ | Boltsmann constant | $\boldsymbol{\sigma}_e$ | Maxwell's stress tensor |
| $k_e$ | Electrical conductivity | $\sigma$ | Normal stress |
| $k_{th}$ | Thermal conductivity | $\tau$ | Shear stress or time constant |
| $l$ | Length | $\omega$ | Frequency of applied stimulus (force, voltage, etc.) |
| $p$ | Perimeter OR pressure | | |
| $q$ | Distributed transverse load OR electric charge | $\omega_n$ | Natural frequency (also called resonance frequency) |
| $q_e$ | Charge of electron | | |
| **r** | Position or distance vector | $\psi_L$ | Line charge density |
| $\hat{\mathbf{r}}$ | Unit vector in the direction of a position or distance vector | $\psi_s$ | Surface charge density |
| | | $\psi_v$ | Volumetric charge density |
| $se$ | Strain energy per unit volume | | |

# Contents

► **CHAPTER 5**

## The Finite Element Method 177

► **CHAPTER 6**

## Modeling of Coupled Electromechanical Systems 245

# Introduction

## LEARNING OBJECTIVES

After completing this chapter, you will be able to:

▶ Get an overview of microsystems and smart systems.

▶ Understand the need for miniaturization.

▶ Understand the role of microfabrication.

▶ Learn about smart materials and systems.

▶ Learn about typical applications of microsystems and smart systems.

Mythology and folk tales in all cultures have fascinating stories involving magic and miniaturization. Ali Baba, in a story in *1001 Arabian Nights*, had to say *Open Sesame* to make the cave door open by itself. We now have automatic doors in supermarkets that open as you move toward them without even uttering a word. Jonathan Swift's fictitious British hero, Lemuel Gulliver, traveled to the island of Lilliput and was amazed at the miniature world he saw there. Gulliver would probably be equally amazed if he were washed ashore into the 21$^{st}$-century world because we now have fabulous miniature marvels undreamt of in Swift's time. Magic and miniaturization are realities today. Arthur C. Clarke, a famous science fiction writer, once said that "a sufficiently advanced technology is indistinguishable from magic." What makes this magic a reality? The answer lies in exotic and smart materials, sensors and actuators, control and miniaturization. Smartness and smallness go hand in hand. Smart systems are increasingly becoming smaller, leading to a magical reality. Small systems are increasingly becoming smarter by integrating sensing, actuation, computation, communication, power generation, control, and more. Like the Indian mythological incarnation Vamana, described as being small and smart, yet able to cover the Earth, sky, and the world beneath in three footsteps, the combination of smallness and smartness has limitless possibilities. This book is about microsystems and smart systems, not about magic. As Nobel Laureate physicist Richard Feynman noted in his lectures, the laws of science as we know them today do not preclude miniaturization, and there is sufficient room at the bottom [1, 2]. It is only a question of developing the requisite technologies and putting them together to make it all happen. Let us begin with the need for miniaturization.

## ▶ 1.1 WHY MINIATURIZATION?

A microsystem is a system that integrates, on a chip or in a package, one or more of many things: sensors, actuators, electronics, computation, communication, control, power generation, chemical processing, biological reactions, and more. You may find it interesting that this definition does not explicitly mention size other than alluding to a chip or a packaged system consisting of chips and other accessories. Miniaturization is essential in achieving this level of integration of a disparate array of components.

There is no doubt that the functionality of such an integrated system will not only be far superior to any other engineered system that we know at the macroscale but will also achieve things beyond those achievable by macrosystems. Think of a big ship, an aircraft, or a power plant: they all serve one primary purpose. But a chip that integrates several components, as already mentioned, *can serve multiple functions*. This is one reason for miniaturization. This does not imply that we cannot integrate many things at the macro scale; it is a question of economy and to some extent functionality. Microsystems technology, by following the largely successful paradigm of microelectronics, remains *economical due to the batch production* of microfabrication processes. You can make plenty of things on one single silicon wafer, and thus, the cost per individual thing comes down drastically. This is another reason for miniaturization. Nothing would better illustrate this than advances in computer technology. Computing systems of today are much more powerful, have many more features, are far less power-consuming and, of course, are significantly cheaper than those available 20 or 30 years ago. Miniaturization and integration approaches have played a significant role in achieving these (see Figure 1.1). Common objects in various size scales are compared in Figure 1.2. In a lighter vein, a popular acronym for the microsystems technology is spelled M€M$: it reads millions of euros and millions of dollars! The market share of *m*icro*e*lectro*m*echanical *s*ystems (MEMS) has exceeded the million-dollar mark some time back and is well poised to cross the billion-dollar mark.

Energy has always been a precious commodity. Today it is increasingly so because of ever-increasing demand coupled with rapidly depleting energy resources. With the human propensity for electronic gadgets, the requirements for batteries are also on the rise, triggering research efforts on high-energy miniature batteries. And the smaller the gadgets (and constituent devices), the *lower* the *energy requirements*, thus adding a further reason for miniaturizing devices and systems.

There are, of course, more technical reasons for miniaturization. *Some phenomena favor miniaturization*. Take optics, for example. If we have a micromechanical device that

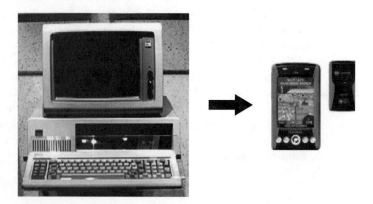

**Figure 1.1** Miniaturization of computer hardware technology.

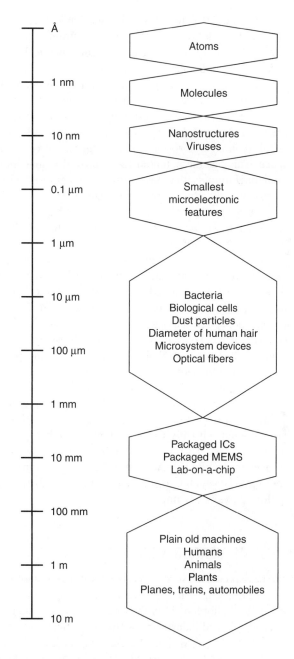

**Figure 1.2** Illustration of objects at various size scales.

can move a micron-sized component and control its movement to a fraction of a micron (the range of light wavelength visible to humans), this opens up numerous new possibilities. There are already commercial products that use optical microsystems, also known as micro-opto-electromechanical systems (MOEMS).

Think of the biological cells that are the basic building blocks of living organisms. They are the workshops where amazing manufacturing, assemblies and disassemblies take place most efficiently. These cells too have features, that is, size, motion, and forces, that are comparable to those of micromechanical structures. So, there is a subfield within microsystems known as "bio-microelectromechanical systems" (bio-MEMS). Furthermore, there are reasons for miniaturizing chemical processing. Controlling process conditions over a small volume is much easier than over a large volume. Hence, the *efficiency of a chemical reaction is greater* in miniaturized systems. It is clear from this perspective why living organisms are compartmentalized into micron-sized units—the cells.

Miniaturization can result in faster devices with *improved thermal management*. Energy and materials requirement during fabrication can be reduced significantly, thereby resulting in *cost/performance advantages*. Arrays of devices are possible within a small space. This has the potential for improved redundancy. Another important advantage of miniaturization is the possibility of integration of mechanical and fluidic parts with electronics, thereby simplifying systems and reducing power requirements. Microfabrication employed for realizing such devices has *improved reproducibility*, and devices thus produced have *increased selectivity and sensitivity*, *wider dynamic range*, *improved accuracy and reliability*.

Integrated microsystems are a collection of microsensors and actuators that can sense their environment and react to changes in that environment by use of a microcircuit control. Such microsystems include, in addition to the conventional microelectronics packaging, integrated antenna structures for command signals, and microelectromechanical configurations for desired sensing and actuating functions. The system may also need micropower supply, microrelay, and microsignal processing units. Such systems with microcomponents are faster, more reliable, more accurate, cheaper, and capable of incorporating more complex and versatile functions than systems used today.

A miniaturized low-power transceiver is an excellent example of a microsystem technology. Figure 1.3(a) shows a simplified block diagram of a transceiver and a board-level implementation of the same consisting of several chips that are basically components with high quality-factors such as radio frequency (RF) filters, surface acoustic wave (SAW) intermediate-frequency (IF) filters, crystal oscillators, and transistor circuits [3]. A possible single-chip implementation of this transceiver using microsystems technology is shown in Fig. 1.3(b). This enables miniaturization as well as low power consumption.

## ▶ 1.2 MICROSYSTEMS VERSUS MEMS

We have presented several reasons for miniaturization from different points of view. Let us now examine how microsystems technology came into existence. In the late 1960s and early 1970s, not long after the emergence of the integrated chip, researchers and inventors in academia and industry began to experiment with microfabrication processes by making movable mechanical elements. Accelerometers, micromirrors, gas-chromatography instruments, etc., were miniaturized during this period. Silicon was the material of choice at that time. A seminal paper by Petersen [4] summarizes these developments. Increased attention to the microsystems field came in late 1980s when a micromachined electrostatic motor was made at the University of California, Berkeley and Massachusetts Institute of Technology, Cambridge. This moving micromechanical entity fascinated everyone and showed the way forward for many other developments. The acronym MEMS was coined during that period. However, this acronym is inadequate today because of the numerous disciplines beyond just mechanical and electrical that have joined the league. A more suitable term is "microsystems;" hence the title for this book.

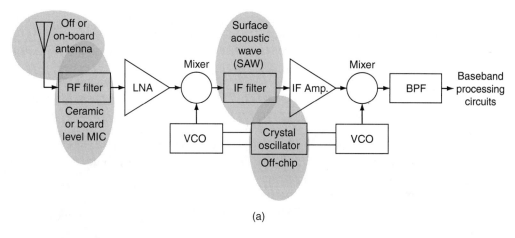

(a)

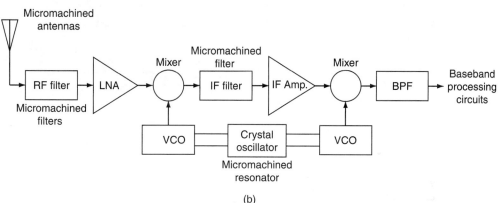

(b)

**Figure 1.3** Integrated radio frequency transceivers: a typical application of microsystems [3]. (*a*) Current approaches (shaded remarks indicate components outside the electronics die). (*b*) Integrated MEMS-based RF receiver on a single chip. LNA, low noise amplifier; RF, radio frequency; VCO, voltage-controlled oscillator; IF, intermediate frequency; BPF, bandpass filter.

## ▶ 1.3 WHY MICROFABRICATION?

Microsystems technology emerged as a new discipline based on the advances in integrated circuit (IC) fabrication processes, by which sensors, actuators and control functions were co-fabricated in silicon. The concepts and feasibility of more complex microsystems devices have been proposed and demonstrated for applications in such varied fields as microfluidics, aerospace, automobile, biomedical, chemical analysis, wireless communications, data storage, and optics.

In microsystems, miniaturization is achieved by a fabrication approach similar to that followed in ICs, commonly known as *micromachining*. As in ICs, much of the processing is done by chemical processing rather than mechanical modifications. Hence, *machining* here does not refer to conventional approaches (such as drilling, milling, etc.) used in realizing macromechanical parts, although the objective is to realize such parts. As with semiconductor processing in IC fabrication, micromachining has become the fundamental technology for the fabrication of microsystems devices and, in particular, miniaturized sensors and actuators. Silicon micromachining, the most mature of the micromachining technologies, allows the fabrication of microsystems that have dimensions in the sub-millimeter to micron range. It refers to fashioning microscopic mechanical parts out of a silicon substrate or on a silicon substrate to make three-dimensional (3D) structures, thereby creating a new paradigm in the design of miniaturized systems. By employing materials such as crystalline silicon, polycrystalline silicon, silicon nitride, etc., a variety of mechanical microstructures including beams, diaphragms, grooves, orifices, springs, gears, suspensions, and a great diversity of other complex mechanical structures have been conceived, implemented, and commercially demonstrated.

Silicon micromachining has been the key factor in the fast progress of microsystems in the last decade of the 20th century. This is the fashioning of microscopic mechanical parts out of silicon substrates and more recently other materials. This technique is used to fabricate such features as clamped beams, membranes, cantilevers, grooves, orifices, springs, gears, chambers, etc., that can then be suitably combined to create a variety of sensors. *Bulk micromachining*, which involves carving out the required structure by etching out the silicon substrate, is the commonly used method. For instance, a thin diaphragm of precise thickness in the few-micron range suitable for inte-grating sensing elements is very often realized with this approach. Figure 1.4 shows the scanning elec-tron microscope (SEM) image of a cantilever beam of $SiO_2$ fabricated by bulk micro-machining (wet chemical etching) of silicon. The dimensions of this cantilever beam are length $= 65\ \mu m$, width $= 15\ \mu m$, and thickness $= 0.52\ \mu m$ ($1\ \mu m = 10^{-6}\ m$).

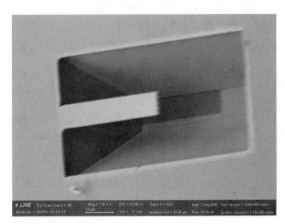

**Figure 1.4** SEM image of $SiO_2$ microcantilever beam prepared by bulk micromachining (wet chemical etching) of silicon.

*Surface micromachining* is an alternate micromachining approach. It is based on patterning layers deposited on the surface of silicon or any other substrate. This approach offers the attractive possibility of integrating the micromachined device with micro-electronics patterned and assembled on the same wafer. In addition, the thickness of the structural layer in this case is precisely determined by the thickness of the deposited layer and hence can be controlled to submicron thickness levels.

Figure 1.5 shows a schematic of a polycrystalline silicon resonating beam that can be made by the surface micromachining technique. Note that the resonator beam is anchored at its ends and that a gap of $d = 0.1\ \mu m$ exists between the resonator and the rigid beam laid perpendicular to it. As a result, the resonator vibrates at the frequency of the ac signal applied to it with respect to the rigid beam and a current flows due to the capacitance variation with time. This current peaks when the frequency of the input signal matches with the mechanical resonance frequency of the vibrating beam. Thus the resonator can be used as a filter.

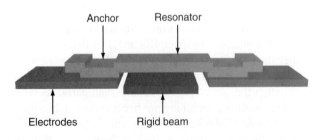

**Figure 1.5** Schematic of a resonator beam [4].

Until about the 1990s, most microsystems devices with various sensing or actuating mechanisms were fabricated using silicon bulk micromachining, surface micromachining, and micromolding processes. More recently, 3D microfabrication processes incorporating other materials (such as polymers) have been introduced in microsystems to meet specific application requirements (e.g. biomedical devices and microfluidics).

It is interesting to note that almost all the microsystems devices employ one or more of the three basic structures—namely, a diaphragm, a microbridge, or a beam—that are realized using micromachining of silicon in most cases and other materials including polymers, metals, and ceramics. These three structures provide feasible designs for microsensors and actuators that eventually perform the desired task in many smart structures. The main issues with respect to implementing these structures are the choice of materials and the micromachining technologies used to fabricate these devices.

## ► 1.4 SMART MATERIALS, STRUCTURES AND SYSTEMS

The area of smart material systems has evolved from the unending quest of mankind to mimic systems of natural origin. The objective of such initiatives is to develop technologies to produce nonbiological systems that achieve optimum functionality as observed in natural biological systems and emulate their adaptive capabilities by an integrated design approach. In the present context, a smart material is one whose electrical, mechanical, or acoustic properties or whose structure, composition, or functions change in a specified manner in response to some stimulus from the environment. In a similar way, one may envisage smart structures that require the addition of properly designed sensors, actuators, and controllers to an otherwise ''dumb'' structure (see schematic in Figure 1.6).

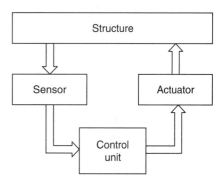

**Figure 1.6** Building blocks of a typical smart system.

As smart materials systems should mimic naturally occurring systems, the general requirements expected from these nonbiological systems include:

1. Full integration of all functions of the system.
2. Continuous health and integrity monitoring.
3. Damage detection and self-recovery.
4. Intelligent operational management.
5. High degrees of security, reliability, efficiency and sustainability.

As one can note, the materials involved in implementing this technology are not necessarily novel. Yet the technology has been accelerating at a tremendous pace in recent years and has indeed been inspired by several innovative concepts.

As mentioned earlier, the structural, physical, or functional properties of smart materials respond to some stimulus from the environment and this response should be repetitive in the sense that the same change in the environment must produce the same response. We know that even without design, most materials do respond to their environment. For example, note the change in dimensions of most materials when heated or cooled. However, what distinguishes a smart material from the rest is that we *design* the material so that such changes occur in a specific manner for some defined objective to be accomplished. Hence, the main feature that distinguishes smart materials is that they respond *significantly* to some external stimuli to which most materials are unresponsive. Furthermore, one would like to enhance such a response by at least one or two orders of magnitude over other materials.

Both active and passive approaches have been attempted in this context. This distinction is based on the requirement to generate power required to perform responses. Hence, an active system has an inbuilt power source. Typically, active sensors and actuators have been favored in designing smart structures. However, in recent years the concept of passive smartness has come to the fore. Passive smartness can be pervasive and even continuous in the structure. Such structures do not need external intervention for their operation. In addition, there is no requirement for a power source. This is particularly relevant in large-scale civil engineering structures. Passive smartness can be derived from the unique intrinsic properties of the material used to build such structures. One common example is the shape-memory-alloy (SMA) material embedded in aerospace composites, designed so that cracks do not propagate. Such smart materials are discussed briefly in Chapter 2.

Besides ''smart system'' or ''smart material,'' another widely used term is *smart structure*. One distinctive feature of smart structures is that actuators and sensors can be embedded at discrete locations inside the structure without affecting the structural integrity of the main structure. An example is the embedded smart structure in a laminated composite structure. Furthermore, in many applications, the behavior of the entire structure itself is coupled with the surrounding medium. These factors necessitate a coupled modeling approach in analyzing such smart structures. The function and description of various components of the smart system in Table 1.1 are summarized in Table 1.1 [5].

Civil engineering systems such as building frames, trusses, or bridges comprise a complex network of truss, beam, column, plate, and shell elements. Monitoring the structural health of bridge structures for different moving loads is an area of great importance in increasing the structural integrity of infrastructures and has been pursued in many countries. Damping of earthquake motions in structures is yet another area that has been taken up for active research in many seismically active countries. Smart devices derived from smart materials are extensively used for such applications. A bridge whose structural health can be monitored is shown in Figure 1.7. In such a bridge, fiberoptic sensors are used as sensing devices, whereas lead zirconate titanate (PZT) or Terfenol-D actuators are normally used for performing actuations such as vibration isolation and control.

**Table 1.1 Purpose of various components of a smart system**

| Unit | Equivalent in Biological Systems | Purpose | Description |
|---|---|---|---|
| Sensor | Tactile sensing | Data acquisition | Collect required raw data needed for appropriate sensing and monitoring |
| Data bus 1 | Sensory nerves | Data transmission | Forward raw data to local and/or central command and control units |
| Control system | Brain | Command and control unit | Manage and control the whole system by analyzing data, reaching the appropriate conclusions, and determining the actions required |
| Data bus 2 | Motor nerves | Data instructions | Transmit decisions and associated instructions to members of structure |
| Actuator | Muscles | Action devices | Take action by triggering controlling devices/units |

**Figure 1.7** The historic Golden Bridge built in 1881 to connect Ankleshwar and Bharuch in Gujarat, India. The inset shows the details of one-time structural health monitoring of a section of the bridge. It had 66 fiber optic strain gauges and micromachined accelerometers. Courtesy: Instrumentation Scientific Technologies Pvt. Ltd., Bangalore, www.inscitechnologies.com.

**Table 1.2 Some sensors and actuators used in smart systems**

| Device | Physical Quantity | Example | Successful Technologies |
|---|---|---|---|
| Sensor | Acceleration | Accelerometer | PZT, Microfabrication |
| | Angular rate | Gyroscope | Fiber optic, Microfabrication |
| | Position | Linear variable differential transformer (LVDT) | Electromagnetic |
| Transducer | Crack detection | Ultrasonic transducer | PZT |
| Actuator | Movement | Thermal | Shape memory alloy |

The beneficiaries and supporters of the smart systems technology have been military and aerospace industries. Some of the proof-of-concept programs have addressed structural health monitoring, vibration suppression, shape control, and multifunctional structural concepts for spacecrafts, launch vehicles, aircrafts and rotorcrafts. The structures built so far have focused on demonstrating potential system-level performance improvements using smart technologies in realistic aerospace systems. Civil engineering structures including bridges, runways and buildings that incorporate this technology have also been demonstrated. Smart system design envisages the integration of the conventional fields of mechanical engineering, electrical engineering, and computer science/information technology at the design stage of a product or a system.

As discussed earlier, smart systems should respond to internal (intrinsic) and environmental (extrinsic) stimuli. To this end, they should have sensors and actuators embedded in them. Some of these devices commonly encountered in the context of smart systems are listed in Table 1.2.

## ▶ 1.5 INTEGRATED MICROSYSTEMS

Integrated microsystems can be classified into three major groups as follows:

**1. Micromechanical structures:** These are non-moving structures, such as microbeams and microchannels.

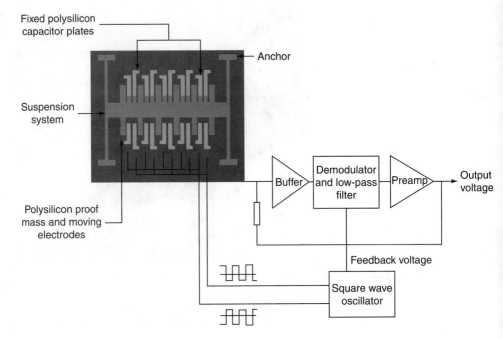

**Figure 1.8** A schematic diagram of ADXL50 accelerometer.

2. **Microsensors:** These respond to physical and chemical signals (such as pressure, acceleration, pH, glucose level, etc.) and convert them to electrical signals.

3. **Microactuators:** These convert electrical or magnetic input to mechanical forms of energy (e.g. resonating beams, switches, and micropumps).

**Microsystems** integrate sensors, actuators and electronics to provide some useful function. The ADXL50, which was released in 1991 and was Analog Devices' first commercial device, is an excellent example of such a microsystem. The block diagram of this system is shown in Figure 1.8 [6]. This microsystem is based on a surface micromachining technology with sensing electronics integrated on the same chip as the accelerometer. Here, the accelerometer is a sensor that responds to the acceleration or deceleration and gives an output voltage to the control circuit, which in turn triggers an actuator to deploy the airbag during a crash, so that the persons seated in the front seat are protected from crashing into the front windshield or the dashboard.

### 1.5.1 Micromechanical Structures

Micromachining is used commercially to produce channels for microfluidic devices and also to fabricate systems referred to as ''labs on a chip'' for chemial analysis and analysis of biomedical materials [7]. Usually such channels are made on plastics or glass substrates. Figure 1.9 (a) shows one such device reported in the literature [8]. Micromachining is also used to make a variety of mechanical structures. Figure 1.9 (b) shows an SEM image of a silicon nanotip fabricated [9] using a ''bulk micromachining'' process. Such microtips find applications in atomic force microscopy (AFM) technology and field emission array for futuristic vacuum electron devices capable of operation in terahertz frequency range [10].

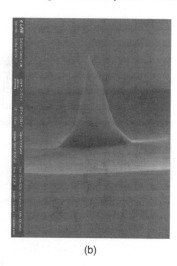

**Figure 1.9** Miniature mechanical structures showing (a) polymer mesopump; (b) silicon nanotip fabricated using bulk micromachining.

(a)                    (b)

## 1.5.2 Microsensors

Several micromachined sensors have evolved over the last two decades. Among them the pressure sensors occupy almost 60% of the market. A schematic isometric cut-away view of a piezoresistive pressure sensor die is shown in Fig. 1.10 (a). Here, we can see the four

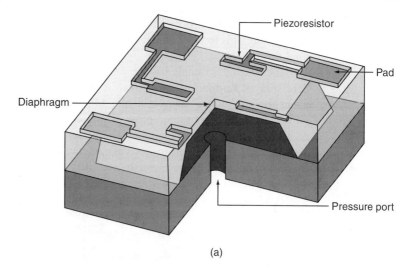

(a)

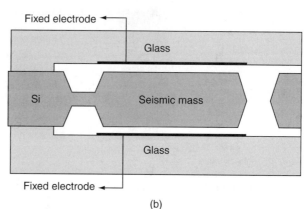

**Figure 1.10** Schematic diagrams of microsensors: (a) cut-away view of a piezoresistive pressure sensor; (b) capacitive-sensing accelerometer.

(b)

pressure-sensitive resistors (piezoresistors) integrated on a micromachined silicon diaphragm. Micromachined accelerometer is yet another device which has received considerable attention from the aerospace, automobile, and biomedical industries. Figure 1.10 (b) shows a schematic cross-sectional view of one such device. The seismic mass responds to acceleration and deflects, thus bringing about a change in the capacitance between the mass and the fixed electrodes. The change in capacitance is a measure of the displacement, which in turn depends upon the acceleration.

## 1.5.3 Microactuators

Over the last few years, micromachined actuators such as RF switches [11], micropumps, and microvalves [12], which can be actuated using the electrostatic or piezoelectric effect, have been reported. In addition, electrostatically actuated tiny micromirror arrays acting as optical switches have been developed by Lucent Technologies for fiber-optic communication [13] and as digital micromirror devices (DMDs) by Texas Instruments in projection video systems [14]. In this section, two examples of actuators are presented to highlight the wide range of applications of microactuators. First, a schematic diagram of the electrostatically actuated bulk micromachined silicon micropump [12] is shown in Figure 1.11. The diaphragm deflects upward when an actuation voltage is applied as shown and the inlet check valve opens due to a fall in chamber pressure, thereby letting in the fluid flow into the pump chamber. When the actuation voltage drops to zero, the diaphragm moves back to its equilibrium position. As the chamber is now full of fluid, the pressure inside is higher than that outside. This forces the inlet valve to close and the outlet valve to open, thus letting the fluid flow out.

The second example is the famous electrostatically actuated micromirror array consisting of tiny mirrors as shown in Figure 1.12. Each mirror is 0.5 mm in diameter, about the size of the head of a pin. Mirrors rest 1 mm apart and all 256 mirrors are fabricated on a 2.5 cm square piece of silicon. The figure on the left-hand side shows how the tilted mirrors switch the optical signals from one fiber to another fiber. Thus, there is no need of making an optoelectronic conversion. This arrangement gives 100-fold reduction in power consumption over electronic switches.

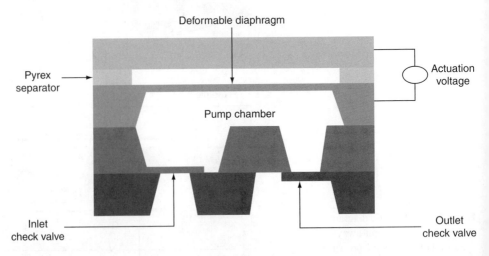

**Figure 1.11** Schematic of a bulk micromachined micropump.

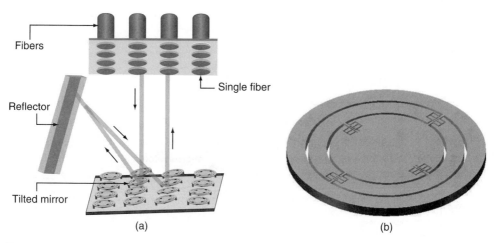

**Figure 1.12** (a) Schematic of micromirrors developed by Lucent Technologies optical switches in fiber-optic communication. (b) Lucent micromirror details.

## ▶ 1.6 APPLICATIONS OF SMART MATERIALS AND MICROSYSTEMS

The area of smart materials and structures is an interdisciplinary field bridging the gap between science and technology and combining knowledge in physics, mathematics, chemistry, material science, electrical and mechanical engineering. Through such a technology, we can build safer cars, more comfortable airplanes, self-repairing water pipes, etc. Smart structures can help us to monitor and control the environment better and to increase the energy efficiency of devices. The applications of microsystems encompass the entire range of civilian, defense, automobile, biomedical, and consumer devices used in day-to-day life. Some of the applications areas of micromachined pressure sensors are illustrated in Figure 1.13. For instance, pressure sensors are used

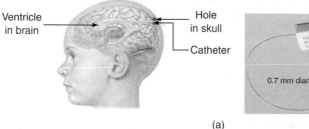

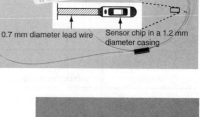

(a)

**Figure 1.13** Some application areas for pressure sensors. (a) Codman (www.codman.com) ICP microsensor showing the dimensions of the packaged transducer. (b) Mapping the pressure on aircraft wings during the development stages of the aircraft.

(b)

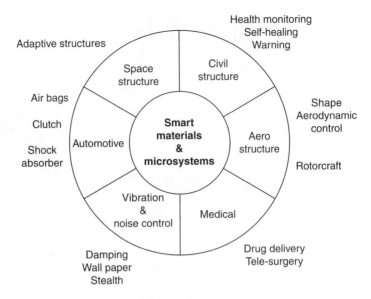

**Figure 1.14** Applications of smart materials and microsystems.

in the intracranial pressure (ICP) monitoring system. This requires pressure sensors of chip size less than 1 mm × 1 mm; also, the packaging material must be biocompatible as the sensor is inserted into the ventricle in the brain using a catheter through a small hole created in the skull. In yet another application, a large number of pressure sensors, over 500, must be laid out on each wing of an aircraft to map the pressure across the aerofoil during the development stages of an aircraft. Similarly, in oceanographic applications, depth monitoring can be done with pressure sensors designed for a wide range of pressures and suitably packaged to protect against the corrosive sea water. Figure 1.14 gives a broad range of applications of smart materials and microsystems.

Furthermore, the applications of microsystems can be conveniently grouped according to major industries as shown in Table 1.3. Though this table lists diverse applications of smart structures [15], it should be clear to the reader that there is a significant scope for developing radically new capabilities in this domain, and that it is only limited by the creativity and innovativeness of the human mind.

**Table 1.3 Applications of smart systems in various areas**

| Application Area | Microcomponent or Smart Component | Purpose |
|---|---|---|
| Machine tools | Piezoceramic transducers | To control chatter and thereby improve precision and increase productivity |
| Photolithography | Vibration control during the process using piezoceramic transducers | To manufacture smaller microelectronic circuits |
| Process control | Shape memory alloy | For shape control, for example, in aerodynamic surfaces |
| Health monitoring | Fiber-optic sensors | To monitor the health of fiber-reinforced ceramics and metal-matrix composites, and in structural composites |

**Table 1.3** (*Continued*)

| Application Area | Microcomponent or Smart Component | Purpose |
|---|---|---|
| Consumer electronics | Piezoceramic and Microaccelerometers and rotation rate sensors; quartz, piezoceramic, and fiber-optic gyros; piezoceramic transducers | For shake stabilization of hand-held video cameras |
| Helicopters and aircraft | Piezoceramic stack actuators; PZT and Microaccelerometers; magnetostrictive mounts | For vibration and twist control of helicopter rotor blades, adaptive control of aircraft control surfaces |
| | Piezoceramic pickups and error sensors; PZT audio resonators and analog voice coils; digital signal processor chips | For active noise control |
| Submarines | Piezoceramic actuators | For acoustic signature suppression of submarine hulls |
| Automotive | Electrochromic | For chromogenic mirrors and windows |
| | Piezo yaw-axis rotation sensors (antiskid, antilock braking); Microaccelerometer (air bag controls) | For navigation and guidance, electronic stability control (four- heel independent auto braking) |
| | Ceramic ultrasonic ''radar'' | For collision avoidance, parking assist |
| | Piezopolymer infrared (IR) sensors; rain monitors; occupant identification; heat, ventilation and air-conditioning (HVAC) sensors; air pollution sensors (CO and $NO_x$) | For smart comfort control systems |
| In buildings | IR, vision and fiber-optic sensors and communications systems | For improved safety, security and energy control systems. Smart windows to reduce heating, ventilation, and air-conditioning costs |
| Biomechanical and biomedical systems | SMA and polymer gel | To develop artificial muscles, active control of *in vivo* drug delivery devices (insulin pumps) |
| | Piezoceramic and other ultrasonic sensors and actuators | Catheter guide wire; surgical tools; imaging devices |
| Computer industry | Piezoceramic and Microaccelerometers and rotation rate sensors | For smart read/write head micropositioners in next-generation data storage devices |
| | Quartz, piezoceramic and fiber-optic gyros | For high-density disk drives |
| | Bimorph-type piezopositioner | To correct for head-motion-related read/write errors |
| | Piezoaccelerometers | |

*Note:* Modified after [15].

## ▶ 1.7 SUMMARY

Miniaturization and integration of a variety of components are the key aspects of micro-systems. Besides giving rise to compact devices and systems, miniaturization leads to the many other benefits discussed in this chapter. Miniaturization is made possible by microfabrication techniques. Integration of sensors, actuators and micromechanical structures on a single chip or package leads to novel solutions in many fields. Smart materials help in devising new types of sensors and actuators. They extend the application domain of microsensors and microactuators to integrated smart systems that can sense their environment and intelligently and spontaneously respond to external stimuli. A few representative applications discussed in this chapter provide a glimpse into the unlimited opportunities provided by the combination of micro and smart systems.

## ▶ REFERENCES

1. Feynman, R.P. (1992) There's plenty of room at the bottom, *Journal of Microelectromechanical Systems*, **1**(1), 60–66.

2. Feynman, R. (1993) Infinitesimal machinery, *Journal of Microelectromechanical Systems*, **2**(1), 4–14.

3. Nguyen, C.T.C. (1999) Frequency-selective MEMS for miniaturized low-power communication devices, *IEEE Transactions on Microwave Theory Techniques*, **47**(8), 1486–1503.

4. Petersen, K.E. (1982) Silicon as a mechanical material, *Proceedings of the IEEE*, **70**(5), 420–57.

5. Akhras, G. (2000) Smart materials and smart systems for the future, *Canadian Military Journal*, **1**(3), 25–31.

6. Analog Devices (1996) ADXL-50 Monolithic accelerometer with signal conditioning, Data sheet, 16 pp.

7. Nagel, D.J. and Zaghloul, M.E. (2001) MEMS: micro technology, mega impact, *IEEE Circuits and Devices*, **17**(2), 14–25.

8. Balaji, G., Singh, A. and Ananthasuresh G.K. (2006) Electromagnetically actuated minute polymer pump fabricated using packaging technology, *Journal of Physics: Conference Series, Institute of Physics Publishing*, **34**, 258–63.

9. Private communication, Center of Excellence in Nanoelectronics, Indian Institute of Science, Bangalore, India.

10. Ive, R.L. (2004) Microfabrication of high-frequency vacuum electron devices, *IEEE Transactions on Plasma Science*, **32**(3), 1277–91.

11. Goldsmith, C.L., Yao, Z., Eshelman, S. and Denniston, D. (1998) Performance of low-loss RF MEMS capacitive switches, *IEEE Microwave and Guided Wave Letters*, **8**, 269–71.

12. Zengerle, R., Kinge, S., Richter, M. and Riscter, A. A bidirectional silicon micropump, Proceedings of the IEEE 1995 MEMS Workshop (MEMS'95, January 29–February 2, 1995, Amsterdam, Netherlands, 19–24.

13. Bishop, D.J., Giles, C.R. and Das, S.R. (2001) The rise of optical switching, *Scientific American*, 88–94.

14. Hornbeck, L.J. Current status of the digital-mirror device (DMD) for projection television Applications, *Technical Digest, International Electron Devices Meeting*, Washington, DC, 1993.

15. Varadan, V.K., Vinoy, K.J. and Gopalakrishnan, S. (2006) *Smart Material Systems and MEMS: Design and Development Methodologies*, John Wiley & Sons, Chichester, London, UK.

# 2

# Micro Sensors, Actuators, Systems and Smart Materials: An Overview

## LEARNING OBJECTIVES

After completing this chapter, you will be able to:

▶ Understand principles of operation of some microsensors and microactuators.

▶ Become familiar with some microsystems.

▶ Learn about some materials and processes used to make microsystem components and devices.

▶ Get an overview of types of smart materials.

Sensors that sense the environment and actuators that provide the force needed to cause intended actions are integral parts of many systems in general and microsystems in particular. Both sensors and actuators are *transducers*—devices that convert one form of energy to another form. A sensor, independent of the parameter it senses, usually provides an electrical output such as voltage or current, while an actuator, in general, reacts to electrical input in order to provide a mechanical output—usually displacement and force. In this sense and from the energy point of view, a sensor may be considered to be primarily in the electrical domain and an actuator in the mechanical domain.

Sensors, actuators, electronics circuitry, power sources and packaging—which interfaces with the environment and protects all the components enclosed by it—comprise a microsystem. At the macroscale, a car, a robot, an air-conditioning unit, a water-purification unit, a television, etc., are all systems. Similarly, there are systems made possible by micromachined components. These components are not just electrical and electronics but mechanical devices fabricated using micromachining techniques with features as small as a few microns. These micromachining techniques are extended to thermal, optical, chemical, biological and other types of devices and components, and they can be packaged with microelectronic/micromechanical components to yield special microsystems.

Before we learn about fabricating, modeling, simulating, controlling and packaging of microsystems, it is useful to get a bird's-eye view of this field. This forms the core of the chapter. We provide salient features of commonly used microsensors, actuators, and systems. All of these are presented in the same format so as to highlight the similarities and differences among various devices and systems. Although one may not understand all the terms, the design details, and the application areas by reading this chapter, we believe that the material of this chapter provides motivation for learning the content presented in the rest of the book. The chapter ends with concise descriptions of various smart materials.

The following lists the devices and systems presented in this chapter. They cover a variety of transduction principles, fabrication processes, disciplines and applications. Although not exhaustive, they collectively represent a sufficiently wide cross-section to make readers familiar and comfortable with the field and its uses.

1. Silicon capacitive accelerometer.
2. Piezoresistive pressure sensor.
3. Conductometric gas sensor.
4. Fibre-optic sensors
5. Electrostatic comb-drive.
6. Magnetic microrelay.
7. Microsystems at radio frequencies
8. Portable blood analyzer.
9. Piezoelectric inkjet print head.
10. Micromirror array for video projection.
11. Micro-PCR systems

## ▶ 2.1 SILICON CAPACITIVE ACCELEROMETER

Summary

| | |
|---|---|
| Category | Sensor |
| Purpose | Measures the acceleration of the body on which the sensor is mounted. |
| Key words | Proof mass, suspension, capacitance |
| Principle of operation | Converts displacement caused by the inertial force on the proof-mass to a voltage signal via a change in capacitance between movable and fixed parts. |
| Application(s) | Automotive, aerospace, machine tools, biomedical, consumer products, etc. |

### 2.1.1 Overview

An accelerometer measures the acceleration of a body on which it is mounted. It becomes a tilt sensor if it measures gravitational acceleration. Almost all types of accelerometers have a basic structure consisting of an inertial mass (also called a *proof mass* or a *seismic mass*), a suspension, and a transducing mechanism to convert the acceleration signal to an electrical signal.

In a capacitive accelerometer, the sensing method is capacitive in that any change in acceleration results in a change of capacitance that is measured electronically. One of the first micromachined accelerometers was reported in 1979 by Roylance and Angell at Stanford University [1]. It used piezoresistive transduction and weighed less than 0.02 g in a $2 \times 3 \times 0.6 \, \text{mm}^3$ package. It took over 15 years for such devices to be accepted as a product for large-volume applications.

A wide variety of micromachined capacitive accelerometers is commercially available today. Some of the manufacturers of this type of accelerometer are Analog Devices,

Honeywell, Texas Instruments, Endevco Corporation, PCB Piezotronics, Freescale Semi-conductors, Crossbow, Delphi, Motorola, etc. Some of these are sold for less than ₹500 and some of them cost more than ₹50,000. This is because of the variation in performance and the applications. Low-cost accelerometers are used in consumer applications where very coarse resolution (even $0.1\,g = 0.981\,\text{m/s}^2$) is enough. Expensive accelerometers can resolve $10^{-6}\,g$ or even $10^{-9}\,g$.

## 2.1.2 Advantages of Silicon Capacitive Accelerometers

Silicon capacitive accelerometers have:

1. Very low sensitivity to temperature-induced drift.
2. Higher output levels than other types.
3. Amenability for force-balancing and hence for closed-loop operation.
4. High linearity.

## 2.1.3 Typical Applications

Typical applications of silicon capacitive accelerometers include:

1. **Consumer:** airbag deployment systems in cars, active suspensions, adaptive brakes, alarm systems, shipping recorders, home appliances, mobile phone, toys, etc.
2. **Industrial:** crash-testing robotics, machine control, vibration monitoring, etc.
3. **High-end applications:** military/space/aircraft industry navigation and inertial guidance, impact detection, tilt measurement, high-shock environments, cardiac pacemaker, etc.

## 2.1.4 An Example Prototype

Figure 2.1(a) shows a photograph of a packaged, two-axis, planar, micromachined capacitive accelerometer with the mechanical sensor element and two Application Specific Integrated Circuit Chips (ASICs) [2]. Figure 2.1(b) shows a close-up view of the sensor element and Figure 2.1(c) shows its schematic details. This device is one of the many accelerometers developed in research laboratories in academia and industry and has the same features that can be found in any capacitive accelerometer. The physical arrangement and shapes of components differ among the different types. Materials and fabrication processes used may also be different.

## 2.1.5 Materials Used

The materials used to form these devices include:

1. Single-crystal silicon to form the physical structure.
2. Silicon dioxide sandwiched in a silicon-on-insulator (SOI) wafer gives electrical isolation.
3. Handle-layer of the SOI wafer is the substrate.
4. Gold for electrodes.

## 2.1.6 Fabrication Process

This is a bulk micromachining. In general, almost any microfabrication process can be used to make an accelerometer.

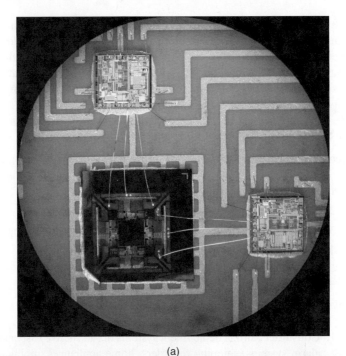

(a)

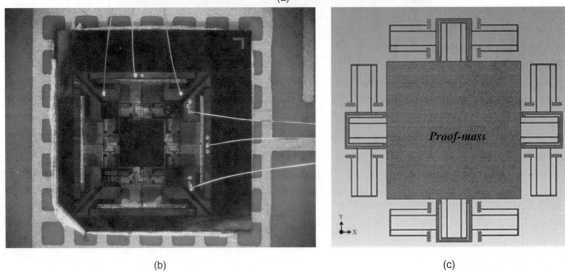

(b)                                                                      (c)

**Figure 2.1** (a) Fabricated two-axis, planar microaccelerometer with the sensor element and two ASICs; (b) a close-up view of the sensor element; and (c) schematic of the sensor element. *Courtesy*: Sambuddha Khan.

## 2.1.7 Key Definitions

The important terms used and their definitions are:

1. **Proof mass:** the inertial mass used in the accelerometer whose displacement relative to a rigid frame is a measure of the influence of external acceleration.

2. **Suspension:** the compliant structure by which the proof mass is suspended from the frame.

3. **Capacitance:** the capacity of a body to hold an electrical charge. Capacitance is also a measure of the amount of electric charge stored for a given electric potential. For a two-plate capacitor, if the charges on the plates are $+q$ and $-q$ and $V$ is the voltage between the plates, then the capacitance is given by $C = q/V$. The international standard (SI) unit of capacitance is the farad (1 farad = 1 coulomb/volt).

4. **Parallel-plate capacitor:** a pair of parallel plates separated by a dielectric (nonconducting substance) medium.

5. **Differential capacitance arrangement:** in this arrangement, there are three plates with a movable middle plate. As the plate moves, the capacitance between one of the pairs will increase while that of the other decreases. This gives a signal that is linearly proportional to the applied acceleration, and hence is the preferred configuration.

6. **Quality factor:** a system's quality factor, $Q$, describes the sharpness of the system's dynamic response.

## 2.1.8 Principle of Operation

An accelerometer can be thought of as a mass suspended by a spring. When there is acceleration, there will be a force on the mass. The mass moves, and this movement is determined by the spring constant of the suspension. By measuring the displacement, we can get an estimate of the acceleration (Figure 2.2). A capacitor may be formed with two plates of which one is fixed while the other moves [Figures 2.3(a), (b)]. In another arrangement, the mass can move in between two plates [Figure 2.3(c)]. In all three, the capacitance changes according to the motion caused by acceleration.

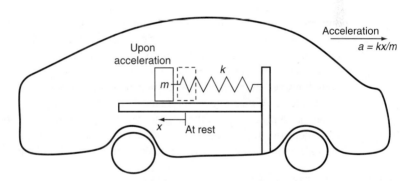

**Figure 2.2** A quasi-static accelerometer model.

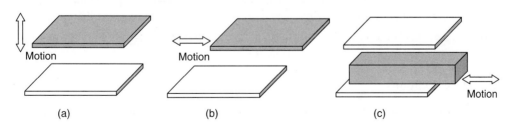

**Figure 2.3** Examples of simple capacitance displacement sensors: (a) moving plate; (b) variable area; and (c) moving dielectric.

▶ **2.2 PIEZORESISTIVE PRESSURE SENSOR**

**Summary**

| | |
|---|---|
| **Category** | Sensor |
| **Purpose** | Measures the pressure, typically of gases or liquids. |
| **Key words** | Piezoresistivity, diaphragm |
| **Principle of operation** | External pressure loading causes deflection, strain, and stress on the membrane. The strain causes a change in the resistance of the material, which is measured using the Wheatstone bridge configuration. |
| **Application(s)** | Automotive, aerospace, appliances, biomedical, etc. |

### 2.2.1 Overview

Pressure measurement is a key part of many systems, both commercial and industrial. In most pressure-sensing devices, the pressure to be measured is applied on one side of a diaphragm and a reference pressure on the other side, thus deforming the diaphragm. This deformation is measured by measuring the change in electrical resistance due to mechanical strain (i.e. piezoresistivity) of the material. The deformation is then related to the pressure to estimate the latter.

Many devices are available for pressure measurement at the macroscale. Liquid column gauges consist of a vertical column of liquid in a tube whose ends are exposed to different pressures. The rise or fall of the column represents the applied pressure. Piston-type gauges counterbalance the pressure of a fluid with a solid weight or a spring. A Bourdon gauge uses a coiled tube that, when it expands due to increased pressure, causes rotation of an arm connected to the tube. This motion is transferred through a linkage connected to an indicating needle. Diaphragm-type pressure sensors include the aneroid gauge, which uses the deflection of a flexible membrane that separates regions of different pressures. The amount of deflection is indicative of the pressure to be determined.

The first micromachined piezoresistive pressure sensors were developed in the early 1960s; however, the first reported practical pressure sensor was in the 1980s. During the 1990s several integrated microelectromechanical pressure sensor products were launched. Pressure sensors presently constitute the largest market segment of microsystems devices. There are many industries that manufacture and sell these sensors, including Motorola, Honeywell, Freescale Semiconductors, Micro Sensor Co. Ltd, American Sensors Technologies, Inc., etc.

### 2.2.2 Advantages of Piezoresistive Pressure Sensors

Micromachined piezoresistive pressure sensors have:

1. Compact size, making them suitable for a variety of applications, including those that use an array of such sensors to measure pressure distribution.
2. Good thermal stability, since thermal compensation can be built into the sensor.
3. Good market potential due to low cost.

### 2.2.3 Typical Applications

Typical applications of piezoresistive pressure sensors include:

1. **Direct pressure-sensing applications:** such as weather instrumentation, combustion pressure in an engine cylinder or a gas turbine, appliances such as washing machines, etc.

2. **Altitude-sensing applications:** such as in aircraft, rockets, satellites, weather balloons, where the measured pressure is converted using an appropriate formula.

3. **Flow-sensing applications**, **manifold pressure sensing:** in automobiles, etc.

## 2.2.4 An Example Commercial Product

Figure 2.4 shows a manifold absolute pressure (MAP) sensor commercially available from Motorola [3]. It is used in automobiles. Its packaged form and the schematic of the die without the outer plastic casing and fluidic fittings are shown in Figure 2.5.

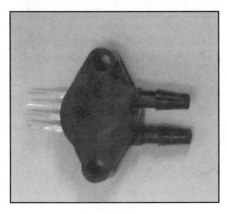

**Figure 2.4** Micromachined pressure sensor.

**Figure 2.5** Schematic of the packaged pressure sensor.

## 2.2.5 Materials Used

The material used to form this device is single-crystal silicon.

## 2.2.6 Fabrication Process

Bulk micromachining with bipolar circuitry plus glass frit wafer-bonding is the fabrication process.

## 2.2.7 Key Definitions

The important terms used and their definitions are:

**Diaphragm:** a thin sheet of a flexible material, anchored at its circumference, over which differential pressure is applied.

**Piezoresistivity:** the dependence of electrical resistivity on mechanical strain. Polysilicon shows substantive piezoresistivity.

## 2.2.8 Principle of Operation

Figure 2.6 shows the cross-section and top view of a pressure sensor die. When the diaphragm deforms due to the pressure to be measured, the resistances of the piezoresistors change. Calculation of piezoresistive responses in real structures must take account of the fact that piezoresistors are typically formed by diffusion and hence have nonuniform doping. They also span a finite area on the device and hence have nonuniform stress. For a

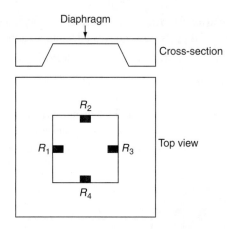

**Figure 2.6** Schematic of a pressure-sensing element.

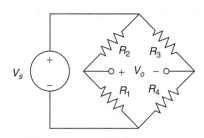

**Figure 2.7** Wheatstone bridge circuit used in sensing the change in resistance of the four resistors in Figure 2.6.

complete representation of all effects, one should solve the Poisson equation for the electrostatic potential throughout the piezoresistor, subject to boundary conditions of applied potentials at its contacts and subject to stress induced by deformation of the structural elements. From this potential, one determines the electric field, then the current density, and finally the total current. However, because the direction of current flow is along the resistor axis and, except at the contacts, parallel to the surface, considerable simplification is possible. Furthermore, the change in resistance due to stress is small, typically 2% or less. The Wheatstone bridge used to measure the change in resistance is shown in Figure 2.7.

## ▶ 2.3 CONDUCTOMETRIC GAS SENSOR

| Summary | |
|---|---|
| Category | Sensor |
| Purpose | It detects and quantifies the presence of a gas, that is, its concentration. |
| Key words | Adsorption, desorption |
| Principle of operation | The principle is that a suitable catalyst, when heated to an appropriate temperature, either promotes or reduces the oxidation of the combustible gases. The additional heat released by the oxidation reaction can be detected. The fundamental sensing mechanism of a gas sensor relies on a change in the electrical conductivity due to the interaction between the surface complexes such as $O^-$, $O_2^-$, $H^+$, and $OH^-$ reactive chemical species and the gas molecules to be detected. |
| Application(s) | Environmental monitoring, automotive application, air conditioning in airplanes, spacecraft, living spaces, sensor networks, breath analyzers, food control applications, etc. |

### 2.3.1 Overview

Gas sensing is concerned with surface and interface interactions between the molecules on the surface and the gas molecules to be detected. Many reactions are possible on the surface of sensors, and they can be accepted as the gas-sensing transduction schemes. However, the dominant reaction is a reversible gas-adsorption mechanism that occurs on the sensor's surface. The adsorbed gas atoms inject electrons into or extract electrons from the semiconducting material, depending on whether they are reducing or oxidizing, respectively. The resulting change in electrical conductivity is directly related to the amount of analyte present in the sensed environment, thus resulting in a quantitative determination of the concentration of the gas present in the environment.

For process control and laboratory analytics, large and expensive gas analyzers are used. Micromachining simplifies this through miniaturization and also reduces the cost. The oldest sensor of this kind is the Pellister, which is basically a heater resistor embedded in a sintered ceramic pellet on which a catalytic metal (platinum) is deposited. A few other methods, such as nondispersive IR methods using pyroelectric IR sensors and solid electrolyte gas-sensing mechanisms, can also be used for sensing.

Commercial gas sensors are made or sold by a number of companies such as: MICS, Applied Sensor, UST, FIS, Figaro, City Tech, New Cosmos, etc. Not all of them use microsystem technology.

### 2.3.2 Typical Applications

Typical applications of conductometric gas sensors include:

1. Environmental monitoring.
2. Exhaust gas sensing in automobiles.
3. Air conditioning in airplanes, spacecrafts, houses, and sensor networks.
4. Ethanol for breath analyzers.
5. Odor sensing in food-control applications, etc.

### 2.3.3 An Example Product Line

Commercial gas sensors of this type are available in the market today. Figure 2.8 shows a schematic of the sensing element of a typical sensor.

### 2.3.4 Materials Used

The materials used for making these are films of metal oxide such as $SnO_2$ and $TiO_2$.

### 2.3.5 Fabrication Process

Gas sensors are fabricated using single-crystalline $SnO_2$ nanobelts. Nanobelts are synthesized by thermal evaporation of oxide powders under controlled conditions in the absence of a catalyst.

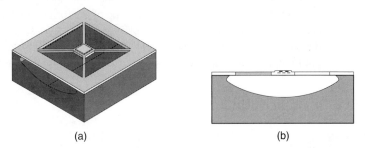

(a)  (b)

**Figure 2.8** A conductometric gas sensor die and its cross-section when isotropic etch is used to release the suspended area of the sensor element.

### 2.3.6 Key Definitions

The important terms used and their definitions are:

1.  **Conductivity:** a material property that quantifies the material's ability to conduct electric current when an electric potential (difference) is applied. It depends on the number of free electrons available.
2.  **Adsorption:** the process of collection and adherence of ions, atoms or molecules on a surface. This is different from absorption, a much more familiar term. In absorption, the species enter into the bulk, that is, the volume; in adsorption, they stay put on the surface.
3.  **Desorption:** the reverse of adsorption; species (ions, atoms or molecules) are given out by the surface.
4.  **Combustion:** a technical term for burning: a heat-generating chemical reaction between a fuel (combustible substance) and an oxidizing agent. It can also result in light (e.g. a flame).

### 2.3.7 Principle of Operation

An active area suspended by four beams off the fixed part of the wafer is the main element of this gas sensor in Figure 2.8. The adsorption or reaction of a gas on this active surface of the semiconducting material induces a change in the density of the conducting electrons in the polycrystalline sensor element. This change in conductivity is detected using electronic circuitry. The chemical reaction can be described in four steps:

1.  Preadsorption of oxygen on semiconducting material surface.
2.  Adsorption of a specific gas.
3.  Reaction between oxygen and adsorbed gas.
4.  Desorption of reacted gas on the surface.

The above process of delivering electrons between the gas and the semiconductor actually represents the sensitivity of the gas sensor. While reacting with the gas, the conductivity of the semiconductor gas sensor decreases when the adsorbed oxygen molecules play the role of the acceptor, whereas the conductivity increases when the adsorbed oxygen molecules play the role of the donor. The principle is based upon initial reversible reaction of atmospheric oxygen with lattice vacancies in the oxide and the concurrent reduction in electron concentration. This reaction generates various oxygen species according to the temperature and oxygen pressures, that is, $O_2, O^-, O_2^-$, which can then react irreversibly with certain combustible species.

Metal electrodes deposited on the top of the formed membrane that contains the active area make the measurement of the resistance of the gas-sensitive layer possible. Generally, the electrodes are located underneath the sensing film. Usually, the electrode materials are gold and platinum and, in some cases, aluminum or tungsten.

## ▶ 2.4 FIBER-OPTIC SENSORS

### 2.4.1 Overview

An optical fiber has a plastic or glass fiber core with a cladding. It is primarily used as a medium for transmitting light signals for communication. It can also be used to sense any quantity that changes the intensity, wavelength, polarization, phase, or transit time of the light passing through it. Typical measurements of a fiber-optic sensor (FOS) are pressure, temperature, electric and magnetic currents, strain, etc. Their simplicity and easy implementation makes fiber-optic sensors particularly suitable for structural health monitoring and remote sensing.

If the fiber is used simply to transmit a measured signal, it is called an extrinsic sensor. The sensor is located outside the fiber in this case. On the other hand, if the fiber assumes the role of sensing in which a characteristic of the light passing through it is changed, it is called an intrinsic sensor. The intensity-based (i.e., intrinsic) sensors are simple in construction, while sensors based on interferometric techniques, necessary for detecting changes in frequency or phase, are more complex but provide better sensitivity than intensity-based sensors. An intrinsic FOS consists of a source of light, a fiber, a photo detector, demodulator, processing and display optics, and electronics for signal conditioning. The sensor can emit, receive, or convert light into an electrical signal.

### 2.4.2 Advantages of Fiber-Optic Sensors

A major advantage of FOSs is that they are free from electromagnetic interference (EMI). Additionally, they have wide bandwidth and are compact, versatile, and economical. Their sensitivity is higher than that of the other types of sensors. They can be deployed as an array for distributed sensing. Specially made fibers can withstand high temperatures and harsh environments. In telemetry and remote-sensing applications, a small segment of the fiber, which is primarily used for communication, can be used for sensing. Additionally, by combining fiber-based signal processing components such as splitters, mixers, multiplexers, filters, delay lines, etc., a complete fiber-optic system can be realized along with the sensors. FOSs are commercially available.

### 2.4.3 An Example Prototype

Figure 2.9 shows a fiber-optic temperature sensor (os4410) made by Micron Optics (www.micronoptics.com). It is capable of sensing temperature at multiple points along its length accurate to less than a degree Celsius from $-40°C$ to $100°C$. The length of the cable can be several kilometers, making it suitable for monitoring temperatures in mines, tunnels, pipelines, and other service areas for fire safety. Its temperature sensors (small bulges in the cable in Figure 2.9), which can be spaced as closely as 0.5 m apart, can respond within seconds. It weighs only about 50 gm.

**Figure 2.9** A fiber-optic temperature sensor from Micron Optics. Redrawn after www.micronoptics.com

### 2.4.4 Materials Used

The exact material used by Micron Optics is not mentioned in the company's data sheet. The fiber material and cladding material are often proprietary to a company's product. In general, the fiber is made of glass or plastic, and the cladding is plastic, possibly coated with a special material for a particular sensing purpose. There is a plastic casing to protect it from harsh environments and to provide the necessary mechanical strength.

### 2.4.5 Fabrication Process

Usually, the optical fibers are bought from a commercial source and the casing and cladding material are removed before exposing it for coating with special materials. The coating can be achieved by spraying, sol-gel binding at a suitable temperature, or simply soaking the cable in a solution. Exposure to lasers of particular wavelength can change the refractive index of the cladding material.

### 2.4.6 Key Definitions

1. **Total internal reflection:** when light is incident at the interface between two materials of different refractive indices, it is completely reflected back if the angle of incidence is greater than the critical angle for the two materials.
2. **Fiber Bragg grating (FBG):** a segment of an optical fiber whose core is modified so that light of a certain wavelength is either reflected back or blocked.
3. **Fabry–Pérot interferometer (FPI):** an instrument used to measure the wavelength of light.
4. **Sol-gel process:** a process in which *sol* (a stable suspension of colloidal particles in a liquid) and *gel* (a porous 3D interconnected network) phases are used to synthesize a material.

### 2.4.7 Principle of Operation

An optical fiber works on the principle of total internal reflection. Some optical fibers, especially those made by doping germanium, are highly sensitive to incident light of a specific wavelength. That is, the refractive index of the fiber changes permanently when these fibers are exposed to light of a specific wavelength. A uniform FBG has a segment of an optical fiber containing periodic modulation of its refractive index. The principle of operation of an FBG sensor is shown in Figure 2.10. As shown in the figure, a grating is created on the fiber core.

When the input light is incident on the grating, some part of it is reflected while the rest is transmitted. An FBG works by measuring the changes in the reflected signal, which is the central wavelength of the back-reflected light from the Bragg grating. This depends on the effective refractive index of the core and the periodicity of the grating. Knowing the grating periodic spacing $d$ and the effective refractive index $r_{\text{eff}}$, the Bragg wavelength is given by

$$\lambda_b = r_{\text{eff}}\delta \tag{2.1}$$

Hence, the wavelength will shift with a change in the effective refractive index or the spacing of the grating. Such changes can be caused when the grating area is subjected to

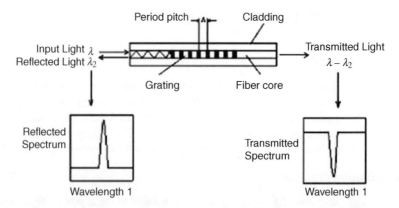

**Figure 2.10** Principle of operation of fiber Bragg grating of an FOS.

mechanical or thermal load. If $\Delta \in$ is the change in mechanical strain due to a mechanical load and $\Delta T$ is the change in temperature due to thermal load, then these changes can be related to the change in the Bragg wavelength as

$$\Delta \lambda_b = \alpha \, \Delta \in + \beta \, \Delta T \tag{2.2}$$

$$\alpha = \lambda_b (1 - p_c) \tag{2.3}$$

$$\beta = \lambda_b (\alpha_A + \alpha_B) \tag{2.4}$$

where $p_c$ is the strain-optic constant, $\alpha$ the thermal expansion coefficient of the fiber, and $\beta$ the thermal-optic coefficient of the fiber. Measurement of changes in the Bragg wavelength is crucial to this technique. This is done by opto-electronic means.

## ▶ 2.5 ELECTROSTATIC COMB-DRIVE

| Summary | |
|---|---|
| Category | Actuator and sensor. |
| Purpose | As an actuator: sets something into motion or applies force on something. <br> As a sensor: measures displacements by quantifying capacitance change. |
| Key words | Linear electrostatic actuator, suspension, fixed and moving comb fingers. |
| Principle of operation | By applying external voltage between fixed and moving combs, an electrostatic force is generated that provides the actuation along the length of the comb fingers. |
| Material(s) used | Silicon and polysilicon. |
| Fabrication process | Silicon etched with reactive ion etching and wet chemical etching. |
| Application(s) | Actuators: resonators, microgrippers, force balancing accelerometers and microscanners. <br> Sensors: automotive, aerospace, machine tools, biomedical, space applications, etc. |

### 2.5.1 Overview

The electrostatic comb-drive is a common microsystems component used in accelerometers, gyroscopes, microengines, resonators and many other applications. It can function as an actuator: when supplied with electric energy, it then sets a microcomponent into motion. It can also be used as a sensor in a microdevice that is actuated by a different source in order to measure the displacements by quantifying capacitance changes. Thus, it can sometimes serve a dual purpose in the same microsystem.

The force output is generally low (on the order of a few μN to a few hundred μN) but its linearity, easy fabrication, and highly predictive behavior make it a popular microactuator. Comb-drives have been used in microengines by combining with a linkage and gear trains.

At the macro level, hydraulic, pneumatic and servo actuators are used for various linear and rotary motions. Electrostatic actuation is not used at the macro level. At the micro level, however, the electrostatic force generation scales favorably. Indeed, electrostatic actuation is the workhorse at the microscale.

The first MEMS electrostatic comb-drive was developed at the University of California, Berkeley around 1989 [4]. It was a lateral comb-drive made of polysilicon and was used in a scanning micromirror application.

Competitors to electrostatic comb-drive at the microscale are:

1. **Piezoelectric actuators:** In these, the force is generated by piezoelectric materials that develop mechanical strain upon application of voltage. They can generate a large force but very low displacement. They are hard to microfabricate. Also, they require large voltages for actuation.

2. **Electrothermal actuators:** These work on the principle of mismatched thermal expansion because of different materials or specific shapes. They are capable of generating a large force and can be designed to have large displacements, require low voltages to actuate, and are easy to build. Their main drawback is that they are slow and take time to heat and cool. Electrostatic actuators can move in microsecond or nanoseconds but electrothermal actuators take milliseconds.

3. **Magnetic microactuators:** These generate force with current-carrying conductors in an externally created magnetic field or with another set of coils. They are hard to microfabricate. They generate moderate forces and displacements.

Applications of comb-drives are many; some of them are:

1. In closed-loop capacitive sensors where a feedback force is to be provided.

2. In opto-mechanical systems to move optical elements such as mirrors and lenses.

3. In resonant sensors and gyroscopes to set a component into motion.

### 2.5.2 An Example Prototype

A photograph of a micromachined electrostatic comb-drive made by Sandia National Laboratories, Albuquerque, NM [5] is shown in Figure 2.11. A simplified schematic is shown in Figure 2.12.

**Figure 2.11** Sandia National Laboratories' recent comb-drive. *Courtesy*: Sandia National Laboratories, SUMMiTTM Technologies, www.mems.sandia.gov.

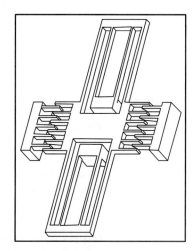

**Figure 2.12** A simplified schematic of a comb-drive. The proof-mass can oscillate along the length of the comb fingers. The mass is suspended by four beams on either side.

### 2.5.3 Materials Used

The materials used for making electrostatic comb-drives are:

1. Polysilicon and single-crystal silicon to form the physical structure.
2. Silicon dioxide as a sacrificial layer material.
3. Silicon nitride for electrical insulation.
4. Gold for electrodes.

### 2.5.4 Fabrication Process

Fabrication is mainly by surface micromachining. Sandia National Laboratories uses a very sophisticated surface micromachining process called SuMMiT (Sandia Ultra Multi-layer Microfabrication Technology) that has five movable and released polysilicon structural layers with four gap layers created by the sacrificial oxide. With this process, not only comb-drives but chain drives, gear trains, rachet drives, and many other micro-mechanical devices have been made.

### 2.5.5 Key Definitions

The important terms used and their definitions are:

1. **Electrostatics:** the phenomena arising from stationary or slowly moving electric charges. Electrostatic phenomena arise from the forces that electric charges exert on one another, as described by Coulomb's law.
2. **Parallel-plate capacitor:** a capacitor consists of two conductors separated by a nonconductive region that may be dielectric medium or air gap. The conductors contain equal and opposite charges on their facing surfaces, and the medium contains an electric field.
3. **Folded-beam suspension:** a planar suspension to suspend a body so that it can move freely with low stiffness in one planar direction but is stiff in the other planar direction as well as in the out-of-plane direction. This is a compliant-design equivalent of a sliding joint.

### 2.5.6 Principle of Operation

A top-view schematic of a comb-drive is shown in Figure 2.13. It consists of a shuttle mass suspended by a planar folded-beam suspension. The shuttle mass has combs attached to it. These moving combs interleave with the combs that are anchored to the substrate. When an electric voltage is applied, the moving comb fingers try to align with the fixed comb fingers. By alternating the voltage on the top and bottom fixed combs while keeping the moving combs at ground voltage, the shuttle mass can be made to move to and fro.

The same arrangement can be used for measuring capacitance. A pair of moving and fixed fingers forms a parallel-plate capacitor. By packing a number of such pairs, we can get sufficiently large capacitance. The capacitance change is larger when the moving comb moves along the gap between itself and the fixed finger rather than along the length. So, when used as a sensor, the folded-beam suspension is turned by $90°$.

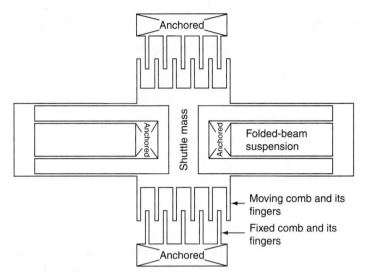

**Figure 2.13** Schematic of a comb-drive and its various parts.

## ▶ 2.6 MAGNETIC MICRORELAY

**Summary**

| | |
|---|---|
| Category | Actuator. |
| Purpose | Acts as a relay for electrical circuits and signal lines. |
| Key words | Magnetomotive force, planar exciting coil. |
| Principle of operation | Magnetization of an iron core due to the current in a wound coil causes the motion of a mechanical element. The contact between the core and the beam defines the ON position for the relay. |
| Applications | Automotive, computer, medical, etc. |

### 2.6.1 Overview

Relays are used in a variety of industrial applications to open or close the connection in an electric circuit. Traditional mechanical relays, although large, slow, and noisy, are still widely used in various industrial control processes. Solid-state relays have much longer lifetime, faster response, and smaller size than mechanical relays. However, solid-state relays generally have high ON resistance and low OFF resistance, resulting in high power consumption and poor electrical isolation, respectively.

Micromachined relays have many advantages, such as small size, fast switching speed, and low cost. Microrelays are usually categorized into groups of electrostatically actuated microrelays, thermally driven microrelays, electromagnetically driven microrelays, etc. The fabrication process of electrostatically actuated microrelays is simple and is suitable for fabricating small devices with low power. However, this device needs driving voltage of tens to hundreds of volts, which is not compatible with the ordinary electric circuit power supply. Therefore, the application of the electrostatic device is limited. Thermally driven microrelays are also seldom researched due to their large power consumption and slow switching speed. Compared with electrostatically driven microrelays and thermally driven microrelays, microelectromagnetic relays are superior because of their lower driving voltage (about 5 V), easy compatibility with ordinary electric circuit voltage, and tolerance of poor working conditions of dust, humidity, and low temperatures.

There are also two problems associated with magnetic actuation at the microscale. The first is the need for coils to generate the magnetic field. A planar spiral-like coil is easy to realize in a microfabrication process; however, making a wound-up coil is difficult, although not impossible. One easy way to overcome the need for a coil is to provide the magnetic field with an external source such as rare-earth type powerful permanent magnets. The second problem associated with microscale magnetic actuation is possible electromagnetic interference.

### 2.6.2 An Example Prototype

Figure 2.14 shows a schematic of a microrelay [6]; this is not a commercial device but rather a prototype made in a research laboratory. It has a planar coil that creates a magnetic field for the core to move. The size of this microrelay is $5 \times 5 \times 0.4 \, mm^3$. The resistance of the exciting coil is about 300 Ω. The switch-on state resistance is about 1.7 Ω at an excitation current of about 50 mA.

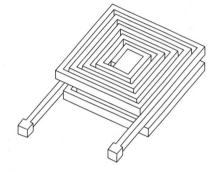

**Figure 2.14** Schematic of a magnetic microrelay with a planar coil.

### 2.6.3 Materials Used

The materials used in the preparation of this magnetic microrelay are:

1. Single-crystal silicon or polysilicon
2. Permalloy
3. Copper–chromium alloy
4. Polyimide as a sacrificial layer

### 2.6.4 Fabrication Process

Usually a customized microfabrication process is required that may consist of lithography, sputtering, electroplating, etching, sacrifice-layer technology, etc. A permanent magnet may also be used to create magnetic field locally.

### 2.6.5 Key Definitions

1. **Magnetomotive force:** the equivalent of voltage in the magnetic domain. It is like electromotive force that occurs because of a magnetic field. It has the strange unit of ampere-turn.
2. **Rare-earth magnet:** a strong permanent magnet made using rare-earth elements. Though small (e.g. a disk of a few mm in diameter and height), these magnets give very high (more than 1 tesla) magnetic fields. They are very brittle and hence are coated with nickel.

### 2.6.6 Principle of Operation

A schematic of a magnetic microrelay is shown in Figure 2.15. The magnetic flux is generated when a current passes through the excited coil. Most of the flux is concentrated in

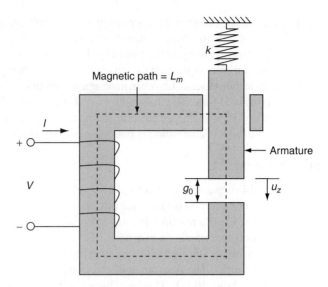

**Figure 2.15** A schematic illustration of a magnetic microrelay.

the permalloy magnetic core. The permalloy cantilever, represented as a spring in the schematic, is attracted by the magnetic force and then bends down to connect to the fixed contact. The active armature is also a conductor; hence, the current will flow from the active armature to the fixed contact when the active armature connects to the fixed contact, and then the relay will stay in the ON state. The active armature is forced to leave the fixed contact by the spring's restoring force when the exciting current is cut off, and then the relay stays in the OFF state. The modeling details of this magnetic actuator are in Section 6.7.

## ▶ 2.7 MICROSYSTEMS AT RADIO FREQUENCIES

### 2.7.1 Overview

The market for wireless personal communication devices has expanded so dramatically in the last two decades that the focus of research in the microwave and millimeter-wave areas has shifted from the more traditional defense-related products toward consumer applications. Processing techniques for microsystems have improved significantly over the past few decades, paving the way for application of microfabricated devices for radio-frequency (RF) systems, popularly known as RF MEMS.

### 2.7.2 Advantages of RF MEMS

The need for microfabricated components in RF systems arises from the inherent limitations of existing architectures. The motivations for incorporating microfabrication technologies in RF systems are of three types.

First, as the frequency increases, the size of these components becomes smaller. Thus for millimeter-wave RF systems, it is imperative that dimensions of most of the components be in the submillimeter range. This calls for fabrication techniques that provide ultra-small feature sizes. A supplementary advantage of this approach is in system integration capabilities. These micromachining-based techniques are therefore preferred for frequencies above 30 GHz (millimeter waves) for fabrication of components such as filters, directional couplers, etc.

Second, many microfabricated RF components are aimed at reducing the insertion loss and increasing the bandwidth. This aspect is of interest for surface-micromachined devices such as RF switches and inductors. Conventional RF switching systems are inefficient at higher frequencies. Microsystems-based RF switches with very low actuation voltages have been reported recently. Similarly, lumped components are preferred for wideband systems.

Third, microfabricated inductors and tunable capacitors lead to the additional feature of integration compatibility. The new research area that has emerged based on these developments is generally called RF MEMS. Several books are available on this topic [11–13].

### 2.7.3 Typical Applications

1. **Microfabricated RF switches:** One of the earliest applications of microsystems technology for RF MEMS was in surface micromachined actuators for switches with very high linearity, low DC standby power, and low insertion loss [14]. The switch design is based on singly or doubly clamped beams with electrostatic attraction employed as the mechanism to pull the switch into position. This electrostatic force is counterbalanced by a mechanical upward restoring force of the beam. The advantage of this approach is that the switch can be designed for standard transmission-line impedance of 50 Ω for a broad range of frequencies and approximately as an open/short circuit when there is no connection. These bridge structures generally utilize a very high capacitance variation to emulate the switching action.

2. **Lumped and variable capacitors:** Another use of microfabrication technology for RF applications is as variable capacitors to replace *varactor* diodes (whose junction capacitance varies with an applied reverse bias voltage) for tuning reconfigurable RF subsystems [15]. Two approaches, one employing lateral capacitance variation and the other parallel-plate capacitance variation, are used to meet this goal. The capacitance variation of this structure is over 3:1, making it very attractive for wideband tuning of monolithic voltage-controlled oscillators (VCOs). The measured quality factor at 2 GHz is over 30. The major limitation of these variable-capacitance circuits is the low-frequency mechanical resonance of the structure that creates spectral sidebands in many oscillator applications.

3. **Micromechanical filters:** High-$Q$ filters are widely used in most communication systems and in radars. For very low frequencies, bulk mechanical filters are common. Their principles have recently been translated to smaller devices at higher frequencies [16]. These micro devices can be used for frequencies up to tens of MHz and can have $Q$ of the order of 100 with proper packaging.

4. **Planar filters:** On thin dielectric membranes, planar filters show low loss and are suitable for low-cost, compact, high-performance radio frequency ICs (RFICs) and monolithic microwave ICs (MMICs). Some of these tunable filters can use micromachined cantilever type variable capacitors [17]. Applying a DC bias voltage on these cantilevers increases the electrostatic force between the plates and pulls the plates closer, resulting in an increase in capacitance.

### 2.7.4 An Example Prototype

Figure 2.16 shows a Raytheon microfabricated switch (www.raytheon.com) that is a capacitive switch that works in the shunt mode. It can be used in X-band and K-band phase-shifters, switched capacitor banks, tunable filters, and microwave switching networks [11]. It consists of a rectangular thin plate that is clamped on two edges while the other two edges

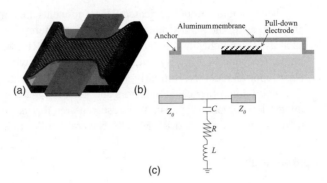

**Figure 2.16** Raytheon micromachined capacitive shunt switch: (a) SEM; (b) cross-section; and (c) electrical model.

are free. The plate is actuated electrostatically to increase the capacitance, thereby shunting the electrical circuit and causing the switching action.

## 2.7.5 Materials Used

As described in [11], the membrane in the Raytheon capacitive switch is made of aluminum. The thickness of the membrane is 0.5 μm with a gap of about 3 μm between the membrane and the transmission line with the tungsten-alloy electrode underneath it.

## 2.7.6 Fabrication Process

The tungsten electrode layer is sputtered and patterned lithographically. A layer of silicon nitride is deposited using plasma-enhanced physical vapor deposition (PECVD) for insulating the electrode. The anchors of the thin plate and the transmission lines are created by evaporating aluminum and then patterning it with wet-etching. A polyimide layer is spin-coated and planarized using a special process. A thin layer of aluminum is deposited and patterned to define the plate. The polyimide is sacrificed to create the gap between the thin plate and the insulated tungsten electrode.

## 2.7.7 Key Definitions

At microwave frequencies, voltages and currents are not known directly, and one usually uses scattering parameters to characterize components and subsystems. They are:

1. **Return loss (S11):** RF power returned back by the device when the switch is ON.
2. **Insertion loss (S21):** RF power dissipated in the device. These power losses may be due to resistive losses, contact loss, or skin depth effect.
3. **Isolation (S21):** isolation between input and output when the switch is in the OFF position. This is caused by capacitive coupling or surface leakage.

Other figures of merit concerning RF MEMS are:

1. **Switching speed:** transition time between ON and OFF states of the switch.
2. **Lifetime:** represented as number of ON/OFF cycles the switch can go through before failing due to structural damage or breakdown. Typically, lifetime of MEMS components is of the order of hundreds of millions of cycles.

### 2.7.8 Principle of Operation

The structure of RF switches is very simple: mere beams that are electrostatically actuated are adequate. Figure 2.17 shows two configurations, one a cantilever beam and the other a doubly clamped beam. The series switch, on the left in Figure 2.17, closes the electrical path when the actuation is applied. In the shunt switch (right) the beam is pulled in toward the bottom electrode, thus shorting the switch.

**Figure 2.17** Simplified schematics of series (left) or shunt (right) RF MEMS switches.

The cantilever design in the figure has the intrinsic advantage of higher isolation in the open state, especially at lower frequencies, due to the very small fringing capacitance. In general, both the cantilever and doubly clamped switches can be designed for less than 1 dB insertion loss from DC to 50 GHz in the ON state [13]. Apart from electrostatic actuation, other schemes such as thermal and magnetic have also been explored for RF switches and microrelays.

## ▷ 2.8 PORTABLE BLOOD ANALYZER

| **Summary** | |
|---|---|
| Category | Lab-on-a-chip or micro-total analysis system (μTAS). |
| Purpose | Lab-on-a-chip technology allows chemical and biological processes to be performed on small glass plate/plastic substrates with fluid channels (microfluidic capillaries). |
| Key words | Biosensor arrays, lab-on-a-chip, microfluidics. |
| Principle of operation | This device has a smart passive microfluidic manipulation system based on the structurally programmable microfluidic system (sPROMs) technology, allowing pre-programed sets of microfluidic sequencing with only an on-chip pressure source and thus making these small quantities of blood available for various tests (done with the help of application-based sensors). |
| Application(s) | Clinical analysis, DNA analysis, proteomic analysis, etc. |

### 2.8.1 Overview

This blood analyzer needs a small blood sample (~0.3 ml). The blood enters the sensing region aided by a smart passive microfluidic manipulation system that includes a microdispensor, a unit multiplexer, and air-busting detonators. Here, application-specific biosensors such as oxygen sensors and glucose sensors perform analysis on a chip. Then, the information is sent to the application-specific integrated circuit (ASIC) and the results are displayed using a liquid crystal display (LCD). All this happens in a small hand-held device. See [7] for details.

Contrast the above with what happens today. Depending upon the tests that have been ordered, a large quantity of blood sample is processed before it is analyzed. Most routine laboratory tests are performed on either plasma or serum. Plasma, the liquid portion of blood, is separated from the cellular portion of blood by rapidly spinning the specimen in a

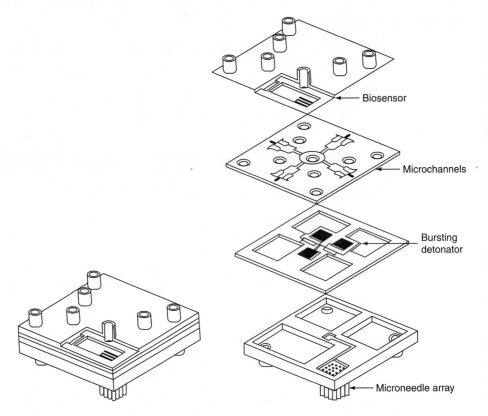

**Figure 2.18** Schematic of a lab-on-a-chip (left) and its components (right). Reservoirs, reaction chambers, channels, microneedles, and tubes for fluidic fittings are shown in the exploded view of the stacked lab-on-a-chip. Redrawn after [7].

centrifuge for several minutes. Then the tests are performed on it. It takes several hours to get the result. Portable clinical analyzers of this type are made by Abbott Laboratories, Abbott Park, IL, U.S.A, [8]. Figure 2.18, a schematic of a lab-on-a-chip, shows how a stack of patterned polymer layers gives a platform for flowing fluids to perform biochemical reactions for sensing and analyzing [7]. A commercial lab-on-a-chip is shown in Figure 2.19. Applications of such devices are in the biomedical and forensic fields.

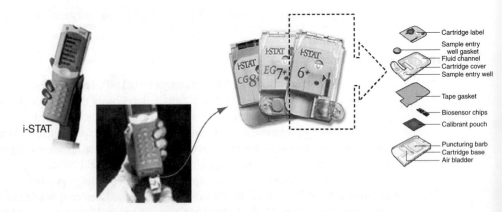

**Figure 2.19** A commercial lab-on-a-chip from Abbott Labs (http://www.istat.com).

### 2.8.2 Advantages of Portable Blood Analyzer

The following are the advantages of a portable blood analyzer:

1. Quick analysis of up to 50 blood parameters within a few seconds.
2. Provisions for recording and transmitting the results automatically.

### 2.8.3 Materials Used

The materials used for making portable blood analyzer are:

1. **Cyclic olefin copolymer (COC):** used for the substrate; this material is preferred over other polymers.
2. **Silicon:** used for the semipermeable membrane of the oxygen sensor because of its high permeability and low signal-to-noise ratio.
3. **Polyurethane:** used for the semipermeable membrane of the glucose/lactate sensor.

### 2.8.4 Fabrication Process

A fabrication process that combines polymer-processing techniques is needed here. Unlike silicon micromachining, these are not yet standardized. Typical processing techniques are: the replaceable mold disk technique, plasma treatment with reactive ion etching, thermoplastic fusion-bonding, spin-coating, and screen-printing.

### 2.8.5 Key Definitions

Important terms used and their definitions are:

1. **sPROMs:** a passive microfluidic control technique in which a set of microfluidic manipulations is carried out in a preprogramed sequence. The microfluidic operations and their sequence are determined primarily by the structural arrangement of the system without the need for an external control signal.
2. **Microdispenser:** a dispenser giving out graduated amounts of blood at a given time.
3. **Multiplexer:** a channel network carrying the required amount of sample from the common source point.
4. **Air-bursting detonators:** As the name suggests, compressed air bursts out from an encapsulated membrane in order to develop a pressure difference across the microfluidic network.

### 2.8.6 Principle of Operation

At the start of the operation sequence, the blood sample is loaded using microneedles. Then, the microdispenser reservoir is filled with the sample. When the air-bursting detonators break open, the liquid sample travels through the multiplexed channels. The divided sample reaches the biosensing reservoir on which the biosensors are mounted. When the sample is loaded into the sensor reservoir, the measurement cycle is initiated and the concentrations of the desired analytes are measured. The transduced signals are processed in the electronics.

▶ **2.9 PIEZOELECTRIC INKJET PRINT HEAD**

| **Summary** | |
| --- | --- |
| Category | Actuator and system |
| Purpose | To produce spherical droplets of the same or varying sizes with high precision and at high drive frequencies. |
| Key words | Piezoelectric, drop-on-demand |
| Principle of operation | A piezoelectric inkjet print head utilizes a voltage-pulse-driven piezoelectric actuator to generate high chamber pressure to force a certain amount of liquid out of the ink chamber through the nozzle. |
| Application(s) | Paper printing, biochip production, etc. |

### 2.9.1 Overview

Piezoelectric *inkjet* print-head technology uses electrical signals to deform a piezoelectric element to eject ink droplets. When a voltage is applied to piezoelectric material (e.g. piezoceramics or quartz), it causes the material to change its size and shape. By controlling the amount and type of movement, ink in an adjacent chamber can be jetted out through an orifice or a nozzle in precisely measured droplets at high speeds.

The first practical inkjet device based on this principle was patented in 1948. This invention was like a galvanometer, but instead of a pointer as an indicator, a pressurized continuous ink stream was used to record the signal onto a passing recording medium. In the early 1960s, Sweet at Stanford University demonstrated that by applying a pressure wave pattern, the ink stream could be broken into droplets of uniform size and spacing. This printing process is known as continuous inkjet. Drop-on-demand (DoD) inkjet systems were developed and produced commercially in the 1970s and 1980s, for example, the Siemens PT-80 serial character printer.

Micromachined piezoelectric technology offers features in ink ejection performance, compatibility with a wide variety of inks, and durability. As a result, this technology is already utilized effectively in various commercial and industrial fields such as photo lab printing, digital printing, textile printing, color filter manufacturing, desktop printing, wide-format printing on paper/film/sheet, package printing, card printing, tile printing, etc.

### 2.9.2 An Example Product

Figure 2.20 shows a piezoelectric print-head system and Figure 2.21 schematically illustrates a piezoelectric print head and its parts as well as the working principle [9]. Figure 2.21(a) shows the nozzle's inner details. In Figure 2.21(b), we can see that one of the orifices of the nozzle is spread wide by piezoelectric actuation in order to eject a drop.

Nozzles

Electrodes

**Figure 2.20** A model of a print head that uses piezoelectric actuation.

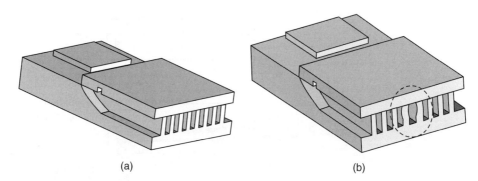

(a)                                          (b)

**Figure 2.21** A schematic of piezoelectric inkjet print head in unactuated and actuated modes. (a) Before actuation the nozzles are narrow; (b) upon actuation the nozzle orifices widen to let the ink out, as shown in the circled region.

### 2.9.3 Materials Used

The materials used for making piezoelectric inkjet print head are:

1. PZT.
2. Metal for electrodes.
3. Plastic for packaging.

### 2.9.4 Fabrication Process

The fabrication process includes precision machining of PZT and lift-off for metal. The channels and the package (which is usually a molded plastic) can be made using a variety of techniques.

### 2.9.5 Key Definitions

The important terms used and their definitions are:

1. **Piezoelectric effect:** a behavior of special materials in which an electrical field causes mechanical strain and vice versa.
2. **DoD:** as opposed to continuous jet of ink, in DoD a drop of liquid is generated and ejected.

### 2.9.6 Principle of Operation

A print head for a piezoelectric inkjet printer includes a piezoelectric actuator in the form of a plate, which lies on one side of a plate. The actuator includes drive electrodes. The plate has pressure chambers each aligned with one of the drive electrodes. The cavity plate also has nozzles, each communicating with one of the chambers. When a voltage is applied, the PZT disk attached to the diaphragm experiences a strain and deforms, as shown in Figure 2.22. When the deformation of the diaphragm begins, the ink in the chamber is pressurized and starts to move through the nozzle. When the diaphragm deflects further, the ink drop separates and is ejected, as shown in Figure 2.22.

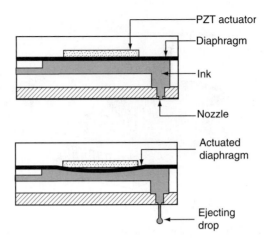

**Figure 2.22** Working principle of a piezoelectric inkjet actuator.

## ▶ 2.10 MICROMIRROR ARRAY FOR VIDEO PROJECTION

Summary

| | |
|---|---|
| Category | Actuator and system |
| Purpose | The orientation of the reflected light can be controlled by tilting the mirror. Individual control of the mirrors of an array, lets them reproduce the images they receive. Hence, they are used for video projection. |
| Key words | Micromirror arrays, torsion bar suspension |
| Principle of operation | By electrostatic actuation of the mirrors, the incoming light can be redirected. With the help of an optical window, light output can be switched ON or OFF. By actuating the mirrors at a very high rate different gray scales can be produced. |
| Application(s) | Commercial display devices especially DLP$^{TM}$ Projectors, HDTV, optical routers in high speed fiber optic communication network |

### 2.10.1 Overview

Standard projectors use various technologies such as cathode ray tubes (CRTs), LCDs, and light-emitting diodes (LED). In a CRT projector, three CRTs—one for each primary color—are used to produce the image. LCD projectors typically send light from a metal halide lamp through a prism that separates and directs light to three panels, one each for the red, green, and blue components of the video signal. As polarized light passes through the panels, individual pixels can be opened to allow the light to pass or closed to block the light. The combination of open and closed pixels can produce a wide range of colors and shades in the projected image. LED projector uses an array of LEDs as the light source. As an alternative to these, researchers in Texas Instruments built an electrostatically actuated micromirror array for video projection.

Each micromirror in the array is suspended using a torsion beam and acts as a light switch. These mirrors can be tilted by applying necessary voltage to the two pairs of electrodes, one of each pair on the mirror and the other one on the substrate. When light from a source falls on the mirror, it is reflected. Now the angle of tilt of the mirror decides whether the light is reflected to the desired location or not. Thus, it can actually switch the light output on the display area OFF and ON. Each micromirror can be actuated well over 1000 times a second, thereby producing a large number of gray-scale image pixels on a screen. Colors can be displayed using a single chip by including a synchronized color wheel or using an LED on the chip. Otherwise, three chips, each corresponding to the primary colors, can be used to display color images.

Digital micro mirror device chips (DMD$^{TM}$ chips), each housing millions of these mirrors along with the associated circuitry, are available in the market. These chips are widely used in the entertainment industry [in projectors, high-definition televisions (HDTVs), movie projection systems, etc.] and in fiber-optic communication (as routers for the optical signal). Also, 3D micromirrors are used as optical tunable filters and also as optical assembly error correctors. Other proposed applications include 3D displays, medical and scientific imaging (medical imaging, controlled radiation for cancer treatment), and industrial applications such water-marking, lithography, engraving, etc.

### 2.10.2 An Example Product

Figure 2.23 shows the digital light processor (DLP) chip made by Texas Instruments. It consists of an array of micromirrors (DMDs). Figure 2.24 illustrates the geometry of micromirrors and how they are actuated [10]. Individual mirrors are tilted to direct the light to or away from the screen using electrostatic actuation. Three DMD chips are used to generate a color image (see Figure 2.25).

### 2.10.3 Materials Used

The materials used are:

1. Aluminum for mirrors.
2. Gold for better mirror-reflectivity.
3. Complementary metal-oxide-semiconductor (CMOS) substrate for electrodes.
4. Silicon dioxide for sacrificial layer.
5. Silicon nitride for electrical insulation.

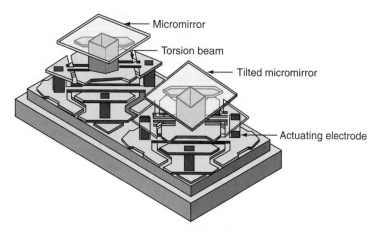

**Figure 2.23** Schematic illustration of micromirrors to direct light as desired through electrostatic actuation. *Courtesy:* Texas Instruments (used with permission).

**Figure 2.24** Schematic illustration of two micromirrors to direct light as desired through electrostatic actuation. One mirror is tilted.

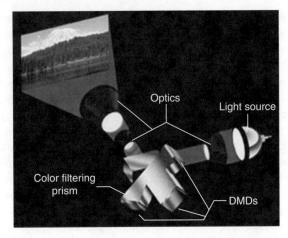

**Figure 2.25** Three DMD chips mounted on three color-filtering prisms to produce a color image. *Courtesy*: Texas Instruments (used with permission).

### 2.10.4 Fabrication Process

This process includes bulk micromachining of metal. $SiO_2$ and aluminum are deposited and patterned on the CMOS substrate electronically. Etching away the photoresist leaves the Al mirror with the associated structures, including the torsion hinge.

### 2.10.5 Key Definitions

The important terms used and their definitions are:

1. **Torsion bar hinge:** a plate held by two bars of rectangular cross-section. When there is a torque on the plate, the bars twist and thus rotate the plate.
2. **Address electrodes:** used to actuate the mirrors. The information about the image is digitized and actuation of the mirrors to reproduce the image is done by a processor on the chip. Based on this, the electrodes attain voltage states corresponding to the ON or OFF position of the mirrors.
3. **CMOS:** micromirrors are fabricated using a CMOS-compatible process. It is actually produced by bulk micromachining of layers of $SiO_2$, a photoresist, and aluminum deposited over a CMOS substrate. Familiarity with and available expertise in CMOS fabrication, coupled with the ease of integrating electronics also on the same substrate, have made this processing technology popular.
4. **S-RAM:** a memory cell fabricated on the CMOS substrate. It provides bias voltage to the electrodes, thereby determining the ON or OFF position.
5. **DMD$^{TM}$chips:** made by Texas Instruments and used in projectors, HDTVs, etc.

### 2.10.6 Principle of Operation

The data for the image to be projected is fed to the chip that is integrated with the micromirror chip. The processor generates the actuation sequence of the mirrors for the particular image. This sequence is fed one by one in to the S-RAM cell beneath the mirror corresponding to each frame. The cell changes the bias voltage of the electrodes, which results in an electrostatic torque. This torque works against the restoring torque of the hinges to produce mirror-and-yoke rotation in the positive or negative direction. The mirror and yoke rotate until the yoke comes to rest against mechanical stops at the same potential as the yoke. Because geometry, not a balance of electrostatic torques, determines the rotation angle, the rotation angle is precisely determined.

Now, the light is fed from a source such as an LED or a halide lamp. It is reflected from the mirror at an angle that depends up on the tilt of the mirror. In the ON position, the mirror reflects the light and lets it pass through the optical window on the chip cover. So the light emerges out of the chip. In the OFF position, the mirror redirects the light to a heat sink inside the chip that absorbs the light and the heat. Now, by controlling the mirrors, different gray scales can be produced depending upon the relative timings of the ON and OFF positions. Thus, they can be used to recreate the image on the screen.

## ► 2.11 MICRO-PCR SYSTEMS

### 2.11.1 Overview

The polymerase chain reaction (PCR) has become an indispensable technique in molecular biology to amplify deoxyribonucleic acid (DNA) and make multiple copies of particular segments of DNA, thereby increasing their concentration in a given sample. A DNA molecule has a double-helix structure made up of sugar molecules and nucleic acid base molecules. There are four types of bases: adenine (A), guanine (G), cytosine (C), and thymine (T). DNA can be thought of as a helically twisted ladder in which the bases join with each other to form the rungs of the ladder. The sequence of the bases is the genetic code where the entire information about the host organism is encoded. It is necessary to make multiple copies of DNA segments in many bio-molecular studies. The PCR technique makes this possible. Invented by K. B. Mullis, who was awarded the Nobel Prize for it [18], the PCR technique has revolutionized molecular biology and has applications in forensic sciences, evolution studies, medical diagnosis, and other fields.

The PCR involves three steps, denaturation, annealing, and extension. These three must happen at specific temperatures. Denaturation, which is splitting the double-stranded DNA into two single strands, takes places at around 90–95 °C. Annealing, where synthetic primer molecules with a specific sequence of the bases are attached to single strands of DNA, occurs between 60 and 70 °C. Finally, extension, by which the nucleotides that are supplied attach to extend the partially attached chain and create two copies of the original DNA segment, happens at 70–75 °C in the presence of a polymerase enzyme. The three steps repeat to make four copies, and then eight, sixteen, etc., as a chain reaction to make many copies in quick succession.

Benchtop PCR instruments consist of many wells into which reagents are supplied as the required thermal cycling takes place. The opportunity for miniaturization arises because of two issues: (i) heating and cooling a large chamber such as that in benchtop instruments and maintaining the precise conditions can be slow, and (ii) large samples are required when the PCR instrument is large. Both issues can be addressed by micro-PCR systems.

### 2.11.2 Advantages of Micro-PCR Systems

Suitably designed micro-PCR systems can vastly improve thermal response times, since the systems become very compact and thus require small sample and reagent volumes, thus reducing the cost. A micro-PCR system can be called a lab-on-a-chip: it is portable and has the potential to reach remote and rural areas where diagnosis is at present difficult.

### 2.11.3 Typical Applications

Applications of PCR include diagnostic instruments for infectious diseases and analysis of specific mutations seen in genetic disorders. Forensic medicine is another broad application. In general, biochemistry and molecular biology research studies can immensely benefit from cost-effective PCRs. Thus, it has applications in medical instrumentation and research.

### 2.11.4 An Example Prototype

Figure 2.26 shows *nanodgx*$^{TM}$, a commercially available micro-PCR kit developed by Bigtec, Inc., Bangalore. It consists of a low-temperature co-fired ceramics (LTCC)-based PCR chip that can cycle DNA/RNA through specific temperature zones for predetermined times in a buffer solution. It has the provision to store, pump and use the necessary primers, bases, and enzymes. A fluorescent dye that binds specifically to double-stranded DNA is used to monitor the reaction.

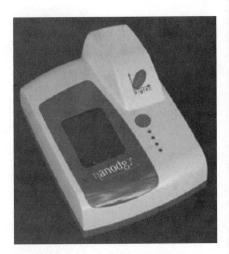

**Figure 2.26** A packaged micro-PCR kit integrated with fluorescent-dye-based detection and capable of diagnosing multiple infectious diseases. *Courtesy*: Bigtec Labs.

The *nanodgx*$^{TM}$ is a battery-operated portable system. Its overall packaged size (Figure 2.26) is less than $20 \times 15 \times 10 \, \text{cm}^3$, while most current laboratory-scale PCR systems are much bigger (typically, $35 \times 45 \times 50 \, \text{cm}^3$ ). The weight of this portable system is less than 1 kg, while other systems weigh more than 30 kg. It takes less than one hour of processing, while laboratory-scale systems take a few hours up to even a day. Another attractive feature of *nanodgx*$^{TM}$ is that it requires minimal sample processing, making it suitable for resource-limited settings that warrant economic viability.

### 2.11.5 Materials Used

A micro-PCR system needs reaction chambers and microchannels with integrated heaters and temperature sensors. Chambers and channels can be easily made in a variety of substrate materials including glass, silicon, LTCC, and polymers. Heating requires a resistive line made of a metal. Temperature sensing can be done in many ways using a variety of materials, from simple thermocouples to fiber-optic temperature sensors.

### 2.11.6 Fabrication Process

Channels and chambers can be etched in glass or silicon using dry or wet etching techniques. In LTCC, individual layers are patterned, laminated, and then fired. Heaters and thermal sensors can be integrated in many ways. In LTCC, it can be achieved with the help of resistive and thermistor pastes that are screen-printed and then co-fired.

## 2.11.7 Key Definitions

1. **DNA:** Deoxyribonucleic acid, a nucleic acid that contains the instructions for the growth and functioning of living organisms.

2. **RNA:** Ribonucleic acid, similar to DNA, that helps in encoding the information from DNA. Some viruses contain only RNA and not DNA.

3. **LTCC:** Low-temperature co-fired ceramics, special ceramic materials prepared first in the green (i.e. unfired) state as sheets, laminated with resistive and conductive pastes after patterning individual sheets, and then fired together (i.e. co-fired) to enable sintering. These are primarily used in multi-chip modules in electronic packaging, and now to make integrated microsystems devices.

4. **Thermal siphoning:** a passive method of exchanging heat through natural convection without using a pump. When this is applied to PCR, it is understood that the flow goes from a hot region to a cooler region without having to cycle the temperature of a reservoir, as in regular PCR.

5. **Primer:** a strand of a nucleic acid that helps in replicating DNA.

6. **Bases:** nucleobases, the molecular parts that help in pairing the two strands of a DNA molecule. There are four bases: cytosine (C), guanine (G), adenine (A), and thymine (T).

7. **Enzymes:** proteins that act as catalysts for biochemical reactions.

8. **Fluorescence dye:** a molecular dye that makes a molecule fluoresce, which means emitting light when subjected to light of certain wavelength,

## 2.11.8 Principle of Operation

All that is needed in a PCR is to subject a liquid sample to a cycle of prescribed temperatures (as in denaturation, annealing, and extension, as noted earlier) repeatedly, supplying the chemical and biochemical reagents at predetermined times. This is done in two ways in micro-PCR systems. One is to have a fixed chamber [see Figure 2.27(a)] where the temperature is cycled. The other (the continuous-flow PCR method) is to flow the sample and the reagents in a long channel with different temperature zones [see Figure 2.27(b)]. The fixed-chamber method is flexible in the sense that the temperatures, heating and cooling rates, and durations at certain temperatures can be changed at will. However, it requires considerable control. The continuous-flow type is inflexible because the channels are designed for particular heating and cooling rates and durations. Naturally, the control required is simple. In either case, there are provisions for introducing reagents and mixing them by active or passive means. Heating is achieved by restive or inductive means [19]. A micro-pump is required to *force* the liquid flow into and out of the chamber or channel. Thermo-siphoning can also be used [20].

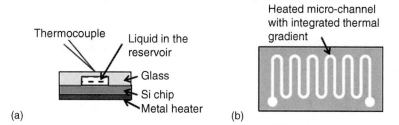

**Figure 2.27** Schematic illustration of PCR techniques: (a) fixed-chamber type and (b) continuous-flow type.

## ▶ 2.12 SMART MATERIALS AND SYSTEMS

Smart materials are usually attached to or embedded in structures or systems so that it is possible to *sense* any disturbance, *process* this information, and then *evoke a reaction* through actuators, possibly to negate the effect of the original disturbance. Thus, we see that smart materials *respond* to environmental stimuli, and for that reason, we could perhaps call them *responsive materials*. In this context, one often talks about *self-healing materials*; such materials have received attention in recent years. For example, self-healing plastics may have the ability to heal cracks as and when they occur. Shape-memory alloys in composite structures can stop propagating cracks by imposing compressive forces resulting from stress-induced phase transformations. SMAs are very useful; for example, they are used in spectacle frames to repair bends. Current research in smart materials aims at developing adaptive, self-repairing materials and structures that can arrest dynamic crack propagation, heal cracks, restore structural integrity and stiffness, and reconfigure themselves.

Responsiveness to external stimuli is probably not sufficient to call a material "smart." A more precise definition would be: a structure or material system may be considered smart if it somehow evaluates external stimuli and takes some action based on these stimuli. This action may be to neutralize the effects of the external stimuli (as described above) or to perform a completely different function. This definition requires the system to have sensor(s), feedback controller, and actuator(s). The selection of sensors may be based on the expected stimuli. The controller may consist of information-processing and storage units. Actuators may depend on the type of action expected of the system. Materials or material systems that can be "programmed" (possibly by tailoring their composition) to behave in a predefined manner in response to an external stimulus may be called *smart*. Thus, smart systems would be required to:

1. Monitor environmental and internal conditions.
2. Process the sensed data according to an internal algorithm.
3. Decide whether to act based on the conditions(s) monitored.
4. Implement a required action (if warranted).
5. Repeat the steps continuously or as required.

As in any other engineering problem, systems designed with these objectives should generally have high reliability, efficiency, and sustainability. It should be possible to integrate such a system with existing platforms by replacing "dumb" counterparts with little or no modification to the rest of the platform. While several areas of this emerging technology—such as the development of new sensing and actuation materials, devices, and control techniques—have received considerable attention, some other areas with immense potential are self-detecting, self-diagnosing, self-correcting, and self-controlling smart materials systems. Therefore, in addition to having sensing and/or actuation properties, smart material also should have favorable:

1. Physical properties (e.g. mechanical, behavioral, thermal, electrical, etc.).
2. Engineering characteristics (e.g. manufacturability, ability to form, weld, etc.).
3. Economic features (e.g. raw material and production costs, availability).
4. Environment friendliness (e.g. free from toxicity, pollution, possibility of reuse or recycling).

Some examples of smart materials are shown in Table 2.1.

**Table 2.1 Examples of materials used in smart systems**

| Development Stage | Material Type | Examples |
|---|---|---|
| Commercialized | SMAs | NITINOL |
| | Polymers | |
| |     Conducting | |
| |     Gels | |
| |     Piezoelectric | PVDF |
| | Ceramics | |
| |     Piezoelectric | PZT-5A, 5H |
| |     Electrostrictive | PMN-PT |
| |     Magnetostrictive | Terfenol-D |
| | Fiber-optic sensor systems | |
| Under development | Chromogenic materials and systems | |
| |     Thermochromic | |
| |     Electrochromic | |
| | Controllable fluids | |
| |     Electrorheologic | |
| |     Magnetorheologic | |
| | Biomimetic polymers and gels | |

Although several natural materials (such as piezoelectric, electrostrictive and magnetostrictive materials) are classified as smart materials, many have limited amplitude responses and certain limitations in operating temperature range. Chemical and mechanical methods may have to be used to tailor their properties for specific smart structure design and application.

## 2.12.1 Thermoresponsive Materials

Smart material alloys (SMAs)—metal alloys that change shape in response to change in temperature—comprise a widely used class of smart materials. Once fabricated to a specified shape, these materials can retain/regain their shape at certain operating temperatures. They are therefore useful in thermostats and in parts of automotive and air vehicles.

The shape-memory effect in materials was first observed in the 1930s by Arne Olander while working with an alloy of gold and cadmium. This Au–Cd alloy was plastically deformed when cold, but returned to its original configuration when heated. The shape-memory properties of nickel-titanium alloys were discovered in the early 1960s. Although pure nickel-titanium has very low ductility in the martensitic phase, this property can be modified significantly by the addition of a small amount of a third element. This group of alloys is known as Nitinol$^{TM}$ (nickel-titanium-Naval-Ordnance-Laboratories). NiTi SMAs are less expensive, easier to work with, and less dangerous than ordinary SMAs.

Commercial products based on SMAs began to appear in the 1970s. Initial applications for these materials were in static devices such as pipe fittings. Later, SMA devices were also used in sensors and actuators. In order to perform well, the SMAs must withstand a cycle of heating, cooling and deformation within a short time span.

Ferroelectric SMAs offer the possibility of introducing strain magnetically. This effect was discovered in the 1990s on SMAs with high magnetocrystalline anisotropy and high magnetic moment (e.g. $Ni_2$ MnGa). These materials produce up to 6% strain at room temperature.

### 2.12.2 Piezoelectic Materials

The piezoelectric effect was initially discovered by Pierre and Jacques Curie in 1880. They discovered a connection between macroscopic piezoelectric phenomena and the crystallographic structure in crystals of sugar and Rochelle salt. The reverse effect, of materials producing strain when subjected to an electric field, was first mathematically deduced from fundamental thermodynamic principles by Lippmann in 1881. Several naturally occurring materials were shown to have these effects, and nickel sonar transducers using them were used in World War I. This application triggered intense research and development into a variety of piezoelectric formulations and shapes. Piezoelectric materials have already found several uses in actuators in diverse fields of science and technology. The converse effect has led to their use as sensors.

The second generation of piezoelectric applications was developed during World War II. It was discovered that certain ceramic materials, called ferroelectrics, showed dielectric constants up to 100 times larger than common cut crystals and exhibited similar improvements in piezoelectric properties. Soon barium titanate and the PZT family of piezoceramics were developed. Some of these began to be used in structural health monitoring and vibration-damping applications. Polymeric materials such as polyvinylideneflouride (PVDF) also exhibit similar characteristics. Intense research is in progress to produce useful and reasonably priced actuators that are low in power consumption and high in reliability and environmental friendliness.

### 2.12.3 Electrostrictive/Magnetostrictive Materials

The electrostrictive effect is similar to piezoelectricity and converts an electrical pulse into a mechanical output. These materials can change their dimensions significantly on application of an electric field. Electrostriction is caused by electric polarization and has a quadratic dependence. The main difference between electrostrictive and piezoelectric materials is that the former show no spontaneous polarization and hence no hysteresis, even at very high frequencies. Electrostriction occurs in all materials, but the induced strain is usually too small to be utilized practically.

Electrostrictive ceramics, based on a class of materials known as relaxor ferroelectrics, show strains comparable to those in piezoelectric materials (strain $\sim 0.1\%$) and have already found application in many commercial systems. New materials such as carbon nanotubes have also been shown to have significant electrostrictive properties. Like piezoelectricity, the electrostrictive effect is also reciprocal. Although the changes thus obtained are not linear in either direction, these materials have found widespread application in medical and engineering fields.

The magnetostrictive effect was first reported in iron by James P. Joule, and the inverse effect was discovered later by Villari. Other materials such as cobalt and nickel also showed small strains. These are quite similar to electrostrictive materials, except that they respond to magnetic fields. Some of the first sonars were built using this principle. Large-scale commercialization of this effect began with the discovery during the 1960s of "giant" magnetostriction in rare-earth alloys. These showed 0.2%–0.7% strain, two orders of magnitude higher than nickel. An alloy of these materials, Terfenol-D (named after its constituents, terbium, iron and dysprosium, and place of invention, the Naval Ordnance Laboratory), exhibits relatively large strains (0.16%–0.24%) at room temperature and at relatively small applied fields. Terfenol-D has now become the leading magnetostrictive material in engineering use. This material is highly nonlinear and has the capacity to produce large strains, which in turn can produce large-block forces. The development of polymer-matrix Terfenol-D particulate composites has further overcome some of the limitations of pure Terfenol-D.

## 2.12.4 Rheological Materials

While the materials described above are all solids, rheological materials are in the liquid phase and can change state instantly through the application of an electric or magnetic charge. Field-responsive fluids have also been known to exist since the 19th century. The effective viscosity of some pure insulating liquids was found to increase when electric field is applied. This phenomenon, originally termed the electroviscous effect, later came to be called the electrorheological (ER) effect. These materials usually consist of suspensions of solid semiconducting materials (e.g. gelatin) in low-viscosity insulating oil (e.g. silicone oil).

In some ER compositions, both Coulomb and viscous damping can be achieved so that a vibration damper can be fabricated. Limitations of most ER fluids include relative low yield stress, temperature dependence, sensitivity to impurities (which may alter the polarization mechanisms), and the need for high-voltage power supplies (which are relatively expensive).

The magnetorheological (MR) effect was discovered by Rabinow in the late 1940s, but due to some difficulties in using MR fluids in actual applications these have not yet become popular. One of the difficulties was the low quality of the early MR fluids, which prevented the particles from remaining suspended in the carrier liquid. Recently, MR fluids have found new potential in engineering applications (vibration control) due to their higher yield stress and lower voltage requirement than ER fluids. They have also been commercially exploited for active automobile suspension systems, controllable fluid brakes for fitness equipment, as shock absorbers, and as dampers for vehicle seats.

## 2.12.5 Electrochromic Materials

Electrochromism is the ability of a material to change its optical properties (e.g. color) when a voltage is applied across it. These materials are used as antistatic layers, electrochrome layers in LCDs, and cathodes in lithium batteries.

## 2.12.6 Biomimetic Materials

Most available engineered materials contrast sharply with those in the natural world, where animals and plants have the clear ability to adapt to their environment in real time. Some interesting features of the natural world include the ability of plants to adapt their shape in real time (e.g. to allow leaf surfaces to follow the direction of sunlight) and limping (essentially a real-time change in the load path through the structure to avoid overload of a damaged region). The materials and structures involved in natural systems have the capability to sense their environment, process the data, and respond instantly. It is widely accepted that living systems have much to teach us about the design of future manmade materials. The field of biomimetic materials explores the possibility of engineering material properties based on biological materials and structures.

## 2.12.7 Smart Gels

Smart gels are gels that can shrink or swell by several orders of magnitude (even by a factor of 1000). Some of these can also be programed to absorb or release fluid in response to a chemical or physical stimulus. These gels are used in areas such as food, drug delivery, and chemical processing.

## ▶ 2.13 SUMMARY

This chapter has presented several micromachined devices and systems and described several smart materials and their potential applications. The purpose of this chapter was twofold: first, to familiarize readers with successful products and applications in the field, and second, to provide basic information on such products to motivate readers to find out more about them. Information is at everyone's fingertips today. Now it is ''*Your Turn*'' to scout for other microsystems and applications of smart materials. Remember that the application of the smart materials and microsystems is limited only by our innovativeness and creativity.

## ▶ REFERENCES

1. Roylance, L.M. and Angell, J.B. (1979) Batch-fabricated silicon accelerometer, *IEEE Transactions on Electron Devices*, **ED-26**, 1911–917.

2. Khan, S., Thejas, Bhat, N., and Ananthasuresh, G.K., ''Design and Characterization of a Micromachined Accelerometer with a Mechanical Amplifier for Intrusion Detection,'' 3rd National ISSS Conference on MEMS, Smart Structures and Materials, Kolkatta, Oct 14–16, 2009.

3. Motorola (Freescale Semiconductor) Manifold Absolute Pressure (MAP) Sensor, http://www.freescale.com/files/sensors/doc/data_sheet/MPX4100A.pdf

4. Tang, W.C., Nguyen, T.-C.H., Judy, M.W. and Howe, RT. (1990) Electrostatic comb drive of lateral polysilicon resonators, *Sensors and Actuators A: Physical*, **21**(1–3), 328–31.

5. Garcia, E.J. and Sniegowski, J.J. (1995) Surface micromachined microengine, *Sensors and Actuators A: Physical*, **48**(3), 203–14.

6. Ruan, M. (2001) Latching microelectromagnetic relays, *Sensors and Actuators A: Physical*, **91**, 346–50.

7. Ahn, C.H., Choic, J.-W., Beaucage, G. and Nevin, J.H. (2004) Disposable smart lab on a chip for point-of-care clinical diagnostics, *Proceedings of the IEEE*, **92**(1), 154–73.

8. http://www.abbottpointofcare.com/testing-products.aspx

9. Brunahl, J. and Grishnin, A.M. (2002) Piezoelectric shear mode drop-on-demand inkjet actuator, *Sensors and Actuators A: Physical*, **101**, 371–82.

10. http://focus.ti.com/pdfs/dlpdmd/117_Digital_Light_Processing_MEMS_display_technology.pdf

11. Rebeiz, G.M. (2002) *RF MEMS: Theory, Design, and Technology*, Wiley Interscience, New York.

12. Varadan, V.K., Vinoy, K.J., and Jose, K. A. (2002) *RF MEMS and Their Applications*, John Wiley & Sons, London, UK.

13. DeLos Santos, H.J. (2002) *RF MEMS Circuit Design for Wireless Communications*, Artech House, Boston.

14. Peterson, K. (1979) Micromechanical membrane switches on silicon, *IBM J. Res. Dev.* 376–385.

15. Young, D.J. and Boser, B.E. (1997) A micromachine-based RF low-noise voltage-controlled oscillator, *IEEE CICC Symposium Digest*, 431–434.

16. Bannon, F.D., Clark, J.R. and Nguyen, C.T.C. (2000) High-Q HF microelectromechanical filters, *IEEE J. Solid State Circuits*, **35**, 512–526.

17. Wu, H.D. et al. (1998) MEMS designed for tunable capacitors, *IEEE MTTS Symposium Digest*, 129–130.

18. Mullis, K.B. (1994) The polymerase chain-reaction (Nobel lecture), *Angewandte Chemie-International Edition in English*, **33**, 1209–1213.

19. Pal D. and Venkataraman, V. (2002) A portable battery-operated chip thermocycler based on induction heating, *Sensors and Actuators A: Physical*, **102**, 151–156.

20. Chen, Z., Qian, S., Abrams, W. R., Malamud, D. and Bau, H. H. (2004) Thermosiphon-based PCR reactor: experiment and modeling, *Analytical Chemistry*, **76**, 3707–3715.

# ▶ EXERCISES[1]

2.1 Which of the following transduction mechanisms can be used to realize a micromachined accelerometer? **(a)** piezoresitivity, **(b)** piezoelectricity, **(c)** Peltier effect, **(d)** Hall effect, and **(e)** photoelectric effect. For those transduction techniques useful in making an accelerometer, draw a sketch and explain how such an accelerometer would work.

2.2 As compared with the size of the other components (e.g. fan, lamp, lens, etc.) in a digital projector, the size of the micromirror array chip is quite small. Argue why it is necessary to make the chip so small. What happens if the size of each mirror is increased?

2.3 Collect data on electrostatic comb-drives to answer the following questions:

    a. Why is a comb arrangement with many interdigitated fingers used?

    b. Is the force vs. deflection characteristic of a comb-drive actuator linear?

    c. How much force and displacement can be generated with a typical comb-drive?

    d. Will a comb-drive work in aqueous environments?

2.4 We discussed a conductometric gas sensor in this chapter. What are other methods that are used to detect gases? Which one has been used in microsystems? Which ones could be used and which ones cannot be used? Support your answers with suitable arguments.

2.5 Find an used inkjet print head. Break it open to see where the chip is and how it is connected to the components that are around it. Do the same for the ink-cartridge. Identify different components that make the microsystems chip in them useful in practice. Comment on the importance of packaging in microsystems in this example.

2.6 Visit a molecular biology laboratory that uses a PCR system and discuss with its users how they use it and what for what purpose. If they are using a system that does not use microsystems components, how would you convince them to switch over to a microsystems-based PCR system?

2.7 This chapter included a description of smart materials. A beam made of a ''normal'' material bends when a load is applied to it. A beam made of a ''smart'' piezoelectric material not only bends but produces an electric charge when a load is applied. Both materials behave in their specific ways. Then, why is that we consider some materials to be smart? Does smartness lie in the way we use the material or in their very nature?

2.8 Search the literature and identify a few devices that use smart materials. Which of these have been miniaturized using microsystems technology?

2.9 Choose a system that you consider ''smart'' and explain why you think it is smart. Is mobile phone a smart system? Is a motorcycle a smart system? How about a washing machine and a home water-purifier?

2.10 Are biological materials smart in the sense we call some materials smart? Explain with examples.

---

[1] Consulting books, the Internet and other reference material is necessary to answer these questions. Knowledge of basic sciences and engineering as well as imagination may also be useful in giving plausible answers.

# 3

# Micromachining Technologies

## LEARNING OBJECTIVES

After completing this chapter, you will be able to:

▶ Understand the importance of silicon as a substrate material.

▶ Get an overview of physical and chemical techniques for thin-film deposition.

▶ Learn about lithography and lift-off techniques for patterning thin films.

▶ Understand dry and wet chemical etching techniques.

▶ Get an overview of bulk and surface micromachining processes for microsystems.

▶ Learn about wafer bonding techniques.

▶ Understand polymeric and ceramic materials and their processing for microsystems.

In the previous chapter we presented the operational principles of several sensors and actuators. Many of these are fabricated using techniques evolved over the past few decades using integrated circuit (IC) fabrication approaches. As they were specially developed for the fabrication of microsystems, these techniques came to be known as *micromachining*— the subject of this chapter.

Materials used for micromachining include semiconductors, metals, ceramics, polymers, and composites. A microsystem typically comprises a substrate with various doped regions for electronics and thin films for mechanical elements as shown in Fig. 3.1.

Semiconductor substrates form the basic materials in micromachining. While their mechanical properties are utilized in realizing structural components, their electrical properties are used for building electronic functions. The most commonly used semiconductor material is silicon, though gallium arsenide (GaAs) is used in some specialized cases. Other common substrate materials include glass, fused quartz, and fused silica. However,

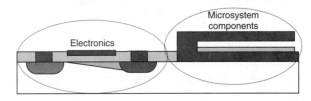

**Figure 3.1** A typical microsystem with electronics and electromechanical sensing part.

silicon is the most sought-after substrate material for integrating electronics; it is discussed in Section 3.1. Thus, silicon, silicon dioxide and silicon nitride are the most common materials, and hence silicon fabrication technologies are the cornerstones of microelectronic fabrication and micromachining. Their deposition and patterning techniques are discussed in Sections 3.2 and 3.3. Like in microelectronics processing, doping is required to modify the electrical and chemical properties of semiconductors. Diffusion and ion implantation methods used for this purpose are discussed in Section 3.4.

Wet etching techniques for building microsystems are covered in Section 3.5. Both isotropic and anisotropic etching methods are discussed in detail. Common methods to terminate the etch process are also discussed in this context. While wet etching facilitates low-cost R&D, industrial production processes largely depend on dry etching techniques which are described in Section 3.6.

Using micromachining techniques discussed in Section 3.7, microscopic mechanical elements are formed on or out of a substrate. Micromachining is classified into *bulk micromachining,* in which components are etched into the bulk of the substrate, and *surface micromachining,* in which layers are formed from thin films deposited on the substrate surface. The basic characteristics of these processes along with fabrication case studies are presented there.

Though silicon is the workhorse material in microsystems devices, polymer- and ceramic-based microcomponents are needed in certain applications. These are introduced in Section 3.8. Although the geometries thus fabricated are functionally different, the processes used typically do not differ greatly from those in silicon ICs. For fabricating high-aspect-ratio micromachined components, the specialized approaches discussed in Section 3.9 have been developed. The *wafer-bonding* processes used in ICs have also been extended for microsystems, often for bonding dissimilar materials. The dissolved wafer process discussed in the section is often considered as an alternative to surface micro-machining. Another multilayer fabrication approach that has received interest in recent years uses *low-temperature cofired ceramics* (LTCC). A laser-based fabrication approach, although used by only a few researchers so far, is also gaining popularity. Soft lithography is another approach gaining wide acceptance. These are also described in this chapter.

## ▶ 3.1 SILICON AS A MATERIAL FOR MICROMACHINING

### 3.1.1 Crystal Structure of Silicon

The most commonly used semiconductor material for microsystems applications is silicon, since it can be modified to alter its electrical, mechanical, and optical properties. Silicon and its compounds are the key materials used for micromachining and fabrication technology, for these materials have reached a state of maturity. In addition, silicon has many desirable mechanical properties that can be exploited in microsystems applications. Some properties of silicon relevant in microsystems are shown in Table 3.1

As one can note, the Young's modulus of silicon is comparable to that of steel ($200 \times 10^9$ N/m$^2$), but the failure strength of silicon is far superior to that of steel. Obviously, it is not possible to fabricate microstructures with steel and silicon by identical processes. Fabrication processes introduced in subsequent sections can be used to realize micro-systems with small parts made of silicon. These include silicon wafer processing, lithography, thin-film deposition, etching (wet and dry), and wafer-bonding processes.

Silicon crystals are made up of two interpenetrating face-centered cubic (FCC) cells. In an FCC cell, atoms are located at the eight corners of the cubic lattice structure and one atom at the center of each face. Because of interpenetration of two FCC unit cells, the unit

**Table 3.1 Typical values for electrical, mechanical and thermal properties of silicon**

| Electrical | |
|---|---|
| Minority-carrier lifetime | 30–300 μs |
| Energy bandgap | 1.1 eV |
| Lattice spacing | 5.43 Å |
| Electron affinity | 4.05 eV |
| Refractive index | 3.42 |
| Dielectric constant | 11.7 |
| Resistivity (B-doped) | 0.005–50 Ω cm |
| Resistivity (P-doped) | 1–50 Ω cm |
| Resistivity (intrinsic) | $3.2 \times 10^5$ Ω cm |
| **Mechanical** | |
| Density | 2.33 gm/cm$^3$ |
| Dislocations | ≪ 500/cm$^2$ |
| Young's modulus | |
| [100] silicon | $130 \times 10^9$ N/m$^2$ |
| [110] silicon | $168 \times 10^9$ N/m$^2$ |
| Poisson's ratio | 0.22–0.28 |
| **Thermal** | |
| Thermal conductivity | 1.57 W/cm°C |
| Thermal expansion | $2.6 \times 10^{-6}$/°C |
| Specific heat | 0.7 J/g/°C |
| Melting point | 1410°C |

*Note:* Modified after [1].

cell of silicon crystal contains four more atoms. Therefore, the silicon unit cell shown in Fig. 3.2 has 18 atoms with eight atoms at the corners, six atoms at the face centers, and four more atoms inside the unit cell (shown as hollow spheres) [2]. Atoms at the boundaries are shared between adjacent cells. A bulk silicon crystal may be considered as stacked layers of periodically repeating FCC unit cells.

The arrangement of atoms in silicon crystal is not rotationally symmetric, and as a result its material properties are anisotropic. Therefore, properties are specified with respect to designated orientations and planes in the crystal. Crystal planes and orientations are designated by *Miller indices* as described next [3].

In Cartesian coordinates, the equation of a plane is given by

$$\frac{x}{a} + \frac{y}{b} + \frac{z}{c} = 1 \qquad (3.1)$$

where $a$, $b$, $c$ are the intercepts of the plane on the $x$-, $y$-, and $z$-axes, respectively. This equation can also be written as

$$hx + ky + mz = 1 \qquad (3.2)$$

where $h = 1/a$, $k = 1/b$ and $m = 1/c$. Miller indices are integer numbers obtained by multiplying the reciprocal of intercepts with their least common multiple (LCM). These integers are given in the form $(hkl)$ to designate the plane and $[hkl]$ to designate the direction normal to the plane.

### Example 3.1

A plane intercepts the $x$-, $y$-, and $z$-axes at 2, 3, 4 respectively. Obtain an equation for the plane. Write down the Miller indices for this plane.

**Solution:**  This plane intersects the crystallographic axes at (2,0,0), (0,3,0), (0,0,4).

**Step 1:** The equation of the plane is $x/2 + y/3 + z/4 = 1$.

**Step 2:** Multiply by 12 to express this with smallest integers.

**Step 3:** (6, 4, 3) are the Miller indices. This is a (6 4 3) plane.

### Example 3.2

A crystal plane in silicon intersects the $x$-, $y$-, and $z$-axes at $x = 2$, $y = 1$, and $z = 1$. Determine the Miller indices of this plane.

**Solution:**  The plane is denoted by (1/2, 1, 1) or properly denoted by (1, 2, 2).

### Problem 3.1

The equation of a plane is given by $7x + 3y = 2$. Obtain the Miller indices for a plane that is normal to this plane.

It may be noted that, due to the cubical symmetry of the silicon crystal, the (100), (010), and (001) planes [cut along three principal directions in Fig. 3.2(b)] are indistinguishable, and are collectively referred to as {100}. Similarly, the family of directions normal to these planes (which are crystallographically equivalent) is indicated as ⟨100⟩.

Various crystal planes for a unit cell of dimension $a$ are shown in Fig. 3.3. The atomic bonds of surface atoms to those within are strongest in the (111) plane, and hence crystal growth is easiest and the etching rate is slowest in the (111) plane.

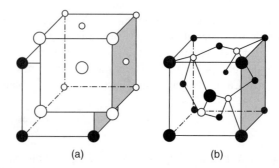

(a)                              (b)

**Figure 3.2** Zinc blende lattice structure for silicon. Atoms from one sublattice are shown by dark spheres, those from the other by hollow spheres. (a) Interpenetrating FCC sublattices; (b) atoms within a cube of side $a$.

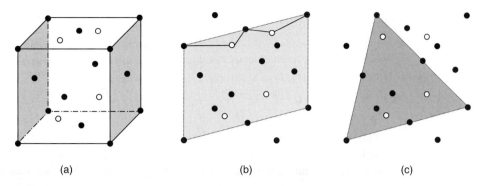

(a)                              (b)                              (c)

**Figure 3.3** Atomic arrangements in various crystal planes in a unit cell of silicon. (a) Any of the sides of the cube forms a 100 plane. Four corner atoms and an FCC atom form a unit cell in each of these planes; (b) one of the possibilities for {110} planes. Six atoms from the original FCC unit cell connected by solid lines to two atoms from the interpenetrating FCC (hollow spheres) are in this plane; and (c) a (111) plane (shaded). Six atoms from the unit cell are in this plane.

## 3.1.2 Silicon Wafer Preparation

Silicon is abundantly available in nature in the form of sand ($SiO_2$) or quartz. It must be converted into high-purity single-crystal silicon (SCS) with impurity level less than 1 ppb (parts per billion) for use in semiconductor and microsystems devices. Sand (high-purity sand, commonly known as *quartzite*) is converted into high-purity SCS by the following series of processing steps.

Quartzite is refined chemically with carbon in an arc furnace at a very high temperature to obtain 90%–99% pure silicon. This material, usually known as metallurgical-grade silicon (MGS), is treated with hydrochloric acid to get trichlorosilane ($SiHCl_3$). At room temperature trichlorosilane is in liquid form. Fractional distillation of $SiHCl_3$ in a hydrogen atmosphere yields electronic-grade silicon (EGS). The polycrystalline EGS obtained by the above process is the basic material for the preparation of silicon wafers. The impurity level at this stage is about 1 ppb. The *Czochralski method* (a method of growing crystals) is adopted to convert polycrystalline EGS into an SCS ingot.

The surface of the single-crystal ingot thus obtained is ground to make it cylindrical. For ingots with diameter less than 8″ (200 mm), one or more *flat regions* are ground along the length of the ingot to specify the crystal orientation and the type of dopant (Fig. 3.4). In manual processing, these flats are also useful in aligning masks with crystal directions during lithography.

Following this, circular wafers (thin sections of the crystal) are sliced off the ingot using a high-speed diamond saw. The thickness of a wafer depends on its diameter and ranges from 100 μm to about 1 mm. To produce perfectly flat, smooth, and damage-free surfaces, the wafers are first lapped to remove irregularities introduced during sawing. They are chemically etched and polished to a mirror finish and uniform thickness.

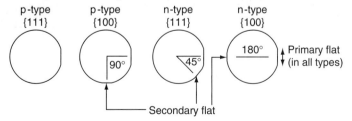

**Figure 3.4** Crystal orientation and dopant type in commercial silicon wafers with small diameters.

## ▶ 3.2 THIN-FILM DEPOSITION

Thin films are an integral part of IC and microsystems fabrication. In ICs thin films are largely used as conductors or insulators. On the other hand, several types of thin films are used in microsystems for various purposes. These may be broadly classified into metals, dielectrics (including ceramics and polymers), and functional materials. Metal films are used to form low-resistance ohmic connections to heavily doped $n^+/p^+$ regions or poly-Si layers, and are also used for rectifying contacts in metal-semiconductor barriers. Common dielectric films include silicon dioxide (referred to as oxide) and silicon nitride. These are typically used as insulating layers between conducting layers in MOS devices, for diffusion and ion implantation masks, and for passivation to protect devices from impurities, moisture, and scratches. Polysilicon (abbreviated as poly-Si or simply poly) is used as a gate electrode in metal-oxide-semiconductor (MOS) devices, as a conductive material for multilevel metallization, and as a contact material for devices with shallow junctions. In addition to these, several smart systems applications require deposition of ceramic thin-film materials such as PZT and barium strontium titanate (BST). Also, plastic materials such as polyethylene (PE), polyvinyl chloride (PVC), polymethyl methacrylate (PMMA), and polydimethylsiloxane (PDMS) are finding innovative applications in the context of microsystems. Applications of some of these materials have already been introduced in Chapter 2.

The source material for depositing these thin films on a substrate can be in the gaseous, liquid, plasma, or solid state. The deposited films are characterized by their grain size, thickness, uniformity, step coverage, adhesion, and corrosion resistance. In fact, some characteristics such as continuity, uniformity, surface properties, and adhesion need to be evaluated to determine the suitability of thin films for a particular application. In the following paragraphs the principles of some of the common deposition processes are explained.

Metal films are useful in forming low-resistance ohmic contacts and conductivity paths for interconnects. The most commonly used materials for these requirements are aluminum and gold. Metal films can be formed by physical vapor deposition (PVD) and chemical vapor deposition (CVD) processes. In these processes, favorable conditions are created to transfer the material from the source (*target*) to the destination (*substrate*). In the PVD process, this transfer takes place by physical means such as evaporation or impact, while in the CVD process, this happens through a chemical reaction.

### 3.2.1 Evaporation

Evaporation is a relatively simple process of deposition of metal films on a substrate. An evaporation system consists of an evaporation chamber, high-vacuum pumping system, substrate holder, crucible/filament, and a shutter (Fig. 3.5). The source material is

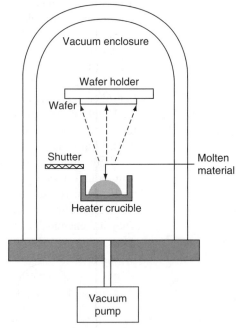

**Figure 3.5** Schematic diagram of a thermal evaporation unit for depositing metals and other materials.

**Table 3.2  Melting point of commonly used metals for microsystems applications**

| Metal | Melting Point (°C) | Preferred Methods of Deposition |
|---|---|---|
| Aluminum | 659 | Thermal evaporation |
| Silver | 957 | Sputtering |
| Gold | 1067 | Electron beam evaporation, |
| Copper | 1083 | Thermal evaporation, sputtering |
| Nickel | 1453 | Electron beam evaporation |
| Palladium | 1552 | Sputtering |
| Titanium | 1677 | Electron beam evaporation, sputtering |
| Platinum | 1769 | Sputtering |
| Chromium(sublimes) | 1887 | Electron beam evaporation, sputtering |
| Tungsten | 3377 | Sputtering |

placed in the crucible/filament. The chamber is evacuated using vacuum pumps to a pressure of $10^{-6}$–$10^{-7}$ torr. The crucible/filament is heated by passing a heavy current or directing an electron beam onto the material. The evaporated metal condenses on the substrate. The thickness of the deposited film depends upon the duration of evaporation and the distance between the source and the crucible, and its purity depends upon contaminations from the source, support material (such as the substrate holder and the crucible), and residual gases. Metal alloy films can be deposited by evaporating constituent elements simultaneously from different sources.

The melting point depends upon the ambient pressure. In Table 3.2 some of the commonly used metals are listed along with their melting points. The purity of the source material is an important factor and normally is of the order of 99.99%. Depending on process conditions, the actual temperature may be different from the melting point shown. As shown in Table 3.2, evaporation by resistive heating is used only for low melting point metals (e.g. Al, Ag, Cu) whereas electron-beam-assisted evaporation is used for high melting point metals such as Cr, Ni and Ti.

### 3.2.2  Sputtering

Sputtering is a physical phenomenon in which ions accelerated through a potential gradient bombard a target which is set to be the cathode. Because of the momentum transfer of the accelerated ions to the atoms near the surface, target atoms are released and are transported in vapor form to the substrate for deposition. Fig. 3.6 shows typical locations of targets and substrate, and indicates the ionization and deposition process in a sputtering system. The vacuum

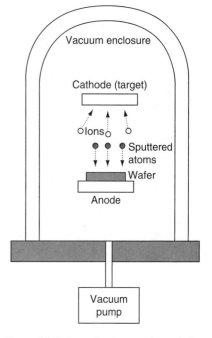

**Figure 3.6** Schematic of sputtering unit for depositing materials.

chamber is evacuated to $10^{-6}$–$10^{-8}$ torr. Argon gas is introduced into the chamber and the pressure is maintained at a few millitorr. Argon is ionized by the application of a dc (for conducting targets) or RF (for insulators) voltage. The ions bombard the target, releasing atoms from it. These atoms are deposited on the substrate surface. The deposition rate depends on the sputtered material, RF power applied, pressure inside the chamber, and spacing between the electrodes.

The sputtering process can be used for depositing one material at a time. However, compound thin films can be deposited by using co-sputtering from multiple targets. Since with sputtering, it is possible to obtain the same stoichiometric composition as that of the source material, alloys too can be used as target materials and deposited as thin films.

Sputtered films usually have good compositional uniformity, adhesion to substrate surface, and grain orientation. The sputtered films are amorphous, but can be made crystalline by suitable annealing. Their mechanical properties and stresses depend on the sputtering conditions.

### 3.2.3 Chemical Vapor Deposition

The CVD process involves convective heat and mass transfer and chemical reaction at the substrate surface. Even though the CVD process is more complex, it gives more effective control on the growth rate and the quality of deposited films. Most CVD processes involve low gas pressures (100–200 mtorr) [4].

The principle of operation involves the flow of a carrier gas with diffused reactants over large substrate surfaces. The energy supplied by the heated surface triggers chemical reaction of the reactants, thus forming thin films over the surface of the substrates. The byproducts of chemical reaction are vented out. A typical CVD reaction chamber is shown in Fig. 3.7.

A CVD process that requires elevated temperature (700–800°C) and near-atmospheric pressure is called atmospheric pressure chemical vapor deposition (APCVD). The low-pressure chemical vapor deposition (LPCVD) and plasma-enhanced chemical vapor

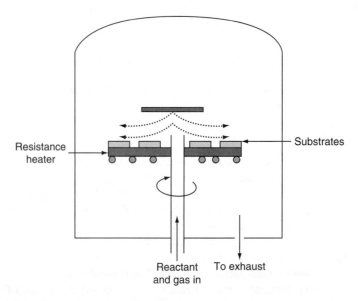

**Figure 3.7** CVD reaction chamber.

**Table 3.3 Typical process conditions for various CVD techniques**

| Process Key | | Temperature | Pressure | Typical Materials |
|---|---|---|---|---|
| Atmospheric pressure | APCVD | 700–800°C | 1 atm (760 torr) | Polysilicon |
| Low pressure | LPCVD | 600–620°C | 0.25–2 torr | Polysilicon, silicon nitride |
| Plasma enhanced | PECVD | 250–300°C | 100–200 mtorr | Silicon nitride, amorphous silicon, silicon dioxide |

deposition (PECVD) processes are used to achieve higher growth rates and better deposited film quality at lower temperatures. Typical reaction conditions for these methods are provided in Table 3.3 for quick comparison. The deposited film is usually uniform. Batch processing of stacked wafers is possible by this method.

PECVD utilizes RF plasma to transfer energy to reactants, with the result that the substrate can remain at lower temperature than in APCVD or LPCVD. Precise temperature control of substrate surface is necessary to ensure good deposited film quality.

Nonsilicon materials such as compound semiconductors can be deposited by appropriate CVD techniques. For example, metal-organic chemical vapor deposition (MOCVD) is a relatively low-temperature (200–800°C) process for epitaxial growth on semiconductor substrates. Metal-organics are compounds in which each atom of the element is bound to one or many carbon atoms of hydrocarbon groups. For precise deposition control, high-purity materials and very accurate control are necessary. For example, trimethyl aluminum, triethyl aluminum, or tri-isobutyl aluminum may be used for depositing aluminum compounds such as AlGaAs and AlGaInP in some photonic devices. Due to the high cost, this approach is used only where high-quality films are required.

Parameters that significantly influence the rate of CVD are:

1. Temperature
2. Pressure of carrier gas
3. Velocity of gas flow
4. Distance along the direction of gas flow

The following sections describe the CVD process for silicon dioxide, silicon nitride, and polysilicon films.

### 3.2.3.1 Silicon Dioxide

The commonly used carrier gas is $O_2$ and the other reactant is silane ($SiH_4$). The chemical reaction involved is:

$$SiH_4 + O_2 \xrightarrow{\text{400°C to 500°C}} SiO_2 + 2H_2$$

This may also be done by decomposing dichlorosilane:

$$SiCl_2H_2 + 2H_2O \xrightarrow{\text{900°C}} SiO_2 + 2H_2 + 2HCl$$

$SiO_2$ can also be deposited from tetraethyl orthosilicate [TEOS or $Si(OC_2H_5)_4$] by vaporizing this from a liquid source. $SiO_2$ is available in various forms as silica, quartz, glass, and borosilicate glass (BSG). Quartz is a single-crystal material that typically has low impurity concentration. Fused quartz is the amorphous form of quartz. On the other hand, glass is an amorphous solid having impurities. BSG is developed specifically for laboratories and heating applications. Some common names are Pyrex$^{TM}$ by Corning and Duran$^{TM}$ by Schott Glass. In these, the dominant component is $SiO_2$, but boron and various other elements are added for improved properties.

### 3.2.3.2 Silicon Nitride

Silicon nitride films can be deposited by a LPCVD process using ammonia as the carrier gas and silane, silicon tetrachloride, or dichlorosilane as other reactants. The chemical reactions are as follows:

$$3SiH_4 + 4NH_3 \xrightarrow{700^\circ C \text{ to } 900^\circ C} Si_3N_4 + 12H_2$$

$$3SiCl_4 + 4NH_3 \xrightarrow{800^\circ C} Si_3N_4 + 12HCl$$

$$3SiH_2Cl_2 + 4NH_3 \xrightarrow{650^\circ C \text{ to } 750^\circ C} Si_3N_4 + 6HCl + 6H_2$$

Typical properties of $Si_3N_4$ deposited by LPCVD include low density, high-temperature strength, superior thermal shock resistance, excellent wear resistance, good fracture toughness, mechanical fatigue and creep resistance, and good oxidation resistance. These films are used as etch masks, gate insulators, thermal insulators, and chemical-resistant coatings in various devices.

Both silicon nitride and silicon dioxide can be deposited at lower temperatures using PECVD process. The properties of thin films vary depending on the process. PECVD $Si_3N_4$ is typically used as a capping and hermetic sealing layer just before packaging ICs and microsystems.

### 3.2.3.3 Polysilicon

Deposition of polysilicon is typically done by a decomposition of silane using an LPCVD process. The following chemical reaction takes place:

$$SiH_4 \xrightarrow{600^\circ C \text{ to } 650^\circ C} Si + 2H_2$$

It may be noted that polysilicon comprises small crystallites of SCS, each with different orientations, separated by grain boundaries. This is a common structural material used in microsystems. In addition, it is also used in microelectronics for electrodes and as a conductor or high-value resistor, depending on its doping level. As with SCS, this must be highly doped to increase the conductivity substantially. When doped, the resistivity can be as low as 500–525 $\mu\Omega$ cm. Therefore, this material is preferred in complementary metal-oxide-semiconductor (CMOS) fabrication for gate electrodes.

## 3.2.4 Epitaxial Growth of Silicon

The method of growing a silicon layer on a substrate maintaining its crystalline orientation is known as the *epitaxial process*. In this, the substrate wafer acts as the seed crystal. The epitaxial process is carried out at a much lower temperature than the Czochralski process, in which the crystal is grown from the melt. Epitaxial layers of silicon can be grown by either vapor-phase epitaxy (VPE) process or molecular-beam epitaxy (MBE) process, the former being more commonly used.

Silicon tetrachloride ($SiCl_4$) is one of the commonly used silicon sources. Silicon is obtained by reducing $SiCl_4$ with hydrogen at around 1200°C:

$$SiCl_4 + 2H_2 \rightarrow Si + 4HCl$$

A mole fraction of 0.03 $SiCl_4$ in hydrogen gives a good-quality silicon epilayer. However, for mole fractions above 0.28, etching of silicon takes place. The MBE process is an epitaxial process involving a thermal beam with the silicon substrate kept under ultrahigh vacuum ($10^{-10}$–$10^{-11}$ torr). Precise control of chemical composition, impurity profiles, single-crystal multilayer structures on atomic scale are possible in MBE.

### 3.2.5 Thermal Oxidation for Silicon Dioxide

Thermal oxidation is a process by which a thin film of silicon dioxide is grown over the native silicon substrate surface. The oxidation equipment is shown in Fig. 3.8. The equipment consists of a resistance heated furnace and a cylindrical fused quartz tube containing silicon wafers vertically loaded in a slotted quartz boat.

Depending upon whether steam or dry oxygen is used, the oxidation process is called *wet oxidation process* or *dry oxidation process*. The chemical reactions involved are:

$$Si + 2H_2O \xrightarrow{900°C \text{ to } 1200°C} SiO_2 + 2H_2 \text{ (wet oxidation)}$$

$$Si + O_2 \xrightarrow{900°C \text{ to } 1200°C} SiO_2 \text{ (dry oxidation)}$$

The wet oxidation process yields faster oxide growth. However, $SiO_2$ films grown by this process are less dense and more porous. Dry oxidation results in much slower oxide growth (typically 1/10 of the growth rate of wet oxidation), producing films that are compact, dense and nonporous. It can be shown from molar calculations that the thickness of silicon consumed is about 0.44 times the oxide thickness.

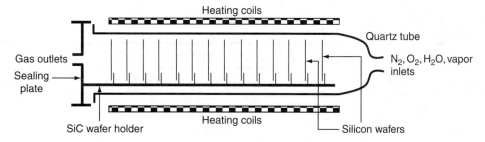

**Figure 3.8** Schematic diagram of a silicon thermal oxidation system.

## ▶ 3.3 LITHOGRAPHY

The preceding section presented processes for creating material layers required in the fabrication of microsystems. The successful development of microsystems involves successive steps of deposition and patterning of various material layers. One of the key steps in patterning is the process of transferring a geometrical pattern on a mask to a radiation-sensitive material called a *resist*. This process is known as *lithography*. As shown later, this geometrical pattern defines the area being protected or removed in a subsequent process step.

The required geometrical pattern of the mask is first prepared in digital form in a computer program. Commands from the computer drive a pattern generator that photo-engraves the particular pattern on an optically flat glass or quartz plate covered with a thin chromium film. This plate containing the patterned chromium layer for the whole wafer is known as a *mask*. The opaque portion of the mask prevents the radiation from reaching the resist on the wafer. A mask is used repetitively to expose selected regions of the resist on the wafer.

Since ultraviolet (UV) radiation (wavelength in the range of 350 to 450 nm) is most commonly used, the process is known as *photolithography*. A mercury vapor lamp with radiation wavelength 300 to 400 nm is a popular UV light source. As the resolution achievable is proportional to the wavelength of the incident radiation, deep UV (<250 nm) and extreme UV (<100 nm) radiation are used to achieve higher resolution than is possible with normal UV light.

### 3.3.1 Photolithography

Resists used in this approach are called *photoresists*. Depending upon their response to UV radiation, photoresists are classified as positive or negative photoresists. In a positive photoresist, the exposed region is removed after developing, while in a negative photo-resist, the exposed region is retained after developing (Fig. 3.9). Examples of positive and negative photoresists along with relevant parameters are given in Table 3.4.

One of the positive photoresists is PMMA. Another is the two-component system made up of diazoquinone ester and phenolic novolak resin. Positive photoresists are sensitive to UV light with maximum sensitivity in the wavelength range 300–400 nm. Most positive photoresists can be developed in alkaline solvents such as potassium hydroxide, ketones, or acetates.

Negative photoresists are typically two-component bisazide rubber resist or azide sensitized polyisotroprene rubber. Negative photoresists are less sensitive to optical and x-ray exposure but are more sensitive to electron beam. Xylene is a common developer for negative resists.

**Table 3.4 Positive and negative photoresists**

| Parameter | Positive Photoresist | Negative Photoresist |
|---|---|---|
| Commercial examples | AZ 1350J, PR120 | Kodak 747, SU8 |
| Adhesion to silicon surface | Fair | Excellent |
| Cost | High | Low |
| Developer process tolerance | Small | Wide |
| Suitable for lift-off process | Yes | Yes |
| Minimum feature size with UV | < 0.5 μm | < 2 μm |
| Opaque dirt on clear portion of mask | Not very sensitive | Causes pinholes |
| Resistance to plasma etch | Very good | Not so good |
| Step coverage | Good | Not so good |

*Note:* Modified after [5].

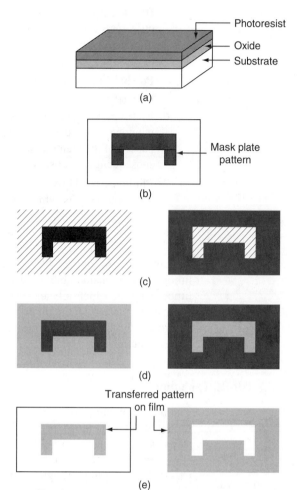

**Figure 3.9** Schematics comparing etches using positive and negative resists. (a) Start wafer with oxide thin film and resist coating; (b) mask plate with the image. (c)–(e) Various stages of pattern transfer for positive resist on the left and negative resist on the right: (c) after exposure, (d) after developing; and (e) after removal (stripping) of resist.

Some of the key processes involved in photolithography are the coating of resist, pattern transfer, and development. The resist material is entirely removed after an underlying layer is etched with the resist as the masking window.

1. **Resist coating**: A few drops of the resist materials are placed on a clean wafer. A high-speed (typically ranging from 2000 to 8000 rpm), high-acceleration spinner is used to produce a uniform coating of photoresist on the substrate surface.

2. **Prebaking**: Before transferring the pattern to the photoresist film, one must ensure that it sticks well to the surface. To improve adhesion of the film, the substrate is prebaked at 75°C to 100°C for about 10 minutes. This step removes organic solvents and releases stress in the film in addition to improving adhesion.

3. **UV exposure**: Pattern transfer by UV exposure is carried out using a mask aligner. For good results, the UV source should have proper intensity, directionality, spectral characteristics, and radiation uniformity across the exposed area.

4. **Developing**: After exposure, developing removes the softened portion of the photoresist. In positive resists, this removes the exposed region; in negative resists, it removes the unexposed region of the resist. The solvent system used for developing depends on the photoresist material.

5. **Postbaking**: Postbaking at 120°C for about 20 minutes removes residual solvents, improves further adhesion, and toughens the remaining photoresist.

6. **Etching**: Etching removes unwanted material from the substrates through the windows opened by developed regions of the photoresist. This step transfers a pattern of the material layer on to the substrate. Details of this step are discussed in subsequent sections.

7. **Photoresist stripping**: After all process steps are performed for pattern transfer, the photoresist is removed by an organic solvent.

Most pattern transfer required for microsystems can be done by photolithography. One of the limitation of this technique is its scalability to nanodevices. Fresnel diffraction of the optical beam places a limit on the minimum feature size in photolithography. Considerable reduction in feature size is possible with electron-beam lithography, in which a beam of electrons (~20 nm diameter) is directed on the wafer to write the pattern on the photoresist. The switching and movement of electron beam is controlled by a pattern generator. However, this writing process usually takes several hours to complete.

### 3.3.2 Lift-Off Technique

The lift-off technique is used to define structural geometry on the substrate. The process steps involved in the lift-off technique are shown in Fig. 3.10. The resist is spin-coated, exposed to radiation through a suitable mask, and developed in a developer. At this stage, selected portions of resist remain on the surface, as shown in Fig. 3.10(c), and this is hardened by postbaking. A thin film of the required material is then deposited above this layer of the resist [Fig. 3.10(d)]. The resist is dissolved in appropriate solvent that detaches the material on top of the resist as well, leaving the film in the position where the photoresist was removed during development [see Fig. 3.10(e)]. The unwanted material is "lifted off" while dissolving the photoresist.

Unlike conventional lithography, in the lift-off process, a photoresist pattern is generated initially on the substrate instead of etching the unwanted material. Fig. 3.11 shows the key differences in process steps in these processes. The basic criterion for the lift-off technique is that the thickness of deposited film be

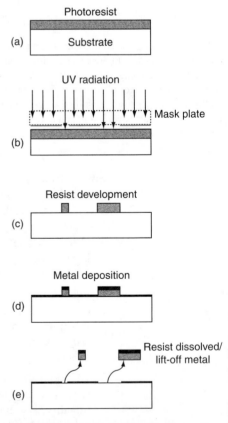

**Figure 3.10** Steps in the lift-off process of patterning.

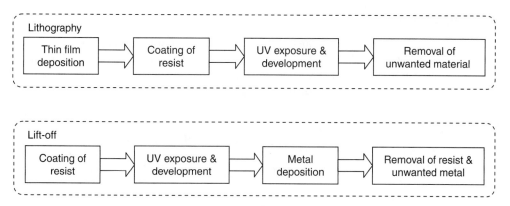

**Figure 3.11** Comparison of major process steps in lithography and lift-off-based patterning.

significantly less than that of the photoresist and that the developed patterns have vertical sidewalls. Metal layers with high resolution can be patterned using lift-off technique. Metals such as gold (which can be etched with *aqua regia*—nitrohydrochloric acid) can be patterned with simple processes by lift-off.

## ► 3.4 DOPING THE SILICON WAFER: DIFFUSION AND ION IMPLANTATION OF DOPANTS

The silicon atom has four valence electrons. Pure silicon is an insulator at 0 K because all four valence electrons are bonded with the four neighboring atoms. However, at higher temperatures the electrons gain sufficient energy to free themselves from the bonds, creating mobile electrons that are free to move in the crystal. These electrons leave behind positive charges due to the vacancies created in equal numbers in the bonds. These vacancies, or positive charges, are called *holes* and can move from bond to bond due to the jump-movement of electrons from one bond to the neighboring bond. In the thermal equilibrium situation, the hole density $p_i$ is equal to the electron density $n_i$. This semiconductor is pure and is called an *intrinsic semiconductor*. In the case of intrinsic silicon, $p_i = n_i = 1.5 \times 10^{10}/\text{cm}^3$ and the resistivity is 300 k$\Omega$.

In order to make useful devices, it is necessary to control the carrier concentration and hence the resistivity of silicon. This is achieved by a process known as doping, which involves adding impurities in controlled quantities. Impurities such as phosphorus and arsenic have five valence electrons (an excess of one electron per atom than in silicon) and hence can donate one free electron per impurity atom to the crystal. Thus if the phosphorus concentration is $10^{16}/\text{cm}^3$, there will be $10^{16}$ free electrons per cm$^3$. As the electron density in this material is several orders of magnitude greater than the hole density, this silicon is called an *n*-type silicon. Similarly, when the silicon is doped with boron, which has only three valence electrons per atom, a vacancy (a hole or a mobile positive charge) is created in silicon. This silicon is called *p*-type because the positive charges are in excess compared to the electrons.

As pointed out in Section 3.1.2, commercially available semiconductor wafers are either *p*-type or *n*-type. Doping either some selected regions or the entire surface of the semiconductor wafer plays an important role in both VLSI and microsystems technology for realizing devices. This is carried out by two different approaches: diffusion and ion implantation, which are briefly described in this section.

## 3.4.1 Doping by Diffusion

In the diffusion process, an oxide of the desired impurity dopant atom is deposited on the silicon wafer surface kept at a high temperature (in the range 900–1200°C) inside a quartz tube furnace. The oxide is reduced to the dopant atom by reaction with silicon. Examples of such reactions for diffusion of $n$-type dopants such as phosphorus and $p$-type dopants such as boron from their oxides are as follows:

$$2P_2O_5 + 5Si \rightarrow 4P + 5SiO_2$$

$$2B_2O_3 + 3Si \rightarrow 4B + 3SiO_2$$

The released phosphorus or boron diffuses into silicon. Similarly, $As_2O_3$ and $Sb_2O_3$ are used for diffusing arsenic (As) and antimony (Sb) as $n$-type dopants. The dopant oxide is deposited by passing a gas mixture containing the desired dopant through the furnace.

The dopant source can be solid, liquid, or gas, the choice depending upon the ease with which it can be incorporated. Halide-bearing liquid sources such as phosphorus oxychloride ($POCl_3$) for phosphorus diffusion and boron tribromide ($BBr_3$) for boron diffusion are preferred because the halide ($Cl_2$ or $Br_2$) is released in the reaction:

$$4POCl_3 + 3O_2 \rightarrow 2P_2O_5 + 6Cl_2$$

$$4BBr_3 + 3O_2 \rightarrow 2B_2O_3 + 6Br_2$$

The halogen assists in reducing any mobile ions into volatile compounds and the quartz tube remains free of these contaminants. A typical liquid source open-tube diffusion system is shown in Fig. 3.12. The temperature of the bubbler through which nitrogen is passed is controlled to keep the vapor pressure constant during the diffusion.

The gas source system uses phosphine ($PH_3$) gas for phosphorus and diborane ($B_2H_6$) gas for boron. Diborane and phosphine are both highly poisonous/toxic, explosive gases and are used with 99.9% dilution in helium, hydrogen or argon. The diffusion systems in this case are similar to those shown in Fig. 3.12 except that the bubbler is replaced by the dopant gas.

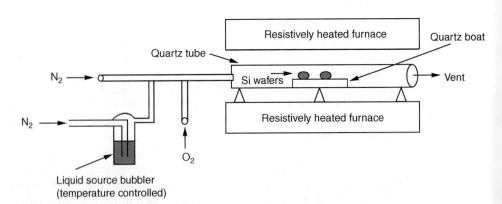

**Figure 3.12** Schematic diagram of a typical liquid source open-tube diffusion system.

Compared to gaseous sources, solid sources are extremely simple to use and hence are often preferred. A very popular solid source for boron diffusion is the boron nitride (BN) disc. These discs are available in the same size as the silicon wafer and are activated in oxygen ambience to convert the outer surface of the BN disc into $B_2O_3$; they are stacked with silicon wafers so that each dopant wafer acts as a source for two adjacent silicon wafers. The gas handling system required for the solid-source diffusion process is extremely simple because all that is needed is a flow of few liters of nitrogen or argon per minute.

When the $P_2O_5$ or the $B_2O_3$ glassy layer is formed on the silicon surface, the concentration of phosphorus or boron atoms on the surface is determined by the solid solubility of the impurity in silicon. For boron, this is in the range $5 \times 10^{19} - 2 \times 10^{20}/cm^3$ at temperatures in the range 900–1100°C, whereas for phosphorus this is in the range $6 \times 10^{20} - 10^{21}/cm^3$. The impurity atom moves into the silicon wafer by the well known diffusion mechanism. In this process the flux $F$ of impurity atoms is given by Fick's law, which says that $F$ is proportional to the concentration gradient of impurities, with the constant of proportionality defined as the diffusion coefficient, $D$. The diffusion coefficient $D$ increases exponentially with temperature, and hence at the diffusion temperatures ($> 900°C$) the dopant atoms move down the concentration gradient to a depth determined by the temperature and time, forming a junction. Thus if the starting Si wafer is $p$-type with a uniform boron doping concentration of $N_A = 10^{15}/cm^3$, a $p$-$n$ junction is formed at $x_j$ where the $n$-type impurity (phosphorus) concentration $N_D(x)$ becomes equal to $10^{15}/cm^3$, as shown in Fig. 3.13(a).

Doping can be done in the selected regions of the wafer surface by protecting the other regions with a layer of $SiO_2$ whose thickness determines the duration for which the diffusion through it can be masked. As the diffusion process is isotropic in nature, that is, taking place equally in all directions in the underlying silicon wafer, the junction is terminated on the surface as shown in Fig. 3.13(b).

In the diffusion process, since the surface concentration depends only on the solid solubility, a wide range of surface concentrations cannot be achieved with constant source diffusion when the source supply is maintained throughout the diffusion process. On the other hand, if the supply of impurities is shut down and the diffusion process is continued at a higher temperature, the surface concentration falls as the dopants move deeper. Also, in

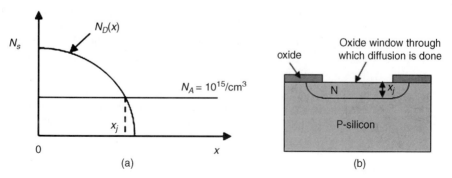

**Figure 3.13** (a) Doping profile of $n$-type impurities diffused into a wafer having $p$-type dopants with boron concentration $N_A = 10^{15}/cm^3$, showing the junction at $x_j$. Note that the $y$-axis is on the log scale to include the wide range of doping density; (b) Two-dimensional picture of the cross section of a $p$-type silicon into which phosphorus is diffused through an oxide window.

the diffusion process, it is very important to use different furnaces for each impurity in order to avoid cross-contamination of the impurities.

## Example 3.3

Phosphorus is diffused into a silicon wafer having boron concentration (i) $N_A = 10^{13}$/cm$^3$ and (ii) $N_A = 10^{17}$ cm$^3$. Determine the junction depth in the two cases. Assume that the phosphorus doping profile is $N_D(x) = N_o \exp(-x/L)$ where $N_o = 10^{20}$/cm$^3$ and $L = 1$ µm.

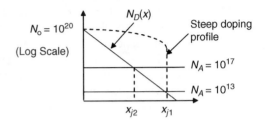

**Figure 3.14** The doping profile $N_D(x)$ in the log scale for two different background acceptor doping concentrations $N_A$. At the junction $x = x_j$ and we have $N_D(x_j) = N_o \exp(-x_j/L) = N_A$

**Solution:** The junction is formed at $x = x_j$ where $N_D(x_j) = N_A$. Note that the exponential profile is shown by a linear plot in the log scale. Therefore

$$x_j = L \ln \frac{N_o}{N_A}$$

Substituting into this equation $L = 1$ µm, $N_o = 10^{20}$/cm$^3$, we obtain $x_j$ for the two cases (i) $x_j = 16.1$ µm and (ii) $x_j = 6.9$ µm. Note that the junction depth depends upon the substrate doping density for the exponentially decaying doping profile for the diffused layer. However, if the doping profile is very steep, as in the arsenic diffusion profile (shown by dotted lines in Figure 3.14 (a) the junction depth is independent of the substrate doping density and depends only on the profile of the dopant that was diffused.

## 3.4.2 Doping by Ion Implantation

Ion implantation is an alternative to diffusion for introducing dopants into semiconductors at room temperature by means of an energetic ion beam of the dopants. Ion energies in the 50–200 keV range are used for implantation into silicon, while higher energies of up to 500 keV are used in GaAs technology. In this process, the ions of the dopants produced in an ionization chamber are accelerated in a high-voltage column, deflected by magnetic fields and a mass separation arrangement, and finally impinge on the substrate. The energetic ions striking the silicon substrate undergo a series of collisions with the host atoms, losing energy in each collision and finally coming to rest. As these collisions are statistical in nature, the implanted dopant profile is Gaussian with its peak located inside the semiconductor at a distance from the surface $R_p$, referred to as the projected range, with standard deviation called the straggle $\Delta R_p$. The ions are neutralized by electrons entering from the negative electrode on which the substrate is placed. The number of ions implanted per unit area, called the implantation dose $N_T$, can be precisely determined by measuring the beam current, which is equal to the electron current flowing out of the negative electrode. Mostly, singly charged ions are used for ion implantation. The parameters of the implanted doping profile are determined mainly by the mass and the implanted energy of the ion. The implantation profile $N(x)$ can be shown to be given by.

$$N(x) = \frac{N_T}{\sqrt{2\pi}(\Delta R_p)} \exp\left[-\frac{1}{2}\left(\frac{x - R_p}{\Delta R_p}\right)^2\right] \tag{3.3}$$

A typical doping profile with ion implantation is shown in Fig. 3.15 (a); Fig. 3.15 (b) shows the two-dimensional cross-section of the $n$-type region implanted through a mask into the $p$-type silicon substrate. Unlike in diffusion, the lateral spread is negligible beneath the mask layer near the surface. However, deeper inside the substrate, lateral spread occurs close to $R_p$ due to lateral straggle.

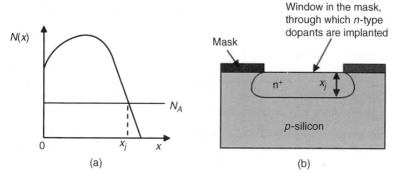

(a)  (b)

**Figure 3.15** (a) Doping profile of ion-implanted $n$-type dopant and the background $p$-type dopant $N_A$; (b) Two-dimensional picture showing cross-section of the $p$-type silicon into which phosphorus is implanted through a window in the mask layer.

A wide range of implantation doses, from $10^{11}$ to $10^{17}/\text{cm}^2$, can be achieved. The projected range and straggle $\Delta R_p$ are dependent on the mass of the implanted ion, in addition to the implantation energy and the substrate onto which implantation is carried out. A typical value of $R_p$ for phosphorus implantation onto silicon is 0.1 $\mu$m/100 keV. As the boron atom is about three times lighter than the phosphorus atom, the projected range for boron implantation onto silicon is higher, approximately equal to 0.3 $\mu$m/100 keV in the energy range of 10–100 keV. The corresponding $\Delta R_p$ at 100 keV is about 0.03 $\mu$m and 0.6 $\mu$m, respectively, for phosphorous and boron. During implantation most of the dopant atoms land in the interstitial sites of the crystal and they are inactive. Furnace annealing at high temperatures of up to 900°C is required to activate the dopants. Alternatively, rapid thermal annealing can be done for very short durations to retain the doping profile.

There are several advantages for doping with ion implantation over diffusion:

1. By monitoring the beam current, the dose can be precisely controlled to within ±1% over a wide range of implantation doses, from $10^{11}$ to $10^{17}/\text{cm}^2$. In contrast, in the diffusion process, at best 5%–10% control can be achieved in the dopant concentration.

2. The mass separation technique in ion implantation ensures a highly pure beam of dopant atoms. Hence, a single machine can be used for a variety of impurity implantations.

3. That ion implantation can be carried out at room temperature makes it possible to use a variety of masking materials for implantation in selected regions on the wafer surface. Thus a standard photoresist patterned by photolithography, as discussed in Section 3.3.1, can be easily used as the masking layer during ion implantation; this is not possible in diffusion, which is carried out at high temperatures.

However, the equipment for ion implantation is highly sophisticated and expensive compared to diffusion systems. Also, the implantation introduces damage to the host semiconductor substrate and hence requires annealing for repair. Batch processing of hundreds of silicon wafers is possible with diffusion using furnaces, while this cannot be done with ion implantation even though this problem has been solved to some extent by cassette loading of wafers.

## Example 3.4

Phosphorus implantation is done at 100 keV onto a silicon wafer doped with $N_A = 10^{17}/\text{cm}^3$. Implantation is done for 16s with a beam current $I = 10$ μA. The implantation area is 10 cm². (a) Determine the dose for the two cases when the implantation ions are (i) singly charged and (ii) doubly charged. (b) Determine the peak concentration of the implanted phosphorus for the above two cases, assuming that the straggle $\Delta R_p = 0.03$ μm and $R_p = 0.1$ μm at 100 keV energy.

**Solution:**   Case (i): (a) When the implanted ions are singly charged, each implanted ion requires the flow of one electron to neutralize the implanted charge. The product of the beam current $I$ and the implantation duration $t$ gives the total charge flow. For the singly charged case, the dose $N_T$ is given by

$$N_{T1} = \frac{1}{q \times \text{area}}(i \times t) = \frac{10^{-6} \times 16}{1.6 \times 10^{-19}} = 10^{14}/\text{cm}^2$$

$$N(x) = N_{p1} \exp\left[-\frac{1}{2}\left(\frac{x - R_p}{\Delta R_{p1}}\right)^2\right] \qquad (b)$$

where

$$N_{p1} = \frac{N_{T1}}{\sqrt{2\pi}\,(\Delta R_{p1})} = \frac{0.4 \times 10^{14}}{\Delta R_{p1}} = \frac{4 \times 10^{13}}{3 \times 10^{-6}} = 1.33 \times 10^{19}/\text{cm}^3$$

Case (ii): When the implantation is done with doubly-charged dopant, two electrons flow to neutralize each implanted dopant and the energy acquired by the implanted charge is 200 keV. Hence, assuming linear dependence of $R_p$ and $\Delta R_p$ on this energy acquired are, respectively, $R_p = 0.2$ μm and $\Delta R_p = 0.06$ μm

(a) For the same value of beam current $I = 10$ μA, the implantation dose $N_{T2} = \frac{N_{T1}}{2} = 5 \times 10^{13}/\text{cm}^2$.

(b) $\Delta R_{p2} = 2\,\Delta R_{p1}$, because the energy acquired is double. Hence the peak doping concentration $N_{p2} = N_{p1}/4 = 3.325 \times 10^{18}/\text{cm}^3$.

**Your Turn:**
Determine the junction depth for the two cases (i) and (ii).

## ▶ 3.5 ETCHING

Etching is used to remove unwanted material from a layer of interest. As discussed earlier in the context of lithography, the etching step is performed after a protection layer, that is, the photoresist, above the target layer is patterned. For deeper etching such as bulk etching of silicon, a hard mask of other deposited material may be required. Etching is a chemical reaction in which no solid precipitate is formed. It can be done either in the liquid or the gaseous phase. In the liquid-phase reaction, known as *wet etching*, the material is removed by chemical reaction between the etching solution and the material on the substrate surface. In *dry etching*, this is done by chemical/physical interaction between atoms of an ionized gas and the substrate.

Etching processes (wet and dry) are characterized by etch rate, etch selectivity, and etch uniformity. *Etch rate* is the thickness of material removed per unit time, whereas *etch selectivity* is the ratio of etch rates of the material to be removed to other material present on the wafer. For example, buffered hydrofluoric (HF) acid removes silicon dioxide much faster than silicon. Thus, buffered HF has high etch selectivity for $SiO_2$ with respect to Si. Etch uniformity is the uniformity in etch depth of the etched surface.

In addition to its use in wafer processing (e.g., polishing wafers to achieve optically flat and damage-free surfaces), wet chemical etching is used in semiconductor device fabrication to delineate patterns and open windows in insulating layers. In most wet etching processes, the etch rate is uniform in all directions, that is, it is isotropic and does not depend on crystal structure/orientation.

The basic steps involved in the wet etching process are:

1. Injection of holes into the silicon substrate to create the $Si^+$ state.
2. Attachment of a negatively charged OH group to positively charged Si.
3. Reaction between hydrated Si and complex agents in etchant solution.
4. Dissolving of reaction products into etchant solution.

Wet etching is carried out by dipping the substrate in an etching solution or spraying the etching solution on its surface. This involves the transport of reactants to the surface of the substrate, a chemical reaction there, and the subsequent removal of the products of this chemical reaction. Noteworthy features of wet etching are:

1. Relatively inexpensive equipment is used.
2. Corrosive alkali and acids are used.
3. Waste products are also corrosive.
4. It is difficult to automate.

Due to these features, the wet etching approach is preferred in prototyping but not so much in industrial fabrication facilities. Most semiconductors, metals, and insulators can be etched by wet etching. In crystalline materials, such as SCS, both isotropic and anisotropic etching are possible by suitably selecting the etching chemicals.

### 3.5.1 Isotropic Etching

In isotropic etching of silicon, the etch rate is independent of crystalline direction, with the result that etching takes place in all directions and may even be uniform in all directions. Because of this non-selective nature, undercuts and rounded patterns are unavoidable

(Fig. 3.16). Precise control of lateral and vertical etch rates is difficult, since both depend upon temperature and agitation. Because of lateral etching, masking is difficult in isotropic etching. A commonly used isotropic etchant for silicon is a mixture of HF (3 ml), nitric acid (HNO$_3$, 5 ml) and acetic acid (CH$_3$COOH, 3 ml), usually referred to as HNA. This mixture etches silicon at an etch rate of about 0.4 μm/min at 300 K.

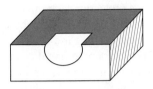

**Figure 3.16** Profile of an isotropically etched wafer.

Silicon etching by HNA may be described by the following chemical processes. Acetic acid in the HNA mixture prevents dissociation of HNO$_3$, which is an oxidant. HNO$_3$ oxidizes Si as indicated by

$$Si + 4HNO_3 \rightarrow SiO_2 + 2H_2O + 4NO_2$$

The resulting silicon dioxide reacts with HF acid and dissociates to form H$_2$SiF$_6$, which dissolves in water:

$$SiO_2 + 6HF \rightarrow H_2SiF_6 + 2H_2O$$

Isotropic etching is used to round off sharp edges so as to avoid stress concentration; to remove roughness after dry/anisotropic etching so as to achieve a smooth polished surface; to thin the silicon substrate; to pattern and delineate junctions; and to evaluate defects. Because of lateral etching, deep etching is generally not carried out by isotropic etching.

Table 3.5 lists commonly used etchants and their respective etch rates for various materials in the fabrication of microsystems. In device processing, etch selectivity is another important parameter to be considered.

### 3.5.2 Anisotropic Etching

Some wet etchants are orientation-dependent, that is, they dissolve a given plane of semiconductor material much faster than other planes (Table 3.6). The more closely the plane is packed, the smaller is the dissolution by the etchant. The {111} planes of a SCS, have the highest packing density and are thus non-etching compared to other planes. (Refer to Miller indices discussed in Section 3.1) The angle between the [100] and [111] planes is 54.74° and that between the [110] and [111] planes can be made 90°.

Although KOH-based etchants have good anisotropy in their etch characteristics, these are not compatible with IC fabrication process because they attack Al bond pads. Hence these

**Table 3.5 Etchants used for some metals and dielectrics and their etch rates at 300 K**

| Material | Etchant | Etch rate (nm/min) |
|---|---|---|
| Si | 3 ml HF + 5 ml HNO$_3$ + 3 ml CH$_3$COOH | 35,000 |
| SiO$_2$ | 28 ml HF + 170 ml H$_2$O + 113 g NH$_4$F | 100 |
| | 15 ml HF + 10 ml HNO$_3$ + 300 ml H$_2$O | 12 |
| Si$_3$N$_4$ | Buffered HF | 15–20 |
| | H$_3$PO$_4$ | 10 |
| Al | 1 ml HNO$_3$ + 4 ml CH$_3$COOH + 4 ml H$_3$PO$_4$ + 1 ml H$_2$O | 35 |
| Au | 4 g KI + 1 g I$_2$ + 40 ml H$_2$O | 10,000 |
| TiW | H$_2$O$_2$ | 5 |

*Note:* Modified after [3].

**Table 3.6 Anisotropic etching characteristics of different wet etchants for SCS**

| Etchant | Temperature (°C) | Etch Rate (μm/h) | | |
|---|---|---|---|---|
| | | $\langle 100 \rangle$ | $\langle 110 \rangle$ | $\langle 111 \rangle$ |
| KOH:H$_2$O | 80 | 84 | 146 | 0.27 |
| EDP | 110 | 75 | 37 | 3.7 |
| N$_2$H$_4$H$_2$O | 118 | 176 | 99 | 11 |
| NH$_4$OH | 75 | 24 | 8 | 1 |
| TMAH | 80 | 20–60 | 1–3 | 0.5–1.5 |

*Note:* Modified from [3].

etchants are not preferred even though these have an etch selectivity of about 400:1 with silicon dioxide. A mixture of ethylene diamine pyrocatechol and water known as EDP is usually preferred for etching silicon when masking materials such as SiO$_2$, Si$_3$N$_4$, Au, Cr or Ag are used. EDP has an etch selectivity with SiO$_2$ of 5000:1 for etching silicon. Incidentally, photoresist is seldom used as an etch mask for deep etching of silicon, as the resist becomes unstable and dissolves or peels off after a short time. Hence, a hard mask made of one of the indicated materials is used in such cases. Alkaline aqueous solutions (KOH, NaOH) and alkaline organics (ethylene diamine) are commonly used as anisotropic etchants. Problems associated with anisotropic etching are the requirement of large separation between two parts (because of the slanted profile after etching) and the fragility of fabricated devices.

A comparison of etch rates along different directions of silicon is provided in Table 3.6. For [100]-oriented silicon, if the mask is oriented along the [110] direction, only [111] planes form as side walls. It may be noted that these ⟨111⟩ directions are orthogonal and hence a rectangular opening is formed even if a nonrectangular etch window is used.

As the etching process continues, truncated pyramids grow in the thickness direction, do not spread in the lateral directions, and close when the depth is 0.7 times the width of the rectangular opening (Fig. 3.17). For larger widths, truncated pyramids can be obtained as shown in Fig. 3.17(c). The sidewalls are [111] plane whereas the bottom is [100] plane, parallel to the wafer surface. A photograph of grooves etched on a silicon wafer appears in Fig. 3.18.

As one may notice, sidewalls of the etch profile are always slanted if [100] wafers are used. As shown in Fig. 3.17(d), two of these may be made vertical if [110] wafers are used.

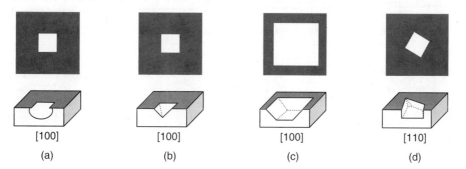

[100]　　　　　[100]　　　　　[100]　　　　　[110]

(a)　　　　　　(b)　　　　　　(c)　　　　　　(d)

**Figure 3.17** Wafer profile of anisotropic etch under various scenarios compared with isotropic etch. (a) Isotropic etch yields rounded profile; (b) anisotropic etch with a small window yields pyramidal profile; (c) for a large window, the profile would be a truncated pyramid if the anisotropic etch is stopped; and (d) anisotropic etch using a [110] wafer yields predominantly vertical sidewalls.

**Figure 3.18** Microphotograph of grooves etched on a silicon wafer.

**Table 3.7 Features of anisotropic etching of (100)- and (110)-oriented silicon wafers**

| (100)-Oriented Si | (110)-Oriented Si |
|---|---|
| Inward sloping wall (54.74°) | Vertical sidewalls possible |
| Sloping walls result in loss of real estate | Narrow trenches with vertical side walls possible |
| Flat bottom parallel to the top surface possible | Multifaceted cavity bottom |
| Underetch of bridges possible (only) in some cases | Bridges cannot be formed. |
| Shape and orientation of diaphragms can be conveniently designed | Shape and orientation of diaphragms are difficult to design |
| Easy to control diaphragm dimensions | Difficult to control diaphragm dimensions |

Note: Modified after [3].

However, care must be taken in aligning the length of the channel along the [111] direction. It can be seen that the other sides are not perpendicular to this (unlike in [110] wafers) and that other slanted [111] planes may also be formed in this case. Features of anisotropic etching of [100]- and [110]-oriented silicon wafers are compared in Table 3.7.

It may be reiterated that the sides of the mask must be aligned with the ⟨110⟩ directions on the wafer surface. Hence, the etch window must have convex rectangular geometries for the mask. If other geometries are used, the etch profile would still be a truncated pyramid whose base would be a rectangle superscribing the etch window and aligned with ⟨110⟩ directions on the wafer surface. In other words, misalignment of the mask results in undercut of the etch window, as pyramidal pits or V-grooves are formed [Fig. 3.19(a), (b)]. This feature can also be exploited to form a free-standing cantilever [Fig. 3.19(c)].

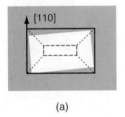
(a)

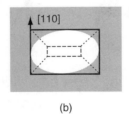
(b)

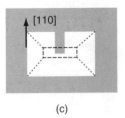
(c)

**Figure 3.19** A few special cases of anisotropic etching: (a) If the masking window is misaligned with the crystal directions, the etch profile is bounded by a superscribing regular geometry; (b) Similar results for nonrectangular etch windows; and (c) A free-standing cantilever of the masking layer results from anisotropic etching with aligned geometry shown.

**Example 3.5**

(a) In a cubic crystal like silicon, determine the angle included between (i) (100) plane and (111) plane; (ii) (100) plane and ($1\bar{1}1$) plane.

(b) Using the above answer, show a 3D picture of the cavity formed when etched though an oxide layer having a square window, whose sides are parallel to the $\langle 110 \rangle$ and $\langle 1\bar{1}0 \rangle$ directions on a (100) wafer.

**Solution:** (a) The angle $\theta$ included between two planes $(u_1 v_1 w_1)$ and $(u_2 v_2 w_2)$ is given by

$$\cos \theta = \frac{u_1 u_2 + v_1 v_2 + w_1 w_2}{\sqrt{\left(u_1^2 + v_1^2 + w_1^2\right)\left(u_2^2 + v_2^2 + w_2^2\right)}}$$

(i) To find the angle $\theta$ included between (100) and (111) planes: In this case, $u_1 = 1$, $v_1 = 0$, $w_1 = 0$ and $u_2 = 1$, $v_2 = 1$, $w_2 = 1$. Substituting these values, we obtain $\cos \theta$ as

$$\cos \theta = \frac{1 + 0 + 0}{\sqrt{1^2 \times (1^2 + 1^2 + 1^2)}} = \frac{1}{\sqrt{3}}$$

Therefore, $\theta = 54.74°$

(ii) To find the angle $\theta$ included between (100) and ($1\bar{1}1$) planes: Note that the ($1\bar{1}1$) plane intercepts the $x$-, $y$-, and $z$-axes at $x = 1$, $y = -1$, $z = 1$. So in this example $u_1 = 1$, $v_1 = 0$, $w_1 = 0$ and $u_2 = 1$, $v_2 = -1$, $w_2 = 1$. Substituting these values, we obtain $\cos \theta$ as

$$\cos \theta = \frac{1 + 0 + 0}{\sqrt{1^2 \times \left[1^2 + (-1)^2 + 1^2\right]}} = \frac{1}{\sqrt{3}}$$

Therefore, $\theta = 54.74°$

(b) The $x$–$y$-plane belongs to the family of (100) planes and hence is denoted by the general symbol {100}. The $\langle 110 \rangle$ and $\langle 1\bar{1}0 \rangle$ directions are perpendicular to each other and lie on the {100} plane, as shown in Fig. 3.20. In the same figure we show that the (111) plane and ($1\bar{1}1$) planes intersect the (100) plane in these directions.

Consider a square window opened in oxide grown on a (100) wafer with the two sides of the window parallel to the 110-line and the other two sides parallel to the $1\bar{1}0$-line, as shown in Fig. 3.20. If this wafer is etched using an anisotropic etchant that delineates the (111) planes, one can get cavities with all the four sides making 54.74° with the wafer surface, as shown in Fig. 3.17 (b) and (c). For example, consider two cases where the etch rate along the $\langle 100 \rangle$ direction is 1 μm/minute and is zero along the $\langle 111 \rangle$ direction.

Case (i): If the square window dimensions are 100 μm × 100 μm and the etch duration is 70.7 minutes, the planes belonging to the {111} family will meet at a depth of 70.7 μm and further etching does not take place. The cutaway view will be as in Fig. 3.17(b).

Case (ii): If the square window is 200 μm × 200 μm and the etch duration is 70.7 minutes, the cutaway view will be as in Fig. 3.17(c), with the etch depth equal to 70.7 μm. The cross sections for the two cases are shown in Fig. 3.21.

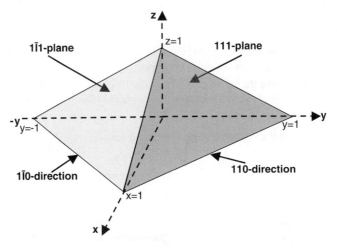

**Figure 3.20** Diagram showing the crystal planes and orientations.

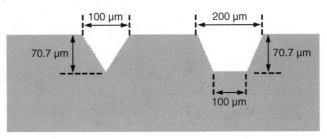

**Figure 3.21** Cross-section of the cavities for the two cases having two different window dimensions.

## Example 3.6

**(a)** Determine the {111} family that is perpendicular to the surface of a silicon wafer having (110) orientation.

**(b)** Examine the possibility of making trenches having vertical side walls on the (110) wafer using anisotropic wet chemical etchant such as KOH or TMAH.

**Solution:** Substituting in the equation governing the angle $\theta$ between two planes $(u_1v_1w_1)$ and $(u_2v_2w_2)$, we obtain the angle between the various {111} planes and (110) as:

**(i)** Angle between (110) and (111) planes: $\cos\theta = \frac{2}{\sqrt{2\times3}} = \sqrt{\frac{2}{3}} \Rightarrow \theta = 35.2°$.

**(ii)** Angle between (110) and $(1\bar{1}1)$ planes: $\cos\theta = \frac{1-1}{\sqrt{2\times3}} = 0 \Rightarrow \theta = 90°$.

**(iii)** Angle between (110) and $(\bar{1}11)$ planes: $\cos\theta = \frac{-1+1}{\sqrt{2\times3}} = 0 \Rightarrow \theta = 90°$.

**(iv)** Angle between (110) and $(11\bar{1})$ planes: $\cos\theta = \frac{1+1}{\sqrt{2\times3}} = \sqrt{\frac{2}{3}} \Rightarrow \theta = 47.78°$.

This shows that only two of the planes from the {111} family, namely $(1\bar{1}1)$ and $(\bar{1}11)$, are perpendicular to the (110) wafer surface. The angle between these two planes $(1\bar{1}1)$ and $(\bar{1}11)$ (using the $\cos\theta$ formula in Example 3.5) turns out to be 109.5°. Hence, we cannot realize square or rectangular trenches on a (110) wafer. However, one can achieve long channels having vertical side walls parallel to each other by terminating (i.e., dicing the wafer) in the direction perpendicular to the channel length.

If a rectangular or square window is opened on the (110) wafer and etched using KOH or TMAH anisotropic etchant, then the cutaway view of the etched pattern will be as in Fig. 3.17(d).

**Table 3.8 Dopant-dependent etch rates of etchants used in wet etching of silicon**

| Etchant (Diluent) | Temperature (°C) | (100) Etch rate (μm/min) for boron doping $\ll 10^{19}$ cm$^{-3}$ | Etch rate (μm/min) for boron doping $\sim 10^{20}$ cm$^{-3}$ |
|---|---|---|---|
| EDP ($H_2O$) | 115 | 0.75 | 0.015 |
| KOH ($H_2O$) | 85 | 1.4 | 0.07 |
| NaOH ($H_2O$) | 65 | 0.25–1.0 | 0.025–0.1 |

## 3.5.3 Etch Stops

Properties that make etchants indispensable in micromachining of planar structures with appropriate thickness are their selectivity and directionality. A region surrounding the etched volume that slows down the etching process is known as an *etch stop*. It is relatively easy to create the masking materials for defining the etch window–this can be achieved by oxide, nitride, or most metal films. However, the depth of etching cannot be controlled precisely even if the process is accurately timed. The uniformity of floor level after etching is controlled by one of the etch-stop techniques described in the following paragraphs.

### 3.5.3.1 Dopant-Selective Etching

Silicon membranes can be fabricated using the etch-stop property of heavily boron-doped layers, known as *dopant-selective etching*. Such thin-doped layers can be grown epitaxially or formed by the diffusion or implantation of boron into a lightly doped substrate. This stopping effect is a useful property of common etching solutions such as KOH, NaOH, and EDP (see Table 3.8).

The main benefits of the boron-doped etch stop are the independence of crystal orientation and the smooth surface finish. In addition, this approach offers the possibility for fabricating released structures with arbitrary lateral geometry in a single-etch step. On the other hand, the high levels of boron required are known to introduce considerable mechanical stress in the material that may cause buckling or even fracture of diaphragms or doubly-clamped beams. Moreover, the introduction of electrical components for sensing purposes into these microstructures, such as the implantation of piezoresistors, is inhibited by the excessive background doping.

### 3.5.3.2 Electrochemical Etch-Stop Technique

The conventional *electrochemical etch-stop technique* is another attractive method for fabricating microsensors and microactuators, since it has the potential for allowing reproducible fabrication of moderately doped *n*-type silicon microstructures with good thickness control. However, a major limiting factor in this process is the effect of reverse-bias leakage current in the junction. Since the selectivity between *n*-type and *p*-type silicon in this process is achieved through the current-blocking action of the diode, any leakage in this diode affects the selectivity. In particular, if the leakage current is very large, etching may terminate well before the junction is reached. In some situations, the etching process may fail completely because of this leakage. A dopant-selective technique that uses pulsed anodizing voltages applied to silicon samples immersed in etching solutions can be used to overcome this problem.

### 3.5.3.3 Bias-Dependent Etching

In *bias-dependent etching*, oxidation is promoted by a positive voltage applied to the silicon wafer, which causes an accumulation of holes at the Si-solution interface. Under these conditions, oxidation at the surface proceeds rapidly while the oxide is readily dissolved by the solution. Holes are transported to the cathode and are released as hydrogen. Therefore, creating excess hole–electron pairs at the silicon surface by optical excitation increases the etch rate.

### 3.5.3.4 Pulsed Potential Anodization Technique

The *pulsed potential anodization technique* is used to selectively etch *n*-type silicon. The difference in dissolution time of anodic oxide formed on *n*-type and *p*-type silicon samples under identical conditions is used for etch selectivity. This dissolution time difference is due to a difference in oxidation rates caused by the limited supply of holes in *n*-type regions. This technique is applicable in a wide range of anodizing voltages, etchant compositions, and temperatures. It differs from the conventional *p-n* junction etch stop in that the performance of the etch stop does not depend on the rectifying characteristics or quality of a diode. Using this technique, *p*-type microstructures of both low and moderate doping can be fabricated. Hence, the pulsed potential anodization technique can also be used for fragile microstructures in *p*-type silicon.

The main problems in the conventional electrochemical etch stop and the pulsed potential anodization techniques are related to the special handling required in contacting the epitaxial layer and protecting the epitaxial side of the wafer from the etchant. Any leakage in these holders interferes with proper operation of the etch stop. Moreover, mechanical stress introduced by the holder substantially reduces production yield. Developing a reliable wafer holder for anisotropic etching with electrochemical etch stop is not straightforward. The process of making contact to the wafer itself can also be critical and difficult to implement. Therefore, single-step fabrication of released structures with either a conventional electro-chemical etch stop or pulsed potential anodization techniques may be troublesome.

### 3.5.3.5 Photovoltaic Electrochemical Etch-Stop Technique

An alternative etch-stop technique that does not require any external electrodes is referred to as the *photovoltaic electrochemical etch-stop technique* (PHET). PHET does not require the high-impurity concentrations of the boron etch stop and also does not require an etch holder, as do the conventional electrochemical etch-stop or pulsed anodization technique. Free-standing *p*-type structures with arbitrary lateral geometry can be formed in a single etch step. In principle, PHET should be seen as a two-electrode electrochemical etch stop where the potential and current required for anodic growth of a passivating oxide are not applied externally, but are generated within silicon itself.

### 3.5.3.6 Buried Oxide Process

The *buried oxide process* generates microstructures by exploiting the etching character-istics of a buried layer of silicon dioxide. After oxygen is implanted into a silicon substrate using suitable ion implantation techniques, high-temperature annealing causes the oxygen ions to interact with the silicon to form a buried layer of silicon dioxide. The remaining thin layer of SCS can still support the growth of an epitaxial layer from a few microns to many tens of microns thick. In micromachining, the buried silicon dioxide layer is used as an etch stop. For example, the etch rate of an etchant such as KOH decreases markedly as the etchant reaches the silicon dioxide layer.

## ▶ 3.6 DRY ETCHING

As noted in previous subsections, wet chemical etching can be used for selectively removing thin films and can be extended to isotropically or anisotropically etching the silicon wafer. However, there are several issues related to wet chemical etching:

1. Wet chemical etching of thin films such as polysilicon, metal, $SiO_2$, silicon nitride, etc., is isotropic in nature. Therefore, if the thickness of film being etched is

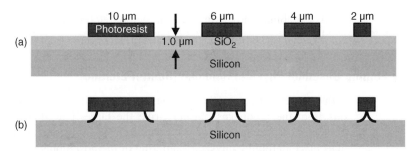

**Figure 3.22** (a) Photoresist patterned on $SiO_2$ grown on silicon; (b) After wet chemical etching of $SiO_2$ using the photoresist as the etch mask.

comparable to the minimum lateral pattern dimension, the undercutting due to such isotropic etching becomes intolerable, as shown in Fig. 3.22. Since this process is isotropic, the $SiO_2$ layer, which is 1 μm thin is almost completely etched when the photoresist pattern width is 2 μm.

2. Anisotropic etching of single-crystal silicon by wet etching techniques such as KOH or TMAH results in slanted etch geometries. As a result, vertical etching cannot be done with this approach, resulting in inadequate usage of silicon material.

3. Disposal of the used chemicals is a major industrial challenge.

Dry etching can serve as a replacement to wet etching in several ways:

1. It offers the capability of non-isotropic (or anisotropic) etching of single-crystal, polycrystal, and amorphous materials such as silicon, oxides, nitrides, and metals.

2. Etch geometries with vertical side walls can be achieved even in single-crystal silicon and other semiconductors.

3. Dry etching also eliminates the important manufacturing disadvantage of handling, consumption, and disposal of the large quantities of dangerous acids and solvents used in wet etching.

4. Dry-etching and resist-stripping operations use relatively small amounts of chemicals.

However, the prohibitive cost of the equipment and other infrastructure required for dry etching usually compels many small laboratories to be content with wet-etching-based processes for their developmental activities. Nevertheless, for fine geometry patterning, dry etching is indispensable.

Dry etching is synonymous with plasma etching because all the dry etching processes use a plasma of either chemically inert or active species. Plasma is a fully or partially ionized gas composed of equal numbers of positive and negative charges and different un-ionized molecules and radicals. Plasma is produced by the collision of electrons energized by an electric field of sufficient magnitude, causing the gas to breakdown and become ionized and is considered the fourth state of matter.

Several types of dry etching processes based on the etching mechanism can be attempted:

1. Physical removal: (a) ion milling or ion-beam etching (IBE); (b) glow-discharge sputtering.

2. Chemical removal: plasma etching.

3. Combination of 1 and 2: reactive ion etching (RIE).

4. Deep reactive ion etching (DRIE), a special case of RIE.

### 3.6.1 Dry Etching Based on Physical Removal (Sputter Etching)

In this process, the gases (e.g. argon) used are chemically non-reactive (inert). The $Ar^+$ ions from the argon gas plasma strike the target placed on the cathode and transfer momentum to sputter (dislodge) atoms from the substrate surface. The etching process takes place purely by physical sputtering of the host atoms either in an ion beam etching (IBE) or a glow-discharge sputtering process.

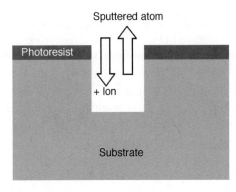

As the process relies predominantly on the physical mechanism of sputtering, the strongly directional nature of incident energetic ions allows substrate materials to be removed in a highly anisotropic manner (see Fig. 3.23) and does not distinguish between different layers, thus making it useful for etching multilayer structures. Due to the poor selectivity of the process (i.e., etching takes place irrespective of the material type), however, it becomes difficult to adopt for selectively etching some layers on the substrate. However, very thick masking photoresist layers have been successfully used for this purpose, as shown in Fig. 3.23. As the excitation energy of the heavy ions is in the range 1–3 keV, radiation damage is possible with this process, which would lead to deterioration of the properties of active devices.

**Figure 3.23** Cross-section of substrate after IBE (sputter etching).

### 3.6.2 Dry Etching Based on Chemical Reaction (Plasma Etching)

In the plasma etching process, an RF glow discharge produces chemically reactive species (atoms, radicals, and ions) from a relatively inert molecular gas. The etching gas is selected to generate species that react chemically with the material to be etched and whose reaction products are volatile so that they can be vented through the exhaust. Thus, the process is based purely on chemical reactions and is hence highly selective against both the mask layer and the underlying substrate layer.

When plasma etching takes place purely by chemical reaction of the radical with the substrate, the etching process is called plasma etching. For instance, fluorine-containing gases such as $SF_6$ or $CF_4$ are used extensively for etching silicon, silicon dioxide ($SiO_2$), silicon nitride ($Si_3N_4$), etc. The most abundant species in $CF_4$ and $SF_6$ plasma are the radicals, such as F, $CF_3$, and $SF_5$, that are formed by the dissociation reactions given by the equations

$$e^- + CF_4 \rightarrow CF_3 + F + e^-$$

$$e^- + SF_6 \rightarrow SF_5 + F + e^-$$

In the $CF_4$ plasma and $SF_6$ gas plasma, in addition to these radicals, ionic species such as $CF_3^+$ are formed by the ionic reactions given by the equation:

$$e^- + CF_4 \rightarrow CF_3^+ + F + 2e^-$$

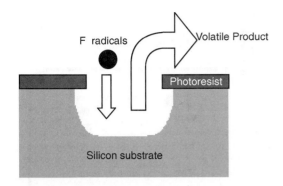

**Figure 3.24** Cross-section of silicon after plasma etching.

which take place along with the dissociation reaction. However, the radicals exist in much higher concentrations than the ions, because they are generated at a faster rate and survive longer than ions. These free radicals are neutral and exist in a state of incomplete chemical bonding, making them chemically very reactive. Thus, when $SF_6$ is used to etch silicon, the F radicals react with silicon and attach to convert Si to a volatile product $SiF_4$ by successive addition of F radicals as follows, until four F radicals are attached:

$$Si + F \rightarrow SiF + F \rightarrow SiF_2 + F \rightarrow SiF_3 + F \rightarrow SiF_4$$

When the four F radicals are attached to silicon, they occupy all silicon's covalent silicon bonds. Hence, the product $SiF_4$ is unstable and gets detached from the silicon, and is pumped out to the exhaust. As this process relies only on chemical mechanisms for etching, the reaction is isotropic, as shown in Fig. 3.24.

The plasma etching process, which depends only on the chemical reaction of the radicals, does not provide a solution to the problem of anisotropy and undercutting. However, it is suitable for photoresist stripping using oxygen gas plasma in the simple barrel-type plasma reactor shown in Fig. 3.25. The oxygen radicals are created in the RF plasma by the following reaction:

$$e^- + O_2 \rightarrow 2\dot{O} + e^- \rightarrow O + O + e^-$$

Oxidation, or burning of photoresist, is caused by these radicals at almost room temperature or at most at 40–50°C. Photoresists of 1 μm thickness can be etched in 5–10 minutes using this plasma etching process.

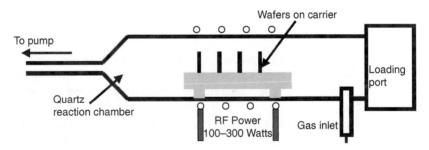

**Figure 3.25** Schematic diagram of RF-excited tubular plasma reactor.

### 3.6.3 Reactive Ion Etching

A plasma etching system that uses a chemical reaction for etching in the presence of energetic ions is called reactive ion etching (RIE). A schematic diagram of the system is shown in Fig. 3.26. In this process the substrate to be etched is placed on an electrode, C (cathode) of area $A_c$ that is driven by the RF power supply. The chamber wall serves as the anode electrode A whose area $A_a$ is large compared to that of the electrode C. This larger electrode is grounded.

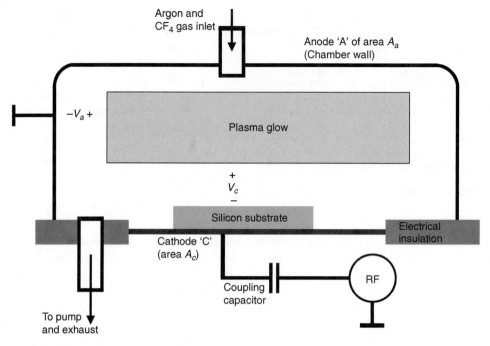

**Figure 3.26** Schematic diagram of RIE system.

The plasma is generated in a gas mixture of the reactive gas and an inert gas like argon. When an RF voltage is applied as shown, electrodes A and C acquire a voltage $V_a$ and $V_c$ due to the relatively fast movement of the electrons and the greater mass of the ions, which mostly become confined to the plasma core region. The voltages developed are related to the area of the electrodes as given by the relation

$$\frac{V_c}{V_a} = \left(\frac{A_a}{A_c}\right)^n$$

As $A_a > A_c$, and in practice, $n$ is between 1 and 2, we can see that $V_c > V_a$. Due to this high sheath voltage, the ions acquire energy and assist in enhancing etch rates produced by the reactive gas in directional etching. This is the reason why this process is called reactive ion etching.

This enhancement in the reaction rate is due to the lattice damage close to the surface by the relatively high energy ($> 50$ eV) of the impinging argon ions. Chemical reaction by the reactant species is higher at these damaged sites than in regions where no damage has occurred. However, the side walls receive a much smaller flux of the bombarding high-energy ions and do not get damaged. Hence, the reaction rate at the side walls is considerably smaller than in the vertical direction where the ion bombardment is received. This difference in the lateral and vertical etch rates makes the RIE process highly anisotropic. Etch rates in the range 40 nm/100 nm to microns per minute can be achieved with this process. Photoresist can be used to some extent as the masking layer to etch the oxide and silicon from selected regions. However, as the etch rate of photoresist in $CF_4$ plasma is comparable to the etch rates of $SiO_2$ and silicon, it is necessary to use photoresist layers with thickness greater than the etch depth required. Alternately, hard masks such as chromium can be used effectively for longer-duration etching of Si, $SiO_2$, and $Si_3N_4$. RIE

**Table 3.9 Etch chemistry used in RIE**

| Material | Gases |
| --- | --- |
| Si | $CF_4$-based, $CF_3Cl$; $CF_4/SF_6/CHF_3$, $SF_4$-based, $Cl_2/H_2$, $C_2ClF_5/O_2$, $NH_3$, $C_2Cl_3F_3$, $CCl_4/He$, $HBr/Cl_2/He$ |
| $SiO_2$ | $CF_4/H_2$, $C_2F_6$, $C_3F_8$, $CHF_3$, $CF_4/O_2$, $CF_4/CHF_3/Ar$, $C_4F_8/CO$, $C_5F_8$, $CH_2F_2$ |
| $Si_3N_4$ | $CF_4/O_2$, $CF_4/H_2$, $C_2F_6$, $C_3F_8$, $SF_6/He$, $CHF_3/O_2$, $CH_2F_2$, $CH_2CHF_2$ |
| Organic solids | $O_2$, $O_2/CF_4$, $O_2/CF_6$ |
| Al | $BCl_3$, $CCl_4$, $SiCl_4$, $BCl_3/Cl_2$, $CCl_4/Cl_2$, $HBr/Cl_2$, $SiCl_4/Cl_2$ |
| Au | $C_2Cl_2F_4$, $Cl_2$ |
| W | $SF_6$, $NF_3/Cl_2$ |

can also replace wet chemical etching of metals using appropriate gas chemistry. Gases used for etching various materials are listed in Table 3.9.

The RIE process, however, is restricted to shallow-trench etching applications, because when used for deeper trenches, lateral etching occurs due to long exposure of the side walls to the reactive species. Improvement in anisotropy can be achieved if a gas mixture of hydrogen and $CF_4$ is used in the RIE system. Alternatively, $CHF_3$ is usually chosen for RIE because the hydrogen available in this gas assists in reducing the pure chemical reaction (which occurs at the sidewalls). Hydrogen reaction with F radicals depletes these radicals and hence lowers the possibility of the pure chemical reaction that is responsible for the lateral etching effect.

For nanoscale applications requiring deep trenches and excellent anisotropy over a long duration of etching, and for fabricating high-aspect-ratio devices for microsystems, RIE is modified into a process called deep RIE (DRIE), discussed briefly in the next section.

### 3.6.4 Deep Reactive Ion Etching (DRIE)

DRIE is a sophisticated version of RIE designed to produce deep and high-aspect-ratio features with near-vertical side walls. This is made possible by using an inductively coupled plasma (ICP) to achieve high plasma density and by employing repeated cycles that consist of an etching step of short duration followed by a coating (passivation) step to protect the side wall. High plasma densities of $10^{11}$–$10^{12}$/cm³, about two orders of magnitude higher than in conventional RIE systems, can be achieved with ICP to create a magnetic envelope inside the etch chamber that reduces the loss of charged species to the surroundings. This DRIE process is based on patents currently owned by Robert Bosch GmbH and Texas Instruments.

In a DRIE process, generally, ICP in a gas mixture of $SF_6$ and argon is used for etching. A negative bias ($-5$ to $-30V$) is applied during the etching step so that the positive ions generated in the plasma are accelerated vertically into the substrate being etched. Next, all the exposed surfaces (side walls and horizontal surfaces) are coated with a Teflon-like polymer layer of thickness 50 nm during the passivation step by switching the gas mixture to trifluoromethane ($CHF_3$) or $C_4F_8$ and argon. During this polymerization step, ion bombardment is used with a small applied bias voltage to prevent polymer formation on the horizontal surfaces of the etched trench. During the next etching step, the side walls are not etched due to the presence of the polymer layer, as shown in Fig. 3.27(a). This is called the Bosch process, named after the company which invented it. The etching and coating cycle

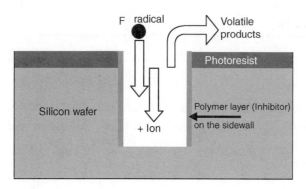 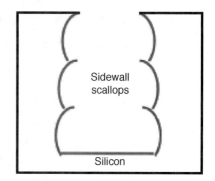

**Figure 3.27** (a) Schematic cross-section of the trench showing the passivating polymer coating on the sidewall during the DRIE process; (b) Schematic cross-section showing the scallops in DRIE as would be seen under a high-magnification SEM.

time is 5–30 seconds to achieve vertical side walls with minimum scallop. The size of the scallops depends on the cycle time; typical scallop depths are 100–200 nm. Fig. 3.27(b) is a schematic diagram of the etched profile similar to that seen with a high-magnification scanning electron microscope (SEM).

With DRIE, silicon etch rate of 5–10 μm per minute can be easily achieved. Maximum etch rates of up to 30 μm per minute in silicon have been reported using a photoresist masking layer. Etch depths of the order of 500–1000 μm can be obtained. Aspect ratios (defined as the ratio of trench height $h$ to width $w$–see Fig. 3.28) up to 30:1 [with side wall angles $(90 \pm 2)°$] can be achieved with DRIE.

**Figure 3.28** Schematic illustrating the concept of aspect ratio of trench in silicon.

Photoresist selectivity of 100:1 and silicon dioxide selectivity of 200:1 can be achieved with a plasma density of $10^{11}$–$10^{12}$ per cm$^3$, which is about two orders of magnitude higher than that in conventional RIE systems. DRIE and RIE are very useful techniques for fabricating microsystems such as accelerometers and gyroscopes using bulk micromachining on silicon and surface micromachining on polycrystalline silicon layers with no undercut etching.

### Example 3.7

Plasma etching of silicon is carried out in a 50 μm × 50 μm square window under the following three plasma conditions at fixed power level for 10 minutes:

(a) Plasma is created in the $CF_4$ gas, without any accelerating ions, so that dry etching takes place due to pure chemical reaction. The etch rate due to this chemical reaction is found to be 0.2 μm per minute.

(b) Plasma is created in a mixture of $CF_4$ gas and argon so that the argon ions can enhance the reaction rate to 1 μm wherever it impinges.

(c) Plasma is created in a gas mixture of $CF_4$ with 40% hydrogen gas so that the pure chemical reaction is zero, and argon gas is added to produce the argon ions.

Draw the cross section of the etched silicon for the above three cases.

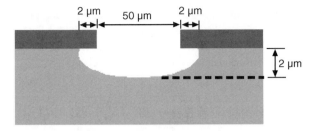

**Figure 3.29** Cross-section showing the isotropic nature of plasma etching in silicon by pure chemical reaction.

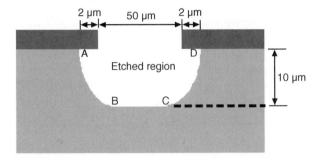

**Figure 3.30** Cross-section showing improved anisotropic nature in silicon in RIE, which uses only chemical reaction in the lateral direction and enhanced etching in the vertical direction by the ion-bombardment-assisted chemical reaction.

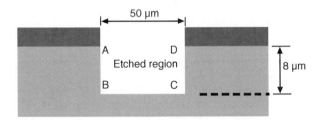

**Figure 3.31** Anisotropic etching showing the silicon being etched anisotropically with no lateral etching as a result of RIE in the plasma containing a gas mixture of $CF_4$, argon, and hydrogen.

**Solution:** **Case (a): Silicon etched in $CF_4$ gas plasma**

In this situation, as the etching is due to pure chemical reaction, it is isotropic. The etch rate being 0.2 $\mu$m per minute, the etch depth at the end of 10 minutes is 2 $\mu$m in all directions. The cross-section of silicon for this case is as shown in Fig. 3.29. The masking layer has a square window opened for plasma etching.

**Case (b): Silicon etched in plasma of a gas mixture containing $CF_4$ and argon**

In this case the presence of argon ions enhances etching in the vertical direction to 1 $\mu$m per minute. However, in the lateral direction the etch rate is governed by the pure chemical reaction and is equal to 0.2 $\mu$m per minute. The undercutting in 10 minutes of etching is 2 $\mu$m. In the vertical direction, the etch depth is 10 $\mu$m. The cross-section is as shown in Fig. 3.30. The region enclosed by ABCD is etched.

**Case (c): Silicon etched in plasma of a gas mixture containing $CF_4$, hydrogen, and argon**

The presence of 40% hydrogen in the $CF_4$ plasma depletes the concentration of fluorine radicals and hence ensures that the pure chemical reaction is negligibly small. Lateral etching can thus be minimized by using about 40% hydrogen in $CF_4$. This hydrogen also lowers the etch rate slightly due the depletion of F radicals. However in the vertical direction, the availability of argon ions and also $F^+$ ions ensure that reasonably high etch rates occur. Thus the etch rate in the vertical direction would be less than 1 $\mu$m/minute [the etch rate in case (b)] by 0.2 $\mu$m/minute [the pure chemical reaction as in case (a)]. Thus we can conclude that the RIE plasma with $CF_4$, hydrogen, and argon, as in case (c), will yield excellent anisotropic etching, resulting in the etched region as shown in Fig. 3.31. The region enclosed by ABCD is the etched portion.

## ▶ 3.7 SILICON MICROMACHINING

Micromachining of silicon can be classified into bulk micromachining and surface micromachining. In bulk micromachining, micromechanical structures are realized within the bulk of a SCS wafer by selective etching. In surface micromachining, structures are built on the surface of a silicon wafer by depositing sacrificial and structural layers and subsequently patterning and removing sacrificial layers. The vertical dimensions of surface-micromachined structures are several orders of magnitude smaller than those produced by bulk micromachining.

Major fabrication steps involved in bulk and surface micromachining are photo-lithography, thin-film deposition, and etching, which are adapted from the IC fabrication processes already described. Fig. 3.32 shows a generalized schematic of these processes. As one may notice, deposition and patterning steps are repeated in microsystems.

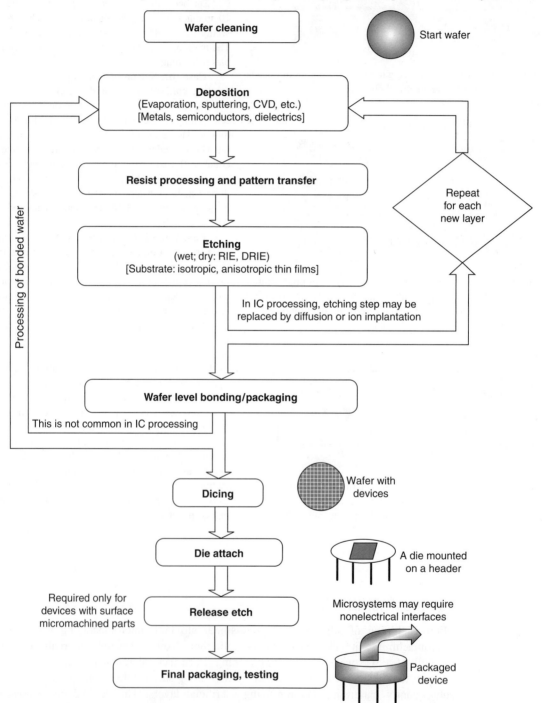

**Figure 3.32** Steps in the fabrication of microsystems.

### 3.7.1 Bulk Micromachining

Bulk micromachining, initiated in the 1960s, has become a mature fabrication technology. Most commercial microsystem devices are fabricated using bulk micromachining. Bulk micromachined structures may have thicknesses in the range from a few microns to 500 μm and lateral dimensions in the range from a few microns to 200 mm. In bulk micromachining techniques, a significant amount of material is removed from the silicon wafer to form membranes, cantilever beams, different types of trenches, holes, and other types of structures.

With bulk micromachining processes and wet chemical etchants, microstructures can be fabricated on SCS by undercutting the silicon wafer. The etch stop may depend on crystal orientation or dopant concentration. However, the type, shape, and size of devices fabricated using wet chemical etch techniques are severely limited.

As an illustration, we describe the fabrication process steps to build a cantilever beam on a silicon wafer by bulk micromachining. A cantilever structure (Fig. 3.33) is realized by forming a beam fixed at one end and a cavity in the silicon substrate so that the other end can move freely. Obviously, the material of the beam must differ from the silicon substrate in which the cavity is formed. The most obvious material for cantilever beam is silicon dioxide ($SiO_2$), which can be conveniently grown on native silicon by thermal oxidation. Anisotropic etching is carried out to form a cavity in the silicon water. The steps involved are:

1. Grow a thick $SiO_2$ layer on a silicon wafer.
2. Spin-coat the photoresist on $SiO_2$ surface.
3. Carry out photolithography to define the cantilever geometry for oxide etch.
4. Etch out oxide in unwanted areas to form a cantilever beam and expose the silicon surface.
5. Do anisotropic etching to form a cavity in the silicon wafer.

All these process steps have been discussed earlier in this chapter. During the last step, a [111] plane is initially formed at all sides of the U-shaped trench. However, the convex corners thus formed (marked in Fig. 3.33) are not stable, as the atoms along this direction are not strongly bonded to other atoms underneath, and as a result are attacked by the etchant. As this process continues, the etch stops with the formation of a complete hollow

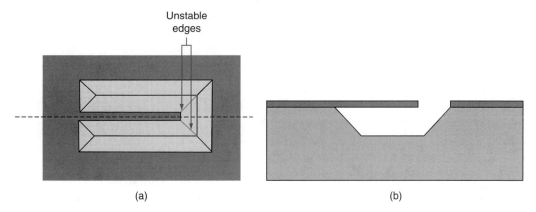

**Figure 3.33** Formation of a cantilever by anisotropic etching of silicon: (a) top view after partial etch; (b) cross-section after complete etch.

pyramid underneath the rectangular etch window. This leaves the part made on the hard mask ($SiO_2$) to remain as a cantilever.

## 3.7.2 Surface Micromachining

In contrast to bulk micromachining, surface micromachining technique builds microstructures on material layers added on the top of the substrate. Polysilicon is commonly used as one of the layers. Silicon dioxide is used as a sacrificial layer that is removed to create the necessary void in the thickness direction. The added layers are not necessarily single crystals or silicon compounds, and their thickness is typically 2–5 μm. The main advantages of surface micromachined structures are the ease with which IC components can be integrated and the possibility of realizing monolithic microsystems components in which mechanical structure and electronic functions are built on the same substrate. As mentioned earlier, the dimensions of surface micromachined components are much smaller than of components fabricated using bulk micromachining techniques.

### 3.7.2.1 Surface Micromachining Process

Surface micromachined microsystem devices have three components: the sacrificial or spacer layer, the microstructure, and the insulator. The sacrificial layer is usually made up of phosphosilicate glass (PSG) or silicon dioxide ($SiO_2$) deposited on the substrate by low-pressure chemical vapor deposition, and its thickness can be 0.1–5 μm. The etch rate of the sacrificial layer must be significantly higher than those of other layers. Both wet and dry etching techniques can be used for surface micromachining.

The basic steps involved in sacrificial layer technology for surface micromachining are illustrated in Fig. 3.34 by a surface micromachining technique used to realize a

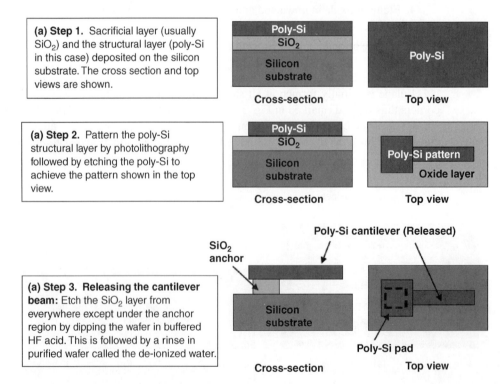

**(a) Step 1.** Sacrificial layer (usually $SiO_2$) and the structural layer (poly-Si in this case) deposited on the silicon substrate. The cross section and top views are shown.

**(a) Step 2.** Pattern the poly-Si structural layer by photolithography followed by etching the poly-Si to achieve the pattern shown in the top view.

**(a) Step 3. Releasing the cantilever beam:** Etch the $SiO_2$ layer from everywhere except under the anchor region by dipping the wafer in buffered HF acid. This is followed by a rinse in purified wafer called the de-ionized water.

**Figure 3.34** Process steps for realizing a cantilever beam having an oxide anchor so that the substrate is electrically isolated from the cantilever.

polysilicon cantilever beam using thermal oxide or deposited oxide as the sacrificial layer and anchor layer.

Note that the polysilicon pad is larger than the width of the cantilever portion. This ensures that the $SiO_2$ anchor is not etched during release of the cantilever. Usually, etch holes are provided on wider structural layers to facilitate easy removal of sacrificial material from underneath. This sacrificial material removal (called the release step) is a very critical step in surface micromachining and is generally carried out after dicing into individual dies, particularly if one is planning to package. Yet when large-area cantilever beams are formed by micromachining, they tend to deflect downward due to surface tension induced by trapped liquid droplets attached to the substrate or isolation layer during final rinsing and drying, a phenomenon is known as *stiction* (short form for *st*atic fri*ction*). Stiction may also be related to residual contamination or van der Waals forces. Several procedures are employed to overcome the problems associated with stiction:

1. Formation of bumps on sacrificial polymer columns along with the oxide. Isotropic oxygen plasma is used to etch away polymer after the oxide etch.
2. Roughening opposite surfaces.
3. Making the silicon surface hydrophobic.
4. Critical-point drying with liquid $CO_2$ at 31°C and 1100 psi.
5. Carrying out release etching after completion of all wet processes.

The main cause for the stiction during evaporation drying of the released structure is the surface tension force pulling the released structure down toward the substrate. If the adhesion force between the contacted areas is larger than the elastic restoring force of the deformed structure, the structure remains stuck to the substrate even after it is completely dried. In order to overcome this problem, several drying techniques have been attempted in addition to the structural changes listed above. All of them use the phase-change release methods adopted to avoid formation of a liquid–gas interface to ensure that surface tension forces do not exist during the drying process. This goal can be achieved in two different ways:

1. A sublimation method in which the final rinsing liquid is solidified and then sublimated.
2. A supercritical drying method in which the final rinsing solution is pressurized simultaneously by heating so that it reaches the supercritical point; it is then vented away. In Fig. 3.35 the arrows show the path of the pressure–temperature state followed by the final rinse during the drying process for evaporation method, sublimation method, and supercritical drying or critical-point drying (CPD).

### 3.7.2.1.1 Evaporation Drying
From this diagram (Fig 3.35) it can be seen that in evaporation drying, the liquid–gas interface is formed during the evaporation from liquid to vapor. So the only solution in this case is to use low-surface-tension liquids such as methanol for the final rinse before evaporation.

### 3.7.2.1.2 Sublimation Method
In this method, after etching the sacrificial oxide layer with HF followed by water rinse, the water is replaced with methanol and then methanol is replaced with tertiary–butyl (t-butyl) alcohol, which melts at 26°C. This is then frozen in a refrigeration system and sublimated

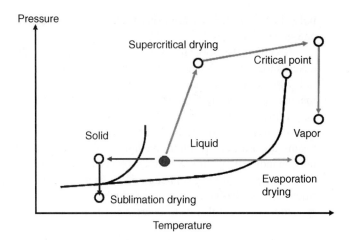

**Figure 3.35** P–T plot for various drying processes.

under relatively low vacuum. A simple vacuum set up with the Peltier chip reduces the operation time. The vapor pressure of t-butyl alcohol is 27 torr at 20°C and hence it sublimates in 15 minutes from a chip of area 1 cm$^2$.

Alternatively, p-dichlorobenzene can be used very easily because it melts at 56°C. This chemical is melted on a hot plate before replacing methanol. Solidification and sublimation are done at room temperature. The vapor pressure of p-dichlorobenzene is very low (1 torr at 25°C) compared to the t-butyl alcohol vapor pressure and hence sublimation takes about three hours from a chip of 1 cm$^2$ area.

### 3.7.2.1.3 Critical Point Drying (CPD) or Supercritical Drying with CO$_2$
The basic idea in this technique is to take advantage of the fact that, when a liquid is heated in a closed system so that the critical pressure is attained at the critical temperature, any visible meniscus (interface between liquid and gas) disappears and the surface tension becomes zero as the distinction between the liquid and gas phase distinction disappears, leading to a continuity of state.

The critical point of water occurs at the very high temperature of 374°C and pressure of 3212 psi, whereas for CO$_2$ the supercritical region occurs for temperatures above 31.1°C and pressures above 72.8 atmospheres (1073 psi). Even though other transitional fluids such as Freon 13 and Freon 23 have still lower critical points, they are not used due to safety concerns. As a result, liquid CO$_2$ is the automatic choice for critical-point drying in surface micromachining, with supercritical drying carried out in the following steps:

1. After HF etch, the structure is rinsed in deionized (DI) water but is not let dry.
2. The DI water is exchanged with methanol by dilution.
3. The wafer is then transferred to a pressure vessel in which the methanol is replaced by liquid CO$_2$ at 25°C and 1200 psi.
4. The contents of the pressure vessel are then heated to 35°C and CO$_2$ is vented at a temperature above 35°C (ensuring that it only exits in gaseous form).

Beams up to 600 μm in length can be released without sticking when the beam width is in the range 10–20 μm and beam thickness and the air gap are 2 μm, compared to a length of only about 80 μm using air drying.

Even though the stiction problem during release can be handled by the various processes described above, stiction during usage is the dominant reliability problem in surface micromachined devices. As these issues are related to issues connected with integration, packaging, and applications, they are presented in Chapter 8.

### 3.7.2.2 Material Systems for Sacrificial Layer Technology

Surface micromachining requires a compatible set of structural materials, sacrificial materials, and etchants. Moreover, structural materials must have mechanical properties such as high yield and fracture strength, minimum creep and fatigue, and good wear resistance. The materials are also required to have good adhesion and low residual stress to avoid device failure by delamination/cracking. The etchants must have excellent etch selectivity and be able to etch off sacrificial layers without affecting other layers. Common materials used in surface micromachining are:

1. **Polysilicon/silicon dioxide**: LPCVD-deposited polysilicon is the structural material and a LPCVD-deposited silicon dioxide layer is the sacrificial material. Silicon dioxide is etched using HF, which does not significantly affect polysilicon.

2. **Polyimide/aluminum**: Polyimide serves as the structural material and aluminum as the sacrificial material. Acid-based etchants are used to etch aluminum.

3. **Silicon nitride/polysilicon**: Silicon nitride is used as the structural material and polysilicon as the sacrificial material. Polysilicon is etched using isotropic etchants.

4. **Tungsten/silicon dioxide**: CV-deposited tungsten is used as the structural material with $SiO_2$ as the sacrificial material. HF can be used to remove $SiO_2$.

### 3.7.2.3 Realizing a Cantilever Structure by Surface Micromachining: A Case Study

So far in this section, the fabrication of a cantilever beam on a silicon substrate using processes described earlier in this chapter such as oxide growth, photolithography, thin-film deposition and etching processes has been described to illustrate the surface micromachining technique. It is clear that the basic steps involved are derived from IC fabrication technology.

The cantilever structure is realized by using silicon dioxide as the sacrificial layer and polysilicon as the structural layer. The silicon dioxide layer is grown on the native Si substrate by thermal oxidation or is deposited by CVD techniques and provides mechanical separation and isolation between the structural layer and the silicon substrate. A poly-silicon layer that acts as the structural layer is deposited on the oxide layer. After formation of the cantilever beam (by photolithography and etching), the sacrificial layer $SiO_2$ is removed by release etching. Removal of the sacrificial layer permits free movement of the structural layer (here the cantilever beam) with respect to the silicon substrate. The steps involved (shown in Fig. 3.36) are:

1. Deposit a thick silicon dioxide layer on the top surface of a Si wafer.

2. Deposit photoresist by spin coating and then transfer pattern by UV lithography.

3. Develop photoresist to expose regions of silicon dioxide for etching. As in lithography, the resist is baked to withstand the etching step that follows.

4. Remove silicon dioxide from those areas where the cantilever beam (polysilicon) is anchored to the Si wafer.

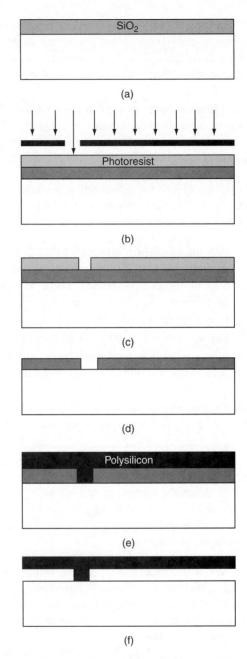

**Figure 3.36** Pictorial representation of surface micromachining to realize cantilever structure: (a) deposit $SiO_2$ on Si wafer; (b) UV lithography; (c) developing; (d) oxide etching; (e) polysilicon deposition and patterning; and (f) removal of oxide (sacrificial layer).

5. Deposit polysilicon on the surface by CVD. This is doped to change its etch characteristics and conductivity. The beam area is defined by patterning steps similar to those followed for etching oxide.

6. Remove the $SiO_2$ sacrificial layer by release etch.

**Table 3.10 Comparison of bulk and surface micromachining processes for microsystems fabrication**

|  | Bulk Micromachining | Surface Micromachining |
|---|---|---|
| Process maturity | Well established | Fairly well established |
| Ruggedness | Structures can withstand vibration and shock | Less rugged |
| Die area | Large | Small |
| IC compatibility | Not fully integrated | IC compatible |
| Structural geometry | Limited options | Wide range possible |
| Maximum structure thickness | Wafer thickness | Typically 2–4 $\mu$m |
| Planar geometry | Rectangular | Unrestricted |
| Lateral dimensions | Minimum 1.4 times depth for (100) wafer | No minimum or maximum; limited by lithography |
| Sidewall features | 54.75° slope for (100) wafer | Limited by etching process |

Table 3.10 compares different aspects of bulk micromachining and surface micromachining processes.

## ▷ 3.8 SPECIALIZED MATERIALS FOR MICROSYSTEMS

As indicated earlier, microsystems use materials other than those used in silicon ICs. Polymers and ceramics are two such specialized materials; in this section we present some details.

### 3.8.1 Polymers

Epoxy resins and adhesives are routinely used in microsystems packaging. Some of the well-established applications of polymers in the area of microsystems are:

1. Etch-stop layers for creating desired patterns for photolithography.
2. As prime molds with desired geometry of microcomponents in the LIGA (see Section 3.9.3.1) process.
3. Conductive polymers as organic semiconductors.
4. Ferroelectric polymers as sources of actuation for micropumping.
5. Coating substances for capillary to enhance electro-osmotic flow in microfluids.
6. Shielding of electromagnetic interference (EMI).
7. Encapsulation of microsensors and packaging.

Some of the features which make polymers suitable for microscale applications are moldability, conformability, ease of deposition as thick and thin films. In addition, some polymers have conducting and semiconducting behaviors, while some others show pyroelectric effects in polymer side chains and these properties can be exploited. Properties of some of the commonly used polymers are listed in Table 3.11.

**Table 3.11 Polymeric materials used in microsystems and their characteristics**

| Polymer | Properties of Interest |
|---------|------------------------|
| Polyethylene (PE) | Excellent chemical resistance, low cost, good electrical insulation properties, clarity of thin films, easy processability |
| Polyvinyl chloride (PVC) | Excellent electrical insulation over a range of frequencies, good fire retardance, resistance to weathering |
| PVDF | Piezoelectric and pyroelectric properties, excellent resistance to harsh environments |
| Polytetra-fluoroethylene (PTFE) | High heat resistance, high resistance to chemical agents and solvents, high anti-adhesiveness, high dielectric properties, low friction coefficient, nontoxicity |
| Polystyrene | Optical property (transparency), ease of coloring and processing |
| Polydimethlysiloxane (PDMS) | High viscoelasticity, biocompatible, easily moldable |

*Note:* Modified after [1].

The required characteristics of polymers used in microsystems are [1]:

**1.** Strong interfacial adhesion between layers.

**2.** Suitable elastic moduli and strength to provide and sustain the needed deformation.

**3.** Dimensional and environmental stability.

These materials are usually deposited as monomers or oligomers and then allowed to polymerize. The energy required for the polymerization reaction may be supplied in various forms, such as photopolymerization, electrochemical polymerization, and vacuum polymerization.

Polymers used in microscale applications are broadly classified in two categories: (a) structural polymers and (b) sacrificial polymers. A structural polymer is usually a UV-curable polymer with urethane acrylate, epoxy acrylate, or acryloxy silane as the main constituent. Processing with automated equipment or manual methods without additional solvents or heat is possible because of the low viscosity. It has the requisite flexibility and resistance to solvents, chemicals, and water. A structural polymer can be used as the backbone structure in building multifunctional polymers. SU8 has emerged as a popular structural polymer for microsystems. Structural polymers can also be used as sensing and actuating components.

PDMS is a commonly used silicone-based organic polymer. It is inert, nontoxic and highly viscoelastic. Its shear modulus varies from 10 kPa to 1 MPa. Its surface chemistry can be modified by plasma oxidation and addition of SiOH groups to the surface to make it hydrophilic. PDMS is widely used as a stamp material in *soft lithography* described in Section 3.9.3.6. It is also used in microfluidic systems.

## 3.8.2 Ceramic Materials

Ceramics are another major class of materials widely used in smart systems. These generally have good hardness and high-temperature strength. Thin and thick ceramic films and 3D ceramic structures are used in specialized microsystems applications; crystalline as well as noncrystalline materials are used. For example, ceramic pressure microsensors have been developed for high-temperature environments, silicon carbide for harsh environments, etc. In addition to these structural ceramics, some functional ceramics, such as ZnO and PZT, have also been incorporated into smart systems.

New functional microsensors and microactuators can be realized by combining ferroelectric thin films having good sensing properties (such as pyroelectric, piezoelectric, and electrooptic effects) with microdevices and microstructures. There are several such ferroelectric materials including oxides and non-oxide materials. One useful ferroelectric thin film studied for microsensors and RF-MEMS is BST. BST is of interest in bypass capacitors, dynamic random-access memories, and RF phase shifters because of its dielectric constant, which can be as high as 2500. Titanates having their Curie temperatures in the vicinity of room temperature are well suited for several sensor and actuator applications. Above the Curie temperature, however, these materials lose their piezo-electric characteristics, and since processing of these ceramics is usually done at high temperatures, they may not be directly compatible with CMOS processes.

# ► 3.9 ADVANCED MICROFABRICATION PROCESSES

## 3.9.1 Wafer Bonding Techniques

The deposition and etch processes described thus far in this chapter can be used in succession to realize some of the structures required in the fabrication of microsystems. However, the overall height of such structures is limited. Furthermore, the conformal nature of most deposition techniques limits the ability to fabricate closed cavities. In such cases, parts of a device can be fabricated in different wafers and combined at the wafer level by bonding. Bonding makes use of the mirror finish of the surface of silicon (and other) wafers used in microsystems processing. This process is further aided by heating, applying electrostatic fields, and/or hydrating the surfaces.

Microsystems components often have complex geometry with a finite thickness and are made of dissimilar materials, making their bonding more complex than that in microelectronic components. Furthermore, these systems may contain fluids or environ-mentally hostile substances, further complicating the bonding process. Bonded dissimilar materials are expected to achieve hermetic sealing, low mounting stress, and low sensitivity to static pressure. Different types of bonding techniques suitable for micro-systems applications have been developed.

### 3.9.1.1 Anodic Bonding (Electrostatic Bonding)

Pyrex (typically pyrex 7740) and silicon wafers are bonded together using anodic bonding. The positive terminal of the power supply is connected to the silicon wafer and the negative terminal is connected to glass (Fig. 3.37), and the voltage applied is in the range 200 to 1000 V. A bonding temperature of 450°C is obtained by placing the anode silicon substrate on a heater. Oxygen ions from the glass migrate into the silicon and form a silicon dioxide layer between silicon and glass, resulting in a strong and hermetic chemical bond. The low temperature ensures the good quality of the metal layers already present. Table 3.12 summarizes the process conditions. An anodic bonding technique

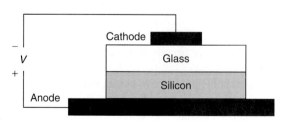

**Figure 3.37** Anodic bonding of glass and silicon wafers.

can also be used to bond two silicon wafers together, using an intermediate sputtered glass layer. Anodic bonding introduces alkali metal ions and hence anodic-bonded devices are usually not allowed for CMOS processing.

**Table 3.12 Typical process parameters in anodic bonding**

| Parameter | Value |
|---|---|
| Temperature | 450°C |
| Voltage | 500 V |
| Electrodes | Point contact |
| Time | Up to 5 minutes |
| Surface unevenness | ~0.1 μm |
| Thermal expansion | Should be matched |
| Bond strength | 10–15 MPa |

### 3.9.1.2 Direct Bonding

Direct bonding or silicon fusion bonding is due to the chemical reaction between OH groups present on the surface of silicon or oxides grown on the surface of silicon wafers. The bonding process involves three steps: surface preparation, contact, and thermal annealing. In surface preparation, the surfaces are cleaned to form a hydrated surface. The wafer surface is required to have a mirror finish, roughness less than 10 Å, and bow less than 5 μm for a 4″ wafer.

After surface preparation in a cleanroom environment, the wafers are aligned and put in contact by gently pressing them together at the center. Intimate contact over the entire wafer surface area is achieved by the surface attraction of two hydrated surfaces. Wet cleaning and plasma hydrophilization assist the bonding process.

The final step in direct bonding involves thermal annealing. Bond strength is increased by more than one order of magnitude by annealing at 450–1200°C. Fusion-bonded silicon wafers can be separated only by breaking. High-temperature annealing (>800°C) may result in broadening of the doping profile, defect generation, contamination, and dissociation in the case of compound semiconductors. Unlike anodic bonding, fusion-bonded wafers can be used for subsequent IC processing.

### 3.9.1.3 Intermediate Layer-Assisted Bonding

In this process, an intermediate layer (metal, polymer, solder, glass, etc.) is used to facilitate bonding. In eutectic bonding, gold is used as the intermediate layer. The stress developed during bonding may be quite significant. Reasonably high strength and low stress are obtained when polymers are used as intermediate layers for bonding; however, hermetic sealing is difficult to achieve. Glasses with low melting temperature as intermediate layers have also been used for bonding in which a layer of glass frit is screen-printed on silicon wafer prior to bonding. This bonding is usually done at low temperatures (<500°C).

A major problem in all bonding processes is the presence of non-contact areas known as *voids* that are due to the presence of particles, organic residues and surface defects and inadequate surface contact. Hence both surfaces must be smooth and clean. Wafer inspection, surface pretreatment, and mechanically controlled and aligned mating in clean room environments ensure minimum voids and good bonding. Because of the dissimilar mechanical properties of the materials to be bonded, the yield of void-free wafer is considerably reduced by wafer bow or defects.

## 3.9.2  Dissolved Wafer Process

One of the early approaches to fabricate microstructures such as beams using silicon makes use of diffusion and bonding techniques discussed in Section 3.9.1 previously. Unlike surface micromachining, this process is not hampered by stiction. However, it requires use of two wafers to realize simple structures such as beams. First, a small trench is formed on a silicon wafer by anisotropic etching as shown in Fig. 3.38. $SiO_2$ may be used as an etch window for this step. The etching depth at this stage is equal to the height of the beam eventually realized. A heavily doped region in the shape of the beam is formed across this trench on the silicon wafer by ion implantation or diffusion. (You may recall that highly doped region in a silicon wafer works as an etch stop.) Boron doping density on the order of $10^{19}$ cm$^{-3}$ is required to achieve the required selectivity.

The wafer is then bonded on to a glass substrate by anodic wafer-bonding methods discussed in Section 3.9.1. Later, the undoped silicon region from the first wafer is removed

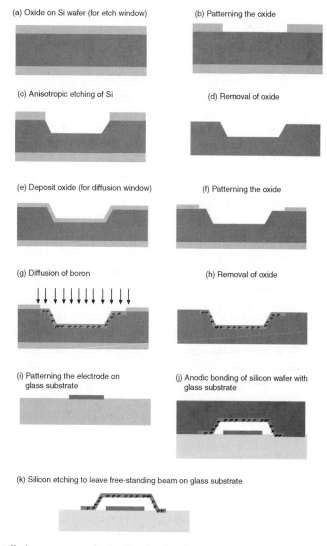

**Figure 3.38** Detailed process steps in the dissolved wafer process to realize beam on a glass substrate.

by wet chemical etching. The microstructure from this part remains unaffected due to the high doping concentration there. The process steps involved in this approach of micro-fabrication are shown schematically in Fig. 3.38.

### 3.9.3 Special Microfabrication Techniques

The preceding sections have explored the extension of silicon IC technologies to micro-machining. However, micromachining often requires certain specialized techniques, especially in fabricating high-aspect-ratio structures. We describe such techniques in the following subsections.

#### 3.9.3.1 LIGA Process

Both the silicon microfabrication techniques discussed in Section 3.5—bulk micromachining and surface micromachining—involve processes that evolved out of conventional microelectronics technologies. Consequently, much of the knowledge and experience as well as equipment used in the production of microelectronics and ICs can be adopted for micromachining with little modification. Unfortunately, these inherited advantages are overshadowed by two major drawbacks: (a) low geometric aspect ratio and (b) use of only silicon-based materials. The geometric aspect ratio of a microstructure is the ratio of the depth dimension to the lateral dimensions. Most silicon-based microsystems use wafers of standard sizes and thicknesses as substrates on which etching or thin-film deposition takes place to form the desired 3D geometry. Severe limitations in the depth dimension are thus unavoidable. The other limitation is on the material itself: silicon-based microsystems often do not use unconventional materials such as polymers and metals for structures and thin films.

The LIGA process for manufacturing microsystems is radically different from the bulk and surface micromachining techniques and does not have the two major shortcomings of silicon-based micromanufacturing techniques. It offers great potential for manufacturing non-silicon-based microstructures. The single most important feature of this process is that it can produce "thick" microstructures that have extremely flat and parallel surfaces such as microgear train motors, generators, and microturbines made of metals and plastics. These unique advantages are the primary reasons for the increasing popularity of LIGA process in the microsystems industry.

The term LIGA is an acronym from the German words for lithography (Lithographie), electroforming (Galvanoformung), and molding (Abformung) [6]. The technique was first developed at the Karlsruhe Nuclear Research Center in Karlsruhe, Germany. The words abbreviated in the term LIGA indeed represent the three major steps in the process, as outlined in Fig. 3.39.

These steps are further illustrated in Fig. 3.40. At the beginning, a seed layer of metal is deposited on the silicon wafer, and a thick photoresist layer is spin-coated above this. X-ray lithography [Fig. 3.40(a)] enables patterning of thick photoresist materials, as x-rays have good penetration depth and low diffraction effects. Using the cavities thus formed [Fig. 3.40(b)], metal structures are realized by electroplating. These have well-defined side walls and can be used to make high-aspect-ratio plastic molds [Fig. 3.40(c)], which in turn can be used in the bulk production of high-aspect-ratio structures [Fig. 3.40(d)].

#### 3.9.3.2 HexSil Process

The HexSil process is used to fabricate micromechanical structures with high aspect ratio. It is a combination of DRIE and surface micromachining techniques [7]. HexSil uses hexagonal honeycomb geometries for making thin-film rigid structures and silicon for micromachining and CMOS electronics. Trenches made by DRIE serve as reusable molds

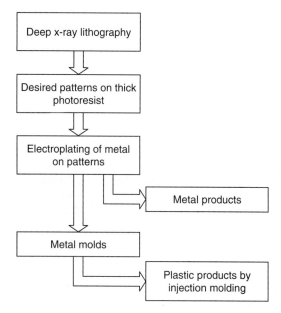

**Figure 3.39** LIGA process steps.

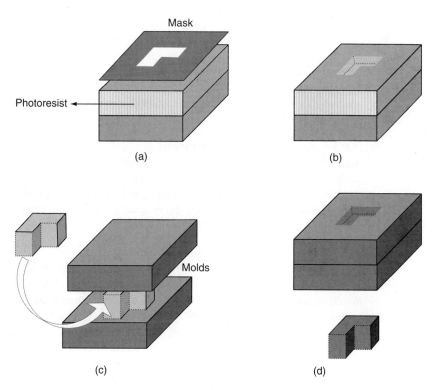

**Figure 3.40** Pictorial representation of LIGA process flow: (a) x-ray lithography on thick resist; (b) patterned resist for electroplating; (c) metal parts for molds; and (d) bulk production of nonmetal parts using molds.

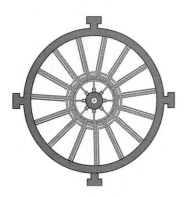

**Figure 3.41** Schematic of a rotary actuator fabricated by the HexSil process.

for sequential filling of polysilicon, sacrificial layers, and silicon dioxide. Patterning and removal of sacrificial layer result in structural parts with large lateral dimensions ranging up to centimeters. Nickel plating can be used to produce highly conducting regions of HexSil plates for contacts and conducting patterns. With the help of the HexSil process, elements of millimeter size can be produced by batch processing of thin films. Using interconnected levers, tiny expansions in polysilicon beams can be multiplied to produce millimeters of motion. Thermal expansion of resistively heated HexSil regions has been used to actuate tweezers that can pick up and place particles ranging from 1 to 2.5 μm.

Another example is the electrostatically actuated polysilicon structure in Fig. 3.41. This uses a reusable silicon mold, and the fabricated structures are solder bonded to a target substrate. As in LIGA fabrication, this makes the mold reusable, thereby reducing the cost per piece.

### 3.9.3.3 Microelectrical Discharge Machining

Microelectrical discharge machining (micro-EDM) is an erosion process in which the material is removed by electrical discharges generated at the gap between two conducting electrodes [8]. Micro-EDM is used to machine microholes, channels and 3D microcavities in electrically conductive materials. The discharge results from a voltage applied between the tool electrode and the workpiece electrode, which are separated by the dielectric fluid in the work tank. An electric arc discharge melts a small amount of material on both electrodes, part of which is removed by the dielectric fluid and the rest solidified on the surface of the electrodes, leaving a small crater on the workpiece and tool electrodes. Hardened steels, carbides, high-strength alloys, and ultrahard conductive materials such as polycrystalline diamond and ceramics can be machined using micro-EDM techniques. Special features of micro-EDM are: (a) machining any conductive or semiconducting material; (b) non-contact machining; (c) high-aspect-ratio machining and (d) high-precision and high-quality machining.

### 3.9.3.4 Laser Mill Micromachining

Laser micromachining is used to fabricate complex patterns. Solid-state lasers with switchable wavelengths are used as an optical energy source. The energy level of the laser beam is about 25 J/cm$^2$ [7]. Apertures less than 2 μm wide can be fabricated using laser mill micromachining, and it can be used for prototype, short-run, R&D and pilot-line development.

### 3.9.3.5 Low-Temperature Cofired Ceramic

LTCC technology is a method for producing multilayer circuits and components using ceramic tapes to apply conductive, dielectric, and resistive parts. The layers are laminated together and fired as one unit. Because of the low-temperature firing (about 850°C), low-resistance materials such as gold and silver can be used. LTCC technology has been recently adopted for the fabrication of microsystems [9].

Ceramic tapes in green form are commercially available in rolls or sheets and are tape-cast and cut to size. Otherwise these can be made by tape casting and slitting. The steps involved in LTCC process, as shown in Fig. 3.42, are:

1. **Via hole punching**: Via holes are drilled using a laser beam or multiple-pin high-speed punching machines.

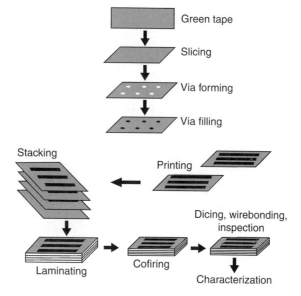

**Figure 3.42** Fabrication process steps in LTCC.

2. **Via filling**: Via holes are filled either by conventional thick-film screen printing or an extrusion via filler. Both methods require a mask or a mylar foil on which the pieces are fixed.

3. **Conductive line printing for interconnects and electrodes**: Cofirable conductors are printed on separate green ceramic sheets using a thick-film screen printer. A porous plate is used to locate the tape in place. After printing, the vias and conductors are dried in an oven. Some pastes may need leveling at room temperature for a few minutes.

4. **Stacking**: Sheets for different layers are stacked or collated one over the other with the help of a digital vision system or positioning pins.

5. **Lamination**: There are two methods for the lamination of types in LTCC:

   a. *Uniaxial lamination:* The tapes are pressed between two plates heated to 70°C. The pressure applied is 200 bar and the duration is 10 min. The tapes are rotated through 180° halfway through. The drawbacks of uniaxial lamination are problems associated with cavities or holes, the requirement for higher shrinking tolerance, and the varying thickness of single parts of each layer.

   b. *Isostatic press:* In this method stacked tapes are vacuum-packed in foil and pressed in hot water. The pressure is about 350 bar.

6. **Cutting into individual pieces**: After laminating, the parts are partially cut into individual pieces that may be smaller in size or different in shape. Alternatively, this step can be performed with a laser cutting tool after cofiring.

7. **Cofiring**: Laminates are cofired in one step on a smooth flat setter tile. The firing follows a specific firing profile. The duration of the firing cycle is between 3 and 8 hours depending upon the material being fired. The temperature is ramped up or down at a specified rate in a closed furnace with controlled gas flow so that thermal interlayer stresses are minimized.

**Table 3.13 Comparison of photolithography and soft lithography**

|  | Photolithography | Soft lithography |
|---|---|---|
| Definition of patterns | Rigid photomask | Elastomeric stamp or PDMS mold |
| Surfaces that can be patterned | Planar | Both planar and non-planar |
| Some materials that can be patterned directly | Photoresists<br><br>Monolayers on Au and $SiO_2$ | Photoresists<br><br>Monolayers on Au, Ag, Cu, GaAs, Al, Pd, and $SiO_2$<br><br>Unsensitized polymers<br><br>Biological macromolecules |
| Structures that can be made patterned | 2D structures | Both 2D and 3D structures |
| Laboratory level limits to resolution | ∼100 nm | ∼1 μm |
| Minimum feature size | ∼100 nm | 10–100 nm |

*Note:* Modified after [10].

### 3.9.3.6 Soft Lithography for Microstructures

Although photolithography is the dominant technology for microfabrication, it is not always the only or best option for all applications. Some of the limitations of photolithography are [10]:

1. Poorly suited for nonplanar surfaces.
2. Limited to only two-dimensional (2D) microstructures.
3. Directly suited to a limited set of photosensitive materials.
4. Not suited to patterns of specific chemical functions on the surface.

Photosensitive materials in devices may not be patterned, since it may be necessary to attach chromophores or add photosensitisers for better adhesion to the substrate surface.To overcome the above limitations, several approaches based on self-assembly and replica molding have been developed for micro- and nanofabrication, which provide convenient, low-cost methods for fabrication of these small structures. In soft lithography, an elastomeric stamp with patterned relief structures on its surface generates patterns and structures with feature sizes ranging from 30 nm to 100 μm. Several techniques have been demonstrated among which microcontact printing (μCP), replica molding, microtransfer molding (μTM), micromolding in capillaries, and solvent-assisted micromolding are popular. Table 3.13 compares the main features of photolithography and soft lithography.

## ▶ 3.10 SUMMARY

A number of basic process steps required for silicon and non-silicon-based microsystems have been discussed in this chapter. The crystal structure of silicon wafer determines its chemical characteristics, as explained by the isotropic or anisotropic etch behaviors of silicon. A series of deposition and patterning steps can be combined to form most silicon-based microsystems.

Physical techniques used in microsystems include evaporation and sputtering. Sputtered films have better uniformity and adhesion characteristics. Various chemical vapor deposition techniques are useful even for better-quality films. One of the most common structural thin-film materials in microsystems is polysilicon, which consists of small crystallites of silicon separated by grain boundaries and has characteristics very similar to those of single-crystal silicon.

Patterning of thin films is achieved by lithography or lift-off process. In the context of microsystems, photolithography therefore plays a key role, as this can be extended to nonplanar geometries by using a series of steps known as surface micromachining. Free-standing microstructures fabricated by surface micromachining can be separated from the wafer surface by up to several micrometers. For geometries with larger vertical dimensions, bulk micromachining can be used in which cavities are formed by etching material from the silicon substrate itself. Acidic etchants are fast but form ill-defined cavities on silicon. However, by judicious choice of etch windows one can utilize the anisotropic nature of some etchants to form well-defined cavities and free-standing thin-film structures. While wet etching processes are cheaper and are widely used, many of their disadvantages can be overcome by the more expensive and industrial dry etching processes involving plasma.

Yet another class of wafer-level process used in silicon microsystems includes anodic, fusion, or eutectic bonding. LIGA and HexSil have been proposed as possibilities for building three-dimensional structures for microsystems. Although silicon is still the most widely used material, ceramic and polymeric materials have niche applications in microsystems. Processes such as soft lithography and low-temperature cofired ceramics (LTCC) are discussed in this context.

Unit processes involved in all the above fabrication approaches have been discussed in this chapter. These are discussed in more specific detail in Chapter 8 on the development of various microsystems.

# ► REFERENCES

1. Varadan, V.K., Vinoy, K.J. and Gopalakrishnan S. (2006) *Smart Material Systems and MEMS: Design and Development Methodologies*, John Wiley & Sons, Chichester, London, UK.

2. Ghandhi, S.K. (1994) *VLSI Fabrication Principles: Silicon and Gallium Arsenide*, 2nd edn, John Wiley & Sons, New York.

3. Madou, M. (1998) *Fundamentals of Microfabrication*, CRC Press, Boca Raton, FL, USA.

4. May, G.S. and Sze, S.M. (2003) *Fundamentals of Semiconductor Processing*, John Wiley & Sons, New York.

5. Gardner, J.W. and Varadan, V.K. (2001), *Microsensors, MEMS and Smart Devices: Technology, Applications and Devices*, John Wiley & Sons, Chichester, London, UK.

6. Beckera, E.W., Ehrfeld, W., Hagmann, P., Maner, A. and Münchmeyer, D (1986) Fabrication of microstructures with high aspect ratios and great structural heights by synchrotron radiation lithography, galvanoforming, and plastic moulding (LIGA process), *Microelectronic Engineering*, **4**(1), 35–56.

7. Horsley, D.A., Cohn, M.B., Singh, A., Horowitz, R. and Pisano, A.P. (1998) Design and fabrication of an angular microactuator for magnetic disk drives, *Journal of Microelectromechanical Systems*, **7**(2), 141–48.

8. Pham, D.T., Dimov, S.S., Bigot, S., Ivanov, A. and Popov, K. (2004) Micro-EDM—recent developments and research issues, *Journal of Materials Processing Technology*, **149**(1–3), 50–57.

9. Thelemann, T., Thust, H. and Hintz, M. (2002) Using LTCC for microsystems, *Micro-electronics International*, **19**(3), 19–23.

10. Xia, Y. and Whitesides, G.M. (1998) Soft lithography, *Annual Review of Materials Science*, **28**, 153–84.

## ▶ FURTHER READING

1. Kovacs, G.T.A. (1998) *Micromachined Transducers Sourcebook*, McGraw-Hill, New York.

2. Campbell, S.A. (2001) *The Science and Engineering of Microelectronic Fabrication*, 2nd ed., Oxford University Press, New York.

## ▶ EXERCISES

3.1 It is required to fabricate a polysilicon cantilever beam whose dimensions are: length × breadth × thickness = 2000 μm × 10 μm × 2 μm. The sacrificial oxide thickness is 1 μm. Discuss the anchor pad size you would choose and explain whether it is possible to realize such a structure by surface micromachining. Give reasons for your answer.

3.2 In one of the designs of an accelerometer, the structure of the seismic mass (500 μm × 500 μm) and the four supporting springs (20 μm wide) were first obtained by etching a silicon layer of thickness 10 μm. The top view of this structure and the anchor pads are shown in Fig. 3.43. It was required to release the mass and the spring while keeping the oxide below the anchor pads (150 μm × 150 μm) intact. The oxide is only 1 μm thick. Suggest any modifications that would be necessary in the structure for its successful release.

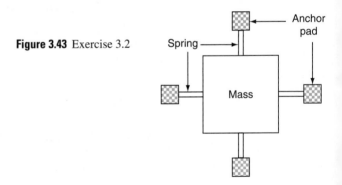

**Figure 3.43** Exercise 3.2

3.3 A silicon wafer has been etched through square a window opening of size 10 μm × 10 μm in the oxide layer. Draw cross-sectional profiles and mark all dimensions of etched silicon for the following cases:
(a) The chemical is isotropic etchant, wafer is ⟨100⟩ silicon, etch depth is 5 μm.
(b) Etchant is 30% KOH solution, wafer is ⟨100⟩ silicon, etch depth is 5 μm.
(c) Isotropic etchant, wafer is ⟨110⟩ silicon, etch depth is 10 μm.
(d) Anisotropic etchant, wafer is ⟨110⟩ silicon, etch depth is 10 μm.

3.4 A (100) silicon wafer of thickness 300 μm is etched using a square window of size 500 μm × 500 μm in the oxide on silicon. The window sides are parallel to ⟨110⟩. The etchant etches in only the ⟨100⟩ direction; the etch rate in the ⟨111⟩ direction is negligible. Draw the cross-section through the wafer, showing the dimensions, after the wafer is etched for a sufficiently long time to make a through hole in the wafer.

3.5 A (100) silicon wafer is 500 μm thick. A mask consists of rectangular window of unknown size. The sides of the window are parallel to ⟨110⟩. After wafer etching a hole size of 50 μm × 80 μm is formed on the other side of the wafer. Find the size of the mask window. The undercut rate is negligible.

3.6 Repeat Problem 3.5, taking into account the undercutting if the etch rate in the ⟨111⟩ direction is only 50 times smaller than that in the ⟨100⟩ direction.

3.7 Thickness of a (100) silicon wafer is 410 μm. A square window of 1000 μm size is opened in the oxide on the front surface of the wafer with the mask edge aligned parallel to the ⟨110⟩ direction. The oxide on the back of the wafer is completely etched. This wafer is subjected to anisotropic etchant whose etch rate along the ⟨100⟩ direction is 50 μm/hour. Determine the diaphragm thickness and size when the etch duration is 4 hours. What is the wafer thickness outside the diaphragm at this stage?

3.8 In a bio-MEMS application involving microchannels (vertical cross-section: 20 μm × 20 μm), it is specified that the walls should make an angle of 90° to the surface (vertical channel). Identify the type of silicon wafer that can be used for the purpose. Which chemical can be used for wet etching of this channel?

3.9 The etch windows shown in Fig. 3.44 are used in 100 silicon. What will be the substrate cross-sections (along lines) after a complete etch in each case? Sketch the top view after the masking layer is removed. In Fig. 3.44(a), the length of the major axis (oriented at 45° to the [110] direction) of the ellipse is 100 μm. In Fig. 3.44(b) a projection of 50 μm × 5 μm in a square area of 150 μm × 150 μm is used.

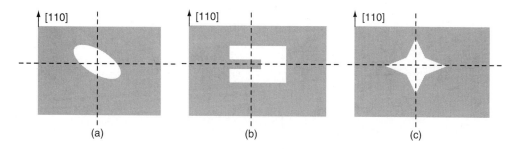

**Figure 3.44** Etchants used: in (a) and (b) KOH; in (c) HNA. The mask is aligned as shown.

3.10 It is required to fabricate a device with a pair of interdigital electrodes by gold thin film. Its fingers are 5 μm wide and are separated by 5 μm. Identify a fabrication scheme that guarantees the high precision required in such situations. Explain the process steps. (Mention generic names of materials involved at each deposition stage and methods for deposition and etching.) Indicate the constraints of this method regarding choice of material and geometry.

3.11 Identify the best deposition/growth techniques for the following thin films: (a) 0.2 μm aluminum, (b) 0.1 μm BST, (c) high-purity silicon nitride, (d) 0.1 μm silicon dioxide for insulation layer, (e) 0.1 1 μm tungsten for conduction layer and (f) 1 μm PSG for sacrificial layer.

3.12 A chemical etches various crystalline planes of silicon with the following rates:

| | Etch rate μm/hr | |
|---|---|---|
| ⟨100⟩ | ⟨110⟩ | ⟨111⟩ |
| 51 | 57 | 1.25 |

The (100) silicon substrate is 200 μm thick and has an etch window of 500 μm × 500 μm aligned to {110} directions. Draw and mark approximate dimensions of the etch profile after (a) 15 minutes and (b) 5 hours of etching.

3.13 Isotropic etching is known to cause rounding off of etch profile. Anisotropic etching, though it yields better-defined shapes, can only form pyramidal openings. Determine (a) The approximate shape for a mask; (b) the type of etch; (c) etch-stop material and (d) a typical etchant required to form a triangular opening for a cavity formed on silicon by wet etching. *Note:* No restriction is imposed on the etch profile.

# CHAPTER 4

# Mechanics of Slender Solids in Microsystems

## LEARNING OBJECTIVES

After completing this chapter, you will be able to:

▶ Understand deformation of bars and beams.

▶ Learn energy methods for computing deflections.

▶ Understand how to compute the stiffness of micromechanical suspensions.

▶ Have an overview of thermal loading and bimorph effects.

▶ Understand the consequences of residual stresses and stress gradients.

▶ Learn about the Poisson effect and anticlastic curvature.

▶ Understand how to handle large displacements and dynamic behavior.

▶ Have an overview of general 2D and 3D governing equations for elastic bodies.

Moving solids and flowing fluids distinguish microsystems from microelectronics. While there is no intended physical movement in electronic circuit elements, in most microsystems the functionality is due to the movement of solids and/or fluids. Hence, microsystems are in fact microdynamic systems. In order to understand how such dynamic micromachined devices perform and how to design them, we need to model the behavior of moving solids and fluids. Our focus in this chapter is on solids.

Movements of solids can be modeled in two fundamentally different ways: as rigid bodies and as deformable bodies. In rigid bodies, there is no relative motion between any two points. A block moving in a slot and a cylindrical pin rotating in a cylindrical hole are examples of rigid-body motion [see Figure 4.1(a)]. In contrast, there is relative motion between points in deformable bodies. Imagine the stretching of a rubber string or a bar or the bending of a beam as shown in Figure 4.2(b). Note that in reality no solid is perfectly rigid; we only assume some solids to be rigid if their deformation is negligibly small.

Both types of movement of solids exist in microsystems, but since deformable solids dominate we focus on them. The deformable solids in microsystems have several forms: bars (also called rods in Chapter 5), beams, membranes, plates, shells, and more complicated geometric shapes. In this chapter, we learn the basics of a few such slender deformable solids that are acted upon by different types of forces.

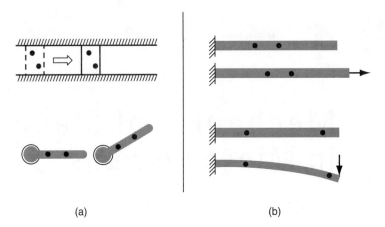

(a)                                              (b)

**Figure 4.1** Rigid-body motion versus deformable motion in solids. Microsystems have both types. (a) In rigid-body motion, no relative motion occurs between points, as in the movement of a block (above) and rotation of a crank (below); (b) Relative motion between points when a bar stretches (above) or a beam bends (below).

**Your Turn:**
Deformable motion is more widely used than rigid-body motion at the microscale. Do you know why? Think about microfabrication. Which is easier to make: a rigid-body joint or a deforming structure?

In general, fluid flow behavior is more involved than the deformation of solids. Another complicating factor is that fluids behave substantially differently in small volumes than in large volumes. For example, the flow in a large pipe is different from that in a pipe of micron dimensions: scaling effects begin to appear in fluid behavior at micron sizes.[1] Microfluidics is an important and challenging part of microsystems (e.g. bioMEMS, lab-on-a-chip). Chapters 6 and 9 briefly introduce this topic in the context of coupled simulations and scaling effects.

It is also important in designing microsystems to consider heat transfer in solids and fluids. In some applications, the parts of a microdevice are intentionally heated. In others, solids in particular, Joule heating is inevitable due to the passage of electric current. Unintended heating leads to two complications. First, most material properties are dependent on temperature. Second, heating causes thermal stresses. Therefore, while modeling one must pay attention to heat transfer aspects. This chapter gives a glimpse into the thermal aspects of microsystems in the context of thermally induced loads.

## ▶ 4.1 THE SIMPLEST DEFORMABLE ELEMENT: A BAR

A bar is a slender structure. For example, a pillar and a post are bars. A taut straight wire of large cross-section area can also be treated as a bar. A simple microfabricated electrical resistor element fixed at its two ends is also a bar. Thus, any slender structure that can stretch or contract without bending is a bar.

The geometric parameters of a bar are its length and the area of cross-section. Note that the shape of the cross-section does not matter for a bar; only its area does. This is because a

---

[1] For all practical purposes in modeling microdevices, the principles of solid mechanics can be applied at the micron scale as at the macroscale.

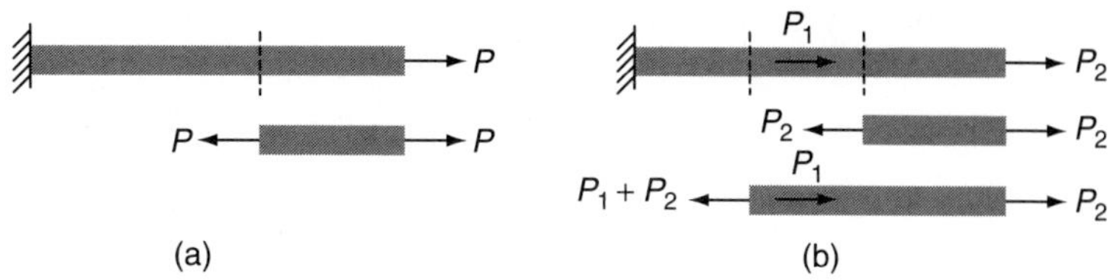

**Figure 4.2** (a) A bar under a single load. Wherever you imagine cutting it, the internal force will be the same, that is, $P$; (b) A bar under two loads. The internal force is either $P_1$ or $(P_1 + P_2)$ depending on where the bar is cut.

bar's stretching or contraction along its length depends only on its area and not on its shape as long as its length is much greater than its cross-section dimensions. A bar can take loads only along its length, which we call its *axis*. When electric current is passed through a bar-like resistor, it is heated due to Joule heating. If the bar is fixed at its two ends, it will not be able to stretch even though the heat causes it to expand; this results in axial stress in the bar. Understanding the notion of stress is important in micromechanical (indeed in any mechanical) systems, as we discuss in subsequent paragraphs.

*Stress* in an elastically deformable solid is defined as the force divided by the area over which it acts. Thus, stress has the same units as pressure. We use N/m$^2$ or Pa (Pascal) as its unit. Let us consider a bar perpendicular to the axis of uniform cross-section area $A$ and length $L$ [Figure 4.2(a)]. Let it be fixed at one end while a force $P$ is applied axially at the other end, as shown in the figure. Now, if we cut the bar perpendicular to the axis anywhere in the middle, it has force $P$ acting on an area $A$. Notice that the separated segment of the bar is in force equilibrium. That is, the vectorial sum of forces acting on the segment is zero. Thus, we can say that the stress, denoted by $\sigma$, acting throughout the length of the bar given by

$$\sigma = \frac{P}{A} \qquad (4.1)$$

In order to know what stress is, we need to imagine cutting an elastic body and then seeing how much internal force acts on a certain area. Consider the same bar under a different loading, as shown in Figure 4.2(b). Imagine cutting this bar at different points and seeing what the internal force would be to keep the separated segment in equilibrium. If we divide this force by the cross-sectional area, we get the stress in it. Now, this stress is not the same throughout the bar because the internal forces are not the same. Similarly, if the cross-sectional area is not the same, the stress is not the same at all points even if the internal force is. A conical tip of an atomic force microscope (AFM) is an example of a bar with variable cross-sectional area. The axial stress in it varies even though the same internal force acts throughout.

The stress in a bar is called a *normal stress* as the internal force in it acts normal (i.e. perpendicular) to the area of cross-section. Furthermore, if the separated segment is being pulled by the forces, the resulting stress is called *tensile* stress. If the stress acts so as to compress the bar, then the stress is said to be *compressive*. The general convention is to treat tensile stress as positive and compressive stress as negative.

Stress is a fictitious quantity defined in order to understand what happens inside an elastic body. We cannot see stress. But what we can see is that the force acting on the bar stretches or contracts it. We may be able to see this with the naked eye, with a sophisticated measurement technique, or with a powerful microscope. Since the value of the stretching or contraction differs in different situations, we define a relative quantity called *strain*. It is denoted by $\in$ and is defined as the ratio of the change in length to the original length. If the bar in Figure 4.2(a) stretches (or contracts) by $\Delta$, its strain is given by

$$\in = \frac{\Delta}{L} \tag{4.2}$$

Stress and strain are related by the well-known *Hooke's law*, a fundamental law in solid mechanics. Hooke's law states that stress is proportional to strain. The constant of proportionality is called *Young's modulus*, which we denote by $Y$. It is a property of the material of the bar. Mathematically, we write Hooke's law as follows:

$$\sigma = Y \in \tag{4.3}$$

This useful relationship enables us to compute the deformation (stretching or contraction) of the bar in Figure 4.2(a) due to the force $P$. We do this by substituting Eqs. (4.1) and (4.2) into Eq. (4.3):

$$\frac{P}{A} = Y\frac{\Delta}{L}$$
$$\Delta = \frac{PL}{YA} \tag{4.4}$$

Now, imagine a bar fixed at both the ends and heated uniformly to temperature $T$ above its original temperature. The bar will be stressed because it cannot stretch or contract since it is fixed at both the ends. Hooke's law helps us compute this stress. Let us imagine that one end of the bar is free to expand. Then, it will simply stretch by the amount $\Delta$ equal to $\alpha TL$, where $\alpha$ is the thermal expansion coefficient. This $\alpha$, which is a property of the material of the bar, gives the change in length of a material segment per unit length per unit temperature rise. Now, since the bar is fixed at both ends but must expand because of the rise in temperature, it develops a stress to oppose this stretching. The strain due to this stress cancels out the thermally induced strain. We can now calculate the strain and thus the stress in the material as

$$\in = -\frac{\alpha TL}{L} = -\alpha T \quad \text{and} \quad \sigma = Y \in = -Y\alpha T \tag{4.5}$$

Thus, the temperature-induced stress in the bar is compressive if there is a rise in temperature. You can reason out what happens if the temperature is reduced by $T$ by dipping the bar in, say, ice-cold water.

It is interesting to imagine what happens if the temperature rises beyond a certain value. Our intuition might tell us that the bar, being unable to stretch, will suddenly bend transversely, that is, in a direction normal to the axis. This is called *buckling*. We can calculate the value of $T$ at which buckling would occur. This requires us to study the transverse deformation and the loads that cause it.

### Problem 4.1

If the temperature increase along the bar is not constant but varies parabolically with zero rise at its fixed ends (Figure 4.3), how does the thermally induced stress vary in the bar? An electrical resistor suspended between two fixed regions with a voltage difference across it presents this scenario.

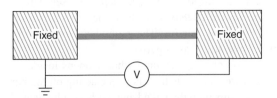

**Figure 4.3** Schematic of a micromachined resistor suspended between two anchors.

## ▶ 4.2 TRANSVERSELY DEFORMABLE ELEMENT: A BEAM

In the preceding section we considered a bar with an axial force. Now we apply a force perpendicular to the axis of the bar, that is, in the *transverse* direction. It is quite easy to see that the bar will bend. We call such an element a *beam*. Let us take the slender structure in Figure 4.1(a) and subject it to a transverse load as shown in Figure 4.4. It will deform as shown. How do we compute the transverse deflection of this beam? What kinds of stresses are induced when a beam deforms? These questions can be answered using the same two simple concepts we used in the case of a bar: force equilibrium and Hooke's law. Before we do that, we look at an application of a beam in a micromachined accelerometer.

**Figure 4.4** A beam under a tip load $F$ in the transverse direction. How much does it deform in the transverse direction? The answer lies in the beam theory discussed in this section.

Figure 4.5(a) is a schematic of a microaccelerometer with mass at the center and four beams attached to it. We can say that the mass is suspended by the four beams. If acceleration $a_x$ is applied in the $x$-direction, as in Figure 4.5(b), each beam deforms and we can calculate the displacement of the mass. The four beams now act like bars because the force equal to $ma_x$, where $m$ is the mass, only stretches or contracts them axially. Each bar acts like a spring. We thus have four springs in parallel and they all experience the same absolute magnitude of deflection because they share the force.

Since all bars have the same length (say, $L$), the same area of cross-section (say, $A$) and are made of the same material, they all have the same *spring constant* $k_a$, where the subscript $a$ indicates an axial deformation mode. The spring constant is the slope of the force–deformation curve. For small deformations, the axial force versus axial deformation

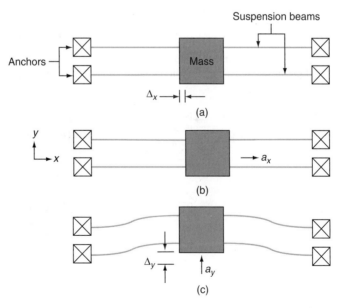

**Figure 4.5** (a) Schematic of a micromachined accelerometer with a mass, four suspension beams, and anchors; (b) axial deformation of the beams, which act as bars because of axial force due to an acceleration in the $x$-direction; (c) Vertical motion of the mass due to an acceleration in the $y$-direction induces bending in the beams. The $z$-axis is perpendicular to the plane of the page.

curve remains straight, that is, the relationship is linear. The axial spring constant $k_a$ is given by [recall Eq. (4.4)]

$$k_a = \frac{P}{\Delta_x} = P\left(\frac{AY}{PL}\right) = \frac{AY}{L} \tag{4.6}$$

Now, since the axial force $ma_x$ is shared equally by all the four springs, the deflection is

$$\Delta_x = \frac{ma_x}{4k_a} \text{ and } k_x = 4k_a = \frac{ma_x}{\Delta_x} = \text{Equivalent axial spring constant} \tag{4.7}$$

### Example 4.1

A mass of 50 μg is attached to two bars, both of length 100 μm but cross-sectional areas $4\,\mu m^2$ and $25\,\mu m^2$, respectively, in three different ways, as shown in Figure 4.6. Compute the axial stiffness and displacement of the mass in each case for $1g = 9.81\ m/s^2$ acceleration. The material is silicon with $Y = 150\,\text{GPa}$.

**Solution:**   The axial stiffness (i.e., spring constants) of the two bars can be computed using Eq. (4.6) as

$$k_{a1} = \frac{A_1 Y}{L_1} = \frac{(4 \times 10^{-12})(150 \times 10^9)}{100 \times 10^{-6}} = 6000\,\text{N/m} \tag{4.8a}$$

$$k_{a2} = \frac{A_2 Y}{L_2} = \frac{(25 \times 10^{-12})(150 \times 10^9)}{100 \times 10^{-6}} = 37,500\,\text{N/m} \tag{4.8b}$$

The force acting on the mass in all cases is given by[2]

$$F = ma = (50 \times 10^{-9})(9.81) = 0.4905 \times 10^{-6}\,\text{N} = 0.4905\,\mu\text{N} \tag{4.9}$$

In cases (a) and (b) in Figure 4.6, the force acting on the mass is shared by the two bars because they move the same absolute distance. Hence, they are in parallel. In that case, the equivalent

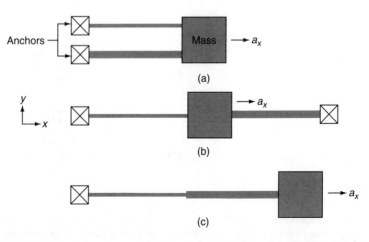

**Figure 4.6** Parallel and serial arrangements of two axially deforming beams depending on how the force is applied.

spring constant is given by $k_{x(a,b)} = k_{a1} + k_{a2} = 43,500$ N/m. Hence, the displacement of the mass, $\Delta_{x(a,b)}$, is

$$\Delta_{x(a,b)} = \frac{ma_x}{k_{x(a,b)}} = \frac{0.4905 \times 10^{-6}}{43,500} = 0.0113 \times 10^{-9} \text{ m} = 0.0113 \text{ nm} \tag{4.10}$$

Note that

$$ma_x = k_{a1}\Delta_{x(a,b)} + k_{a2}\Delta_{x(a,b)}$$
$$k_{x(a,b)}\Delta_{x(a,b)} = (k_{a1} + k_{a2})\Delta_{x(a,b)} \tag{4.11}$$
$$k_{x(a,b)} = (k_{a1} + k_{a2})$$

On the other hand, in case (c) of Figure 4.6, the bars experience the same force but undergo different displacements. Therefore,

$$k_{a1}\Delta_{x1(c)} = ma_x \quad \text{and} \quad k_{a2}\Delta_{x2(c)} = ma_x$$
$$\Delta_{x1(c)} = \frac{ma_x}{k_{a1}} \quad \text{and} \quad \Delta_{x2(c)} = \frac{ma_x}{k_{a2}} \tag{4.12}$$

Now, the total displacement of the mass is $\Delta_{x(c)} = \Delta_{x1(c)} + \Delta_{x2(c)}$, where $\Delta_{x1(c)}$ and $\Delta_{x2(c)}$ are the elongation of springs 1 and 2, respectively. If we define the equivalent spring constant as $k_{x(c)} = (ma_x/\Delta_{x(c)})$, we get

$$\Delta_{x(c)} = \Delta_{x1(c)} + \Delta_{x2(c)}$$
$$\frac{ma_x}{k_{x(c)}} = \frac{ma_x}{k_{a1}} + \frac{ma_x}{k_{a2}} \tag{4.13}$$
$$\frac{1}{k_{x(c)}} = \frac{1}{k_{a1}} + \frac{1}{k_{a2}}$$

Hence, $k_{x(a,b)} = k_{a1} + k_{a2}$ is equal to 444 N/m and the displacement of the mass in case (c) is 0.0948 nm:

$$k_{x(c)} = \frac{k_{a1}k_{a2}}{k_{a1} + k_{a2}} = \frac{(6000 \times 37,500)}{6000 + 37,500} = 5,172.4 \text{ N/m}$$
$$\Delta_{x(c)} = \frac{ma_x}{k_{x(c)}} = \frac{490.5 \times 10^{-6}}{5,172.4} = 0.0948 \text{ nm}$$

Note that the equivalent spring constant for parallel springs is larger than the largest, whereas the equivalent spring constant for serial springs is smaller than the smallest. This insight is useful in design when we want to get large or small spring constants.

Now, imagine what happens if the acceleration $a_y$ acts in the $y$-direction as shown in Figure 4.5 (c). The four beams bend as shown in the figure. How do we get $k_y$, the spring constant for the bending mode of deformation in the $y$-direction? Note that, just like the axial springs, the four bending springs act in parallel here. That is, they too share the force $ma_y$ equally while having the same deflection. Let us use the force equilibrium and Hooke's law to derive an expression for the bending spring constant, $k_b$, for beams.

The deformation in a beam is primarily due to bending. We now introduce a concept called the bending *moment*. Just as we imagined cutting a bar [see Figures 4.2(a) and (b)], we imagine cutting the beam at a distance $x$ from the fixed end and consider its equilibrium, as shown in Figure 4.7(a). For static equilibrium, we have a vertical force in the $y$-direction at the cut section. We call it a vertical *shear force* because its effect is to shear the beam vertically. Note that a shear force is parallel to the surface on which it acts, unlike a normal stress or force that acts perpendicular to the surface. The shear force $V$ and the applied force $F$ are the balancing forces in the vertical ($y$) direction. If we take moments about the $z$-axis at any point in the cut portion of the beam (the easiest point would be

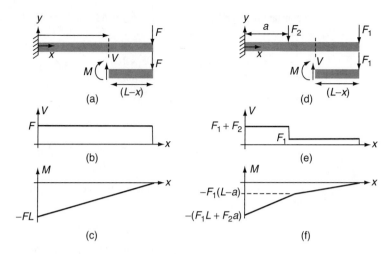

**Figure 4.7** (a) A cantilever beam with a tip load that gives rise to a vertical shear force, $V$, and bending moment, $M$; (b) shear force diagram; (c) bending moment diagram; (d) cantilever beam with two loads; (e) shear force diagram; and (f) bending moment diagram.

the point where we have cut), we can conclude that there must be a bending moment $M$ equal to $V(x - L)$ and this in turn is equal to $F(x - L)$ at the cut point in order to maintain the moment equilibrium. Clearly, the bending moment varies along the length (i.e. along the axis) of the beam. Vertical shear force and bending moment diagrams for this beam are shown in Figures 4.7(b) and (c), respectively. The vertical shear force is constant in this case, but it need not be so in general. As an illustration, see Figures 4.7(d)–(f), where a cantilever beam is subjected to multiple loads. The expressions for vertical ($y$-axis) shear force and bending moment are obtained by considering the static equilibrium of vertical forces and moments about the $z$-axis. Now, let us see how the bending moment can be used to compute the deflection.

Imagine a small segment of a beam that has a bending moment $M$ throughout its small length. Such a segment is said to be in pure bending. In Figure 4.8, beginning at the top, we see that a straight segment[2] is bent into a circular arc with enclosed angle $\theta$. As the bending moment is constant, the rate of bending (i.e. the rate of change of $\theta$ with respect to the length along the segment considered) is also constant. Hence the arc can only be circular! Now, think about several lines parallel to the axis of the beam in its undeformed or straight condition. The lines closer to the top must contract when the beam bends and the lines closer to the bottom must elongate. Clearly, there must be a line somewhere in the middle that neither contracts nor elongates. We call this the *neutral axis*. The plane formed by the neutral axis perpendicular to the $xy$ plane is called the *neutral plane*. The distance between any pair of points in the neutral plane remains constant.

If we take the radius of the circular arc as $\rho$, the length of the neutral axis and hence the length of the undeformed beam segment is $\rho\theta$. Now, imagine a line at a distance $y$ from the neutral axis. As it bends, its length will be $(\rho - y)\theta$. If $y$ is positive, the length of the deformed arc is smaller than that of the neutral axis, and if $y$ is negative the length is greater.

The axial strain can be calculated from Eq. (4.2). Obviously, the largest axial strain occurs when $y$ has the greatest value and this is at $c$, where $2c$ is the thickness of the beam [see Eq. (4.14 a)]. Now, we can write the strain along the vertical cross section of a beam as $(-\epsilon_{max}y/c)$, as given in Eq. (4.15 a). The maximum stress from Hooke's law then becomes $(-Y\epsilon_{max}y/c)$ as given in Eq. (4.16 a). We now raise the question: can we locate the neutral axis in the vertical section? The answer is: yes, we can do this with the help of force equilibrium on the vertical section; this is shown

---

[2] Readers familiar with straight beam bending may want to look at Figure 4.10 where the bending of initially curved thick beams is described. Thin curved beam analysis, as discussed later, does not differ from that for straight beams. But thick curved beams, as shown here, do differ. Both thin and thick curved beams are used in micromachined and smart systems structures.

in Figure 4.9, along with the stress acting on it. Notice that this stress varies linearly with $y$ [from Eq. (4.16a)]. As expected, the stress is zero at the neutral axis because the neutral axis does not stretch or contract. Now, as shown in Eq. (4.17), summation of the $x$ forces on the cross-section shows that $\int y\,dA = 0$. This expression defines the neutral axis. Recall the definition of the centroid of an enclosed area: the point such that the area moment about it is zero. Clearly, for a straight beam, the neutral axis passes through the centroid.[3]

However, we still do not know $\epsilon_{max}$ and $\sigma_{max} = Y\epsilon_{max}$ because they are in terms of an unknown $\rho$ [see Eq. (4.14a)]. In order to resolve this unknown, we use the moment equilibrium about the $z$-axis of the normal force acting on the cross-section of the beam. The normal force is $\sigma_x$ multiplied by the cross-sectional area. We take the moment of this force about the point where the neutral axis crosses the cross-sectional area. As shown in Eq. (4.18a), the summation of the moments about the $z$-axis gives an expression for the bending stress, $\sigma_x$.

Note that we have defined a quantity $I = \int y^2 dA$. This is called the *area moment of inertia* and is a physical property of the cross-sectional area of the beam. Its value depends on the shape of the cross section. Thus, unlike in bars, the shape of the cross-sectional area is important for beams, since different cross sections of the same area have different moments of inertia. Figure 4.11 shows some typical cross-sections encountered in micromachined structures along with their area moments of inertia. Square and rectangular cross-sections are common in surface-micromachined beams. Trapezoidal cross-sections are obtained with anisotropic wet chemical etching. The T or I sections are also possible in structures consisting of multiple structural layers of surface micromachining or wafer-bonding processes.

---

### Straight Beam in Pure Bending

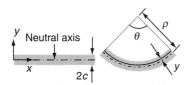

**Figure 4.8** A short straight segment with constant bending moment before (left) and after (right) bending. Lines above the neutral axis contract and those below stretch.

Axial strain is given by

$$\epsilon_x = \frac{(\rho - y)\theta - \rho\theta}{\rho\theta}$$

$$= -\frac{y}{\rho}$$

$$\epsilon_{max} = c/\rho \tag{4.14a}$$

$$\epsilon_x = -y\epsilon_{max}/c \tag{4.15a}$$

$$\sigma_x = -Yy\epsilon_{max}/c = -y\sigma_{max}/c \tag{4.16a}$$

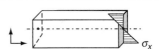

**Figure 4.9** The bending stress in the axial direction varies linearly in the $\sigma_x$ direction as per Eq. (4.16a).

$$\sum F_x = 0$$

$$\int \sigma_x dA = 0$$

$$-\frac{\sigma_{max}}{c}\int y\,dA = 0 \tag{4.17}$$

$$\int y\,dA = 0$$

$$\sum M_z = 0$$

$$\int y\sigma_x dA + M = 0$$

$$-\frac{\sigma_{max}}{c}\int y^2 dA = M \tag{4.18a}$$

$$-\frac{\sigma_{max}}{c}I = M$$

$$\sigma_{max} = -\frac{Mc}{I}$$

---

[3] Readers familiar with straight beam theory should look at the derivation for the initially curved beams and note two things: the stress is not linear along the vertical cross section, and the neutral axis is not located at the centroid (Figure 4.10).

Therefore,

$$\sigma_x = \frac{My}{I} \qquad (4.18b)$$

**Initially Curved Beam in Pure Bending**

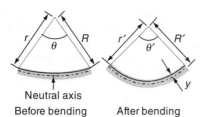

Neutral axis
Before bending    After bending

**Figure 4.10** A segment of a curved beam.

Let $\theta' = \theta + \Delta\theta$. Since the neutral axis length does not change,

Now,
$$R\theta = R'\theta' \qquad (4.14b)$$

$$\epsilon_x = \frac{r'\theta' - r\theta}{r\theta} = \frac{(R' - y)\theta' - (R - y)\theta}{(R - y)\theta}$$

$$\qquad (4.15b)$$

$$= -\frac{y\Delta\theta}{(R - y)\theta}$$

$$\sigma_x = -\frac{Yy\Delta\theta}{(R - y)\theta}. \qquad (4.16b)$$

Note that the stress is not linear in $y$!

**Your Turn:**

By following the method outlined for a straight beam, that is, by integrating the stress times the area over the cross-section and equating it to zero, show that the distance between the neutral axis and the centroidal axis for a curved beam is

$$\left(\frac{1}{A}\int r\,\mathrm{d}A - \frac{A}{\int \mathrm{d}A/r}\right)$$

Also find expressions for this quantity when the cross-sections are circular and rectangular.

Obtain an expression for the bending stress by summing the bending moment across the area of cross-section.

Can you think of an application of curved beam theory in microsystems or smart systems?

**Problem 4.2**

In Figure 4.11, the area of moment of inertia is intentionally omitted for the inverted T-section. By using the definition [Eq. (4.18a)], derive the moment of inertia.

*Hint*: Note that you first need to compute the location of the neutral axis, which is at the centroid for a straight beam.

We have thus calculated the bending stress using static equilibrium and Hooke's law (and nothing more!). Now, let us see how we can calculate the transverse deflection of a beam. We need to look at only the geometry of pure bending already considered in

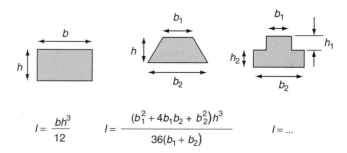

$$I = \frac{bh^3}{12} \qquad I = \frac{(b_1^2 + 4b_1 b_2 + b_2^2)h^3}{36(b_1 + b_2)} \qquad I = \dots$$

**Figure 4.11** Area of moment of inertia, $I$, for typical cross-sections encountered in micromachined structures.

Figure 4.8. Equation (4.14a) gives

$$\epsilon_{max} = \frac{c}{\rho}$$

$$\frac{1}{\rho} = \frac{\epsilon_{max}}{c}$$

(4.19)

By replacing $\epsilon_{max}$ with $\sigma_{max}/Y$ and substituting for $\sigma_{max}$ using Eq. (4.18a), we get

$$\frac{1}{\rho} = \frac{\epsilon_{max}}{c} = \frac{\sigma_{max}}{Yc} = -\frac{Mc}{I} \cdot \frac{1}{Yc} = -\frac{M}{YI}$$

$$\frac{1}{\rho} = -\frac{M}{YI}$$

(4.20)

Equation (4.20) is called the *Euler–Bernoulli theorem* for the bending of beams. Recall from geometry that $\rho$, the radius of curvature, is given in terms of the first and second derivatives of a curve. Here, the curve is the bent profile of the neutral axis of the beam. Let us denote the deflection of the beam at any point $x$ by $w(x)$. The definition of the radius of curvature[4] gives us

$$\frac{1}{\rho} = \frac{d^2w/dx^2}{\left\{1 + (dw/dx)^2\right\}^{3/2}}$$

(4.21a)

which for small deflections becomes

$$\frac{1}{\rho} \approx \frac{d^2w}{dx^2} \quad \text{because } (dw/dx)^2 \approx 0$$

(4.21b)

Now, Eqs. (4.20) and (4.21b) give us a differential equation to solve for the deflection of beams:

$$\frac{d^2w}{dx^2} = -\frac{M}{YI}$$

(4.22)

Once we know the bending moment $M(x)$, we can compute $w(x)$ by integrating Eq. (4.22) twice and using the end conditions of the beam to solve for the two constants of integration. That is,

$$\frac{dw}{dx} = -\int \frac{M}{YI}dx + C_1$$

$$w = -\int \left(\int \frac{M}{YI}dx\right)dx + C_1x + C_2$$

(4.23)

---

[4] Refer to an elementary analytical geometry book. Alternatively, you can compute the curvature using calculus from its definition: the rate of change of the tangent vector as we move along the curve.

Let us see how this works for the cantilever with a tip load shown in Figure 4.4. Since $M = F(x - L)$ in that example, we get

$$w = -\int \left( \int \frac{F(x - L)}{YI} \, dx \right) dx + C_1 x + C_2$$

$$= -\int \left( \frac{Fx^2}{2YI} - \frac{FLx}{YI} \right) dx + C_1 x + C_2 \qquad (4.24)$$

$$= -\frac{Fx^3}{6YI} + \frac{FLx^2}{2YI} + C_1 x + C_2$$

The end conditions for the cantilever beam require that the deflection and the slope both be zero at the fixed end. That is,

$$\left. \begin{array}{c} w = 0 \\ \dfrac{dw}{dx} = 0 \end{array} \right\} \quad \text{at} \quad x = 0 \qquad (4.25)$$

Substituting the values obtained in Eq. (4.25) into Eq. (4.24) yields the following at $x = 0$:

$$w = C_2 = 0$$
$$\frac{dw}{dx} = C_1 = 0 \qquad (4.26)$$

Thus, for a cantilever beam with a tip load, the deflected curve of the neutral axis is given by

$$w = -\frac{Fx^3}{6YI} + \frac{FLx^2}{2YI} \qquad (4.27)$$

Consequently, for a cantilever beam with a tip load, the magnitudes of the downward deflection and its slope at the loaded tip are

$$w|_{x=L} = \frac{FL^3}{3YI} \quad \text{and} \quad \left. \frac{dw}{dx} \right|_{x=L} = \frac{FL^2}{2YI} \qquad (4.28)$$

Therefore, for a cantilever beam with a tip load, we can write the bending spring constant as

$$k_{b\_cantilever} = \frac{F}{|w|_{x=L}} = \frac{3YI}{L^3} \qquad (4.29)$$

In the preceding paragraphs, we described a method of calculating bending spring constant of beams. However, we have not yet derived $k_y$ for the accelerometer suspension in Figure 4.6(c) for vertical ($y$-axis) acceleration. Observe that the bending moments will be different when the end conditions are different, and hence the solution

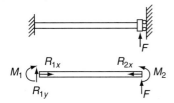

**Figure 4.12** A fixed–guided beam with a transverse load at the guided end, and its reaction forces and moments. This is a *statically indeterminate* beam.

of Eq. (4.22) will also be different. The end conditions of each of the four beams in Figure 4.6(c) are described as *fixed* and *guided*. That is, one end is fixed as in a cantilever and the other is guided along a vertical line or the *y*-direction (see Figure 4.12). If this beam is separated from its supports, we must show the reaction forces and reaction moments acting on it by using a *free-body diagram* (Figure 4.12).

The concept of free-body diagrams is almost indispensable in the analysis of mechanical systems. The box item below explains the significance of free-body diagrams. In the present context, when we look at a physical connection of a beam to a mass, we must be able to understand the reactions forces at the points of connection. For this, we draw the free-body diagram of the beam. In essence, free-body diagrams help us understand how forces and moments are transferred from one portion to another portion of a mechanical assembly of rigid and elastic bodies.

## ▶ *FREE-BODY DIAGRAMS*

A mechanical system is an assemblage of individual elements of different shapes. The elements may be connected with joints such as hinges and sliders or rigidly fastened to one another. Whenever a single element or a subassembly is separated as a free body, we need to show the reaction forces and moments at the points of connection with the other members. This may seem easy at first, but careful examination is warranted to see what kinds of reaction forces and moments act at the points of connections. The general rule is that whenever a particular type of motion is restricted, a corresponding force or moment exists as a reaction. Likewise, when a motion is free, there is no reaction force or moment. Figure 4.13 illustrates some easy cases in 2D. The same notion can be extended to 3D and to cases in which there is line or point contact, possibly with friction as well.

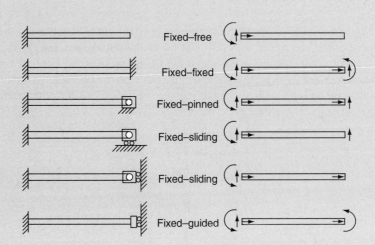

**Figure 4.13** Different end conditions for 2D beams and reaction forces/moments. All reaction forces and moments are shown in their positive directions. If they are in opposite directions, their magnitudes become negative in the calculations.

The procedure presented for computing the deflection of a cantilever beam cannot be readily applied to the fixed–guided beam because we cannot determine the unknown reaction forces and moments ($R_{1x}$, $R_{1y}$, $M_1$, $R_{2x}$ and $M_2$) from the three static equilibrium equations we have. Such an elastic system is called *statically indeterminate*. In order to determine the reactions in such a system, we need to use more than statics; we need kinematic conditions at the end. In this example, such a kinematic end condition is given by the fact that the $x$-direction displacement and the slope at the guided end are zero. This gives two more equations, and that suffices to solve for all the five unknown reactions and moments. Once we know these, we can obtain the bending moment and axial force as functions of $x$ and then solve for the deflections. However, there is a catch here: how do we know the end reactions before solving for the deflections? This can be resolved in more than one way. Here, we use an energy method that comes in handy in many other situations arising in micromachined mechanical structures.

## ▶ 4.3 ENERGY METHODS FOR ELASTIC BODIES

A deformed elastic body stores energy, as is evident from a simple spring. This is analogous to an electrical capacitor, which also can store and release energy. The energy stored in an elastic body, called the *strain energy*, is denoted by $SE$. The strain energy per unit volume, called the *strain energy density* and denoted by $se$, of a body in equilibrium is given by

$$se = \int_0^\epsilon \sigma \, \mathrm{d}\epsilon \qquad (4.30a)$$

This is simply the area under the stress-strain curve. By virtue of Hooke's law, the relationship between them is linear. Therefore, we get

$$se = \int_0^\epsilon \sigma \, \mathrm{d}\epsilon = \int_0^\epsilon (Y\epsilon) \, \mathrm{d}\epsilon = \frac{Y\epsilon^2}{2} = \frac{\sigma\epsilon}{2} = \frac{\text{Stress} \times \text{Strain}}{2} \qquad (4.30b)$$

Now, in order to compute the strain energy of the entire elastic body, we need to integrate $se$ over the volume of the body. Let us do this for the bar and the beam.

For a bar, we can write the stress as the ratio of the internal force and the cross-sectional area ($\sigma = p/A$) and the strain as $\epsilon = \mathrm{d}u/\mathrm{d}x$ (the latter is simply the definition of strain for an infinitesimally small segment of bar of length $\mathrm{d}x$ that stretches by $\mathrm{d}u$). Thus, we write the strain energy of the bar, $SE_{bar}$:

$$SE_{bar} = \int_0^L \frac{\sigma\epsilon}{2} A \, \mathrm{d}x$$

$$= \int_0^L \frac{\sigma^2}{2Y} A \, \mathrm{d}x = \int_0^L \frac{p^2}{2AY} \, \mathrm{d}x \qquad (4.31)$$

$$= \int_0^L \frac{Y\epsilon^2}{2} A \, \mathrm{d}x = \int_0^L \frac{YA}{2} \left(\frac{\mathrm{d}u}{\mathrm{d}x}\right)^2 \mathrm{d}x$$

Note that we have written the strain energy in terms of only the internal force denoted as p(x) (which can be obtained from the free-body diagrams) and the deflection (actually, its derivative). Note also that the expression for *SE* is valid even when the cross-sectional area *A* varies along the bar. Needless to say, the internal force and Young's modulus may also change with *x* and the expression in Eq. (4.31) is still valid when both *A* and *Y* become functions of *x*.

By using the definitions of bending stress and strain for a beam, and with Eqs. (4.18b) and (4.22), the strain energy of a beam can be written as:

$$
SE_{beam} = \int\limits_{Volume} \frac{\sigma \in}{2} \, dV
$$

$$
= \int\limits_{Volume} \frac{\sigma^2}{2Y} \, dV = \int\limits_{Volume} \frac{M^2 y^2}{2YI^2} \, dV = \int\limits_0^L \left\{ \frac{M^2}{2YI^2} \left( \int\limits_{Area} y^2 dA \right) \right\} dx = \int\limits_0^L \frac{M^2}{2YI} \, dx \quad (4.32)
$$

$$
= \int\limits_0^L \frac{\left( YI \frac{d^2 w}{dx^2} \right)^2}{2YI} \, dx = \int\limits_0^L \frac{1}{2} YI \left( \frac{d^2 w}{dx^2} \right)^2 dx
$$

Armed with the strain energy definition and its expressions for a bar and a beam, we state two important energy theorems known as *Castigliano's first* and *second theorems*.

**Castigliano's First Theorem**

If the strain energy of a linearly elastic body is expressed in terms of its displacements, then the force at a point is equal to the partial derivative of the strain energy with respect to the displacement in the direction of the force at the point where that force is applied. Mathematically, we write this as follows:

$$
F = \frac{\partial SE(u)}{\partial u} \quad (4.33)
$$

**Castigliano's Second Theorem**

If the strain energy of a linearly elastic body is expressed in terms of forces, then the force at a point is equal to the partial derivative of the strain energy with respect to the force at the point where that displacement is considered in the direction of the displacement. Mathematically, we write this as[5]:

$$
u = \frac{\partial SE(F)}{\partial F} \quad (4.34)
$$

The two Castigliano theorems are general enough that the force can be replaced by moment and the displacement by rotation. We can now return to our main problem of the fixed-guided beam.

---

[5] It is more accurate to state this force using complementary strain energy ($SE_c$), to be introduced in Section 6.2.1 (Chapter 6). But for linear elastic situations, $SE_c = SE$.

At the end of Section 4.2 and in Figure 4.12, we had five unknowns ($R_{1x}$, $R_{1y}$, $M_1$, $R_{2x}$ and $M_2$) and five equations to solve for them. The first three were static equilibrium equations:

$$\sum F_x = R_{1x} - R_{2x} = 0 \tag{4.35}$$

$$\sum F_y = R_{1y} + F = 0 \tag{4.36}$$

$$\sum M_z = -M_1 + M_2 + FL = 0 \tag{4.37}$$

The remaining two were kinematic equations based on the end conditions of the guided end:

$$u_{x=L} = \text{Axial displacement} = 0 \tag{4.38}$$

$$\left.\frac{dw}{dx}\right|_{x=L} = \text{Slope} = 0 \tag{4.39}$$

Equations (4.38) and (4.39) can now be put into mathematical form using the Castigliano's second theorem as follows:

$$u_{x=L} = \frac{\partial SE}{\partial R_{2x}} = 0 \tag{4.38a}$$

$$\left.\frac{dw}{dx}\right|_{x=L} = \frac{\partial SE}{\partial M_2} = 0 \tag{4.39a}$$

By noting that the axial internal force, $p(x)$, for this beam is equal to $R_{1x} = R_{2x}$ and that the bending moment is $M = -M_2 - F(L - x)$, we can compute the total strain energy for this beam using Eqs. (4.31) and (4.32):

$$
\begin{aligned}
SE = SE_{\text{axial}} + SE_{\text{bending}} &= \int_0^L \frac{p^2}{2YA}\,dx + \int_0^L \frac{M^2}{2YI}\,dx \\
&= \int_0^L \frac{R_{2x}^2}{2YA}\,dx + \int_0^L \frac{[-M_2 - F(L-x)]^2}{2YI}\,dx \\
&= \frac{R_{2x}^2 L}{2YA} + \frac{M_2^2 L}{2YI} + \frac{M_2 F L^2}{2YI} + \frac{F^2 L^3}{6YI}
\end{aligned}
\tag{4.40}
$$

Now, when we use $SE$ from Eq. (4.40) in Eq. (4.38a), we find that $R_{2x} = 0$. Since $R_{1x} = R_{2x}$, we conclude that axial reaction forces have no influence in this problem and Eq. (4.39a) gives

$$
\begin{aligned}
&\frac{\partial SE}{\partial M_2} = 0 \\
&\frac{M_2 L}{YI} + \frac{FL^2}{2YI} = 0 \Rightarrow M_2 = -\frac{FL}{2}
\end{aligned}
\tag{4.41}
$$

Equation (4.37) then gives $M_1 = FL/2$, but we do not need this. What we need is the transverse deflection at the guided end. We get this by using Castigliano's second theorem once again:

$$w_{x=L} = \frac{\partial SE}{\partial F}$$

$$\frac{M_2 L^2}{2YI} + \frac{2FL^3}{6YI} = -\frac{FL^3}{4YI} + \frac{FL^3}{3YI} = \frac{FL^3}{12YI} \tag{4.42}$$

This is the result we need to obtain the $y$-direction spring constant, $k_y$, for the accelerometer suspension of Figure 4.6(c). From Eq. (4.42), we get

$$k_y = 4k_b = 4\frac{F}{w_{x=L}} = \frac{48YI}{L^3} \tag{4.43}$$

where we used the fact that all four fixed-guided beams are like four springs in parallel.

In this long digression we discussed beam bending theory and two energy methods for solving the practically useful problem of determining the spring constant of an accelerometer suspension. What we have learned is useful in solving many problems encountered in microsystems and smart systems A few are presented as solved example problems in the next section. However, there is more to learn about beams in the context of microsystems and smart systems components.

## ▶ 4.4 EXAMPLES AND PROBLEMS

**Example 4.2:** A Two-Axis Micromirror

Figure 4.14 shows the schematic of a two-axis micromachined mirror (developed by Agere Systems, a Bell Labs spin-off now with LSI) made of polycrystalline silicon using surface micromachining. The mirror is suspended by four serpentine springs and a ring from an outer square frame. The square frame is fixed. The ring can rotate about the $x$-axis and the mirror rotates about the $y$-axis relative to the ring. Find the rotational stiffness of the serpentine spring whose details are shown in Figure 4.14. The material of the spring has Young's modulus $Y$.

**Solution:** The torsional stiffness of the serpentine spring is calculated on the basis of the bending of the vertical beams when a torque is applied about the axis as shown in Figure 4.15(a). Since the horizontal beams in the figure are very small, their twist is neglected (we learn about twisting of beams later in the chapter).

There are two vertical beams of length $p$ and four of length $2p$. The angular rotation of the serpentine spring when subjected to a torque $T$ results in an angular rotation of the vertical beams, which per Castigliano's second theorem is given by

$$\theta = \frac{M_0 L}{YI} = \frac{Tp}{YI} \quad \text{or} \quad \frac{T(2p)}{YI} \tag{4.44}$$

where the area moment of inertia $I = qt^3/12$ (see Figure 4.11). To derive the result of Eq. (4.44), consider a cantilever beam with applied moment $M_0$ at the tip, as shown in Figure 4.15(b). From

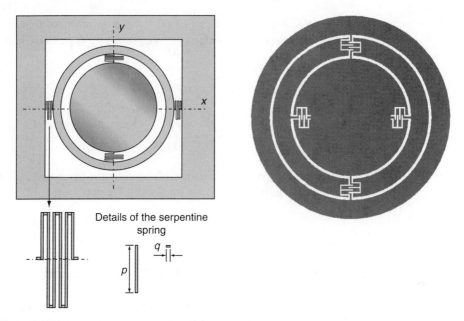

**Figure 4.14** A two-axis micromirror for deflecting light beams in optical fiber communication applications. The circular mirror is supported by four serpentine beam structures. The mirror can tilt about two axes in its plane when the beams in the serpentine structure bend. Here $q$ = in-plane width, $p$ = length of vertical beams, and $t$ = out-of-plane thickness (not shown). The figure at the top is a schematic while the one at the bottom is closer to the actual implementation.

Eq. (4.32), we can write its strain energy and then use Castigliano's second theorem to get the angular rotation at the free end:

$$SE = \int_0^L \frac{M^2}{2YI} \, dx = \int_0^L \frac{M_0^2}{2YI} \, dx = \frac{M_0^2 L}{2YI}$$

$$\theta_{x=L} = \frac{\partial SE}{\partial M_0} = \frac{M_0 L}{YI}$$

(4.45)

Now, the total angular displacement for torque $T$ (which is equal to $M_0$) in Figure 4.15(a) is the sum of the rotations of all the six (four long and two short) beams. Then the angular stiffness constant $\kappa$ can be calculated as follows:

$$\theta_{\text{total}} = \frac{2Tp}{YI} + \frac{4T(2p)}{YI} = \frac{10Tp}{YI}$$

$$\kappa = \frac{T}{\theta_{\text{total}}} = \frac{YI}{10p}$$

(4.46)

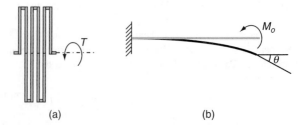

(a)                                    (b)

**Figure 4.15** (a) Modeling the twisting of the serpentine spring in terms of bending the beams in it. (b) One beam under an applied bending moment and the angle of rotation at the free tip.

Since there are two serpentine springs, one on either side, the stiffness will be $2\kappa$ for the spring constant for the rotation about the $x$-axis and for the $y$-axis as well.

## Example 4.3: A Folded-Beam Suspension of an Electrostatic Comb-Drive Microactuator

Figure 4.16 shows the suspension of an electrostatic comb-drive microactuator. Except for the darkly-shaded region marked as fixed, the rest of the structure is free to move above the substrate separated by a gap.

**(a)** Obtain the linear stiffness of the suspension in the vertical direction (i.e. along the axis of symmetry shown) in terms of the length $L$, in-plane width $w$, Young's modulus $Y$, and out-of-plane layer thickness $t$.

**(b)** Obtain the expression for the maximum stress.

**(c)** Calculate also the maximum vertical deflection possible without exceeding the maximum permissible stress $S$.

Given $Y = 150$ GPA, $S = 900$ MPa, $L = 200$ μm, $w = 5$ μm and $t = 3$ μm.

**Solution:** The suspension is symmetric about the vertical ($y$) axis. Hence, we consider only the left half. This part consists of four beams, all of which are fixed at one end and guided at the other. Figure 4.17(a) shows the deformed configuration of the suspension. From Eq. (4.42), we can write the deflection as

$$\Delta = \frac{FL^3}{12YI} = \frac{FL^3}{12Y(tw^3/12)} = \frac{FL^3}{Ytw^3}$$

$$k = \frac{F}{\Delta} = \frac{Ytw^3}{L^3}$$

(4.47)

Thus, for the numerical values given, we get

$$k = \frac{Ytw^3}{L^3} = \frac{(150 \times 10^9)(3 \times 10^{-6})(5 \times 10^{-6})^3}{(200 \times 10^{-6})^3} = 7.0313 \text{ N/m}$$

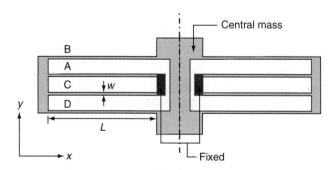

**Figure 4.16** A folded-beam suspension. Only the shaded portion (marked Fixed) is immovable. The central mass has low stiffness in the vertical direction but very high stiffness in the horizontal direction.

If we denote each fixed–guided beam as a spring of stiffness $k$, then in the left half of the comb-drive suspension there are four such springs, arranged as shown in Figure 4.17(b). The boxes in dashed and dot-dash lines in Figures 4.17(a) and (b) show the correspondence between the beams in Figure 4.17(a) and the lumped springs in Figure 4.17(b). The correspondence exists because beams A and B (and also C and D) have the same force and the springs can be considered to be in series. The A–B and C–D spring pairs have the same displacement and hence can be considered to be in parallel. From the spring schematic in Figure 4.17(b), we can observe that the stiffness of the left half of the suspension is $k$ (note that the A–B and C–D spring pairs each have $k/2$ so together they have $k/2 + k/2 = k$). The left and right halves of the suspension are in parallel (since they have the same displacement). Hence, the total stiffness of the suspension is $k + k = 2k = 14.0626\,\text{N/m}$.

(a)  Now let us calculate the stress as the suspension is deformed. From Eq. (4.18b) and using $M = M_2$ from Eq. (4.41), we get the maximum stress in the beam:

$$\sigma = \frac{Mc}{I} = \frac{M_2 c}{I} = \frac{(FL/2)(w/2)}{w^3 t/12} = \frac{3FL}{w^2 t} \tag{4.48}$$

By substituting for $F$ in terms of $\Delta$ from Eq. (4.47), we get $\sigma = 3Yw\Delta/L^2$. Note that $\Delta$ for one fixed–guided beam is half the displacement of the central mass in Figure 4.16, that is,

$$\Delta = \frac{\Delta_{\text{central mass}}}{2}$$

Thus, the maximum stress is

$$\sigma = \frac{3Yw\Delta_{\text{central mass}}}{2L^2}$$

(b)  If $\sigma = S = 900\,\text{MPa}$ is the maximum permitted stress before failure occurs, the maximum permitted shuttle displacement is given by

$$\Delta_{\text{central mass}}^{\max} = \frac{2L^2\sigma_{\max}}{3Yw} = \frac{2(200 \times 10^{-6})^2 900 \times 10^6}{3(150 \times 10^9)(5 \times 10^{-6})} = 32 \times 10^{-6}\,\text{m} = 32\,\mu\text{m}$$

Then the maximum deflection is $16\,\mu\text{m}$ per beam, which is well above the limit for which the linear deflection analysis is valid. Thus, this should be thought of as merely an estimate of the maximum

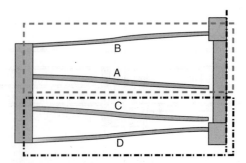
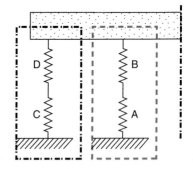

**Figure 4.17** (a) Simulated deformation of the left half of the folded-beam suspension; (b) equivalent spring representation.

deflection. When a beam undergoes large deflections, geometrically nonlinear analysis is necessary. A beam such as this experiences increased stiffness due to the axial stretching (recall that linear analysis assumes that the length of the beam along the neutral axis does not change). These factors can also be modeled analytically, but we need more concepts and more mathematical modeling. Instead, in practice, engineers and scientists use finite element analysis (FEA, see Chapter 5) that helps us compute nonlinear deformations. Even in FEA, one must ensure that the nonlinear option is enabled rather than the default linear option.

**Your Turn:**

Note that movable planar micromachined structures have stiffness that is of the order of N/m or tens of N/m. Compare their stiffness with those you are familiar with at the macroscale. What might be the spring constant of the spring in your ballpoint pen?

Recall that the axial stiffness of the beams considered in Example 4.1 was high (a few thousand N/m). This may seem high relative to the bending stiffness of micromachined beams, but it is quite low relative to the axial stiffness of macro-sized beams. Estimate the axial stiffness of a beam in a building for comparison with that in Example 4.1.

Note that actuators at the microscale can move tens of microns. The force required to achieve this displacement is tens to hundreds of micro-Newtons. Why? Because the stiffness is of the order of N/m or tens of N/m.

Examples 4.2 and 4.3 are covered here as illustrations and should not be considered exhaustive. Readers (clearly your turn!) are encouraged to attempt the other structures presented in Problems 4.3 to 4.5.

### Problem 4.3 A Gyroscope Suspension

A gyroscope is a device that measures the angular rotation rate of an automobile, an aircraft, a ship, or any object on which it is mounted. The suspension of a gyroscope should have equal stiffness in the horizontal and vertical directions. For the schematic shown in Figure 4.18, derive the formula for the stiffness in the horizontal and vertical directions using Castigliano's second theorem.

### Problem 4.4 An Orthoplanar Compliant Microplatform

Figure 4.19(a) shows what is called an *orthoplanar compliant microplatform*. It consists of four folded beams attached to a fixed outer ring and movable central disk. When a force is applied, the central disk moves up and down with low stiffness. However, it has high stiffness in the planar directions. Derive expressions for the stiffness in terms of the symbols indicated in the figure. Compute the stress as in Example 4.3. These equations help you intelligently design such a device.

### Problem 4.5 Bent-Beam Thermal Microactuator

The principle of a bent-beam thermal microactuator is shown schematically in Figure 4.19(b). The length of each beam

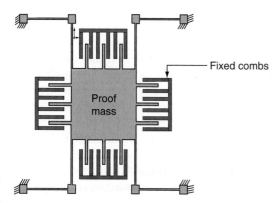

**Figure 4.18** Schematic of a micromachined gyroscope. The central mass is suspended by four L-shaped beams. The fixed comb fingers detect the motion of the mass by measuring the change in capacitance between the moving and fixed fingers. The same combs can also be used for actuation.

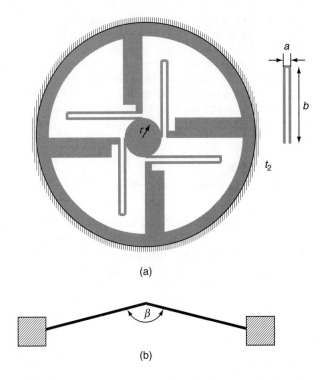

(a)

(b)

**Figure 4.19** (a) An orthoplanar compliant micromachinable platform; (b) a bent-beam thermal actuator.

segment is $L$. Use the usual symbols for the cross-sectional properties and material properties. In the complete actuator, there will be an array of bent beams connected so that their corners lie on a vertical beam connecting them all. Determine an expression for the vertical displacement of the corner of the bent beam for a uniform temperature rise, $T$, if the two ends are mechanically fixed.

## ▶ 4.5 HETEROGENEOUS LAYERED BEAMS

The preceding section demonstrated how bar and beam theories and Castigliano's theorems enable us to analyze micromachined mechanical structures. In this and the next section, we learn a few more things about beams. First, recall that micromachined structures are mostly layered structures. Each layer can be of a different material. Figure 4.20 shows one such case: a long cantilever beam with a short beam on the top and at the left end. Let us say that the short beam is a patch of piezoelectric material. Let the total length of the beam be $l_1$ and that of the piezopatch be $l_2$; let their respective thicknesses be $t_1$ and $t_2$ and widths $w_1$ and $w_2$. In order to analyze this structure, let us first ask the question: when we consider a heterogeneous layered beam, what changes are necessary in our analysis of a beam?[6]

---

[6] We do not consider the piezoelectric effect here, but merely analyze the additional mechanical stiffness due to the piezoelectric patch.

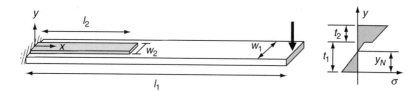

**Figure 4.20** A silicon beam of length $l_1$ with a short piezoelectric beam of length $l_2$ on top at the fixed end.

First, the neutral axis does not remain at the same level all along the beam's axis because there is a sudden change in the thickness on only one side (here, the top side). Furthermore, the neutral axis does not lie at the centroid of the cross-section in that part of the beam where layers of different materials exist. In order to determine the location of the neutral axis, we need first to find out how the stress varies along the vertical ($y$) cross-sectional axis.

As shown in Figure 4.20, the stress distribution along the $y$-axis of the cross-section in the part containing the two materials is linear in the respective parts with a sudden change in the value at the interface. Let $Y_1$ and $Y_2$ be the Young's moduli of the two materials. The stress can then be written as

$$\sigma = \begin{cases} \sigma_1 = \dfrac{Y_1}{\rho}(y - y_N) & \text{for } y \leq t_1 \\[3mm] \sigma_2 = \dfrac{Y_2}{\rho}(y - y_N) & \text{for } t_1 < y \leq t_1 + t_2 \end{cases} \tag{4.49}$$

where $y_N$ is the height of the neutral axis from the bottom of the beam (Figure 4.20). In pure bending, there is no axial force acting at the cross-section. Therefore, the stress integrated over the area should be equal to zero. This enables us to find $y_N$ as follows:

$$\int_0^{t_1} \sigma_1 w_1 \, dy + \int_{t_1}^{t_1+t_2} \sigma_2 w_2 \, dy = 0$$

$$\Rightarrow y_N = \frac{Y_1 A_1 y_1 + Y_2 A_2 y_2}{Y_1 A_1 + Y_2 A_2} \tag{4.50}$$

where $y_1 = (t_1/2)$ and $y_2 = [t_1 + (t_2/2)]$ are the heights from the bottom of the cross-section to the centroids of layers 1 and 2, and $A_1 = w_1 t_1$ an $A_2 = w_2 t_2$ are the cross-sectional areas, respectively. Next, we take the moment of force ($\sigma A$) about the $z$-axis at the point where it meets the neutral axis. This moment should be equal to the bending moment at that cross-section. This was the reasoning we followed for a homogeneous beam [see Eq. (4.18a)]:

$$\int_0^{t_1} \sigma_1 w_1 \, y \, dy + \int_{t_1}^{t_2} \sigma_2 w_2 \, y \, dy = M \tag{4.51}$$

$$M = c\{Y_1[I_1 + y_1 A_1(y_1 - y_N)] + Y_2[I_2 + y_2 A_2(y_2 - y_N)]\} = c(YI)_e$$

where $I_1$ and $I_2$ are the area moments of inertia, respectively. As shown in the rightmost expression of Eq. (4.51), the effective *beam modulus* $(YI)_e$ can be obtained from the individual geometric and material properties.

If there are many layers one on top of the other, the location of the neutral axis and the equivalent beam modulus can be derived in exactly in the same way as indicated here. The formulae are given for easy reference with usual definition of $y_i$, $Y_i$, $A_i$ and $I_i$:

$$y_N = \frac{\sum_{i=1}^{n} Y_i A_i y_i}{\sum_{i=1}^{n} Y_i A_i} \qquad (4.52)$$

$$(YI)_e = \sum_{i=1}^{n} Y_i[I_i + y_i A_i(y_i - y_N)] \qquad (4.53)$$

**Example 4.4:** Mechanical Analysis of an RF Relay

Figure 4.21 shows isometric and cross-section views of a surface-micromachined RF relay (only the left symmetric half is shown in the cross-section image). Compute the deflection of the point of symmetry for the force indicated. The width is uniform throughout and the thicknesses of the polysilicon and silicon dioxide layers are $t_{Si}$ and $t_{SiO_2}$, respectively.

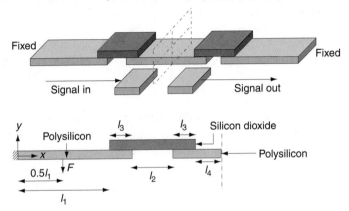

**Figure 4.21** Isometric and cross-section views of an RF relay.

**Solution:** It suffices to solve the left symmetric half of the full relay by imposing the symmetric boundary condition at the point of symmetry. Thus, the slope is zero at the point of symmetry. This left symmetric half becomes a fixed-guided beam comprising five segments, each of which has a different equivalent beam modulus. They can be written as follows with the notation used so far (i.e. $y$, $Y$, $A$ and $I$) with subscripts Si for polysilicon and SiO$_2$ for silicon dioxide:

$$(YI)_e = \begin{cases} Y_{Si} I_{Si} & \text{for} & 0 \leq x < l_1 \\ Y_{Si}[I + yA(y - y_N)]_{Si} + Y_{SiO_2}[I + yA(y - y_N)]_{SiO_2} & \text{for} & l_1 \leq x < l_1 + l_3 \\ Y_{Si} I_{Si} & \text{for} & l_1 + l_3 \leq x < l_1 + l_2 + l_3 \\ Y_{Si}[I + y_1 A(y - y_N)]_{Si} + Y_{SiO_2}[I + y_1 A(y - y_N)]_{SiO_2} & \text{for} & l_1 + l_2 + l_3 \leq x < l_1 + l_2 + 2l_3 \\ Y_{Si} I_{Si} & \text{for} & l_1 + l_2 + 2l_3 \leq x < l_1 + l_2 + 2l_3 + l_4 \end{cases} \qquad (4.54)$$

Note that this involves locating the neutral axis, $y_N$, for the portions that are two-layer sandwiches. This can be done using Eq. (4.50). After that, the bending moment can be determined to get an expression for the strain energy and thereby apply Castigliano's second theorem to get the deflection. These remaining steps are left as an exercise; it will be instructive to complete it.

## ▶ 4.6 BIMORPH EFFECT

Heterogeneous beams have some advantages in thermal actuation. Imagine a heated sandwich cantilever beam made of two materials of disparate coefficients of thermal expansion. The two layers expand by different lengths but they cannot expand freely because they are adhering to each other. Thus each layer develops a thermal strain as given by Eq. (4.5). Such a sandwich beam relieves its thermal strain by bending, as shown in Figure 4.22. The layer with higher coefficient of expansion will be on the outer side. This effect, known as the *bimorph* effect, is used in thermal microactuators and is also widely used in thermostats to close or open a switch turning an air-conditioner ON/OFF when a certain temperature is reached. The bimorph effect is also used in smart materials such as piezoelectrics: the smart material patch expands/contracts along the length of the beam due to an electrical stimulus, and this effect can be converted to a large deformation in the transverse direction.

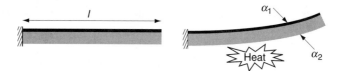

**Figure 4.22** Thermal bimorph effect.

However, the bimorph effect is disadvantageous when out-of-plane deformation of heterogeneous beams is not desired. In micromachined structures, we often have thin layers of different materials one on top of the other. Either during fabrication or use, this heterogeneous beam might become heated, and the mismatch between the coefficients of thermal expansion would cause unwanted deformation. Let us now examine the size of this deformation so that we can use it in actuation or reduce it when we do not want it.

Referring to Figure 4.22, assume that the two layers have thicknesses $t_1$ and $t_2$, coefficients of thermal expansion $\alpha_1$ and $\alpha_2$, and Young's moduli $Y_1$ and $Y_2$, respectively. Let this two-layer heterogeneous beam be heated by a temperature $\Delta T$. It will then curl up as shown in Figure 4.22. The curled-up beam will be under pure bending. The free-body diagram of a small piece of that beam will look like that in Figure 4.23. Since there are no external forces or moments on it, force equilibrium and moment equilibrium give the following:

$$P_1 = P_2 = P$$
$$P\frac{(t_1 + t_2)}{2} = M_1 + M_2 \tag{4.55}$$

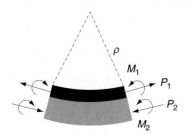

**Figure 4.23** Free-body diagram of a piece of a bent bimorph beam.

Here we have taken the moment about a point on the boundary between the two layers at the end. Let $\rho$ be the radius of curvature of the bent beam. Using Eq. (4.20), we write

$$M_1 = -\frac{Y_1 I_1}{\rho} \text{ and } M_2 = -\frac{Y_2 I_2}{\rho} \tag{4.56}$$

By substituting for $M_1$ and $M_2$ from Eq. (4.56) into Eq. (4.55), we get

$$P\frac{(t_1 + t_2)}{2} = -\frac{(Y_1 I_1 + Y_2 I_2)}{\rho} \tag{4.57}$$

The above equation has two unknowns, $P$ and $\rho$, and we need an additional equation to solve for them. We get this additional equation by noting that the axial elongations at the boundaries between the layers must be the same: if they were not, the layers would not be adhering to each other and would separate. The elongation of the layers has three components: a thermal expansion component, an axial deformation component due to the force $P$, and an elongation component due to bending. These are written for each layer by using Eqs. (4.5), (4.4,) and (4.19), respectively. Dropping the length as it appears on both sides, we get

$$\alpha_1 \Delta T + \frac{P}{Y_1 t_1 b} + \frac{(t_1/2)}{\rho} = \alpha_2 \Delta T - \frac{P}{Y_2 t_2 b} - \frac{(t_2/2)}{\rho} \tag{4.58}$$

Note the minus signs for the last two terms on the right-hand side. They arise here because the lower layer has a compressive force due to $P_2 = P$ and has compressive normal stress due to bending. By substituting for $P$ from Eq. (4.57) into Eq. (4.58), we can solve for $\rho$:

$$\rho = \frac{[(t_1 + t_2)/2] + [2(Y_1 I_1 + Y_2 I_2)/(t_1 + t_2)][(1/Y_1 t_1 b) + (1/Y_2 t_2 b)]}{(\alpha_1 - \alpha_2)\Delta T} \tag{4.59}$$

Note that when the difference between the coefficients of expansion is large or the temperature rise is large, the radius of curvature will be small (i.e. there will be a lot of bending).

The stress in the curled-up beam will have two components: axial stress due to $P$ and bending stress:

$$\sigma = \frac{P}{A} + \frac{My}{I} = \frac{P}{t_1 b} + \frac{Y_1 y}{\rho} \tag{4.60}$$

where we have written the stress for a point in the top layer of the bimorph beam that is at height $y$ from its neutral axis. For maximum bending stress, $y = t_1/2$. Note also that we have used Eq. (4.19) to eliminate $M$ in Eq. (4.60).

As noted here, the mismatched expansion need not occur only due to heating of two layers of different thermal expansion coefficients in a bimorph beam; it can also occur when one layer is of a piezoelectric material and the other of a normal material. A piezoelectric material contracts axially when a voltage is applied, while the normal material does not, so that there is mismatched expansion between the layers. Another reason for mismatched expansion is residual stress induced during fabrication. This is discussed next.

## ▶ 4.7 RESIDUAL STRESSES AND STRESS GRADIENTS

Residual stresses are common when thin films are deposited on a substrate. Since deposition is usually done at an elevated temperature, a stress ensues when the fabricated structures are brought back to room temperature. This residual stress is due, once again, to the different coefficients of thermal expansion. However, residual stress can also arise in microfabricated structures due to processing steps such as oxidation, substitutional doping, ion implantation, etc. In such processes, new atoms or species enter a material and push on the existing ones. In epitaxial growth, mismatch of lattice spacing also produces a residual stress. Rapid material-addition processes also result in residual stresses because more atoms try to occupy a limited space.

Residual stress need not be the same in all directions and it need not even be uniform throughout a structure. Hence, we need to consider *stress gradients*, that is, stress variation across a dimension. A stress gradient along the thickness direction is common when a thin film is deposited on a substrate. Let us analyze the effects of residual stresses and stress gradients in micromechanical structures.

### 4.7.1 Effect of Residual Stress

Imagine a beam or a bar fixed at one end and free at the other (Figure 4.24(a)). Let us assume that there is a residual stress along the axial direction. When the beam or bar is not constrained at the right side, it will simply stretch or contract, depending on whether the stress is compressive or tensile. Compressive (negative) residual stress can be cancelled by stretching to induce tensile (positive) stress, and vice versa. Now, as shown in Figure 4.24 (b), consider a beam attached on the bottom to a substrate. Let us assume that the beam has tensile residual stress. It is prevented from contracting due to the structure below it. Only the right edge (marked AB in the figure) is free, but its bottom point is attached to the substrate and cannot move. Since B is free to move and the effect of the tensile stress is to make it contract, the edge AB will rotate. This results in *shear stress* in the region near the edge AB. If both ends of the beam are fixed, the beam can neither stretch nor contract and must hold that stress within itself.

Now, let us remove the bottom substrate and add a fixed support at the right as in Figure 4.24(c). It cannot stretch or contract, so it still contains the residual stress. Now the question is: will it behave in the same way as a normal beam? The answer is: certainly not! Such a beam will have altered bending stiffness. If the stress is compressive, it will have reduced bending stiffness; if tensile, increased bending stiffness. This can be reasoned intuitively: If you try to apply a force in the transverse direction on a wire stretched taut between its two ends, you will feel that it resists bending. Similarly, if you heat a wire held between two ends, it elongates and hence becomes loose between the ends. Now, if you apply a transverse force, it will not resist as much. Since a wire has little bending stiffness, it simply becomes loose by elongating, whereas a beam cannot. Nevertheless, heating reduces the beam's bending stiffness because thermal expansion leads to compressive stresses in a constrained structure. To take this into account in the beam deformation analysis, we need to take a small detour and talk further about the transverse deformation of a beam.

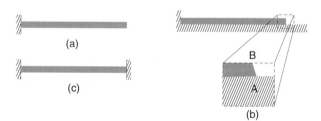

(a)

(c)

B

A

(b)

**Figure 4.24** Effect of residual stress along the length direction (a) a fixed–free beam: it simply stretches or contracts; (b) a fixed–free beam adhering to a substrate at the bottom: it has a shear force on the free vertical edge at the right end; and (c) a fixed–fixed beam with no constraint on the bottom side: it holds the stress within itself.

We return to Eq. (4.22) (repeated here for convenience)

$$YI\frac{d^2w}{dx^2} = -M$$

Differentiating it twice with respect to $x$, we get

$$\frac{d}{dx}\left(YI\frac{d^2w}{dx^2}\right) = -\frac{dM}{dx} = -V \tag{4.61a}$$

$$\frac{d^2}{dx^2}\left(YI\frac{d^2w}{dx^2}\right) = -\frac{dV}{dx} = q \tag{4.61b}$$

Here, we have introduced two new relationships, namely,

$$\frac{dM}{dx} = V \tag{4.62a}$$

$$\frac{dV}{dx} = -q \tag{4.62b}$$

Here, $V(x)$ is the vertical shear force [defined and used earlier in this chapter; see Figures 4.7(a), (b)] and $q(x)$ is the transverse loading per unit length of the beam. These relationships come about if we consider a small piece of bent beam as in Figure 4.25, identify the vertical shear forces and bending moments on it, and consider the static equilibrium of the forces in the vertical direction and the moment equilibrium about the axis perpendicular to the element's face.

For the static equilibrium of the forces in the vertical direction of the small piece of a bent beam, we can write the following:

$$V - qdx - (V + dV) = 0 \Rightarrow \frac{dV}{dx} = -q$$

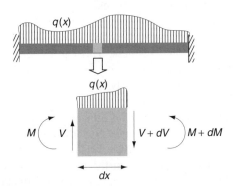

**Figure 4.25** A small piece of a bent beam under the distributed transverse load $q(x)$, the force per unit length.

By taking moment equilibrium and neglecting second-order quantities, we get

$$(M + dM) - M - (V + dV)dx - \frac{qdx}{2}dx = 0 \Rightarrow \frac{dM}{dx} = V$$

Now, we can return to the general differential equation (Eq. (4.61b)) for determining beam deformation when transverse loading $q(x)$ is given. When $Y$ and $I$ are not functions of $x$, we can write Eq. (4.61b) as follows:

$$YI\frac{d^4w}{dx^4} = q \tag{4.63}$$

After this short detour, we now return to the analysis of the fixed-fixed beam of Figure 4.24(c) with residual stress.

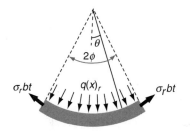

**Figure 4.26** A bent beam with longitudinal tensile residual stress needs an imaginary transverse distributed load $q_r(x)$.

We have already indicated that a beam with positive (i.e. tensile) residual stress will have increased bending stiffness for transverse loads. Let the transverse load be $q_r(x)$. This load would cause the beam to bend. When it bends, the axial force due to the longitudinal residual stress will no longer be perfectly balanced. (For this, see Figure 4.26, which shows the force due to the residual stress on both the ends of a small piece of a beam bent into a circular arc.) While the horizontal components of the forces cancel, the vertical components add up to $2(\sigma_r \sin\phi)bt$. To cancel this, we need an imaginary load $q_r(x)$, the distributed transverse load per unit length of the beam. The net effect of the horizontal component of this load is zero because of how the load is oriented—always normal to the bent circular beam. The vertical component can be integrated to get

$$\int_{-\phi}^{\phi} q_r \cos\theta\, \rho d\theta = 2q_r\rho \sin\phi \tag{4.64}$$

For equilibrium, the two forces must be equal; this helps us calculate the imaginary load, $q_r$:

$$2(\sigma_r \sin\phi)bt = 2q_r\rho \sin\phi$$

$$q_r = \frac{bt\sigma_r}{\rho} = bt\sigma_r \frac{d^2w}{dx^2} \tag{4.65}$$

where we have substituted the small-deflection approximation to the radius of curvature with the second derivative of the transverse displacement of the bent beam. Now, returning to Eq. (4.63), we have an "additional" load that is equivalent to the longitudinal residual stress. This additional load is the negative of $q_r$ because it is this load that takes into account the effect of the residual stress:

$$YI\frac{d^4w}{dx^4} = q - q_r \tag{4.66}$$

This equation shows that the deflection of the beam will be less when $q_r$ is positive, as for instance, when the curvature and residual stress are both positive—in other words, compared with the case of no residual stress, the bending stiffness is higher. We now substitute for $q_r$ from Eq. (4.65) and write a more general differential equation than Eq. (4.63):

$$YI\frac{d^4w}{dx^4} + \sigma_r bt\frac{d^2w}{dx^2} = q \tag{4.67}$$

This differential equation can be solved numerically for any given $q$ and other constants, including $\sigma_r$. As can be observed from our preceding discussion, the effect of residual stress warrants a deeper understanding of beam deflection theory. Interestingly, the residual stress gradient also gives us an opportunity to learn something new, as discussed next.

## 4.7.2 Effect of the Residual Stress Gradient

Consider a fixed-free (i.e. cantilever) beam that has a residual stress gradient in the thickness direction given by

$$\sigma_r = \sigma_0 + \frac{2\sigma_1}{t}y \tag{4.68}$$

indicating that the stress varies linearly with $y$, where $y$ is measured from the neutral axis in the upward transverse direction. This stress gradient causes a bending moment at each cross section. This bending moment can be calculated by taking the moment of the force induced by the stress from a point on the neutral axis:

$$M = \int_{-t/2}^{t/2} \left(\sigma_0 + \frac{2\sigma_1}{t}y\right) by\, dy = \frac{1}{6}bt^2\sigma_1 \tag{4.69}$$

We substitute this $M$ into Eq. (4.20), determine the deflection, and conclude that the curvature is constant:

$$\frac{1}{\rho} = -\frac{M}{YI} = -\frac{bt^2\sigma_1}{6YI} = -\frac{2\sigma_1}{Yt} \tag{4.70}$$

A curve that has constant curvature is a circular arc, and hence a beam with a residual stress gradient bends into a circular arc of radius $Yt/2\sigma_1$. Figure 4.27 shows an array of cantilever beams curled up due to a processing-induced residual stress gradient along the thickness direction. In fact, by measuring the deflection, if any, of a cantilever beam, we can monitor the residual stress gradient in the process. Thus, it may be good practice to add a few cantilever beams to lithography mask layouts if we suspect that processing may lead to a residual stress gradient.

Figure 4.28 shows another array: a polar array of curled-up cantilever beams acting as a cage to capture biological cells. By deforming the flexible membrane (not shown) on

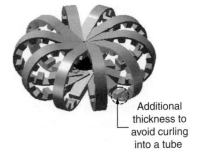

Additional thickness to avoid curling into a tube

**Figure 4.27** Schematic of an array of cantilever beams fixed at the right end curled up due to the residual stress gradient along their thickness direction.

Source: Redrawn after Trimmer (ed.) *Micromechanics and MEMS: Classic and Seminal Papers to 1990.*

**Figure 4.28** A cell cage made of a polar array of cantilevers curled due to the residual stress gradient. The circled patch and ones similar to it have increased thickness to avoid curling along the beam width. Due to the residual stress gradient and anticlastic curvature effect, they curl into tube-like structures without the additional bending stiffness provided by increased thickness along the width direction.

Source: Redrawn after a microcage developed by C.J. Kim's group at the University of California at Los Angeles.

which the cantilevers are fixed, we can open or shut the cage at will. However, observe that there are some patches along the length of the beams. There is a good reason for them. This leads us to an important concept in elasticity—the Poisson effect—which we discuss in the next section.

## ▶ 4.8 POISSON EFFECT AND THE ANTICLASTIC CURVATURE OF BEAMS

We began discussing the deformation of solids with axially deforming bars in Section 4.1. We talked about the strain arising in the axial direction when we apply an axial load on it [see Eq. (4.2)]. However, we did not discuss the effect of this load on the cross-section of the solid. It is our everyday experience that when we stretch a rubber band, it becomes thinner. That is to say, the strain in the direction perpendicular to the axis has opposite sign to that of the strain in the axial direction. The strain in the perpendicular direction depends on what is called *Poisson' ratio*, $\nu$. With reference to Figure 4.29, we write the strains in the perpendicular directions as

$$\epsilon_y = -\nu\epsilon_x$$
$$\epsilon_z = -\nu\epsilon_x$$

(4.71)

Poisson's ratio varies from $-1$ to $0.5$ for isotropic materials, although negative values are rare. Most often we find that it varies beween $0$ and $0.5$. For plastics it is close to $0.5$, whereas for metals, polysilicon, and silicon compounds it is between $0.2$ and $0.3$.

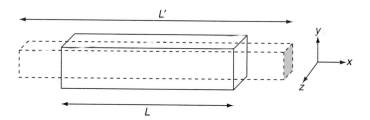

**Figure 4.29** Poisson effect for an axially deforming bar (exaggerated). When there is a strain in one direction, there is a negative strain in the perpendicular directions. The strain in the perpendicular directions is $\nu$ times the strain in the axial direction and with opposite sign.

We discuss two consequences of Poisson's ratio here. First, it gives us an indication of the compressibility of the material. Since $\epsilon_x$ is defined as the ratio of the change in length to the original length, we can write the new length of the bar in Figure 4.29, $L'_x$, as

$$\Delta x = \epsilon_x L_x$$
$$L'_x = L + \Delta x = L_x(1 + \epsilon_x)$$

(4.72)

From the Poisson effect, we can write the new lengths for the other directions:

$$L'_y = L_y(1 - \nu\epsilon_x)$$
$$L'_z = L_z(1 - \nu\epsilon_x)$$

(4.73)

Then, the change in volume of the bar is given by

$$\Delta V = L'_x L'_y L'_z - L_x L_y L_z = L_x L_y L_z \{(1 + \epsilon_x)(1 - \nu \epsilon_x)(1 - \nu \epsilon_x) - 1\} \qquad (4.74)$$

which for small strains (i.e. neglecting second-order small terms) can be approximated as:

$$\Delta V \approx L_x L_y L_z (1 - 2\nu)\epsilon_x \qquad (4.75)$$

Since $V = L_x L_y L_z$, we can write the relative change in volume as

$$\frac{\Delta V}{V} = (1 - 2\nu)\epsilon_x \qquad (4.76)$$

By writing $\epsilon_x$ as stress divided by Young's modulus (Hooke's law!) and denoting stress as pressure $p_x$ (after all, stress is force per unit area) we can rewrite Eq. (4.76) as:

$$\frac{\Delta V}{V} = (1 - 2\nu)\frac{p_x}{Y}$$

If, in addition, there are forces in the y- and z-directions also due to pressures $p_y$ and $p_z$ applied in those directions, we get

$$\frac{\Delta V}{V} = (1 - 2\nu)\frac{1}{Y}(p_x + p_y + p_z) \qquad (4.77)$$

If the pressures in all the directions are equal to $p$ (we call this *hydrostatic pressure*, since it is like the uniform pressure an object immersed in water would experience), we get

$$\frac{\Delta V}{V} = 3(1 - 2\nu)\frac{1}{Y}p$$
$$\frac{p}{(\Delta V/V)} = \frac{Y}{3(1 - 2\nu)} = K \qquad (4.78)$$

where we have defined a new material property, $K$, the *bulk modulus*, as the ratio of the hydrostatic pressure to the relative change in volume. Note that when the Poisson ratio approaches 0.5, the bulk modulus approaches infinity, that is, the relative change in volume approaches zero. Such a solid material is called *incompressible*. The reciprocal of the bulk modulus indicates compressibility and materials with $\nu$ less than 0.5 are compressible.

A second consequence of Poisson's ratio considered here is for the case of beams. From Eqs. (4.14a) and (4.15a), we write

$$\epsilon_x = -\frac{y}{\rho} \qquad (4.79a)$$

Now, due to the Poisson effect, we have

$$\epsilon_z = -\nu \epsilon_x = \frac{\nu y}{\rho} \qquad (4.79b)$$

What does this mean? It means that when a beam bends in one direction, $x$ here, there will be bending in the perpendicular directions too. We show this *anticlastic curvature* for the $z$-direction in Figure 4.30.

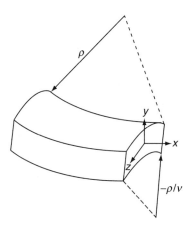

**Figure 4.30** Anticlastic curvature of a beam. The radius of curvature for bending in the $x$-direction is $\rho$ and that in the $z$-direction is $-\rho/v$.

Now, in view of the anticlastic curvature, think about what happens to the cell cage in Figure 4.28. As the cantilever beams bend into semicircles in their length direction, they also tend to curve into tubes in the width direction when there is a high residual stress gradient. To prevent this from happening, the thickness is increased in patches, as seen in the circled region of Figure 4.28. One might ask: Why not make it thick all through? If we do this, then the beam will have increased bending stiffness along the length direction, too, making it bend into much less than a semicircle.

The anticlastic curvature effect is used in many interesting ways. First, it can be a good technique for measuring the residual stress gradient in plates (which are simply wide beams). Another interesting application is in cantilever beams used to detect cancer and other viruses as well as biological molecules in general. For this, the top surface of a cantilever beam is coated with an antigen (a biological molecule with affinity for an antibody produced by our body to tackle a virus). If there are antibodies in the body fluid in which the coated cantilever is immersed, they bind to the antigens. This creates a surface stress (and strain) in the width direction. Due to the Poisson effect, there will be a strain in the length direction, and the cantilever bends. This is measured using lasers. Did you know that a modest microcantilever beam can detect a cancer virus? It is a mix of biology, mechanics, and sophisticated measurement techniques that makes this possible.

**Your Turn:**

You have work to do now: go and read up about antigens, antibodies, surface stress, laser-based measurements, and how they all add up in cancer diagnostics. Well, this is a concept that has been demonstrated in research labs, but no commercial diagnostic tools of this kind exist in the market at this time. You may investigate why this is so and how those problems might be overcome.

**Problem 4.6**

A cylindrical microcage is needed for trapping biological cells to allow them to be queued up for mechanical characterization. A designer wants to exploit the residual stress gradient in a surface-micromachined polysilicon layer with a sacrificial oxide layer underneath. The cylindrical microcage is shown schematically in Figure 4.31, where the cantilever beams have curled into almost complete circles. Assume that the spherical cells to be queued up have a diameter of 200 μm. The thickness of the polysilicon layer is fixed at 2 μm by the process engineer, who is willing to adjust the stress gradient per your specifications. As a designer of this cage, you need to determine the following:

a. The width and the length of the cantilever.

b. The process-induced stress gradient you want.

c. The spacing between the cantilever beams.

Describe your steps in arriving at all these and compute the numerical values. Note that the answer is not unique. So, if you make any assumptions, please state them clearly. Consider everything pertinent to the actual fabrication of this device.

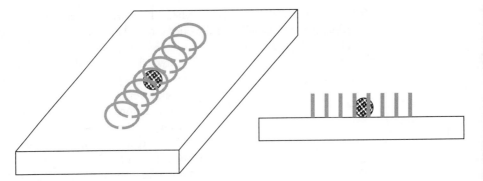

**Figure 4.31** Perspective and orthographic views of a cylindrical cell cage made of curled-up cantilever beams with residual stress gradient.

## ▶ 4.9 TORSION OF BEAMS AND SHEAR STRESSES

So far, we have discussed only how beams bend. But beams can also twist when a torque is applied. Consider a micromirror used in a projector. Its working principle was discussed in Chapter 2, Section 2.10; Figure 4.32 shows a schematic. Here, when an electrode on one side of the torsional beams is turned ON, the central part attached to the beams tilts that side as the beams twist. A micromirror-based projector will contain millions of them and each mirror can be tilted individually. Can you imagine any other engineered system, large or small, where a million components are movable and are individually controllable? Most sophisticated ships, aircrafts, or manufacturing automation machinery might have million of parts if you count all nuts and bolts. But most parts in them do not move, and even if they do, their motions are not individually controlled. This clearly shows that miniaturization provides some unique opportunities in building complex systems not built or known before. However, our task here is to discuss the torsion analysis of beams.

The twisting of a beam of circular cross-section is shown in Figure 4.33(a). In micromechanical structures, circular cross-section beams are rare; most often, as a consequence of the layered microfabrication process steps, we have rectangular or trapezoidal cross-sections and occasionally T-sections. However, since the torsion (i.e. twisting under torque loads) analysis of noncircular cross-section beams is complicated, we discuss here only the torsion of circular cross-sections and then give some empirical formulae for a few other cross-sections.

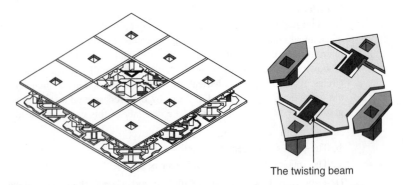

The twisting beam

**Figure 4.32** Texas Instruments' single-axis micromirror array for projection display applications. The torsion hinge is shown. The two beams twist to tilt the mirror.

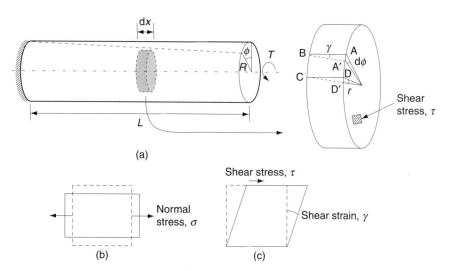

**Figure 4.33** (a) Twisting a beam of circular cross-section under a torque load; (b) a differential element under normal stress; and (c) a differential element under shear stress.

The beam in Figure 4.33(a), with length $L$ and radius $R$, is fixed at the left end while a torque $T$ is applied at the free right end. As to be expected, the beam twists. Let this angle of twist be $\phi$. Our interest is not only in the value of $\phi$ but also in understanding how much stress of what type is created in the beam. For this, we isolate a small disk of radius $r$ and length $dx$ and analyze its deformation. In the isolated segment (enlarged at right in Figure 4.33(a)), let us consider the rectangle ABCD. Because of the relative twist $d\phi$ between the segment's left and right ends, it moves to A′BCD′. We have already discussed the kind of strain (and stress) that would appear when an element is stretched or contracted. This normal strain (stress) is shown in Figure 4.33(b) once again. But now, as shown in Figure 4.33(c), when only one edge translates, two rotate, and one remains the same, we call it a *shear*, and the resulting strain (stress) is called shear strain (stress) and denoted by $\gamma$ (shear strain) and $\tau$ (shear stress). As shown in Figure 4.33(c), $\gamma$ is simply the angle of shear. Shear stress and shear strain are related linearly (as in Hooke's law for $\sigma$ and $\in$) but with a different material property, called *shear modulus* or *rigidity modulus*, denoted by $G$ and given by

$$\tau = G\gamma \tag{4.80}$$

The variable $G$, for an isotropic material, is given in terms of Young's modulus and Poisson's ratio as[7]

$$G = \frac{Y}{2(1+\nu)} \tag{4.81}$$

---

[7] Note that the relationship among the rigidity modulus, Young's modulus, and Poisson's ratio can be derived from Hooke's law relating normal stress and normal strain. It is left as an exercise to the readers.

We apply the definition of shear strain (stress) to the rectangle ABCD in Figure 4.33(a):

$$BC = A'D'$$
$$AA' = DD'$$
$$\gamma dx = r d\phi$$

(4.82)

Now, by using Eq. (4.80), we can extend Eq. (4.82) as

$$r d\phi = \gamma dx = \frac{\tau}{G} dx$$
$$\tau = G \frac{d\phi}{dx} r$$

(4.83)

We note that shear stress is linearly proportional to the radius of the point of interest within the beam and to the rate of twist along the length of the beam. But, at this point, we know neither $\tau$ nor $\phi$. To solve the problem at hand, we need another equation. This comes from the equation of equilibrium. Recall that all we used for the normal strain (stress) case of stretching and bending were Hooke's law and the equilibrium equation. Here, we have already used the Hooke's law equivalent [Eq. (4.80)] and we now use equilibrium under the torque load. Torque is like a moment except that it acts about the longitudinal axis of the beam. We therefore balance the torque $T$ with the moment due to the shear stress $\tau$ over the cross section of the small disk (enlarged in Figure 4.33(a)):

$$T = \int_{\substack{\text{Cross-section} \\ \text{of the circle A}}} (\tau dA) \, r = \int_{A} \left( G \frac{d\phi}{dx} r \, dA \right) r = G \frac{d\phi}{dx} \int_{A} r^2 \, dA = GJ \frac{d\phi}{dx}$$

(4.84)

where we have $J = \int_A r^2 dA$. The variable $J$, called the *polar moment of inertia* of a cross-section, is the counterpart of the area moment of inertia of bending. By assuming that the rate of twist of the beam along the axis is linear, we can replace $d\phi/dx$ with $\phi/L$. Thus, we get

$$T = \frac{GJ}{L} \phi$$
$$\phi = \frac{TL}{GJ}$$

(4.85)

From Eq. (4.83), we get the expression for the shear stress

$$\tau = G \frac{\phi}{L} r = \frac{T}{J} r$$

(4.86)

Note that the shear stress does not depend on the material properties. We may recall that the normal stress for an axially deforming bar also did not depend on the material properties.

Also note that, $J = \int_A r^2 dA$ for a circular cross-section is equal to $J = \pi R^4 / 2$. This can be extended to annular circular cross-sections. If the inner and outer diameters are given by $R_i$ and $R_o$, $J$ for such a section is equal to $\pi (R_o^4 - R_i^4)/2$.

The preceding analysis does not apply to noncircular cross-sections. As noted earlier, analysis becomes complicated for other cross-sections because twisting is not limited to the cross-sectional area but will also extend in the axial direction. However, an approximation can be used to compute the twist by replacing $J$ in Eq. (4.85) with $J_{eq}$ as follows:

$$\phi = \frac{TL}{GJ_{eq}}$$

$$J_{eq} = \frac{A^4}{4\pi^2 J}$$

(4.87)

where $A$ is the cross-sectional area. If we solve the problem of the torsion hinges in the micromirror in Figure 4.32, we need to use this approximation since we have a rectangular cross-section.

---

**Example 4.5:** Helical Spring

Everybody knows a helical spring [Figure 4.34 (a)]; it is present in a ballpoint pen. It has a remarkable design in that its force-displacement characteristic is linear over a large range of its deflection. That is why we use $F = kx$ without a second thought. But how do we obtain this $k$, the spring constant? Obviously, it depends on the geometric parameters of the spring and the material it is made of. But how? It turns out that the twisting of a circular beam helps here, which may appear counterintuitive. You might have thought that the coils of a helical spring bend when you pull or push at its ends. They actually twist. To appreciate this, let us do a thought experiment. Imagine a thread; hold one end of it between thumb and index finger of your left hand, keep it taut with your right hand, and twist the other end by many turns. Now, when you bring your two hands together, the twisted thread curls into a helix. If you spread the hands apart, the helix expands or unwinds. Likewise, in a reverse process to the thought experiment, a helical spring too twists when we push on its ends.

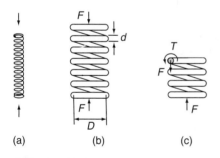

To see this from a mechanics viewpoint, let us consider Figures 4.34 (b) and (c). Figure 4.34 (b) shows a sketch of a helical compression spring (the same analysis applies to extension springs too!). Its geometric parameters are: coil diameter, $D$, wire diameter, $d$, and number of turns, $N$. Figure 4.34 (c) shows a cut-off section of this spring and its free-body diagram. To keep the circular cut-off section in equilibrium, there must be two forces—a vertical force $F$ in the downward direction and a torque $T$. Torque, as discussed earlier, results in a twisting action. If we consider the unwound spring as a shaft of length $N\pi D$, this torque would twist the entire length of the shaft. It is given by

**Figure 4.34** (a) Helical compression spring; (b) sketch of the spring showing geometric parameters and forces; and (c) free-body diagram of a cut-off spring.

$$T = \frac{FD}{2}$$

(4.88)

Our interest is in finding $k$. For this, we need to know the axial deflection of the spring. That is, if we hold one end fixed, how much does the other end move along the axis of the helix? We see that we have a torque load. A way out of this is to use Castigliano's second theorem. We write the strain energy stored in the spring when a force $F$ is applied. The strain energy due to a torque load is equal to half the product of torque and twist (recall that $kx^2/2 = Fx/2$ is the strain energy stored in a spring):

$$SE = \frac{T\phi}{2}$$

(4.89)

From Eq. (4.85), we get

$$SE = \frac{T\phi}{2} = \frac{T^2 L}{2GJ} \tag{4.90}$$

Here, $T = FD/2$, $L = N\pi D$, and $J = \pi R^4/2 = \pi d^4/32$. Hence, we get

$$SE = \frac{1}{2} \cdot \frac{F^2 D^2}{4} (N\pi D) \frac{1}{(G\pi d^4/32)} = \frac{4NF^2 D^3}{Gd^4} \tag{4.91}$$

Now, Castigliano's second theorem gives the axial deflection $\delta$:

$$\delta = \frac{\partial SE}{\partial F} = \frac{8NFD^3}{Gd^4}$$
$$k = \frac{F}{\delta} = \frac{Gd^4}{8ND^3} \tag{4.92}$$

Does this formula for the spring constant agree with our intuition? Check: if the wire diameter is large, the spring is stiff; if the coil diameter is small, the spring is stiff; if the shear modulus of the material is large, the spring is stiff; and if there are more turns, the spring is less stiff; and so on.

## Problem 4.7

In the preceding analysis, we considered only the torque load acting on the cut-off circular section in computing the strain energy. But there is also a shear force $F$. If we consider that too, show that the spring constant becomes

$$k = \frac{Gd^4}{8ND^3} \left(1 + \frac{D^2}{2d^2}\right)^{-1}$$

*Hint:* Add the strain energy due to the shear force and re-apply Castigliano's second theorem.
Do you think there are helical springs in microsystems? Of course. They can be and have been made. Look up the literature or the Internet to find them.

## Example 4.6: Silicon Spiral Spring

Silicon is a brittle material. However, in microsystems, we see a lot of flexible structures made of silicon. The strength of such structures under deformation may be questioned. An interesting experiment can be done to convince ourselves that flexibility is a matter of geometry and is not just dependent on the material type.

Imagine etching a spiral through-cut in a silicon wafer starting from a point slightly away from the center of the wafer. Perhaps deep reactive ion etching (DRIE) can be used to achieve such a cut. Let us suppose that a circular rim is left uncut at the periphery of the wafer, as in Figure 4.35.

Figure 4.36 (a) shows such a cut in an acrylic sheet instead of a silicon wafer. Like silicon, acrylic is brittle. Now, imagine that the central disk to which a vertical rod is attached [see Figure 4.36 (b)] is lifted with a vertical load. The self-weight is itself the load in this case. It can be observed that the spiral left in the sheet deforms into a tapering helical spring. Our intention in this example is to determine the stiffness of such a spring.

Central disk

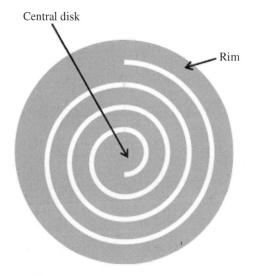

**Figure 4.35** Schematic of a spiral cut in a silicon wafer.

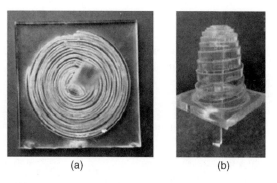

(a)                    (b)

**Figure 4.36** (a) Acrylic sheet with a spiral cut made with a $CO_2$ laser machine; (b) the spiral deformed under its own weight.

By assuming that the spiral is given in polar coordinates where the radius for any angle $\theta$ is $r = a\theta$, with $2\pi \leq \theta \leq 50\pi$, the in-plane width of the spiral $w$, and wafer thickness $t$, compute the stiffness under a vertical load. Assume the following numerical values:

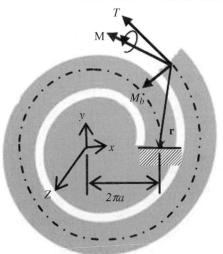

**Figure 4.37** Free-body diagram of a segment of a spiral with one end fixed to the central disk.

$a = 191 \times 10^{-6}$

$w = 3$ mm

$t = 1.1$ mm

Young's modulus, $Y = 169$ GPa
Poisson's ratio, $\nu = 0.3$

**Solution:** Note that this problem is similar to the helical spring in Example 4.5. In order to see this, we consider the free-body diagram of a segment of the spiral and the forces and moments acting on it; see Figure 4.37. Notice that, at the cut-end of the spiral, there will be a vertical reaction force (i.e. in the $z$-direction) as well as a twisting torque and a bending moment. We compute the torque and bending moment using vector notation. Let $\hat{\mathbf{i}}$, $\hat{\mathbf{j}}$, and $\hat{\mathbf{k}}$ denote the unit vectors in the $x$-, $y$-, and $z$-directions.

Note that the position vector from a point on the neutral axis of the spiral beam to the anchored paint is given by

$$\mathbf{r} = (2\pi a - a\theta \cos\theta)\hat{\mathbf{i}} - (a\theta \sin\theta)\hat{\mathbf{j}} \qquad (4.93)$$

The reaction force at the cut-end is given by $\mathbf{F} = -F_z\hat{\mathbf{k}}$. Then, the vector of the moment caused by this force (which has as components the torque and bending moment, as shown in Figure 4.37) can be written as

$$\mathbf{M} = \mathbf{r} \times \mathbf{F} \qquad (4.94)$$

Now, the torque (i.e. moment about the unit tangent vector to the spiral) and the bending moment (i.e. moment about the unit normal vector to the spiral) can be computed as

$$T = \mathbf{M} \cdot \hat{\mathbf{t}} \qquad (4.95a)$$

$$M_b = \mathbf{M} \cdot \hat{\mathbf{n}} \qquad (4.95b)$$

with tangent vector $= \mathbf{t} = (a\cos\theta - a\theta \sin\theta)\hat{\mathbf{i}} + (a\sin\theta + a\theta \cos\theta)\hat{\mathbf{j}} \qquad (4.95c)$

and normal vector $= \mathbf{n} = -(a\sin\theta + a\theta \cos\theta)\hat{\mathbf{i}} + (a\cos\theta - a\theta \sin\theta)\hat{\mathbf{j}} \qquad (4.95d)$

We can now compute the strain energy of the spiral due to the torque and bending moment as follows:

$$SE = \int_{2\pi}^{50\pi} \left( \frac{M_b^2}{2YI} + \frac{T^2}{2GJ} \right) d\theta \tag{4.96}$$

where $I = wt^3/12$ and $J = \{(wt^3/12) + (tw^3/12)\}$. The deflection $\delta_z$ in the $z$-direction can be found using Castigliano's second theorem:

$$\delta_z = \frac{d(SE)}{dF_z} \tag{4.97}$$

Then, the stiffness in the $z$-direction is given by

$$k_z = \frac{F_z}{\delta_z} \tag{4.98}$$

One may need to use symbol manipulation software such as Maple™ or Mathematica™ to evaluate the expressions in the preceding discussion to avoid the tedium of writing out long expressions. We give the final numerical results for the data specified in the problem statement to help the reader verify the results:

$$SE = 0.3397F_z^2$$
$$\delta_z = 0.6794F_z$$
$$k_z = 1.4718 \, \text{N/m}$$

Notice that this is a very soft spring, one that deflects by 1 m for less than 1.5 N force! This shows that a soft spring can be made out of a brittle material. By computing the stresses in it, one can be further convinced that the stress levels are also low in this structure. This removes concerns about this spring (and the many other micromechanical structures made of silicon) breaking due to large deformation it undergoes. Note that a similar problem is solved using 3D FEA in Section 10.2 using ABAQUS FEA software.

---

## Example 4.7: Multi-Axial Stiffness of a Micromechanical Suspension

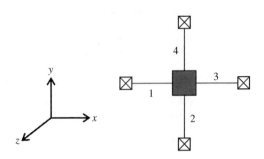

**Figure 4.38** A micromachinable suspension for a mass.

Consider the suspension for a square mass shown in Figure 4.38. It consists of four beams, each of which is connected to the mass at one end and anchored to the substrate at the other. Let the beam's dimensions be: length $= l$, in-plane width $= w$, and out-of-plane thickness $= t$. Let the side of the square mass be $s$. Find the stiffness of this suspension in the $x$-, $y$-, and $z$-directions using $Y$ for Young's modulus.

**Solution:** We first note that there is symmetry in the $x$- and $y$-directions, so that we expect the stiffness in these directions to be the same. Thus, treating one of them suffices.

Imagine a force $F_x$ acting in the $x$-direction. This causes axial loads in beams 1 and 3. Per Eq. (4.6), the axial stiffness of these two beams is given by

$$k_{x:1,3} = \frac{AY}{L} = \frac{wtY}{l} \tag{4.99}$$

The force $F_x$ causes a transverse load on beams 2 and 4. We can see from Figure 4.38 that these two beams are fixed at one end and are guided transversely at the other end. Per Eq. (4.43), the corresponding stiffness is given by

$$k_{x:2,4} = \frac{12YI}{L^3} = \frac{12Yw^3t}{12l^3} = \frac{w^3tY}{l^3} \tag{4.100}$$

Since the four beams share the applied load $F_x$, we need to add the $x$-direction stiffnesses of all of them to get the total stiffness of the suspension in the $x$-direction:

$$k_x = k_{x1} + k_{x2} + k_{x3} + k_{x4} = \frac{2wtY}{l} + \frac{2w^3tY}{l^3} = \frac{2wYt}{l}\left(1 + \frac{w^2}{l^2}\right) \tag{4.101}$$

As noted above, $k_y = k_x$.

Now, to determine $k_z$, imagine a force $F_z$ in the $z$-direction. This force causes all four beams to behave like fixed-guided beams. Once again, per Eq. (4.43), we can get the individual stiffness of each beam as

$$k_{z:1,4} = \frac{12YI}{L^3} = \frac{12Ywt^3}{12l^3} = \frac{wt^3Y}{l^3} \tag{4.102}$$

Since all four beams share the force $F_z$, we need to add their stiffnesses to get

$$k_z = \frac{4wt^3Y}{l^3} \tag{4.103}$$

---

**Problem 4.8**

Consider the suspension in Figure 4.39, a slight variant of the suspension shown in Figure 4.38. Use the same symbols as in Example 4.7 to find the multiaxial stiffness of this suspension. (*Hint*: Think about how beams move under different loads to determine the boundary conditions of the beams.)

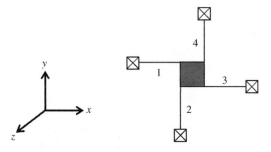

**Figure 4.39** A micromachinable suspension for a mass—a variant of the suspension shown in Figure 4.38.

## ► 4.10 DEALING WITH LARGE DISPLACEMENTS

So far, we have restricted ourselves to small displacements and made approximations by neglecting the second-order quantities. This is not always valid in reality, certainly not in many microsystems and smart-systems components. There are three effects of large displacements: large strain, large rotations and stress stiffening (note that we have encountered all the three in this chapter). We reduced the expression for bending-induced radius of curvature to a simpler expression in Eq. (4.21b) by assuming that the slope of the beam ($dw/dx$) is small enough that we could neglect its square. When the slope is not small, we cannot do so. The expression for strain [see Eqs. (4.79a) and (4.21a)] must be written as

$$\epsilon_x = -\frac{y(d^2w/dx^2)}{\left\{1 + (dw/dx)^2\right\}^{3/2}} \qquad (4.104)$$

Likewise, for deflection of beams, we need to consider the following instead of Eq. (4.22):

$$M = -\frac{YI(d^2w/dx^2)}{\left\{1 + (dw/dx)^2\right\}^{3/2}} \qquad (4.105)$$

Similarly, while discussing bars, we defined the strain as simply the change in length over the original length, but we took only the first-order term. For an infinitesimally small segment of a bar, we wrote

$$\epsilon_x = \frac{du}{dx} \qquad (4.106)$$

and used this in Eq. (4.31) (see the last line). But when strains are large, we need to include the second-order term, rewriting as

$$(\epsilon_x)_{\text{large}} = \frac{du}{dx} - \frac{1}{2}\left(\frac{du}{dx}\right)^2 \qquad (4.107)$$

Things are now nonlinear. We can thus conclude: small displacement assumptions lead to linear behavior, while the large displacements experienced in practice lead to nonlinearities.

The second effect of large displacements arises because of large rotations; its implications lie in the reference configuration we take to write our force and moment equilibrium equations. We had conveniently skipped this nuance when we showed the free-body diagrams of beams in Figure 4.13 and then applied and reaction forces and moments in the undeformed configuration of the beams. This is all right when the beam bending is small. But when the beam bends a lot, it rotates a lot, and hence we must show force and moment equilibrium in that rotated configuration. Let us take the simple example of a cantilever beam.

As shown in Figure 4.40, when the displacements and rotations are small, the reaction bending moment at the left end, $M_S$, is equal to the force at the tip times the total beam length. However, for large displacements and rotations, the reaction bending moment $M_L$ is smaller than $M_S$. This is because the moment arm has decreased as the right tip has moved closer to the left end, that is, the fixed end. In a general structure, too, similar difference

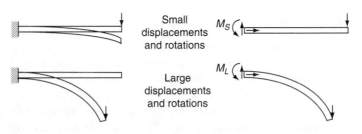

**Figure 4.40** A cantilever beam undergoing a large displacement and rotation.

arises between small and large displacements and rotations. The point to keep in mind is that equilibrium is to be considered in the deformed configuration. Alternately, we can map the forces and moments in the deformed configuration back to the undeformed configuration. This subtlety will become clear if we consider slightly advanced concepts in nonlinear elasticity.

The third nonlinear effect—stress stiffening—arises in beams because as beams bend, their neutral axis stretches. We may recall that we had assumed otherwise and indeed defined in Section 4.2 the neutral axis of the beam as the line that does not contract or stretch. However, that assumption is not valid when the bending is large. A stretched beam is stressed and becomes stiffer. Imagine a relaxed string versus a taut string: more force is required to pluck a taut string than a relaxed string. The same applies to beams undergoing large displacements.

In Eq. (4.67), we wrote the governing equation for a beam with residual stress. The same equation can be used for analyzing the stress-stiffening effect by replacing the residual stress with the bending stress. It then becomes a nonlinear equation. Usually, numerical methods are used to solve such equations. One such method is FEA, which is discussed in Chapter 5.

Nonlinearities can also occur in structures when material behavior is nonlinear (when Hooke's law does not remain linear or the material becomes plastic and deforms permanently) or when a part of a structure touches itself or some other object. The latter, known as *contact nonlinearity*, becomes important in micromachined RF switches. While we do not consider the analysis of such nonlinearities here, we are now ready to think beyond beams.

## ▶ 4.11 IN-PLANE STRESSES

Figure 4.41(a) shows a bulk-micromachined pinching device made using single-crystal (100) silicon. Its purpose is to pinch an object held in the gap to the right of the anchored portion when the forces are applied as shown. Many such miniature grasping and manipulating devices are used in studying biological cells and tissues. The device in Figure 4.41(a) consists of mostly slender beam-like segments, and beam-type analysis might give reasonably accurate results. But a more accurate analysis would require us to think beyond beams. We now consider these segments as general planar structures that deform within the plane. As a case in point, Figures 4.41(b) and (c) show a macroscale replica of a similar device. This device is farther from beam modeling than that in Figure 4.41(a). A question to ask is: what kinds of stresses exist in such planar structures that deform within a plane? To answer this, consider the configuration in Figure 4.42.

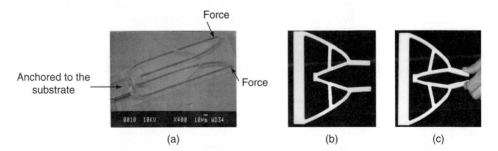

(a)　　　　　　　　　　(b)　　　　　　　(c)

**Figure 4.41** (a) A micromachined pinching device made of single-crystal silicon; (b) and (c) a similar macroscale compliant pinching device in its underformed and deformed configurations.

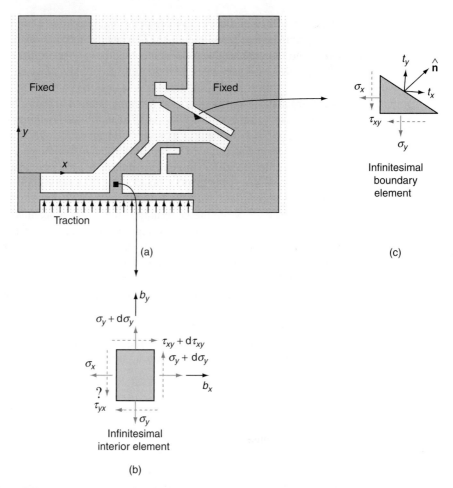

Infinitesimal boundary element

(a)

(c)

(b)

**Figure 4.42** (a) A planar deformable structure; (b) an infinitesimally small piece taken from the interior of a planar structure that deforms within its plane. Three stresses are shown on the element along with body force. (c) Another infinitesimal structure taken from the boundary with its stresses and the force acting on the boundary.

We stress that we need no new concepts in what follows; we simply use force and moment equilibrium and Hooke's law with the Poisson effect, which are all now familiar to us. We do, however, need some new notation for different stresses and forces here and there.

Figure 4.42(a) shows a planar elastically deformable structure. Let its thickness be unity. We say that this structure, after deforming, is in a *plane-stress* condition: that is, any stress involving the $z$-direction is zero. So far, we have considered stresses in one direction only because this is all that matters the most in bars and beams. However, in a general structure, there are stresses in all directions. In the planar structure in Figure 4.42(a), we isolate an infinitesimally small square of size $dx \times dy$ and show the three stresses acting on it [Figure 4.42(b)]. These are: two normal stresses, $\sigma_x$ and $\sigma_y$, and one shear stress, $\tau_{xy}$. Note that $\tau_{xy}$ and $\tau_{yx}$ must be equal for static moment equilibrium about the $z$-axis (the axis perpendicular to the $xy$-plane). Even though we show the two shear stresses separately, we tacitly assume that they are equal except when there is an external moment load on the element. Since such a moment load within the body is rare, we take $\tau_{xy}$ as equal to $\tau_{yx}$.

Since the three in-plane stresses, $\sigma_x$, $\sigma_y$, and $\tau_{xy}$, vary throughout the structure, we add a small differential change, $d(\cdots)$, for all three and show them on the infinitesimal interior element [Figure 4.42(b)]. We also show the body forces that may be present. Body forces are the forces that exist throughout the interior; examples include the force of gravity, centrifugal force for a rotating body, etc.

Now consider the static force equilibrium of this element. (Recall that this is similar to what we did for a beam in Figure 4.25.) Static force equilibrium in the $x$-direction gives

$$(d\sigma_x)(dy \cdot 1) + (d\tau_{xy})(dx \cdot 1) + (b_x)(dx \cdot dy \cdot 1) = 0$$

$$\frac{d\sigma_x}{dx} + \frac{d\tau_{xy}}{dy} + b_x = 0 \tag{4.108}$$

where the stress is multiplied by the area to get the force while the body force, which is defined per unit volume, is multiplied by the volume to get the force. Recall that the thickness is assumed to be unity. Similarly, force equilibrium in the $y$-direction gives

$$(d\sigma_y)(dx \cdot 1) + (d\tau_{xy})(dy \cdot 1) + (b_y)(dx \cdot dy \cdot 1) = 0$$

$$\frac{d\sigma_y}{dy} + \frac{d\tau_{xy}}{dx} + b_y = 0 \tag{4.109}$$

Equations (4.108) and (4.109) apply to every interior point of the deformed planar structure, but not to the points on the boundary. For these, we take an infinitesimally small wedge element on the boundary [Figure 4.42(b)] and consider the forces acting on it. Let the hypotenuse of the wedge element be $ds$ and the edges in the $x$- and $y$-directions be $dx$ and $dy$. Trigonometry then tells us that $dx = n_y ds$ and $dy = n_x ds$, where the unit normal $\hat{n}$ to the hypotenuse of the wedge element [Figure 4.42(c)] has $n_x$ and $n_y$ as components in the $x$- and $y$-directions. We also show the boundary force with its two components $t_x$ and $t_y$ in the $x$- and $y$-directions. We call the boundary force *traction*, a fancy term used in mechanics literature. Traction is a familiar term in the context of an automobile tire: if the tire has no traction it will slip and skid. We have indicated the force acting on the deformable structure of Figure 4.42(a) as traction [Figure 4.42(a) bottom].

Now, let us write the force equilibrium in the $x$-direction:

$$-(\sigma_x)(dy \cdot 1) - (\tau_{xy})(dx \cdot 1) + (t_x)(ds \cdot 1) = 0$$

$$-(\sigma_x)(n_x ds \cdot 1) - (\tau_{xy})(n_y ds \cdot 1) + (t_x)(ds \cdot 1) = 0 \tag{4.110}$$

$$\sigma_x n_x + \tau_{xy} n_y = t_x$$

Similarly, for the $y$-direction we can write

$$-(\sigma_y)(dx \cdot 1) - (\tau_{xy})(dy \cdot 1) + (t_y)(ds \cdot 1) = 0$$

$$-(\sigma_y)(n_y ds \cdot 1) - (\tau_{xy})(n_x ds \cdot 1) + (t_y)(ds \cdot 1) = 0 \tag{4.111}$$

$$\sigma_y n_y + \tau_{xy} n_x = t_y$$

Now, by collecting the end results of Eqs. (Eqs. (4.108)–(4.111)) in one place (because they relate to the interior and boundary of the same elastic body), we can write the governing equations for a general planar elastic structure as

$$
\left.
\begin{aligned}
\frac{\partial \sigma_x}{\partial x} + \frac{\partial \tau_{xy}}{\partial y} + b_x = 0 \\[2mm]
\frac{\partial \sigma_y}{\partial y} + \frac{\partial \tau_{xy}}{\partial x} + b_y = 0
\end{aligned}
\right\} \text{ For the interior}
$$

$$
\left.
\begin{aligned}
\sigma_x n_x + \tau_{xy} n_y = t_x \\[2mm]
\sigma_y n_y + \tau_{xy} n_x = t_y
\end{aligned}
\right\} \text{ For the boundary}
$$

(4.112)

We are not done yet! We have yet to invoke Hooke's law and the Poisson effect. In Eq. (4.112), we must deal with stresses directly but they depend on deformations of the structure. Additionally, we want to know these deformations as well. Let the deformations be denoted by $u$ and $v$ for displacement of the structure in the $x$- and $y$-directions. Note that $u$ and $v$ are functions of $x$ and $y$; that is, $u(x, y)$ and $v(x, y)$, as they vary throughout the structure. In order to express the governing equations in terms of $u$ and $v$, we need to define strains and then use Hooke's law and the Poisson effect.

Since there are three stresses in a planar structure at every point, there should naturally be three corresponding strains. We denote them by $\epsilon_x$, $\epsilon_y$, and $\gamma_{xy}$. The first two, $\epsilon_x$ and $\epsilon_y$, are the normal strains whereas the third, $\gamma_{xy}$, is the shear strain. We have already defined normal strain at the very start of the chapter [Eq. (4.2)] and we have used the concept of shear strain in Eq. (4.80). Let us define all three once again for the planar case.

We see in Figure 4.43 how the rectangle ABCD (an infinitesimally small element of size $dx \times dy$) is first stretched in the two directions and then sheared as well. Point A moves to A′ by $u$ in the $x$-direction and $v$ in the $y$-direction. Before shear, point C moves by $u + (\partial u/\partial x)dx$ in the $x$-direction and $v + (\partial v/\partial y)dy$ in the $y$-direction. So the normal

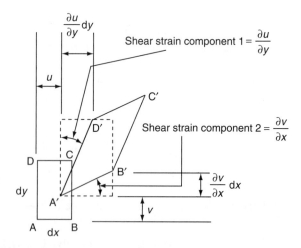

**Figure 4.43** Definition of strains in a planar structure. An infinitesimal element of size $dx \times dy$ is considered (thick solid rectangle). The dashed rectangle shows the form when it is expanded in both directions, while the solid parallelogram shows the form when sheared in two ways. The shear strain has two components, which are also shown.

strains are given by

$$\epsilon_x = \frac{\partial u}{\partial x} \text{ and } \epsilon_y = \frac{\partial v}{\partial y} \tag{4.113}$$

where we have simply used the definition of normal strain (i.e. the change in length over the original length).

Because of shear, point B moves up by $(\partial v/\partial x)dx$ and point D moves to the right by $(\partial u/\partial y)dy$. The shear strain components, which are really angles, turn out to be as shown in Figure 4.43. By adding the two components, we get the shear strain of the infinitesimal element:

$$\gamma_{xy} = \frac{\partial u}{\partial y} + \frac{\partial v}{\partial x} \tag{4.114}$$

The next step is to relate the stresses and strains using Hooke's law and the Poisson effect:

$$\epsilon_x = \frac{\sigma_x}{Y} - v\frac{\sigma_y}{Y}$$

$$\epsilon_y = -v\frac{\sigma_x}{Y} + \frac{\sigma_y}{Y} \tag{4.115}$$

$$\gamma_{xy} = \frac{\tau_{xy}}{G}$$

Note that Hooke's law applies to shear strain too, with the proportionality constant between shear stress and shear strain being $G = Y/2(1 + v)$. In Eq. (4.115), we assume that Young's modulus is the same in the $x$- and $y$-directions. A material that exhibits such a property is called an *isotropic* material. Of course, not all materials are isotropic. In fact, single-crystal silicon is an *anisotropic* material: its properties are different in different directions [see Section 3.1, Chapter 3]. This fact must be kept in mind when we simulate a silicon device. We do not discuss anisotropy or its effects here.

When we invert the three equations in Eq. (4.115) (a symbol manipulation exercise in inverting a matrix), we get the stress–strain relationship for the planar case:

$$\begin{Bmatrix} \sigma_x \\ \sigma_y \\ \tau_{xy} \end{Bmatrix} = \frac{Y}{1 - v^2} \begin{bmatrix} 1 & v & 0 \\ v & 1 & 0 \\ 0 & 0 & (1 - v)/2 \end{bmatrix} \begin{Bmatrix} \epsilon_x \\ \epsilon_y \\ \gamma_{xy} \end{Bmatrix} \tag{4.116}$$

Equation (4.116) is called the *constitutive equation* and is specific to the constitution of a material. Here, as earlier, it is specific to an isotropic material that is also linear and it is applicable to the plane-stress condition.

Now, if we substitute the stresses obtained from Eq. (4.116) into Eq. (4.112), and then substitute for strains in terms of displacements using Eqs. (4.113) and (4.114), we get the governing equations for the interior and the boundary in terms of $u\,(x, y)$ and $v\,(x, y)$. These are formidable partial differential equations that can be solved analytically only in

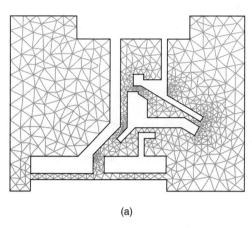

(a)

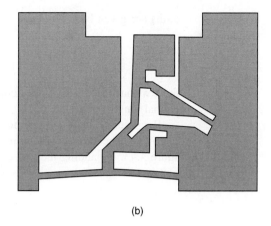

(b)

**Figure 4.44** (a) Finite element mesh for deformation and stress analysis of the deformable structure in Figure 4.42 (a); (b) the deformed geometry of the structure.

the simplest cases. For a general structure such as the one shown in Figure 4.41(a), we must resort to numerical techniques; one such technique is the *finite element method* (FEM), described in Chapter 5.

As a prelude to Chapter 5, consider the finite-element-method-based solution of the elastic structure in Figure 4.42(a). The mesh and the deformed pattern with stress distribution are shown in Figures 4.44(a) and (b). Details of this method are given in Chapter 5.

 **Your Turn:**

The plane-strain condition exists in other types of structures that can be considered to be planar for the purposes of analysis. Consider a long pipe where the stresses do not change along the length of the pipe. In such pipes, the strains associated with the third direction (rather than stresses, as in the plane-stress condition) are zero. For such a case, show that the constitutive equation is the following:

$$\begin{Bmatrix} \sigma_x \\ \sigma_y \\ \tau_{xy} \end{Bmatrix} = \frac{Y}{(1+\nu)(1-2\nu)} \begin{bmatrix} 1-\nu & \nu & 0 \\ \nu & 1-\nu & 0 \\ 0 & 0 & (1-\nu)/2 \end{bmatrix} \begin{Bmatrix} \epsilon_x \\ \epsilon_y \\ \gamma_{xy} \end{Bmatrix}$$

(4.117)

The method used in the preceding analysis to derive the governing equations and the constitutive relationship for a planar structure can be readily extended to three dimensions. The equations would be

$$\left.\begin{aligned} \frac{\partial \sigma_x}{\mathrm{d}x} + \frac{\partial \tau_{xy}}{\partial y} + \frac{\partial \tau_{zx}}{\partial z} + b_x &= 0 \\ \frac{\partial \tau_{xy}}{\partial x} + \frac{\partial \sigma_y}{\partial y} + \frac{\partial \tau_{yz}}{\partial z} + b_y &= 0 \\ \frac{\partial \tau_{zx}}{\partial x} + \frac{\partial \tau_{yz}}{\partial y} + \frac{\partial \sigma_z}{\partial z} + b_z &= 0 \end{aligned}\right\} \text{For the interior}$$

$$\left.\begin{aligned} \sigma_x n_x + \tau_{xy} n_y + \tau_{zx} n_z &= t_x \\ \tau_{xy} n_x + \sigma_y n_y + \tau_{yz} n_z &= t_y \\ \tau_{xz} n_x + \tau_{yz} n_y + \sigma_z n_z &= t_z \end{aligned}\right\} \text{For the boundary}$$

$$\begin{Bmatrix} \sigma_x \\ \sigma_y \\ \sigma_z \\ \tau_{xy} \\ \tau_{yz} \\ \tau_{zx} \end{Bmatrix} = \frac{Y}{(1+\nu)(1-2\nu)} \begin{bmatrix} 1-\nu & \nu & \nu & 0 & 0 & 0 \\ \nu & 1-\nu & \nu & 0 & 0 & 0 \\ \nu & \nu & 1-\nu & 0 & 0 & 0 \\ 0 & 0 & 0 & 0.5-\nu & 0 & 0 \\ 0 & 0 & 0 & 0 & 0.5-\nu & 0 \\ 0 & 0 & 0 & 0 & 0 & 0.5-\nu \end{bmatrix} \begin{Bmatrix} \epsilon_x \\ \epsilon_y \\ \epsilon_z \\ \gamma_{xy} \\ \gamma_{yx} \\ \gamma_{zx} \end{Bmatrix} \quad (4.118)$$

Although the equations look complex and formidable at first sight, once we understand the concepts, it is not so difficult to derive them. Try it!

## ► 4.12 DYNAMICS

So far, we have discussed only static deformation of the slender elastic bodies found in microsystems. However, as noted at the beginning of this chapter, micromechanical components are *dynamic*: they move and change their configurations with time. The mass of an accelerometer continues to move in the presence of external forces that vary with time; the torsion beam in a tilting micromirror must twist and untwist continuously—thousands or millions of times per second—to redirect the light falling on it as required; the diaphragm of a pressure sensor deforms whenever the pressure changes; the beam in an RF-MEMS switch or a flow-control valve must bend and straighten to close or open it. The list is endless! We consider such time-varying behavior in this section. The study of dynamics of rigid and elastic bodies is vast and complex. Nonlinearities abound in it, sometimes even *chaos*—the unexpected strange motion that is unpredictably sensitive to initial conditions. However, we restrict ourselves to slender elastic bodies by considering simple motions as we introduce and work with dynamics.

The starting point for dynamics is Newton's second law of motion: $F = ma$. The acceleration $a$ may vary with time. In fact, the external force $F$ and mass $m$ may also change with time in some systems. Let us first consider the simplest dynamic model of an elastic solid: the so-called *spring-mass-damper* in Figure 4.45(a). Its free-body diagram is shown in Figure 4.45(b). The four forces acting on it balance each other at every instant of time to give the equation

$$m\,\ddot{x}(t) + b\,\dot{x}(t) + k\,x(t) = F(t) \quad (4.119)$$

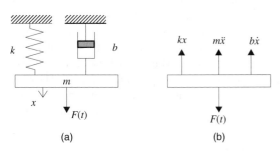

**Figure 4.45** A spring-mass-damper model and its free-body diagram.

where $\dot{x} = \frac{dx(t)}{dt} = v(t) =$ velocity and $\ddot{x} = \frac{dx(t)^2}{dt^2} = a(t) =$ acceleration of the mass $m$; $m\,\ddot{x}(t)$ is the inertial force, $b\,\dot{x}(t)$ is the damping force, $k\,x(t)$ is the spring force, and $F(t)$ is the applied external force. The Appendix describes the solution of this equation when $F(t)$ is zero and when it varies sinusoidally with time. The time and frequency responses of this system, the concepts of amplitude and phase, natural frequency, resonance, time-constant, and quality factor are also discussed in the Appendix.

**Example 4.8:** A Micromachined Accelerometer: A Single-Degree-of-Freedom Dynamic Model

A micromachined accelerometer has a proof mass of dimensions 50 μm × 50 μm × 2 μm. It is held by a suspension with a spring constant of 2.5 N/m in the direction perpendicular to its lateral plane (i.e. the square of 50 μm × 50 μm area). Compute its sensitivity and bandwidth. Assume $Y = 150$ GPa and $\rho_m = 2{,}300$ kg/m$^3$.

**Solution:** If we neglect the mass of the suspension, the contribution to $m$ in Eq. (4.119) comes primarily from the proof mass and is given by

$$m = \rho_m \left(50 \times 50 \times 2 \times 10^{-18}\right) = 11.5 \times 10^{-12} \text{ kg}$$

The force on the right-hand side of Eq. (4.119) is $F(t) = m\,a(t)$, where $a(t)$ is the applied acceleration. In order to compute the sensitivity, let us consider the steady-state situation where only the stiffness and force terms remain in Eq. (4.119):

$$k\,x = m\,a \Rightarrow 2.5\,x = 11.5 \times 10^{-12}a$$

If we assume that $a = 9.81$ m/s$^2$, the displacement for unit $g$ acceleration can be computed as 45.12 pm. This is too small a displacement to cause a measurable change in, say, the capacitance of the accelerometer. Hence, either the suspension must be made more flexible or the mass of the proof mass must be increased. Note that such a simple analysis is sufficient at the beginning stages of design.

The resonance frequency of this accelerometer can be computed as (see Eq. (A.3c) in the Appendix)

$$\omega_n = \frac{1}{2\pi} \sqrt{\frac{k}{m}} = \frac{1}{2\pi} \sqrt{\frac{2.5}{11.5 \times 10^{-12}}} = 74.2 \text{ kHz}$$

As noted in Section A.3 of the Appendix, about one-third of the resonance frequency gives a rough indication of the bandwidth of the accelerometer, that is, 25 kHz. This very high bandwidth is indicative of a very stiff accelerometer. This example shows the compromise between sensitivity and bandwidth.

## 4.12.1 A Micromachined Gyroscope: Two-Degree-of-Freedom Dynamic Model for a Single Mass

A micromachined vibrating gyroscope is an interesting dynamic device. It is shown schematically in Figure 4.18 above, in which a mass is suspended so that it can move in different directions. Its suspension is designed so that its stiffness is low in x- and y-translational motion and very high in z-translation and the three rotational directions. This gyroscope measures the angular rate about the z-axis of the frame on which it is mounted. If the gyroscope is mounted on a car to measure its gentle roll about its longitudinal axis, the car is considered as the *frame*. This gyroscope is a *vibrating* gyroscope because its mass is continuously set into vibration by applying a force. The electrostatic combs are present in Figure 4.18 because they are needed for applying the force.

Let us consider a slightly modified sketch of Figure 4.18 without the combs but with the added frame in Figure 4.46(a) to analyze its dynamics. Consider the intentionally driven vibration of the mass in only the x-direction, for the lumped model in Figure 4.46(b). The added frame is shown as a wafer disk in this figure because it is the diced wafer that will be affixed to a chip-frame, which in turn will be glued to the frame (the car in this case).

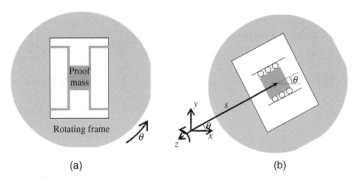

**Figure 4.46** (a) The sketch of a micromachined vibrating gyroscope; (b) a lumped two-degree-of-freedom model of the motion of the proof mass in the $x$-direction.

Unlike the spring-mass-lever model of Figure 4.45, the mass in Figure 4.46(b) has two degrees of freedom because its position needs specification of two quantities relative to a fixed frame: $s(t)$, the position of the mass $m$ with respect to the rotating disk, and $\theta(t)$, the angular position of the disk. [We have used the term "frame" a few times in this and the preceding paragraph with and without the prefix "fixed." We should note that a reference frame is a crucial concept in dynamics: whenever we consider motion, we must know the reference frame with respect to which we consider the motion. In the example of the car on which the gyroscope is mounted, the road is the *fixed* frame and the car is the *rotating* frame (or, in general, *moving* frame). In Figure 4.46(b), the rotating frame is the wafer-disk, which is rotating about the $z$-axis relative to the reference that is assumed to be fixed.] We now see how more terms are added to Eq. (4.119) when we write the velocity and acceleration of the mass in Figure 4.46(b).

Let the position of the mass in Figure 4.46b) be written as a vector $\mathbf{p}$ in Cartesian coordinates as:

$$\mathbf{p} = s\,\cos\theta\,\hat{\mathbf{i}} + s\,\sin\theta\,\hat{\mathbf{j}} \tag{4.120}$$

where $\hat{\mathbf{i}}$ and $\hat{\mathbf{j}}$ are the unit vectors attached to the fixed frame (i.e. the car) in the $x$- and $y$-directions. Note that, for brevity, we do not explicitly indicate the dependence of $s$ and $\theta$ on time from now on; it is understood that $s$ means $s(t)$ and $\theta$ means $\theta(t)$. We differentiate Eq. (4.120) with respect to time, once to get the velocity vector $\dot{\mathbf{p}}$ and twice to get the acceleration vector $\ddot{\mathbf{p}}$:

$$\begin{aligned}
\dot{\mathbf{p}} &= \left(\dot{s}\,\cos\theta - s\,\dot{\theta}\sin\theta\right)\hat{\mathbf{i}} + \left(\dot{s}\,\sin\theta + s\,\dot{\theta}\cos\theta\right)\hat{\mathbf{j}} \\
&= \dot{s}\left(\cos\theta\,\hat{\mathbf{i}} + \sin\theta\,\hat{\mathbf{j}}\right) + \dot{\theta}s\left(-\sin\theta\,\hat{\mathbf{i}} + \cos\theta\,\hat{\mathbf{j}}\right) \\
&= \dot{s}\,\hat{\mathbf{p}} + s\,\dot{\theta}\,\hat{\mathbf{p}}_{\perp}
\end{aligned} \tag{4.121}$$

where we have used

$$\hat{\mathbf{p}} = \left(\cos\theta\,\hat{\mathbf{i}} + \sin\theta\,\hat{\mathbf{j}}\right) \tag{4.122}$$

to indicate the unit vector in the direction of $\mathbf{p}$ and

$$\hat{\mathbf{p}}_{\perp} = \left(-\sin\theta\,\hat{\mathbf{i}} + \cos\theta\,\hat{\mathbf{j}}\right) \tag{4.123}$$

to indicate the unit vector perpendicular to $\mathbf{p}$. So, the final expression for $\dot{\mathbf{p}}$ in Eq. (4.121) can be recognized to have a radial component of velocity $\dot{s}\,\hat{\mathbf{p}}$, and a tangential component of velocity $s\dot{\theta}\,\hat{\mathbf{p}}_\perp$. Differentiation of $\dot{\mathbf{p}}$ with respect to time gives:

$$
\begin{aligned}
\ddot{\mathbf{p}} &= \left(\ddot{s}\,\cos\theta - 2\dot{s}\,\dot{\theta}\sin\theta - s\,\ddot{\theta}\sin\theta - s\dot{\theta}^2\cos\theta\right)\hat{\mathbf{i}} \\
&\quad + \left(\ddot{s}\,\sin\theta + 2\dot{s}\,\dot{\theta}\cos\theta + s\,\ddot{\theta}\cos\theta - s\dot{\theta}^2\sin\theta\right)\hat{\mathbf{j}} \\
&= \ddot{s}\left(\cos\theta\,\hat{\mathbf{i}} + \sin\theta\,\hat{\mathbf{j}}\right) + 2\dot{s}\,\dot{\theta}\left(-\sin\theta\,\hat{\mathbf{i}} + \cos\theta\,\hat{\mathbf{j}}\right) \\
&\quad + s\ddot{\theta}\left(-\sin\theta\,\hat{\mathbf{i}} + \cos\theta\,\hat{\mathbf{j}}\right) - s\dot{\theta}^2\left(\cos\theta\,\hat{\mathbf{i}} + \sin\theta\,\hat{\mathbf{j}}\right) \\
&= \ddot{s}\,\hat{\mathbf{p}} + 2\dot{s}\,\dot{\theta}\,\hat{\mathbf{p}}_\perp + s\ddot{\theta}\,\hat{\mathbf{p}}_\perp - s\dot{\theta}^2\,\hat{\mathbf{p}}
\end{aligned}
\tag{4.124}
$$

where we see that there are four components of acceleration: (i) the easily understandable radial sliding acceleration due to the motion of the mass relative to the moving frame (i.e. the wafer-disk) $\ddot{s}\,\hat{\mathbf{p}}$, (ii) a slightly less familiar acceleration called *Coriolis acceleration* $2\dot{s}\,\dot{\theta}\,\hat{\mathbf{p}}_\perp$ (see the box item below on the *Coriolis effect*), (iii) the familiar centrifugal acceleration $s\ddot{\theta}\,\hat{\mathbf{p}}_\perp$, and (iv) another familiar centripetal acceleration $(-s\dot{\theta}^2\,\hat{\mathbf{p}})$. Therefore, there will be additional inertial forces corresponding to each of these components.

Now, consider a more accurate model of the gyroscope in Figure 4.48 with the stiffness and damping elements in the $x$- and $y$-directions. When the wafer-disk rotates about the $z$-axis relative to a fixed frame, the equation of motion of the mass is given by the following vector equation, which is a vector analog of Eq. (4.119):

$$
m\left(\ddot{s}\,\hat{\mathbf{p}} + 2\dot{s}\,\dot{\theta}\,\hat{\mathbf{p}}_\perp + s\ddot{\theta}\,\hat{\mathbf{p}}_\perp - s\dot{\theta}^2\,\hat{\mathbf{p}}\right) + b\left(\dot{s}\,\hat{\mathbf{p}} + s\dot{\theta}\,\hat{\mathbf{p}}_\perp\right) + 2k_x\,(\delta s)\,\hat{\mathbf{p}} + 2k_y\,(\delta s)_\perp\hat{\mathbf{p}}_\perp = \mathbf{0}
\tag{4.125}
$$

where $(\delta s)$ and $(\delta s)_\perp$ are the extensions of the springs in the respective directions relative to their force-free lengths. In view of Eqs. (4.122) and (4.123), the component form of Eq. (4.125) in $\hat{\mathbf{i}}$ and $\hat{\mathbf{j}}$ directions is given by

$$
\left\{m\ddot{s} - ms\dot{\theta}^2 + b\dot{s} + 2(\delta s)k_x\right\}\cos\theta - \left\{2m\dot{s}\,\dot{\theta} + ms\ddot{\theta} + b\dot{s} + 2(\delta s)_\perp k_y\right\}\sin\theta = 0 \tag{4.126a}
$$

$$
\left\{m\ddot{s} - ms\dot{\theta}^2 + b\dot{s} + 2(\delta s)k_x\right\}\sin\theta + \left\{2m\dot{s}\,\dot{\theta} + ms\ddot{\theta} + b\dot{s} + 2(\delta s)_\perp k_y\right\}\cos\theta = 0 \tag{4.126b}
$$

Thus, we get two coupled ordinary differential equations in terms of $s$ and $\theta$. By denoting the small oscillations of the mass relative to an initial configuration (where the springs in Figure 4.48 are force-free) by $x$ and $y$, Eqs. (4.126a) and (4.126b) can be re-written as

$$
m\ddot{x} + b\dot{x} - \left(2m\dot{\theta} + b\right)\dot{y} - \left(m\dot{\theta}^2 - 2k_x\right)x - \left(m\ddot{\theta} + 2k_y\right)y = 0 \tag{4.127a}
$$

$$
m\ddot{y} + b\dot{y} - \left(2m\dot{\theta} + b\right)\dot{x} - \left(m\ddot{\theta} + 2k_x\right)x - \left(m\dot{\theta}^2 - 2k_y\right)y = 0 \tag{4.127b}
$$

In this gyroscope, shown in Figure 4.51, let the mass be set into vibration in the $x$-direction and let its motion be denoted by

$$
x = x_0\cos\left(\omega_d t\right) \tag{4.128}
$$

where $x_0$ is the amplitude and $\omega_d$ is the driving frequency. Since the motion in the $x$-direction is fixed as given in Eq. (4.128), the two-degree-of-freedom system represented by Eqs. (4.127a) and (4.127b) reduces to a single-degree-of-freedom system governed by

$$m\ddot{y} + b\dot{y} - \left(m\dot{\theta}^2 - 2k_y\right)y = \left(2m\dot{\theta} + b\right)\dot{x} + (m\ddot{\theta} + 2k_x)x \tag{4.129a}$$

$$\Rightarrow m\ddot{y} + b\dot{y} - \left(m\dot{\theta}^2 - 2k_y\right)y = -\left(2m\dot{\theta} + b\right)x_0\omega_d \sin{(\omega_d t)} + (m\ddot{\theta} + 2k_x)x_0 \cos{(\omega_d t)} \tag{4.129b}$$

The preceding equation is important because the gyroscope senses the angular motion by measuring the motion of the mass in the $y$-direction. In a rate gyroscope (a commonly used type of vibrating gyroscope), the mass and stiffnesses (i.e. $m$, $k_x$, and $k_y$) as well as the amplitude and the driving frequency (i.e. $x_0$ and $\omega_d$) are chosen so that the driving term involving $\dot{\theta}$ is the dominant cause of the motion of the mass in the $y$-direction. Evidently, the analysis and design of a gyroscope is more complicated than an accelerometer.

## ► *THE CORIOLIS EFFECT AND ITS APPLICATION IN RATE GYROSCOPES*

The Coriolis acceleration is given mathematically by the vector $2(\mathbf{\Omega} \times \mathbf{V})$, where $\mathbf{V}$ is the velocity of a body relative to a moving frame and $\mathbf{\Omega}$ is the angular velocity of the moving frame relative to a fixed reference frame. Note that both $\mathbf{\Omega}$ and $\mathbf{V}$ are vectors. Hence, the direction of the Coriolis acceleration is perpendicular to both because of the cross product in $2(\mathbf{\Omega} \times \mathbf{V})$.

To understand the effect of this acceleration, consider the Foucault pendulum shown in Figure 4.47. It is a simple pendulum with the difference that the frame from which it is hanging can rotate about an axis. When the frame rotates while the pendulum is oscillating, the plane of the pendulum starts to shift because of Coriolis acceleration. It does so because a force acts perpendicular to the oscillating pendulum's direction and that of the axis of rotation of the frame. The rate of change of angle of rotation of the pendulum's plane of oscillation is directly proportional to the angular rate (i.e., angular velocity) of the frame. The lumped model in Figure 4.46 (b) applies to the Foucault pendulum too. Here also we have two degrees of freedom for the pendulum to vibrate—one in the usual plane of oscillation and the other in the plane perpendicular to it. If we consider small oscillations, motion along $x$- and $y$-axes directly fits the model. Gravity provides the spring constants, as is usual in a pendulum.

An important concept to understand in the Coriolis effect is that the energy from one mode of oscillation is transferred to another mode. The natural frequencies in

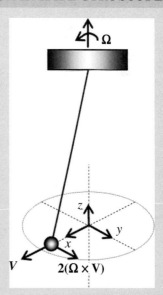

**Figure 4.47** A Foucault pendulum. When the support of the pendulum begins to rotate about the $z$-axis as shown, the ball oscillating along the $x$-axis experiences Coriolis acceleration along the $y$-axis.

the two modes must be the same for this to happen most effectively. In a Foucault pendulum, the natural frequencies of the pendulum in the two perpendicular planes of oscillation are the same, and this should be true in a vibrating rate gyroscope as well.

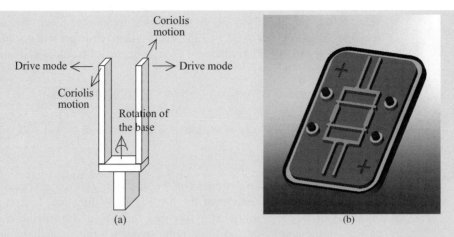

**Figure 4.48** (a) A tuning fork; (b) a micromachined tuning fork gyroscope.

Many configurations are possible in order that the two frequencies corresponding to the two modes are the same. A familiar example is a tuning fork [Figure 4.48(a)]. Its two tines can vibrate in-phase (the first mode) and out-of-phase (the second mode). When the frame of the tuning fork's base rotates while the tuning fork is vibrating in its second mode, its tines start vibrating in the plane perpendicular to their original plane of vibration—just as in the Foucault pendulum. Since their original oscillations (the driven mode) are out of phase with each other, the Coriolis effect causes motion of one tine in one direction and the other in the opposite direction. This principle is used in a micromachined gyroscope shown in Figure 4.48(b).

Just like the tuning fork, one can have a dual-mass gyroscope. Here, too, the masses are constantly driven in-plane in resonance vibration but out of phase with each other. When the frame rotates about an in-plane perpendicular direction, one mass starts oscillating up and down while the other down and up because of the Coriolis effect. This motion can be sensed capacitively or piezoresistively and then correlated to the angular rate. Figures 4.49(a) and (b) show a micromachined dual-mass gyroscope in drive and sense modes. The gyroscope in Figure 4.46(a) has a single mass but its suspension allows it to vibrate in two modes at the same frequency.

A disk held on a double-gimbal arrangement can also act as a gyroscope; it is used at the macro scale and

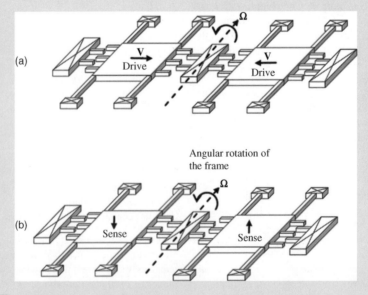

**Figure 4.49** A dual-mass vibrating gyroscope: (a) drive mode; (b) sense mode.

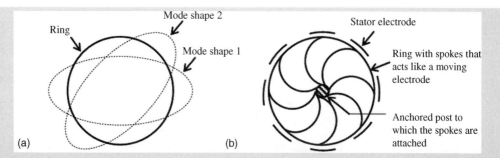

**Figure 4.50** (a) A ring with two degenerate elliptical mode shapes; (b) schematic of a ring gyroscope.

can be realized at the micro scale too. See Figure 2.14 showing a two-axis micromirror. Think about how it could be used as a gyroscope and what conditions need to be satisfied for it to do so.

Shown in Figure 4.50 (a) is a circular ring of uniform cross-section with two mode shapes of the same frequency. Its two *degenerate mode shapes* (i.e. two resonant modes of vibration with the same frequency) are ellipses at 45° to each other. The micromachined ring gyroscope in Figure 4.50 (b) does not have a free ring but a ring supported symmetrically by eight semicircular spokes attached to a central post anchored to the substrate, that is, the frame. Because of this, its degenerate elliptical mode shapes will be slightly distorted but will be present

nevertheless. When the substrate starts rotating relative to a fixed frame while the ring is driven into resonant oscillation in one of its distorted elliptical modes, the Coriolis effect makes the ring vibrate in the other mode. The driving of the ring is accomplished with the electrostatic force in Figure 4.50 with the help of the anchored electrodes at the 12, three, six, and nine o'clock positions. Any rotation of the frame will cause the ring to vibrate in the 45° elliptical mode. Thus measuring the capacitance between the ring electrode and anchored electrodes at the 1:30, 4:30, 7:30, and 10:30 positions senses any rotation of the frame. This concept of an electrostatically actuated and capacitively measured micromachined gyroscope is used in a commercial product.

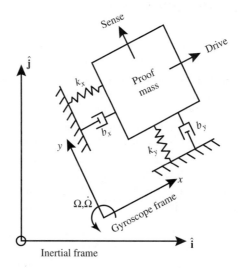

**Figure 4.51** A lumped two degree-of-freedom model of a single-mass gyroscope with spring and damping elements.

An important fact about a vibrating rate gyroscope is that it is the Coriolis acceleration term (i.e. $2\dot{s}\,\dot{\theta}\,\hat{\mathbf{p}}_{\perp}$) that is made dominant. This term is the one that contributes the dominant driving factor to the motion of the mass in the $y$-direction due to the angular rotation of the frame in the presence of forced sinusoidal motion in the $x$-direction. It is worth noticing that this driving factor is linear in $\dot{\theta}$—a good thing for a sensor that senses $\dot{\theta}$!

**Example 4.9**

Assume that the numerical data given in Example 4.8 for an accelerometer is are the same for a dual-mass gyroscope. Let its in-plane driving velocity be 0.002 m/s. Compute the resulting out-of-plane Coriolis acceleration if the substrate rotates about the axis perpendicular to it at the rate of 1 rad/s.

**Solution:** The magnitude of the Coriolis acceleration is given by $2\Omega v = 2 \times 1 \times 0.002 = 0.004\,\text{m/s}^2$. Hence, the Coriolis force (mass times the acceleration) is given by

$$F_{\text{Coriolis}} = (11.5 \times 10^{-12}) \times 0.004 = 0.046 \times 10^{-12}\,\text{N} = 46\,\text{fN}$$

Note that this is an extremely small force, which with an out-of-plane spring constant of 2.5 N/m will give rise to an 18.4 fm displacement. It is interesting to note that 1 rad/s is actually a very fast rotation. Even with that, we get a very small displacement to be sensed. Clearly, this is not a suitable design for a rate gyroscope. As in the accelerometer example considered earlier in the chapter (see Figure 4.5), the proof mass needs to be made bigger and the suspension more flexible. Another option in a gyroscope is to increase the driving velocity.

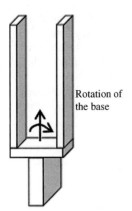

**Figure 4.52** A tuning fork used in a gyroscope.

Rotation of the base

**Problem 4.9**

Consider the tuning fork in Figure 4.52 for use in a micromachined gyroscope. Let the two beams making up the tines of the tuning fork to be 100 μm long with square cross-section of 5 μm × 5 μm. Compute the two natural frequencies of this tuning fork by assuming a Young's modulus of 169 GPa and a mass density of 2,300 kg/m³. Compute also the Coriolis force on the tines for a base-rotation of 0.01 rad/s if the driving in-plane amplitude of vibration of the tines gives a velocity of 0.001 m/s.

### 4.12.2 A Micromechanical Filter: Two-Degree-of-Freedom Dynamic Model with Two Masses

Consider the micromechanical filter shown in Figure 4.53(a), which can serve as a band-pass filter and yet occupy as little space on the chip as two fixed-fixed beams connected by a third small beam do. One of its fixed-fixed beams receives the input signal as a sinusoidal electric potential difference applied between it and the electrode underneath. Consequently, the electrostatic force makes the beam oscillate. Because of the small coupling beam, the other fixed-fixed beam also oscillates. When the frequency of the input signal matches that of the natural frequency of the filter structure, both the beams oscillate the most because of resonance. However, this system has more than one natural frequency. In order to see this, we need to construct the lumped model that is shown in Figure 4.53(b). A lumped model can have as many degrees of freedom as is required by the problem at hand. Here, we need two degrees of freedom because we are interested in the motion of both the fixed-fixed beams.

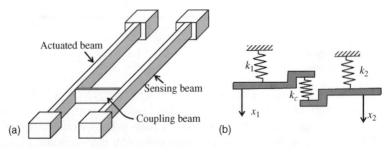

**Figure 4.53** (a) Schematic of a micromechanical filter; (b) its two-degree-of-freedom lumped model.

Both beams have stiffness and mass of their own: $k_1$, $k_2$, $m_1$, and $m_2$. We ignore damping for now. As the two beams are coupled by the small beam, we have a coupling spring of stiffness $k_c$ connecting the two lumped masses. We neglect the mass of the coupling beam as it is much smaller than that of the fixed-fixed beams.

In order to write the equations of motion of this two-degree-of-freedom lumped model, we can use either the method of force-balance and apply Newton's second law or Lagrange's method. In the force-balance method, we consider the free-body diagram of each mass and equate all the forces to the inertial force in the respective directions. The reader may use this method to write the equations of motions for the two-mass system in Figure 4.53(b). However, we will use the Lagrange method here to introduce an often more convenient way to write equations of motion than the force-balance method.

Lagrange's method is an energy-based method, While one would need to consider internal and intermediate forces, such as the force of the coupling spring in Figure 4.53(b), to apply Newton's second law, that is not necessary in Lagrange's method, as we will see.

In Lagrange's method, we write what is called a Lagrangian $L$ as

$$L = KE - PE = \text{kinetic energy} - \text{potential energy} \qquad (4.130)$$

The kinetic energy $KE$, by definition, is half the mass times the square of velocity. Since it is a scalar, the $KE$s of individual masses or parts can simply be added to get the $KE$ of the whole system:

$$KE = \frac{1}{2} \sum_{i}^{n} m_i v_i^2 \qquad (4.131a)$$

for $n$ discrete masses $m_{i=1\cdots n}$ with velocities $v_{i=1\cdots n}$, and

$$KE = \frac{1}{2} \int_V \rho_m v^2 dV \qquad (4.131b)$$

for a continuous system with density $\rho_m$ and velocity $v$. Also, the potential energy $PE$, by definition, is equal to the difference of strain energy and the work done by the external forces:

$$PE = SE - W_{ext} \qquad (4.132)$$

Once we have $KE$, $PE$, and the Lagrangian $L$, the equations of motion for a discrete system can be written as follows:

$$\frac{d}{dt}\left(\frac{\partial L}{\partial v_i}\right) - \frac{\partial L}{\partial x_i} = 0 \qquad (4.133)$$

where $x_{i=1\cdots n}$ and $v_{i=1\cdots n}$ are the positions and velocities of masses $m_{i=1\cdots n}$, respectively.

Let us now use Lagrange's method for writing the equations of motion for the two masses in Figure 4.53(b):

$$KE = \frac{1}{2}\left(m_1 v_1^2 + m_2 v_2^2\right) \qquad (4.134)$$

$$PE = \left\{\frac{1}{2}k_1 x_1^2 + \frac{1}{2}k_2 x_2^2 + \frac{1}{2}k_c(x_1 - x_2)^2\right\} - F_1 x_1 - F_2 x_2 \qquad (4.135)$$

$$L = \frac{1}{2}\left(m_1 v_1^2 + m_2 v_2^2\right) - \left\{\frac{1}{2}k_1 x_1^2 + \frac{1}{2}k_2 x_2^2 + \frac{1}{2}k_c(x_1 - x_2)^2\right\} + F_1 x_1 + F_2 x_2 \qquad (4.136)$$

Note that $v_1 = \dot{x}_1$ and $v_2 = \dot{x}_2$. With substitutions from Eqs. (4.134)—(4.136) into Eq. (4.133), we obtain the equations of motion for the filter's lumped model as

$$m_1\ddot{x}_1 + (k_1 + k_c)x_1 - k_c x_2 - F_1 = 0 \qquad (4.137a)$$

$$m_1\ddot{x}_2 + (k_2 + k_c)x_2 - k_c x_1 - F_2 = 0 \qquad (4.137b)$$

Note that the preceding equations have the same form as the dynamic equation of the single-degree-of-freedom case, that is, Eq. (A.1) in the Appendix. Hence, the same analysis applies to Eqs. (4.137a) and (4.137b). Specifically, the concept of natural frequency applies here too. In fact, for a two-degree-of-freedom system, there will be two natural frequencies. To see this, let us write Eqs. (4.137a) and (4.137b) in the matrix form with external forces set to zero to analyze the free-vibration case:

$$\begin{bmatrix} m_1 & 0 \\ 0 & m_2 \end{bmatrix} \{ \ddot{x}_1 \; \ddot{x}_2 \} + \begin{bmatrix} k_1 + k_c & -k_c \\ -k_c & k_2 + k_c \end{bmatrix} \begin{Bmatrix} x_1 \\ x_2 \end{Bmatrix} = \begin{Bmatrix} 0 \\ 0 \end{Bmatrix} \qquad (4.138)$$

Just as in Section A.1 of the Appendix where we consider the free-vibration solution, if we assume harmonic responses $x_1 = A_1\sin(\omega_n t) + B_1\cos(\omega_n t)$ and $x_2 = A_2\sin(\omega_n t) + B_2\cos(\omega_n t)$, we get

$$\begin{bmatrix} k_1 + k_c - \omega_n^2 m_1 & -k_c \\ -k_c & k_2 + k_c - \omega_n^2 m_2 \end{bmatrix} \begin{Bmatrix} x_1 \\ x_2 \end{Bmatrix} = \begin{Bmatrix} 0 \\ 0 \end{Bmatrix} \qquad (4.139)$$

The preceding equation has a trivial solution, $x_1 = x_2 = 0$, which is of no interest because it corresponds to the motionless case. It has a nontrivial solution only if the determinant of the matrix is singular. Equating the determinant to zero yields a quadratic equation in $\omega_n^2$. Its two solutions, the *eigenvalues*, give the natural frequencies of the system: $\omega_{n1}$ and $\omega_{n2}$. We also now define the concept of *eigenvectors* and *mode shapes*.

Eigenvectors are the solutions $\{ x_1 \;\; x_2 \}^T$ of Eq. (4.139) when we substitute $\omega_{n1}$ and $\omega_{n2}$ for $\omega_n$. This system has two eigenvectors because it has two degrees of freedom. Another name for eigenvectors in the context of dynamics of solids is *mode shapes*. They are called mode shapes because they indicate the mode in which the system would vibrate when left free after an initial disturbance. The mode shapes form the basis for any motion of the system. That is, any motion $\{ x_1 \;\; x_2 \}^T$ of the system can be expressed as a linear

**Example 4.10**

Derive the natural frequencies and mode shapes of the two-degree-of-freedom model of the micromechanical filter in Figure 4.53(b).

**Solution:** Because the two beams of the micromechanical filter in Figure 4.53(a) are identical, we can assume that $k_1 = k_2 = k$ and $m_1 = m_2 = m$. Then, Eq. (4.139) can be rewritten as

$$\begin{bmatrix} k + k_c - \omega_n^2 m & -k_c \\ -k_c & k + k_c - \omega_n^2 m \end{bmatrix} \begin{Bmatrix} x_1 \\ x_2 \end{Bmatrix} = \begin{Bmatrix} 0 \\ 0 \end{Bmatrix} \qquad (4.140)$$

Let $k = 5$ N/m, $k_c = 0.4$ N/m, and $m = 3 \times 10^{-10}$ kg. Then, Eq. (4.140) gives

$$
\begin{bmatrix} 5.4 - 3 \times 10^{-10}\omega_n^2 & -0.4 \\ -0.4 & 5.4 - 3 \times 10^{-10}\omega_n^2 \end{bmatrix} \begin{Bmatrix} x_1 \\ x_2 \end{Bmatrix} = \begin{Bmatrix} 0 \\ 0 \end{Bmatrix} \tag{4.141}
$$

By equating the determinant of the matrix in the preceding equation to zero (as is necessary to get the nontrivial solution), we get a quadratic equation in terms of $\omega_n^2$:

$$
\left(5.4 - 3 \times 10^{-10}\omega_n^2\right)^2 - (0.4)^2 = 0 \Rightarrow 9 \times 10^{-20}\omega_n^4 - 32.4 \times 10^{-10}\omega_n^2 + 29 = 0 \tag{4.142}
$$

Thus, $\omega_n^2 = 1.6667 \times 10^{10}$ or $1.9333 \times 10^{10}$, and the two natural frequencies are $\omega_{n1} = 0.1291 \times 10^6$ rad/s and $\omega_{n2} = 0.1390 \times 10^6$ rad/s.

When we substitute $\omega_n = \omega_{n1} = 0.1291 \times 10^6$ rad/s (or 20.55 kHz) into the equation given by the first or the second row of Eq. (4.141) we get

$$
s_{12} = s_{11} \tag{4.143a}
$$

where $x_1 = s_{11}$ and $x_2 = s_{12}$. Similarly, with $\omega_n = \omega_{n2} = 0.1390 \times 10^6$ rad/s (or 22.12 kHz) in the first or the second row of Eq. (4.141), we get

$$
s_{22} = -s_{21} \tag{4.143b}
$$

where $x_1 = s_{21}$ and $x_2 = s_{22}$. By taking $s_{11} = s_{21} = 1$, Eqs. (4.143a) and (4.143b) can be used to write the *mode shapes* of this two-degree-of-freedom system as:

$$
\mathbf{s}_1 = \begin{Bmatrix} s_{11} \\ s_{12} \end{Bmatrix} = \begin{Bmatrix} 1 \\ 1 \end{Bmatrix} \text{ and } \mathbf{s}_2 = \begin{Bmatrix} s_{21} \\ s_{22} \end{Bmatrix} = \begin{Bmatrix} 1 \\ -1 \end{Bmatrix}
$$

From the mode shapes $\mathbf{s}_1$ and $\mathbf{s}_2$, we can notice that the two masses move together in the same direction in the first mode shape, whereas they move in opposite directions in the second mode shape. Thus, mode shapes indicate two distinct ways in which the masses can vibrate. Now, any motion of the two masses, that is, $\{ x_1 \quad x_2 \}^T$, can be written as:

$$
\begin{Bmatrix} x_1 \\ x_2 \end{Bmatrix} = q_1 \mathbf{s}_1 + q_2 \mathbf{s}_2 = q_1 \begin{Bmatrix} 1 \\ 1 \end{Bmatrix} + q_2 \begin{Bmatrix} 1 \\ -1 \end{Bmatrix} \tag{4.144}
$$

where $q_1$ and $q_2$ are called the *modal amplitudes*. This implies that any motion of the micromechanical filter can be decomposed into its two mode shapes and hence can be written as a *linear combination* of its mode shapes

combination of the two mode shapes. Let us consider the micromechanical filter example to understand this concept.

As an extension of the analysis in Example 4.10, we can consider an *n*-degree-of-freedom system that has *n* natural frequencies and *n* mode shapes corresponding to them. A continuous system will have infinite frequencies and mode shapes because it has infinite degrees of freedom. Determining the natural frequencies and mode shape in the manner of Example 4.10 is called *modal analysis*. It is a useful tool in dynamic analysis because it enables us to transform the original motion coordinates ($x_1$ and $x_2$ in the case of the filter) to *modal coordinates* (i.e. modal amplitudes $q_1$ and $q_2$).

**Your Turn:**
We noted in this section that modal analysis helps us transform the original motion coordinates to modal coordinates and that this is a useful tool in dynamic analysis. However, we did not discuss why and how it is useful. In order to see that, we need to know the properties of mode shapes. They are orthogonal to each other and form the *basis* for the motion of the system. The modal coordinates decouple the equations of motion so that the dynamic equation of a modal coordinate does not involve other modal coordinates. Thus the analysis in the Appendix for the single-degree-of-freedom can be readily used for the entire system of a many-degree-of-freedom system. Or, we can only select a few modes of interest when we know that the response of a system is dominant around a few natural frequencies. Read any standard textbook on vibrations ([1], [2]) to understand the subtleties and uses of modal analysis.

We now return to the dynamic analysis of the micromechanical filter to explain how it works as a band-pass filter. The preceding discussion of modal analysis tells us that the free vibration of a two-degree-of-freedom model of the filter in Figure 4.53 (b) consists of two terms corresponding to its two natural frequencies. When this system is set into forced vibration with a sinusoidal actuation force, we will see that the system's motion is very strong whenever the frequency of the actuation is close to the system's natural frequency (see Figure A.4 in the Appendix). Using the analysis in Section A.3 of the Appendix, we will see that the motion of the second mass of the filter's lumped model of Figure 4.53 (b) will be very strong at two natural frequencies of this system whose first mass is excited with a sinusoidal force. If we add the frequency responses [see Figure (A.4)], we get a combined frequency response as shown in Figure 4.54. This response is simply a sum of two individual frequency responses

$$\frac{x_{\text{output}}}{(F_0/k)} = \frac{1}{\sqrt{\left[1 - (\omega/\omega_{n1})^2\right]^2 + [2\zeta(\omega/\omega_{n1})]^2}} + \frac{1}{\sqrt{\left[1 - (\omega/\omega_{n2})^2\right]^2 + [2\zeta(\omega/\omega_{n2})]^2}} \quad (4.145)$$

whose terms are defined in the Appendix and where the modal amplitudes corresponding to the two modes are assumed equal. The natural frequencies from Example 4.10 were used to draw the frequency response plot in Figure 4.54 showing the normalized output. The damping factor $\zeta$ was 0.04. We can see from the figure that the motion of the second mass is significantly stronger than the rest over only a small band of frequencies of excitation of the first mass. Thus, this simple system of three beams can act as a band-pass filter!

**Your Turn:**
Notice in Figure 4.54 that the there are ripples in the total response of the filter within the passband range of the frequency. This is common in all filters. The parameters of the system need to be adjusted to minimize the ripples. Notice also that the passband too depends on the parameters of the system. In practice, one wishes to have a large passband with minimal ripples. Is it possible to achieve the best of both? Explore using Eq. (4.145). Think beyond what is discussed here and try to understand the tradeoffs in designing band-pass filters.

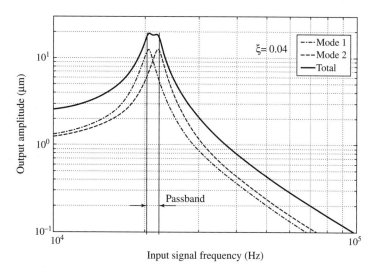

**Figure 4.54** Frequency response of the micromechanical filter in Example 4.10.

We have used the vibrating gyroscope and the micromechanical filter as examples to illustrate some important concepts and techniques of dynamic modeling. We now introduce the continuum view of dynamics so that interested readers can relate the continuum modeling to lumped modeling, which is an important focus of this chapter.

### 4.12.3 Dynamics of Continuous Elastic Systems

We briefly introduce here the governing equations for bars and beams in axial and bending modes. This will help the reader relate the static aspects discussed earlier in the chapter to the dynamic situations. It will also help in understanding the dynamic finite element equations described in Chapter 5.

Just as we wrote the equation of motion, Eq. (4.119), for the single-degree-of-freedom lumped model in Figure 4.45 by simply equating the forces acting on it, we can do the same for the bar. In a continuous system, we consider a differentially small segment along the bar and write the force–balance equation for it. We will have an inertial force because of the dynamic effect in addition to the axial external force, $f(x)$, and internal forces (stress times area of the cross section, i.e. $\sigma A$) on either side of the segment; see Figure 4.55). The inertia force is given by $(\rho_m A\, dx)\ddot{u}$ where $\rho_m$ is the mass density and $\ddot{u}$ the axial acceleration of the differential segment. Damping is neglected in this analysis.

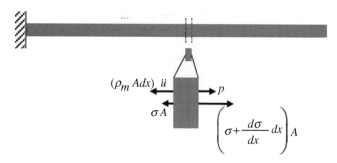

**Figure 4.55** A bar of uniform cross-section under axial loading with forces on its differential element.

The balance of all the forces shown in Figure 4.55 gives

$$(\rho_m A\,dx)\ddot{u} + \sigma A = p + \left(\sigma + \frac{d\sigma}{dx}dx\right)A \tag{4.146a}$$

which, upon substituting $\sigma = Y\frac{du}{dx}$ (recall Hooke's law and the definition of strain), gives

$$\rho_m A\ddot{u} - YA\frac{d^2u}{dx^2} - p = 0 \tag{4.146b}$$

The preceding equation is the differential equation that governs the dynamics of the bar. For free vibrations, $p = 0$. If we assume that $u(x, t) = s(x)\,q(t)$, that is, that the spatial and time-dependent parts are separable, Eq. (4.146b) gives

$$\rho_m As(x)\ddot{q}\,(t) - YAq(t)s''(x) = 0 \Rightarrow \frac{\ddot{q}(t)}{q(t)} = \frac{YAs''(x)}{\rho_m As(x)} = \text{constant} \tag{4.147}$$

where $s''(x)$ denotes the second derivative of $s(x)$ with respect to $x$. We have shown that the ratio in the preceding equation is constant because one ratio is entirely time-dependent while the other is dependent on a spatial variable, $x$. Thus, the ratio must be constant. Let this constant be $(-\omega^2)$. With this, Eq. (4.147) gives

$$\frac{\ddot{q}(t)}{q(t)} = -\omega^2 \Rightarrow \ddot{q}(t) + \omega^2 q = 0 \tag{4.148a}$$

$$\frac{Ys''(x)}{\rho_m s(x)} = -\omega^2 \Rightarrow Y s''(x) + \omega^2 \rho_m\, s(x) = 0 \tag{4.148b}$$

Equation (4.148b) has infinite solutions of the form $\sin\left(\frac{\omega\sqrt{\rho_m}}{\sqrt{Y}}x\right)$ or $\cos\left(\frac{\omega\sqrt{\rho_m}}{\sqrt{Y}}x\right)$ which are indeed the mode shapes of the bar. The values of $\omega$ are the natural frequencies. Since a continuous system has infinite degrees of freedom, it has infinite natural frequencies and corresponding mode shapes. Equation (4.148b) is called the *eigenequation* for a bar.

Similarly, we can first write the equation of motion for a beam and then derive its eigenequation by considering free vibrations. This is discussed in Chapter 5 along with finite element techniques for solving it.

## 4.12.4 A Note on the Lumped Modeling of Inertia and Damping

We have discussed the lumped stiffness of a continuous slender structure consisting of axially deforming bars and transversely bending and twisting beams. Such lumped stiffness values are needed for the $k$'s, the spring constants, in the lumped model dynamic equation in Eq. 4.119. However, that equation also has $m$ and $b$ in it. A question that naturally arises is: how do we derive the spring constants for continuous structures? To answer this question, we should look at the lumped stiffness, $k$, slightly differently.

When we compute $k$ as the force divided by the displacement it caused at the same point, we are essentially equating the strain energy stored in the elastic continuum (be it a bar or a beam) with that stored in the lumped spring. In other words, for a general elastic continuum,

$$\frac{1}{2}k\Delta^2 = \int\limits_{dV} \frac{\sigma \in}{2}\,dV \tag{4.149}$$

The reasoning underlying this argument is that *the strain energy stored in an elastic system is half of the work done by the external forces at static equilibrium*. This is known as the *Clayperon's theorem* in elastic mechanics. One may readily verify this for a linear spring: $(0.5\,F\Delta) = (0.5\,k\Delta^2)$ because $F = k\Delta$.

---

**Your Turn:**

We have used quite a few energy methods in this chapter and Chapter 6 of this book. *The calculus of variations* is the mathematical concept that underlies energy methods. Read up on this, as you will find it rewarding in more ways than one. For example, proving *Clayperon's theorem* and other energy theorems used in this book become easy once you understand the calculus of variations.

---

An alternative way of thinking about lumped $k$ from the energy viewpoint enables us to do the same for other energies. If we do that for the kinetic energy, we get the lumped mass $m$ (lumped inertia, in general). When we consider dissipated energy we get $b$, the lumped damping coefficient. Similar arguments work for electrical resistance, capacitance, inductance, magnetic reluctance, thermal resistance, capacitance, and so on. We discuss these in Chapter 6. In particular, lumped inertia modeling is discussed in Section 6.4.2 and lumped damping modeling in Sections 6.4.3 and 6.5.

## ▶ 4.13 SUMMARY

We have discussed a wide variety of modeling concepts for solids in this chapter. The discussion has been limited to only issues relevant for microsystems and some smart materials systems. A great deal more is needed to model the behavior of moving solids accurately. We included here only those concepts that can be solved using simple algebra and calculus. Indeed, the techniques presented here are adequate for a quick stiffness analysis of most microsystems problems. We started with a humble axially deforming bar and ended with the governing equations for the deformation of 3D solids. Yet, what we have seen it is only the tip of the iceberg!

As we saw in Section 4.11, partial differential equations are inevitable when we consider structures of general shape. While we are able to solve most of the equations in this chapter analytically, it is not possible to solve the general equations (mainly partial differential equations) analytically. Hence, we may have no choice but to resort to numerical methods. Gear up to learn one such method in the next chapter.

## ▶ REFERENCES

1. Thomson, W.T. (1988) *Theory of Vibration*, Prentice Hall, Upper Saddle River, NJ USA.
2. Meirovitch, L. (1986) *Elements of Vibration Analysis*, McGraw-Hill, New Delhi.

## ▶ FURTHER READING

1. Timoshenko, S. (1983) *Strength of Materials*, Parts 1 and 2, 3rd ed., Krieger Publishing Co., Malabar, FL, USA.
2. Den Hartog, J.P. (1961) *Strength of Materials*, Dover Publications, Mineola, NY, USA.
3. Chandrupatla, T. R.and Belegundu, A.D. (1997) *Introduction to Finite Elements in Engineering*, 2nd ed., Prentice Hall, Upper Saddle River, NJ, USA.

▶ **EXERCISES**

4.1 Figure 4.56 shows three variants of a bar of square cross section in a resonant-mode force sensor. An axial force of 100 μN acts on it in all the three cases. Find the elongation of the bar and the maximum stress and strain in all three cases if the length of the bar is 200 μm. Take Young's modulus of silicon to be 150 GPa.

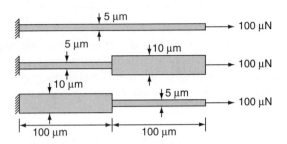

**Figure 4.56** A bar under axial load with three variants.

4.2 Suspension of an accelerometer is schematically shown in Figure 4.57. There are three beams on either side of the square proof mass of 50 μg mass. The beams are identical. They are 150 μm long and have an in-plane width of 8 μm and an out-of-plane thickness of 2 μm. They are made of polysilicon whose Young's modulus is 169 GPa.
  (a) How much does the proof mass move in the x-direction for 1 g (9.8 m/s²) acceleration in that direction?
  (b) How much does the proof mass move in the y-direction for 1 g (9.8 m/s²) acceleration in that direction?
  (c) How much does the proof mass move in the x-direction and y-direction for 1 g (9.8 m/s²) acceleration at an angle of 45° to the x-direction?

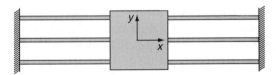

**Figure 4.57** Suspension of the proof mass by six identical beams in an accelerometer.

4.3 Figure 4.58 shows a spring steel cantilever beam (Young's modulus = 210 GPa) over which there is a piezopatch (Young's modulus = 300 GPa). The lengths of the beam and the piezopatch are shown in the figure. The in-plane width (25 μm) and out-of-plane thickness (5 μm) of the beam and piezopatch are the same. If the piezopatch contracts axially by 1%, compute the deflection of the free tip of the spring steel beam in the axial and transverse directions.

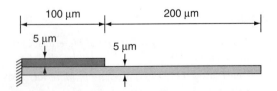

**Figure 4.58** A spring steel beam with a piezopatch over one-third of its length.

4.4 Figure 4.59(a) shows a rigid mass suspended from one cantilever and Figure 4.59(b) shows the same with two cantilevers. If the same force of 1 mN acts in both cases, compute the deflection of the mass in the direction of the force. Sketch the deformed profiles schematically in both the cases. The beams are 250 µm long and have out-of-plane thickness of 3 µm and in-plane width of 4 µm. The cross-sections of the beams are rectangular. Take $Y = 169$ GPa.

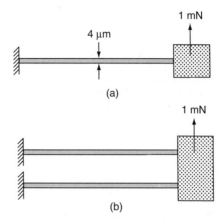

**Figure 4.59** (a) A cantilever beam with an end load applied on a rigid mass attached to the beam, (b) two beams attached to a rigid mass on which a force is applied.

4.5 A research lab tried to make a 314 µm long cantilever beam with diamond-like-carbon (DLC) thin-film technology. The beam had rectangular cross section with 15 µm in-plane width and 5 µm out-of-plane thickness. In its released condition, it was noticed that it had curled into a semicircular form. The Young's modulus and Poisson's ratio of DLC were measured to be 100 GPa and 0.3, respectively.
(a) What is the residual stress gradient that caused this curling?
(b) What is the anticlastic curvature along the in-plane width direction? Draw that shape.

4.6 A silicon nitride beam of rectangular cross-section in a microscale flow-meter is fixed at both the ends. It has length 800 µm, in-plane width 10 µm, and out-of-plane thickness 2 µm. There is an aluminum resistor over this beam all along its length, but the thickness of the aluminum beam is 1 µm and its width is only 5 µm. It is placed symmetrically along the in-plane width direction as shown in Figure 4.60. If there is a distributed transverse load of 20 N/m along the length of this composite beam, what is the transverse deflection of its midpoint? (Take the Young's modulus of silicon nitride to be 200 GPa and that of aluminum to be 70 GPa.)

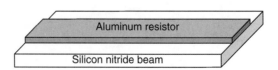

**Figure 4.60** A narrow aluminum beam over a silicon nitride beam.

4.7 A diamond-shaped micromirror is suspended between two torsional aluminum beams of square cross-section as shown in Figure 4.61. The length of each torsional beam is 100 µm and the side of the cross-section is 2 µm. When there is force in the direction perpendicular to the plane of the diamond-shaped beam at one of its free vertices, it tilts by 5° about the axis of the torsional beams. What is the magnitude of the force? The Young's modulus of aluminum is 70 GPa and Poisson's ratio is 0.25.

**Figure 4.61** Schematic of a diamond-shaped micromirror suspended by two torsional beams.

4.8 In a plane-strain condition, all the strains involving the $z$-direction are zero. Derive the stress-strain relationship between stresses and strains in the $x$- and $y$-directions.

4.9 A permanent magnet is attached at the midpoint of a fixed-fixed beam (see Figure 4.62 ). Due to an external magnetic field, there is a couple of $100\,\mu N \cdot \mu m$ on the magnet about the direction perpendicular to the plane of the magnet. If the beam is $500\,\mu m$ long and has circular cross section of diameter $10\,\mu m$, what and where is the maximum stress induced in the beam?

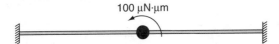

**Figure 4.62** A magnetic couple acting at the midpoint of a fixed–fixed beam.

4.10 A microvalve has a circular poppet (shown filled in Figure 4.63) suspended by four semicircular beams of radius $50\,\mu m$ and square cross-section $2\,\mu m$/side. Compute the stiffness of the poppet in the direction perpendicular to the plane of the disk. It is made of steel with a Young's modulus of $210\,GPa$.

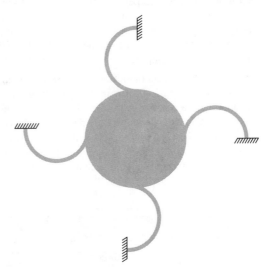

**Figure 4.63** A circular poppet suspended by four semicircular beam.

# CHAPTER 5

# The Finite Element Method

## LEARNING OBJECTIVES

After completing this chapter, you will be able to:

▶ Understand numerical methods for solving partial differential equations.

▶ Get an overview of the variational method and weighted residual method.

▶ Understand the finite element formulation of rod, beam, and plane stress elements.

▶ Learn finite element modeling for piezoelectric structures.

▶ Learn finite element modeling for the analysis of time-dependent problems.

The behavior of most micro and smart dynamic systems is governed by partial differential equations, which normally are called the *equilibrium equations* in mechanics terminology. As described in Chapter 4, defining partial differential equations can be formidable and is normally not amenable to analytical solutions, save in the simplest of cases. Hence, we seek an alternative solution philosophy in which we assume a solution involving many unknown constants, and then determine these constants so that the governing equilibrium equations are satisfied in an approximate sense. However, the assumed solution is required to satisfy a set of conditions called the *boundary conditions* and *initial conditions*. The boundary conditions are of two types: natural boundary conditions and kinematic boundary conditions. The former are also called force boundary conditions (in mechanics) or *Neumann boundary conditions* (in general) and give the value of the derivatives of the unknown variable at the boundary. The kinematic boundary conditions, also called the essential boundary conditions (in mechanics) or *Dirichlet boundary conditions* (in general), and give the value of the unknown variable at the boundary.

However, the governing partial differential equation of the system is not amenable to this solution methodology due to the requirement for higher-order continuity of the assumed solution. Hence, one needs an alternative statement of equilibrium equations that is more suited for numerical solution. This is normally provided by the variational statement of the problem. Many numerical methods can be designed based on the variational statement, and the finite element method (FEM) is one such numerical solution method.

## ▶ 5.1 NEED FOR NUMERICAL METHODS FOR SOLUTION OF EQUATIONS

Some of the differential equations common in microsystem design are discussed next. A smart system based on either bulk or microactuation usually involves the design of a cantilever plate or a beam, which is governed by the beam or plate equations used in mechanics. For example, the response of a cantilever subjected to electrical excitation perpendicular to the axis of the beam is governed by

$$YI \frac{\partial^4 w(x,t)}{\partial x^4} + q(x,t) = \rho A \frac{\partial^2 w(x,t)}{\partial t^2} \tag{5.1}$$

Note that the left-hand side of Eq. (5.1) was derived in Chapter 4 [see Eq. (4.63)] and the procedure for deriving this equation is given in Section 4.7.1. Compared to Eq. (4.63), Eq. (5.1) has an additional term; this is due to the mass of the beam and indicates the inertial force acting on the beam.

Another difference in this equation is that $w(x,t)$, the transverse displacement of the beam, is shown to be dependent on both position and time; $YI$ the bending rigidity of the beam, which is a product of the Young's modulus $Y$ and area moment of inertia of the beam $I$; $\rho$ is the density of the beam material, and $A$ is the cross-sectional area of the beam. The term $q(x,t)$ in Eq. (5.1) is the loading on the beam, which can come from a mechanical load, electric field, magnetic field, or a combination of all these. This equation is of fourth order in spatial coordinates and second order in temporal coordinates; hence, we need four boundary conditions and two initial conditions for its solution.

We derive a procedure using variational methodology to identify the boundary conditions for a given differential equation in Section 5.3. The two initial conditions that this equation must satisfy are that displacement $w(t=0) = 0$ and velocity $\dot{w}(t=0) = dw/dt(t=0)$. It is clear that solving the above problem is not straightforward and obtaining an exact solution to Eq. (5.1) is very difficult. This is the reason why we need numerical methods. The different governing equations arising in microsystem designs that require numerical solution are discussed in Chapters 4 and 6.

Modeling systems with smart sensors/actuators is more complex. The additional complexity arises from the electric and/or magnetic fields that are coupled to the dependent variable in the governing equation of a smart system. The coupling gives rise to additional terms that need to be handled differently. This is addressed in Section 5.7 of this chapter.

### 5.1.1 Numerical Methods for Solution of Differential Equations

Many numerical methods have been reported in the literature for different classes of problems involving microsystems and smart systems. For fluid-related problems as in microfluidic devices or in problems involving design of RF devices, which require the solution of Maxwell's equation, the most common numerical method used is the *finite difference method* (FDM). In this method, all the differential quantities in a differential equation are converted to a difference equation using a suitable finite difference scheme. First, the domain is split into many subdomains or cells, then the differential equation is converted to a difference equation over each of these cells, then the converted difference equation is solved, and finally the solution is marched until the whole domain is covered.

Finite difference schemes can be implicit or explicit. The explicit scheme is a single-step scheme that does not require solution of matrix equations. The most commonly used explicit FDM is the *central difference method*, explained and derived in Section 5.2.3. The main disadvantage of explicit schemes is that a bound is put on the size of the step, since otherwise the solution obtained through such a scheme will be highly divergent. These issues are covered in texts suggested at the end of this chapter for further reading. On the other hand, the implicit scheme is a multistep method involving the solution of the matrix equation at each step. The simplest implicit scheme is the *trapezoidal rule*, whereas the most extensively used implicit scheme in FEM, especially for time-dependent problems, is the *Newmark-β* method. The major advantage of this scheme is that it is unconditionally stable. This method is not discussed in this chapter.

The most common numerical method is the *finite element method* (*FEM*). This method assumes a solution that satisfies only the kinematic or essential boundary conditions. There are different flavors of FEM, such as *Galerkin FEM*, *least-square FEM*, *extended FEM* (*XFEM*), etc., based on the way the assumed solution is constructed to satisfy the governing differential equation. The FEM is explained in detail in the next section.

The *spectral FEM* (*SFEM*) is yet another method based on the FEM principle. This method is normally used to solve wave-propagation problems. In this method, the unknown variable in the governing equation is transformed into the frequency domain using *fast Fourier transform* (FFT), which is a numerical version of the *discrete Fourier transform* (DFT). The finite element (FE) procedure is then applied to the transformed governing equations in the frequency domain. In essence, it can be categorized as a type of FEM. Details of this method can be found in [1].

For problems involving discontinuous domains such as those involving cracks and holes, solution of the governing equation by the *boundary element method* (BEM) is quite popular. It is well known that the domain over which a numerical method is employed comprises of a boundary and a volume. Here, the governing equation is transformed into an equation involving only the boundary of the domain, using the divergence theorem and the Stokes theorem (see [2] for more details). The boundary is then discretized as in FEM. The volume part of the differential equation is forced to zero, and its solution yields the *fundamental solution*. The main advantage of this method over FEM is the reduction in the problem dimension—a 2D problem becomes one-dimensional (1D) and the 3D problem becomes 2D—which significantly reduces the computational effort. The main disadvantage of this method is that the fundamental solution, that is the solution of the volume integral part, is available only for a few selected problems. This method is described in [3].

In addition to these methods, there are other methods in which the domain of interest is never discretized. Instead, the nodes are scattered all over the domain and the governing equation is solved by FEM in the least-square sense. This method is normally referred to as *meshless method*. See [4] for details of this method.

The mother of all the methods mentioned above is the *weighted residual technique* (WRT) *method*. The basic idea is the following: as in all the methods, a solution is assumed. Since the solution is only approximate, substituting it in the governing equations will not be exact and we are left with a residue. This residue is now minimized using a variational statement to obtain the final solution. The variational statement involves certain weights, and the choice of these weights determines the type of the numerical solution. Most of the methods described earlier can be deduced from WRT. The details of the method are explained in Section 5.2.3.

## 5.1.2 What is the Finite Element Method?

The FEM is a stiffness-based method in which the entire problem domain [Figure 5.1(a)] is divided into subdomains called *elements*. The set of elements can make up the domain, as shown in Figure 5.1(b). Each element is described by a set of *nodes* whose connectivity completely describes the element in FE space.

The domain in Figure 5.1(a) has volume *V*. The dependent variable here is $u(x,y,z,t)$, which represents displacements in mechanics problems, temperature in heat transfer problem, voltages or currents in smart actuation problems, or velocities and pressures in microfluidic problems. The domain here has two boundaries, namely $\tau_1$, where the

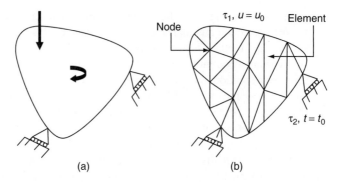

**Figure 5.1** (a) Original problem domain; (b) problem domain discretized into subdomains called elements.

dependent variable $u(x,y,z,t) = u_0$ is specified, and $\tau_2$, where the derivatives of the dependent variable represented by $t = t_0$ are specified. The physical meaning of the variable $t$ is that it represents *traction* in mechanics, temperature gradient in heat transfer problems, pressure gradient in microfluidic analysis, or electric (or magnetic) flux in microsensing/actuation analysis.

The FE procedure begins with the assumption of the displacement field in each of the elements, and this assumed field is then substituted into the weak form of the differential equation. The *weak form* is the integral form of the original differential equilibrium equation. If the gradients (or the derivative of the dependent variable) are high (for instance, near the crack tip in a cracked structure), then one needs very fine mesh discretization. Incidentally, there are other considerations in selecting the size of the mesh. In wave-propagation analysis (e.g. electromagnetic analysis in RF devices), many higher-order modes are excited due to the high-frequency content of the input signal. At these frequencies, the wavelengths are small; the mesh size should be of the order of the wavelengths so that the mesh edges do not act as free boundaries and start reflecting waves from these edges. These small mesh sizes increase the problem size enormously. However, it is to be noted that the size of the mesh is an important parameter in determining the accuracy of the solution.

The variational statement (explained in the next section) is the equilibrium equation in integral form, which is often referred to as the *weak form* of the governing equation. The alternate statement of equilibrium for structural systems is provided by the energy functional governing the system. According to the *calculus of variation*, any differential equation that is *self-adjoint* (refer to [5]–[8])[1] has a quadratic functional. However, in most physical systems governed by the laws of physics, the governing differential equation will have a functional. In microsystems, it is called the *energy functional*, as it describes the energy stored by the system due to external forces acting upon the system. For example, in structural systems, the energy functional is given as the sum of the total energy, which includes the strain energy (explained in the next section), the work done by the external forces on the boundary, and the other concentrated loads. The objective here is to obtain an approximate solution of the dependent variable $u$ (which can represent displacement in

---

[1] Before defining the self-adjoint property, we consider two functions $p$ and $q$. We define the *inner product* between these two functions as $(p, q) = \int_V pq \, dV$, where $V$ is the volume of the domain. Any differential equation can be represented as $Lu = f$, where $L$ is the differential operator, $u$ the dependent variable, and $f$ the forcing function. We consider another variable $v$ that could also satisfy the differential equation $Lv = f$. We define the differential operator $L$ as *self-adjoint* if the following inner product rule is satisfied: $(Lu, v) = (u, Lv)$.

structural systems, velocities in fluid systems or electric field in electrical or electro-magnetic systems) in the form

$$u(x, y, z, t) = \sum_{n=1}^{N} a_n(t) \psi_n(x, y, z) \tag{5.2}$$

where $a_n(t)$ ($n = 1$ to $N$) are the unknown time-dependent coefficients to be determined through some minimization procedure, and $\psi_n(x,y,z)$ are the spatially dependent functions that normally satisfy the kinematic boundary conditions and not necessarily the natural boundary conditions. (Observe that we have expressed dependent variables as functions of all three spatial coordinates and time.) Different energy theorems give rise to different variational statements of the problem, and hence, different approximate methods can be constructed. As explained earlier, the basis for formulation of the different approximate methods is WRT, wherein the residual (or error) obtained by substituting the assumed approximate solution in the governing equation is weighted with a weight function and integrated over the domain.

The different approximate methods again are too difficult to use in situations where the structure is complex. To some extent, methods like the Rayleigh–Ritz [5] that involve minimization of the total energy to determine the unknown constants in Eq. (5.2) cannot be applied to some complex problems. The main difficulty here is to determine the functions $\psi_n$, called *Ritz functions*. However, if the domain is divided into a number of subdomains, it is easier to apply Rayleigh–Ritz method over each of these subdomains (each subdomain is called an element of the FE mesh) and solutions of each are pieced together to obtain the total solution. This in essence is the FEM.

The entire FE procedure can be explained by the flowchart shown in Figure 5.2. FEM converts the governing partial differential equation into a set of algebraic equations with nodal dependent variables as basic unknowns in the static case and into a set of highly coupled ordinary differential equations for time-dependent problems. The use of the Ritz function over a small subdomain (we call this the *finite element*) gives us two different matrices, the stiffness matrix and the mass matrix. In the static case, only the stiffness matrix need be generated. The procedure for developing these matrices is given in Section 5.4.

First, a FE mesh of the domain of interest needs to be created; this is fed into the FE code along with the relevant loads and the boundary constraints. Then stiffness and mass matrices are generated for each element and assembled together. The assembly of matrices ensures that the contribution of stiffness and masses from a node that is common to one or more elements are taken into account. If the problem is static in nature, we need solve only the algebraic set of equations. The algebraic equations are solved using some standard solvers such as Gaussian elimination, Cholesky decomposition, etc., the details of which can be found in [9] and [10]. Once the nodal dependent variables are obtained, secondary quantities such as derivatives of the dependent variables are obtained by post-processing the obtained nodal values of dependent variables. If the problem is time-dependent, the method of solution is quite different. There are two methods of solving dynamic equations in FEM, one based on the normal mode method and the second based on time-marching schemes. See [10] for details on the different solution methods for dynamic problems.

Although FEM is explained here as the assembly of *Ritz solutions* over each subdomain, in principle, all the approximate methods generated by the WRT can be applied to each subdomain. Hence, in Section 5.2, the complete WRT formulation and various other energy theorems are given in detail. These theorems are then used to derive

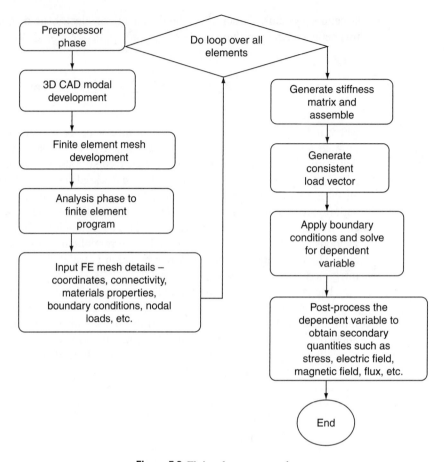

**Figure 5.2** Finite element procedure.

the discretized FE governing equation of motion. This is followed by the formulation of basic building blocks used in FEM, namely, the *stiffness* and *mass matrices*. The main issues relating to their formulation are also discussed.

## ▶ 5.2 VARIATIONAL PRINCIPLES

This section begins with some basic definitions of work, complementary work, strain energy, complementary strain energy, and kinetic energy. These are necessary to define the energy functional, which is the basis of any FE formulation. This is followed by the complete description of WRT and its use in obtaining many different approximate methods. Next, some basic energy theorems such as the principle of virtual work (PVW), principle of minimum potential energy (PMPE), Rayleigh–Ritz procedure and Hamilton's theorem for deriving the governing equation of a system and their associated boundary conditions are explained. Using Hamilton's theorem, FE equations are derived, followed by the derivation of stiffness and mass matrices for some simple finite elements.

### 5.2.1 Work and Complementary Work

Consider a body under the action of a force system described by components of forces $F_x$, $F_y$ and $F_z$ in the three coordinate directions. These components can also be time-dependent.

Under the action of these forces, the body undergoes infinitesimal deformations $du$, $dv$ and $dw$ in the three coordinate directions, where $u$, $v$, and $w$ are the components of displacements in the three coordinate directions, respectively. The work done by the force system in moving the body is given by

$$dW = F_x du + F_y dv + F_z dw \tag{5.3}$$

The total work done in deforming the body from the initial to the final state is given by

$$W = \int_{a_1}^{a_2} \left( F_x du + F_y dv + F_z dw \right) \tag{5.4}$$

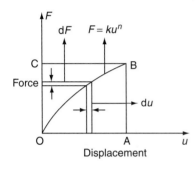

**Figure 5.3** Schematic representation of work (area OAB) and complementary work (area OBC).

where $a_2$ is the final deformation and $a_1$ the initial deformation of the body. To understand this better, consider a 1D system under the action of a force $F_x$ and having initial displacement zero. Let the force vary as a nonlinear function of displacement ($u$) given by $F_x = ku^n$, where $k$ and $n$ are known constants, as shown graphically in Figure 5.3. To determine the work done by the force, a small strip of length $du$ is considered in the lower portion of the curve. The work done by the force is obtained by substituting the force variation in Eq. (5.3) and integrating:

$$W = \int_0^u F_x du = \int_0^u ku^n du = \frac{ku^{n+1}}{n+1} = \frac{F_x u}{n+1} \tag{5.5}$$

Alternatively, work can also be defined by taking the upper portion of the curve in Figure 5.3. In this case, the work done can be defined as

$$W^* = \int_{F_1}^{F_2} \left( u\, dF_x + v\, dF_y + w\, dF_z \right) \tag{5.6}$$

where $F_1$ and $F_2$ are the initial and final applied force, respectively. This alternative definition is normally referred to as *complementary work*. Again considering the 1D system with the same nonlinear force–displacement relationship ($F_x = ku^n$), we can write the displacement $u$ as $u = (1/k)F_x^{(1/n)}$. Substituting this in Eq. (5.6) and integrating, the complementary work can be written as

$$W^* = \frac{F_x^{(1/n+1)}}{k^n (1/n + 1)} = \frac{F_x u}{(1/n + 1)} \tag{5.7}$$

Obviously, $W$ and $W^*$ are different, although they were obtained from the same force–displacement curve. However, for the linear case ($n = 1$), they have the same value $W = W^* = F_x u/2$, which is merely the area under the force–displacement curve.

A similar analogy can be made for work done in a general capacitor. The charge in a capacitor ($Q$) is related to voltage across the capacitor ($V$) through the relation $Q = CV$, where $C$ is the capacitance of the capacitor. This relationship is quite similar to the force–displacement relationship $F_x = ku^n$ with $n = 1$. Hence, in this case, we can define the work or complementary work as

$$W = W^* = \frac{1}{2}QV \tag{5.8}$$

## 5.2.2 Strain Energy and Kinetic Energy

To derive the expression for the strain energy, consider a small element of volume $dV$ of the structure in a 1D state of stress, as shown in Figure 5.4. Let $\sigma_{xx}$ be the stress on the left face and $\sigma_{xx} + (\partial \sigma_{xx}/\partial x)dx$ the stress on the right face. Let $B_x$ be the body force per unit volume along the $x$-direction. Let us first equate the forces on the two opposite sides of a cube:

$$-\sigma_{xx}\mathrm{d}y\,\mathrm{d}z + \left(\sigma_{xx} + \frac{\partial \sigma_{xx}}{\partial x}\mathrm{d}x\right)\mathrm{d}y\,\mathrm{d}z + B_x\mathrm{d}x\,\mathrm{d}y\,\mathrm{d}z = \left(\frac{\partial \sigma_{xx}}{\partial x} + B_x\right)\mathrm{d}x\,\mathrm{d}y\,\mathrm{d}z = 0$$

or $\hspace{8cm}$ (5.9)

$$\frac{\partial \sigma_{xx}}{\partial x} + B_x = 0$$

Equation (5.9) is the equilibrium equation along the $x$-direction (i.e. horizontal). The strain energy increment $dSE$ due to stresses $\sigma_{xx}$ on face 1 (right face) and $\sigma_{xx} + (\partial \sigma_{xx}/\partial x)\,dx$ on face 2 (left face), with infinitesimal deformations $du$ on face 1 and $d[u + (\partial u/\partial x)\,dx]$ on face 2, is given by

$$\mathrm{d}SE = -\sigma_{xx}\mathrm{d}y\,\mathrm{d}z\,\mathrm{d}u + \left(\sigma_{xx} + \frac{\partial \sigma_{xx}}{\partial x}\mathrm{d}x\right)\mathrm{d}y\,\mathrm{d}z\,\mathrm{d}\left(u + \frac{\partial u}{\partial x}\mathrm{d}x\right) + B_x\mathrm{d}y\,\mathrm{d}x\,\mathrm{d}z\,\mathrm{d}u$$

Simplifying and neglecting the higher-order terms, we get

$$\mathrm{d}SE = \sigma_{xx}\,\mathrm{d}\left(\frac{\partial u}{\partial x}\right)\mathrm{d}x\,\mathrm{d}y\,\mathrm{d}z + \mathrm{d}u\,\mathrm{d}x\,\mathrm{d}y\,\mathrm{d}z\left(\frac{\partial \sigma_{xx}}{\partial x} + B_x\right)$$

The last term within the brackets is the equilibrium equation [Eq. (5.9)], which is identically equal to zero. Hence, the incremental strain energy becomes

$$\mathrm{d}SE = \sigma_{xx}\,\mathrm{d}\left(\frac{\partial u}{\partial x}\right)\mathrm{d}x\,\mathrm{d}y\,\mathrm{d}z = \sigma_{xx}\mathrm{d}\varepsilon_{xx}\mathrm{d}V \tag{5.10}$$

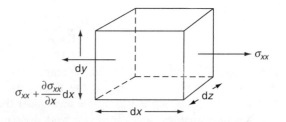

**Figure 5.4** Elemental volume for computing strain energy.

Now, we introduce the term *incremental strain energy density* defined as

$$dS_D = \sigma_{xx} d\varepsilon_{xx}$$

Integrating the above expression over a finite strain, we get

$$S_D = \int_0^{\varepsilon_{xx}} \sigma_{xx} d\varepsilon_{xx} \qquad (5.11)$$

Using the expression in Eq. (5.11) in Eq. (5.10) and integrating it over the volume, we get

$$SE = \int_V S_D dV \qquad (5.12)$$

Similar to the definition of work and complementary work, we can define complementary strain energy density and complementary strain energy as

$$SE^* = \int_V S_D^* dV, \quad S_D^* = \int_0^{\sigma_{xx}} \varepsilon_{xx} d\sigma_{xx} \qquad (5.13)$$

Kinetic energy must also be considered while evaluating the total energy when inertial forces are important. Inertial forces are dominant in time-dependent problems, where both loading and deformation are functions of time. The kinetic energy is given by the product of mass and the square of velocity. This can be mathematically represented in integral form as

$$T = \int_V \rho(\dot{u}^2 + \dot{v}^2 + \dot{w}^2) dV \qquad (5.14)$$

where $u$, $v$, and $w$ are the displacements in the three coordinate directions, respectively; the dots on the characters represent the first time derivative and in this case are the three respective velocities.

## 5.2.3 Weighted Residual Technique

Most dynamical systems that obey the laws of physics are governed by a differential equation of the form

$$Lu = f \qquad (5.15)$$

where $L$ is the differential operator of the governing equation, $u$ the dependent variable of the governing equation, and $f$ the forcing function. For example, let us consider the beam equation [Eq. (5.1)]. It can be written in the form of Eq. (5.15), where $L$ and $f$ are

$$L = \left( EI \frac{\partial^4}{\partial x^4} - \rho A \frac{\partial^2}{\partial t^2} \right), \quad f = -q(x, t)$$

The system may have two different boundaries $\tau_1$ and $\tau_2$ as shown in Figure 5.1, where the displacements $u = u_0$ and tractions $t = t_0$ are specified. In most approximate methods, we seek an approximate solution for the dependent variable $u$, say $\bar{u}$ (in 1D), as

$$\bar{u}(x, t) = \sum_{n=1}^{N} \alpha_n(t)\phi_n(x) \tag{5.16}$$

where $\alpha_n$ are some unknown constants that are time-dependent in dynamic situations, and $\phi_n$ are some known spatially dependent functions. When we use discretization in the solution process as in FEM, $\alpha_n$ will represent the nodal coefficients. In general, these functions satisfy the kinematic boundary conditions of the problem. When Eq. (5.16) is substituted into the governing equation [Eq. (5.15)], we get $L\bar{u} - f \neq 0$, since the assumed solution is only an approximate one. We can define the error functions associated with the governing equations and boundary conditions as

$$e_1 = L\bar{u} - f, \quad e_2 = \bar{u} - u_0, \quad e_3 = \bar{t} - t_0 \tag{5.17}$$

In these expressions, $e_2$ and $e_3$ are the errors associated with the kinematic and natural boundary conditions. The objective of any WRT is to make the error function as small as possible over the domain of interest and also on the boundary. Different numerical methods vary in the distribution of these errors and thus result in new approximate solution methods.

Let us now consider a case where the boundary conditions are exactly satisfied, that is, $e_2 \equiv e_3 \equiv 0$. In this case, we need to distribute the error function $e_1$ only. This can be done through a weighting function $w$ and integration of the weighted error $e_1$ over the domain $V$ as

$$\int_V e_1 w \, dV = \int_V (L\bar{u} - f)w \, dV = 0 \tag{5.18}$$

The choice of the weighting function determines the type of WRT. The weighting functions used are normally of the form

$$w = \sum_{n=1}^{N} \beta_n \psi_n \tag{5.19}$$

When Eq. (5.19) is substituted in Eq. (5.18), we get

$$\sum_{n=1}^{N} \beta_n \int_V (L\bar{u} - f)\psi_n = 0$$

where $n = 1, 2, 3, \ldots, N$. Since $\beta_n$ are arbitrary, we have

$$\int_V (L\bar{u} - f)\psi_n = 0$$

where $n = 1, 2, 3, \ldots, N$. This process ensures that the number of algebraic equations after using Eq. (5.16) for $\bar{u}$ is equal to the number of unknown coefficients chosen.

We now demonstrate how different numerical schemes can be constructed by using different weighting functions. We demonstrate how FDM, the method of moments and the Galerkin method can be constructed as illustrations by using WRT on the following differential equation and its associated boundary conditions:

$$\frac{d^2u}{dx^2} + 4u + 4x = 0, \quad u(0) = u(1) = 0 \tag{5.20}$$

Here $u$ is the dependent field variable.

### 5.2.3.1 Finite Difference Method

For FDM, we choose all $\psi_n$ as Dirac delta functions, normally represented by $\delta$. These are spike functions that have unit value only at the point at which they are defined; at all other points they are zero. They have the following properties:

$$\int_{-\infty}^{\infty} \delta(x - x_n)dx = \int_{x-r}^{x+r} \delta(x - x_n)dx = 1$$

$$\int_{-\infty}^{\infty} f(x)\delta(x - x_n)dx = \int_{x-r}^{x+r} f(x)\delta(x - x_n)dx = f(x_n)$$

Here $r$ is any positive number and $f(x)$ any function that is continuous at $x = n$.

To demonstrate this method, consider a three-point line element as shown in Figure 5.5. Let the points $n - 1$, $n$, and $n + 1$ be located at $x = 0$, $x = L/2$, and $x = L$, respectively, where $L$ is the domain length.

The dependent variable field can be expressed as a three-term series in Eq. (5.16) as

$$\bar{u} = u_{n-1}\phi_1 + u_n\phi_2 + u_{n+1}\phi_3 \tag{5.21}$$

where $u_{n-1}$, $u_n$ and $u_{n+1}$ are the three displacements at $x = 0$, $x = L/2$ and $x = L$, respectively. Now the next step is to construct the functions $\phi_1$, $\phi_2$, and $\phi_3$ that satisfy the kinematic boundary conditions of the given problem. These functions are constructed in exactly the same way as we construct *shape functions* for FE formulation, as explained in in Section 5.4.1. We briefly show the procedure for constructing these functions here. Normally, polynomials are chosen as assumed functions for most problems, and we see from Figure 5.5 that there are three nodes located at points $n-1$, $n$, and $n+1$. Hence, the function to be assumed for $\bar{u}$ should be quadratic and can be taken as

$$\bar{u} = a_0 + a_1x + a_2x^2$$

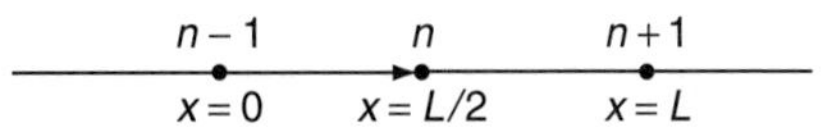

**Figure 5.5** Finite difference as WRT.

Substituting $\bar{u}(x = 0) = u_{n-1}, \bar{u}(x = L/2) = u_n, \bar{u}(x = L) = u_{n+1}$ in this equation and simplifying, we can write the resulting equation in the form of Eq. (5.21), with the functions $\phi_1$, $\phi_2$ and $\phi_3$ given by

$$\phi_1 = \left(1 - \frac{x}{L}\right)\left(1 - \frac{2x}{L}\right), \quad \phi_2 = \left(\frac{4x}{L} - \frac{4x^2}{L^2}\right), \quad \phi_3 = \frac{x}{L}\left(\frac{2x}{L} - 1\right) \tag{5.22}$$

The weighting function can now be taken as

$$w = \beta_1\delta(x - 0) + \beta_2\delta(x - L/2) + \beta_3\delta(x - L) = \sum_{n=1}^{3}\beta_n\delta_n \tag{5.23}$$

Now, let us use this scheme in Eq. (5.20). Here, the independent variable $x$ has limits between 0 and 1. Using Eq. (5.21) in Eq. (5.20), we find that the error function or residue $e_1$ is given by

$$e_1 = \left(\frac{d^2u}{dx^2} + 4u + 4x\right)_n = \left(\frac{1}{L^2}u_{n-1} - \frac{2}{L^2}u_n + \frac{1}{L^2}u_{n+1}\right) + 4u_n + 4x_n \tag{5.24}$$

where $L = 1$ is the domain length. If we now substitute the weighting function Eq. (5.23) and integrate, using the properties of the Dirac delta function, we get

$$\left(\frac{1}{L^2}(u_{n-1} - 2u_n + u_{n+1})\right) + 4u_n + 4x_n = 0 \tag{5.25}$$

Equation (5.25) is the equation for the *central finite differences*. The first term in Eq. (5.25) within the parentheses represents the second-order derivative in $u$, expressed in terms of its value at steps $n - 1, n$, and $n + 1$. Note that this equation is traditionally derived using Taylor series expansion (see [2] for more details), but we have derived the same equation here using WRT.

### 5.2.3.2 Method of Moments

Next, we derive the *method of moments* from WRT. This can be done by assuming weight functions of the form

$$w = \beta_1 + \beta_2x + \beta_3x^2 + \beta_4x^3 + \cdots = \sum_{n=0}^{N}\beta_nx^n \tag{5.26}$$

Let us again consider the problem in Eq. (5.20). In what follows, we consider only the first two terms in the above weighting function series, that is, $w = \beta_1 + \beta_2x$, and also assume the dependent field variable $u$ to be

$$\bar{u} = \alpha_1x(1 - x) + \alpha_2x^2(1 - x) \tag{5.27}$$

Note that the functions $x(1 - x)$ and $x^2(1 - x)$ both satisfy the boundary conditions of the problem, namely, $u(x = 0) = 0$ and $u(x = 1) = 0$. In fact, any function that satisfies the above boundary conditions is acceptable. For example, $\bar{u} = \alpha_1\sin\pi x + \alpha_2\sin3\pi x$ is also an acceptable solution as it satisfies the boundary conditions exactly. The unknown coefficients $\alpha_1$ and $\alpha_2$ are evaluated through a weighting process, as explained now. Substituting Eq. (5.27) in Eq. (5.20) yields the following error function:

$$e_1 = \alpha_1(-2 + 4x - 4x^2) + \alpha_2(2 - 6x + 4x^2 - 4x^3) + 4x \tag{5.28}$$

Next, we outline the procedure for evaluating the unknown constants $\alpha_1$ and $\alpha_2$. We need to weight the error obtained from Eq. (5.28) with each term of the assumed weight function ($w = \beta_1 + \beta_2 x$), which is 1 associated with $\beta_1$ and $x$ associated with $\beta_2$. If we weight this residual, we get the following equations:

$$\int_0^1 1 \cdot e_1 dx = 2\alpha_1 + \alpha_2 = 3$$

$$\int_0^1 x \cdot e_1 dx = 5\alpha_1 + 6\alpha_2 = 10$$

Solving these two equations, we get $\alpha_1 = 8/7$ and $\alpha_2 = 5/7$. Substituting these in Eq. (5.27), we get the approximate solution to the problem as

$$\bar{u} = \frac{8}{7}x(1 - x) + \frac{5}{7}x^2(1 - x)$$

The exact solution to Eq. (5.20) is given by

$$u_{\text{exact}} = \frac{\sin 2x}{\sin 2} - x$$

Comparing the results, say, at $x = 0.2$, we get $\bar{u} = 0.205$, $u_{\text{exact}} = 0.228$. The percentage error involved in the solution is approximately 10, which is very good considering that only two terms were used in the weight function series. The accuracy of the solution can be increased by using more terms in Eqs. (5.26) and (5.27).

### 5.2.3.3 The Galerkin Method

The procedure for deriving the *Galerkin technique* from the weighted residual method is now outlined. In this case, the weighting function variation is the same as that of the dependent variable variation [Eq. (5.16)], that is

$$w = \beta_1\phi_1 + \beta_2\phi_2 + \beta_3\phi_3 + \cdots \tag{5.29}$$

Let us now consider the same problem [Eq. (5.20)] with the assumed displacement field given by Eq. (5.27) and solve the problem by the Galerkin technique. The weight function variation being same as the dependent field variable variation, we need to consider the first two terms in the series for $w$ as

$$w = \beta_1\phi_1 + \beta_2\phi_2 = \beta_1 x(1 - x) + \beta_2 x^2(1 - x) \tag{5.30}$$

The residual $e_1$ is same as that in the previous case [Eq. (5.28)]. Weighting this residual with the weighting function by Eq. (5.29) yields the following equations:

$$\int_0^1 \phi_1 \cdot e_1 \, dx = 6\alpha_1 + 3\alpha_2 = 10$$

$$\int_0^1 \phi_2 \cdot e_1 dx = 21\alpha_1 + 20\alpha_2 = 42$$

Solving the above equations, we get $\alpha_1 = 74/57$ and $\alpha_2 = 74/57$. The approximate Galerkin solution then becomes

$$\bar{u} = \frac{74}{57}x(1-x) + \frac{42}{57}x^2(1-x)$$

The result obtained for $x = 0.2$ is 0.231, which is very close to the exact solution (only 1.3% error). We can design various approximate schemes in a similar manner by assuming different weighting functions.

#### 5.2.3.4 FEM as WRT

FEM is one such WRT in which the displacement variation and the weight functions are same. That is, the solution obtained through the Galerkin technique is essentially the FE solution. However, for FEM, the WRT solution is converted to a weak form of the governing equation by integrating the WRT equation [Eq. (5.18)] by parts; the resulting integral equation is called the *weak form* of the governing equation. In this form the dependent variable variation [Eq. (5.16)] is substituted and minimized using the variational theorem. This process is repeated to all the elements or subdomains, which are synthesized to obtain an algebraic set of equations in terms of nodal variables. The solution to these equations yields an approximate solution over the domain. That is, FEM reduces a differential equation to a set of algebraic equations in the case of loads that do not depend on time. These are illustrated in the next section.

### 5.2.4 Variational Symbol

In most approximate methods based on variational theorems, including the FE technique, it is necessary to minimize the functional, and this minimization process is normally represented by a variational symbol (normally referred to as a *delta operator*) represented as $\delta$. (Although same symbol is used for both the Dirac delta function and the variational operator, their meaning is different. The meaning of the Dirac delta function was explained in Section 5.2.3.1.)

Consider a function represented as $F(w, w', w'')$ that is a function of the dependent variable $w$ and its derivatives, where primes (') and ('') indicate the first and the second derivatives, respectively. For a fixed value of the independent variable $x$, the value of the function depends on $w$ and its derivatives. If the dependent variable $w$ changes to $\alpha p$, where $\alpha$ is a constant and $p$ a function of independent variable $x$, this change is called the *variation* of $w$ and is denoted by $\delta w$. That is, $\delta w$ represents the admissible change in $w$ for a fixed value of the independent variable $x$. At the boundary points where the value of dependent variables is specified, the variation is zero. In essence, the variational operator acts like a differential operator, and hence all the laws of differentiation are applicable here.

**Your Turn:**

At this point, it is interesting to see how the solution changes if different solutions are assumed for some of the methods defined in the preceding sections for solving Eq. (5.20). Let us again consider the finite difference method and the method of moments. Two possibilities exist: either the order of the solution is increased or entirely new solutions can be assumed. For example, if the order of solution is increased, then the solution can be written as

$$\bar{u} = \alpha_1 x(1-x) + \alpha_2 x^2(1-x) + \alpha_3 x^3(1-x)$$

or

$$\bar{u} = \alpha_1 x(1-x) + \alpha_2 x^2(1-x)^2 + \alpha_3 x^3(1-x)^3$$

Both the above assumptions for the dependent variable are possible solutions as they both satisfy the boundary conditions. The reader is strongly urged to derive the solutions from the above methods and compare them with the exact solution. We expect that the solution error would be drastically reduced by using higher-order approximations for the dependent variable.

Alternatively, a completely different solution such as

$$\bar{u} = \alpha_1 \sin \pi x + \alpha_2 \sin 3\pi x$$

could be assumed, which is again a possible solution as it satisfies the boundary conditions exactly. It would be interesting to see what kind of solutions the above approximation yields.

## 5.3 WEAK FORM OF THE GOVERNING DIFFERENTIAL EQUATION

The variational method gives us the alternative statement of the governing equation—normally called the *strong form* of the governing equation—which is essentially an integral equation. It is obtained by weighting the error function of the governing equation with a weighting function and integrating the resulting expression. This process gives not only the *weak form* of the governing equation but also the associated boundary conditions (both essential and natural boundary conditions). We explain this procedure by again considering the governing equation of an elementary beam [Eq. (5.1)] without inertial loads. The strong form of the beam equation is given by

$$YI \frac{d^4 w}{dx^4} + q = 0$$

Now we are looking for an approximate solution $\bar{w}$ of form similar to that in Eq. (5.16). The error function becomes

$$YI \frac{d^4 \bar{w}}{dx^4} + q = e_1$$

If we weight this with another function $v$ (which also satisfies the boundary condition of the problem) and integrate over the domain of length $l$, we get

$$\int_0^L \left( YI \frac{d^4 \bar{w}}{dx^4} + q \right) v \, dx$$

Integrating the above expression by parts, we get

$$\left(YI\frac{d^3\bar{w}}{dx^3}v\right)\Big|_{x=0}^{x=L} - \left[\int_0^L YI\frac{d^3\bar{w}}{dx^3}\frac{dv}{dx} + qv\right]dx$$

The first term is the boundary term representing the shear force normally expressed by $V = -YI(d^3\bar{w}/dx^3)$. It corresponds to transverse displacements in the kinematic boundary conditions in a beam. Integrating again by parts, we get

$$\left(YI\frac{d^3\bar{w}}{dx^3}v\right)\Big|_{x=0}^{x=L} - \left(YI\frac{d^2\bar{w}}{dx^2}\frac{dv}{dx}\right)\Big|_{x=0}^{x=l} + \int_0^L \left[EI\frac{d^2\bar{w}}{dx^2}\frac{d^2v}{dx^2} + qv\right]dx$$

where the second term is the second boundary condition in terms of moment resultants $M = YI(d^2\bar{w}/dx^2)$, and the corresponding slope $(\theta = dv/dx)$. We see that both the kinematic and natural boundary conditions always occur in pairs.

We get the following expression by expanding the above equation:

$$v(0)V(0) - v(l)V(l) - \theta(l)M(l) + \theta(0)M(0) + \int_0^L \left[YI\frac{d^2\bar{w}}{dx^2}\frac{d^2v}{dx^2} + qv\right]dx \qquad (5.31)$$

Equation (5.31) is the weak form of the differential equation, since it requires reduced continuity compared to the original differential equation. That is, the original equation is a fourth-order equation and requires functions that are third-order continuous, whereas the weak form requires solutions that are just second-order continuous. This aspect is exploited in the next section.

## ▶ 5.4 FINITE ELEMENT METHOD

FEM uses the weak form of the governing equation to convert an ordinary differential equation to a set of algebraic equations in the case of static analysis and a set of coupled second-order differential equations in the case of dynamic analysis. Although we have indicated several different approximate methods, they are very difficult to apply to problems involving complex geometry or complicated boundary conditions. However, if we take the approach of subdividing the domain into many subdomains, then in each of these subdomains (a) we assume a solution of the type

$$\bar{u}(x, y, z, t) = \sum_{n=1}^{N} a_n(t)\phi_n(x, y, z) \qquad (5.32)$$

and (b) we use any of the approximate methods described earlier to obtain the solution within the subdomain and (c) then combine (synthesize) to obtain the overall approximate solution to the problem. In FEM, these subdomains are called *elements* and are normally in the shapes of line elements for 1D structures such as rods and beams, rectangles, quadrilaterals or triangles for 2D structures, and bricks or tetrahedrons for 3D structures. Each element has a set of nodes that may vary depending on the order of the functions $\phi_n(x, y, z)$ in Eq. (5.32) used to approximate the displacement field within each element. These nodes have unique IDs that fix their position in the space of a complex structure. In Eq. (5.32), $a_n(t)$ normally represent the time-dependent nodal displacements, whereas $\phi_n(x, y, z)$ are the spatially dependent functions, normally referred to as *shape functions.*

The entire FE procedure for obtaining a solution for a complex problem can be summarized as: it uses the weak form of the governing equation of a system, together with an assumed solution of the form Eq. (5.32), to obtain *stiffness* and *mass matrices* (note that mass matrices are required only when the structures are subjected to inertial loads). The mass matrix formulated through the weak form of the equation is called the *consistent mass matrix.* There are other ways of formulating the mass matrix. That is, the total mass of the system can be distributed appropriately among all degrees of freedom. Such a mass matrix is diagonal and is called a *lumped mass matrix.*

The *damping matrix* is normally not obtained through the weak formulation. For linear systems, it is obtained through a linear combination of the stiffness and mass matrices. A damping matrix obtained through such a procedure is called the *proportional damping matrix.* Minimization is done using an energy theorem called *Hamilton's principle;* the process of deriving FE equations is outlined in Section 5.4.2

FEM is a stiffness method in which the dependent variables (say displacements for structural systems) are the basic unknowns; compatibility of displacements across the element boundaries is automatically satisfied as we begin the analysis with the displacement assumption. Although we start with the weak form of the governing equation, this assures only that the system is in equilibrium within the element, since the assumed displacement variation is valid only for an element.

For the body to be in equilibrium, it is necessary that forces at the interelement boundaries be in equilibrium. This is not ensured automatically in FEM and must be enforced by assembling the stiffness, mass, and damping matrices by adding the element of the stiffness (or mass or damping matrices for time-dependent problems) matrix that corresponds to a particular degree of freedom coming from the contiguous elements. Similarly, the force vectors acting on each node are assembled to obtain a global force vector. If the load is distributed on a segment of the complex domain, then, using the concept of equivalent energy, it is split into concentrated loads acting on the respective nodes making up the segment. The size of the assembled stiffness, mass and damping matrices is equal to $n \times n$, where $n$ is the total number of degrees of freedom in the discretized domain.

After the matrix assembly, the displacement boundary conditions are enforced, which can be homogeneous or nonhomogeneous. If they are homogeneous, then the corresponding rows and columns are eliminated to get the reduced stiffness, mass, and damping matrices. In static analysis, the matrix equation obtained involving the stiffness matrix is solved to obtain the nodal displacements. In dynamic analysis, we get a coupled set of ordinary differential equations that is solved by either modal method or time-marching scheme. The reader is advised to look at [9] and [10] for more details.

## 5.4.1 Shape Functions

The spatially dependent functions $\phi_n(x, y, z)$ in Eq. (5.32) are called the *shape functions* of the element. These functions are normally derived from a polynomial whose order depends on the nodal degrees of freedom that an element can support. Normally denoted by $N$, they relate the nodal-dependent variable with the assumed field. We now show how to derive the shape functions of several elements starting from rod element in Figure 5.6.

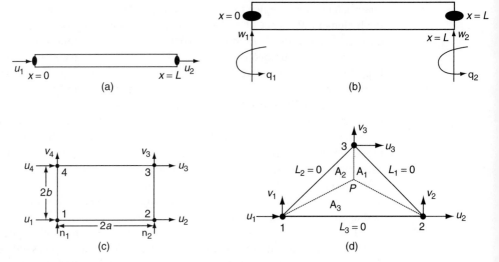

**Figure 5.6** Different finite elements: (a) rod element; (b) beam element; (c) rectangular element; and (d) triangular element.

### 5.4.1.1 Rod Element

Figure 5.6(a) shows a rod FE having length $L$ and axial rigidity $YA$. A rod element can support only axial motion and hence can have only two nodes, with each node supporting one axial motion, as shown in Figure 5.6(a). Clearly, we require a function that is only first-order continuous (elements that are first-order continuous are normally called $C^0$*continuous elements*). Hence, we can assume a displacement field with two constants corresponding to two degrees of freedom. That is,

$$u = a_0(t) + a_1(t)x \tag{5.33}$$

This equation also happens to satisfy the governing static differential equation of a rod, which is given by $YA d^2u/dx^2 = 0$. Equation (5.33) is now expressed in terms of nodal coordinates by substituting $u(x = 0) = u_1$ and $u(x = L) = u_2$ in it. This enables us to write the constants $a_0$ and $a_1$ in terms of nodal displacements $u_1$ and $u_2$. Eliminating these constants and simplifying, we can write Eq. (5.33) as

$$u(x) = \left(1 - \frac{x}{L}\right)u_1 + \left(\frac{x}{L}\right)u_2 \tag{5.34}$$

In Eq. (5.34), the two functions inside the brackets are the two shape functions of the rod corresponding to the two degrees of freedom, namely $u_1$ and $u_2$. Hence, the displacement field can be written in matrix form as

$$u(x) = [N_1(x)\ N_2(x)]\begin{Bmatrix} u_1 \\ u_2 \end{Bmatrix} = [N]\{u\} \tag{5.35}$$

The shape function $N_1$ takes value 1 at node 1, whereas it is 0 at node 2. Similarly, $N_2$ is 0 at node 1 and 1 at node 2. Also, the sum of the shape functions $N_1 + N_2$ is always equal to one. In fact, the displacement for any element can be written in the form of Eq. (5.35).

### 5.4.1.2 Beam Element

One can similarly derive the shape functions for a beam element. The beam element in Figure 5.6(b) has two nodes, and each node has two degrees of freedom, transverse displacement $w$ and rotation $\theta = dw/dx$. Hence the nodal degrees of freedom vectors are given by $\{u\} = \{ w_1 \quad \theta_1 \quad w_2 \quad \theta_2 \}^T$, which require at least a cubic polynomial having four constants for dependent transverse displacement variation. Also, since the slope $\theta$ is derived from the transverse displacement $w$, the polynomial assumed should be second-order continuous. That is, at the interelement boundaries, it is required that both the transverse displacement and its derivative (which is the slope of the beam) must be continuous, which is not the case for rods. Such elements are called $C^1$ *continuous elements*. The cubic interpolating polynomial for the beam is given by

$$w(x,t) = a_0(t) + a_1(t)x + a_2(t)x^2 + a_3(t)x^3 \tag{5.36}$$

As in the case of rods, the solution in Eq. (5.36) happens to be the exact solution of the governing beam equation [Eq. (5.1)]. Now, we substitute $w(0,t) = w_1(t)$, $\theta(0,t) = dw(0,t)/dx = \theta_1(t)$, $w(L,t) = w_2(t)$ and $\theta(L,t) = dw(L,t)/dx = \theta_2(t)$ to get

$$
\begin{Bmatrix} w_1 \\ \theta_1 \\ w_2 \\ \theta_2 \end{Bmatrix} =
\begin{bmatrix} 1 & 0 & 0 & 0 \\ 0 & 1 & 0 & 0 \\ 1 & L & L^2 & L^3 \\ 0 & 1 & 2L & 3L^2 \end{bmatrix}
\begin{Bmatrix} a_0 \\ a_1 \\ a_2 \\ a_3 \end{Bmatrix} = \{u\} = [G]\{a\}
$$

Inverting this matrix, we can write the unknown coefficients as $\{a\} = [G]^{-1}\{u\}$. Substituting the values of the coefficients in Eq. (5.36), we get

$$w(x,t) = [N_1(x) \quad N_2(x) \quad N_3(x) \quad N_4(x)]\{u(t)\} \tag{5.37a}$$

where

$$N_1(x) = 1 - 3\left(\frac{x}{L}\right)^2 + 2\left(\frac{x}{L}\right)^3, \quad N_2(x) = x\left(1 - \frac{x}{L}\right)^2$$

$$N_3(x) = 3\left(\frac{x}{L}\right)^2 - 2\left(\frac{x}{L}\right)^3, \quad N_4(x) = x\left[\left(\frac{x}{L}\right)^2 - \left(\frac{x}{L}\right)\right] \tag{5.37b}$$

These shape functions will take the unit value at the nodes and zero elsewhere. Note here that the shape functions $N_2(x)$ and $N_4(x)$ correspond to the rotations $\theta_1$ and $\theta_2$ and hence $dN_2(x)/dx$ takes the value of 1 at $x = 0$ and $dN_4(x)/dx$ takes the value of 0 at $x = 0$. Similarly, $dN_4(x)/dx$ takes the value of 1 at $x = L$ and $dN_2(x)/dx$ takes the value of 0 at $x = L$.

Before proceeding further, we highlight the necessary requirements on an interpolating polynomial of an element, especially from the convergence point of view. These can be summarized as follows:

1. The assumed solution should be able to capture the rigid body motion. This can be ensured by retaining a constant part in the assumed solution.

2. The assumed solution must be able to attain the constant strain rate as the mesh is refined. This can be assured by retaining the linear part of the assumed function in the interpolating polynomial.

3. Most second-order systems require only $C^0$ continuity, which is easily met in most FE formulations. However, for higher-order systems such as Bernoulli–Euler beams or elementary plates, one requires $C^1$ continuity, which is extremely difficult to satisfy, especially for plate problems where interelement slope continuity is very difficult to satisfy. In such situations, one can use shear-deformable models, that is, models that also include the effect of shear deformations. In such models, slopes are not derived from the displacements and are independently interpolated. This relaxes the $C^1$ continuity requirement. However, when such elements are used in thin-beam or plate models, where the effects of shear deformations are negligible, the displacement magnitude predicted are many orders smaller than the correct displacements. Such problems are called the *shear locking* problems.

4. The order of the assumed interpolating polynomial is dictated by the highest order of the derivative in the energy functional. That is, the assumed polynomial should be at least one order higher than what appears in the energy functional.

### 5.4.1.3 Rectangular Elements
We now determine the shape functions for 2D elements. Let us now consider a rectangular FE of length $2a$ and width $2b$, as shown in Figure 5.6(c). This element has four nodes and each node can support two degrees of freedom, namely, the two displacements $u(x, y)$ and $v(x, y)$ in the two coordinate directions. Since there are four nodes, we can take the interpolating polynomial as

$$u(x,y) = a_0 + a_1x + a_2y + a_3xy$$
$$v(x,y) = b_0 + b_1x + b_2y + b_3xy \tag{5.38}$$

These functions have linear variation of displacement in both the coordinate directions, and hence are normally called *bilinear* elements. In the interpolating polynomial above, we substitute $u(-a,b) = u_2$, $v(-a,b) = v_2$, $u(-a,-b) = u_1$, $v(-a,-b) = v_1$, $u(a,-b) = u_4$, $v(a,-b) = v_4$. These help us to relate the nodal displacements to the unknown coefficients as $\{u\} = [G]\{a\}$. Inverting the above relation and substituting for unknown coefficients in Eq. (5.38), we can write the displacement field and the shape functions as

$$u(x,y) = [N]\{u\} = [N_1(x,y) \quad N_2(x,y) \quad N_3(x,y) \quad N_4(x,y)]\{u\} \tag{5.39a}$$

and $\quad v(x,y) = [N]\{v\} = [N_1(x,y) \quad N_2(x,y) \quad N_3(x,y) \quad N_4(x,y)]\{v\}$

where

$$\{u\} = \{u_1 \quad u_2 \quad u_3 \quad u_4\}^T, \quad \{v\} = \{v_1 \quad v_2 \quad v_3 \quad v_4\}^T$$

$$N_1(x,y) = \frac{(x-a)(y-b)}{4}, \quad N_2(x,y) = \frac{(x-a)(y+b)}{4},$$

$$N_3(x,y) = \frac{(x+a)(y+b)}{4}, \quad N_4(x,y) = \frac{(x+a)(y-b)}{4} \tag{5.39b}$$

Note that these shape functions have the property of taking the value unity for nodes at which they are evaluated and zero at all other nodes.

### 5.4.1.4 Triangular Elements

One can similarly write the shape functions for a triangular element. However, it is more convenient to use the area coordinates for the triangle. Consider a triangle having coordinates of the three vertices $(x_1, y_1)$, $(x_2, y_2)$ and $(x_3, y_3)$ [Figure 5.6(d)]. Consider an arbitrary point $P$ inside the triangle. This point splits the triangle into three smaller triangles of areas $A_1$, $A_2$ and $A_3$, respectively. Let $A$ be the total area of the triangle, which can be written in terms of nodal coordinates as

$$A = \frac{1}{2} \begin{vmatrix} 1 & x_1 & y_1 \\ 1 & x_2 & y_2 \\ 1 & x_3 & y_3 \end{vmatrix} \tag{5.40}$$

We define the area coordinates for the triangle as

$$L_1 = \frac{A_1}{A}, \quad L_2 = \frac{A_2}{A}, \quad L_3 = \frac{A_3}{A} \tag{5.41}$$

Thus, the position of point $P$ is given by the coordinates $(L_1, L_2, L_3)$. These coordinates, which are normally referred to as area coordinates, are not independent; they satisfy the relation

$$L_1 + L_2 + L_3 = 1 \tag{5.42}$$

These area coordinates are related to the global $x$–$y$ coordinate system through

$$x = L_1 x_1 + L_2 x_2 + L_3 x_3 \quad \text{and} \quad y = L_1 y_1 + L_2 y_2 + L_3 y_3 \tag{5.43}$$

where

$$L_i = \frac{a_i + b_i x + c_i y}{2A} \quad (i = 1, 2, 3) \tag{5.44}$$

$$a_1 = x_2 y_3 - x_3 y_2, \quad b_1 = y_2 - y_3, \quad c_1 = x_3 - x_2$$

The other coefficients are obtained by cyclic permutation. Equation (5.44) is required to compute the derivatives with respect to the actual coordinates $x$ and $y$. Now, one can write the shape functions for the triangle as

$$u = N_1 u_1 + N_2 u_2 + N_3 u_3 \quad \text{and} \quad v = N_1 v_1 + N_2 v_2 + N_3 v_3 \tag{5.45a}$$

where

$$N_1 = L_1, \quad N_2 = L_2, \quad N_3 = L_3 \tag{5.45b}$$

These shape functions also follow the normal rules. That is, at point $A$, where $L_1 = 1$, the shape functions take the value 1. At the same point, $L_2 = L_3 = 0$. Similarly, at the other two vertices, $L_2$ and $L_3$ take value 1, whereas the other two go to 0.

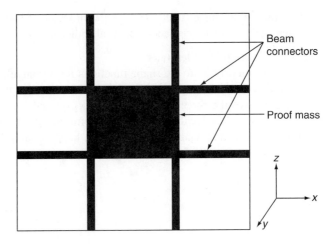

**Figure 5.7** A plan view of a proof-mass accelerometer.

In summary, we can express the displacements of all the elements in terms of shape functions and write the nodal displacements as $u = \sum_{n=1}^{N} N_n u_n$ or $[N]\{u\}$. This spatial discretization is used in the weak form of the governing equation [e.g. Eq. (5.31) for beams] to obtain the FE governing equation, as shown in the next section.

The next question to answer is: is it possible to use combination of elements? The answer is yes. This can be done in the FE code by proper assembly of the stiffness matrix coming from, say, the rod and beam elements, the beam and plate elements, etc. This requires detailed description of typical FE code and the displacement transformation matrix from local to global coordinates.

At this point, let us look at how to perform FEA for a system undergoing multiple motions. We consider the proof-mass accelerometer in Figure 5.7. The accelerometer consists of a proof mass attached to beam connectors that are oriented in both the $x$- and $y$-directions. The induced voltage on the proof mass in the direction perpendicular to plane of the page (the $z$-direction) causes motion of the beam connectors in all three directions. However, the beam FE model (presented in Section 5.4.5 below) lets us model the beam motion only in the $z$-direction (the $x$–$z$ plane). In order to model three-directional motions, it is necessary to generate the elemental stiffness matrix for the axial motion (see the rod stiffness matrix given in Section 5.4.4), the beam motion in the $x$–$y$ plane, and the beam motion in $x$–$z$ plane. The stiffness matrix for beam motion in the $x$–$z$ plane is exactly same as that of the beam motion in the $x$–$z$ plane with area moment of inertia about the $y$–$y$ axis required for the stiffness matrix of the $x$–$z$ plane motion replaced with the area moment of inertia about the $z$–$z$ axis. Note that the axial stiffness matrix will be of size $2 \times 2$, corresponding to two axial degrees of freedom ($u$ along the $x$-direction); the bending motion in $x$–$z$ plane will have stiffness matrix of size $4 \times 4$, where the degrees of freedom at each node will be $w$ (corresponding to $z$-directional motion) and the slope about the $z$-axis is $\theta_z$; and the bending motion in the $x$–$y$ plane will again have stiffness matrix of size $4 \times 4$, where the degrees of freedom will be $v$ (corresponding to $y$-directional motion) and the slope about the $y$-axis is $\theta_y$. Hence, each node will have three translational degrees of freedom ($u$, $v$ and $w$) and two rotational degrees of freedom $\theta_z$ and $\theta_y$, giving a total of 10 degrees of freedom per element.

The stiffness matrices are first constructed in the beam connecters local coordinates, and an updated stiffness matrix of size $10 \times 10$ is created corresponding to the 10 degrees of freedom. This updated stiffness matrix is then transformed to global coordinates using a suitable coordinate transformation, which is then used for coupled analysis. The details of coordinate transformation and solution details can be found in [9] and [10].

## 5.4.2 Derivation of the Finite Element Equation

Consider a body of volume $V$ under the action of surface tractions in the three coordinate directions $\{t_s\} = \{\, t_x \quad t_y \quad t_z \,\}^{\mathrm{T}}$ on the boundary $S$ and body force vector per unit volume $\{B\} = \{\, B_x \quad B_y \quad B_z \,\}^{\mathrm{T}}$. Let the displacement vector be expressed as $\{d(x, y, z, t)\} = \{\, u(x, y, z, t) \quad v(x, y, z, t) \quad w(x, y, z, t) \,\}^{\mathrm{T}}$, where $u$, $v$, and $w$ are the displacement variations in the three coordinate directions, respectively. To derive the FE equation, we need suitable energy theorems, most of which are based on the principle of virtual work (PVW). For problems involving static analysis, principle of minimum potential energy (PMPE) is normally used, which is essentially the PVW in the variational form. For time-dependent problems, a modified form of PMPE called Hamilton's principle is used. More details on the energy theorems are available in [5] & [8].

We now invoke Hamilton's principle, whose physical meaning is as follows: a body under the action of forces deforms and hence does some work that is stored as energy. The path traced by the body in deforming from one state at time $t_1$ to the next state at time $t_2$ is such as to make the total energy of the system stationary. Alternatively, Hamilton's principle can also mean that, if the path traced by the body between time $t_1$ and $t_2$ is such as to minimize the total energy, then such a process ensures that the body is in equilibrium.

Mathematically, we can write Hamilton's principle as follows:

$$\delta \int_{t_1}^{t_2} (T - SE + W_{nc})\, \mathrm{d}t = 0 \tag{5.46}$$

where $T$ is the kinetic energy, $SE$ is the strain energy, and $W_{nc}$ is the work done by the non conservative forces. The kinetic energy $T$ is given by

$$T = \frac{1}{2} \int_V \rho(\dot{u}^2 + \dot{v}^2 + \dot{w}^2)\mathrm{d}V$$

Taking the first variation and integrating, we get

$$\int_{t_1}^{t_2} \delta T\mathrm{d}t = \int_{t_1}^{t_2}\int_V \rho\left(\frac{\mathrm{d}u}{\mathrm{d}t}\frac{\mathrm{d}(\delta u)}{\mathrm{d}t} + \frac{\mathrm{d}v}{\mathrm{d}t}\frac{\mathrm{d}(\delta v)}{\mathrm{d}t} + \frac{\mathrm{d}w}{\mathrm{d}t}\frac{\mathrm{d}(\delta w)}{\mathrm{d}t}\right)\mathrm{d}V\,\mathrm{d}t$$

Integrating by parts and noting that the first variation vanishes at times $t_1$ and $t_2$, we get

$$\int_{t_1}^{t_2} \delta T\mathrm{d}t = -\int_{t_1}^{t_2}\int_V \rho(\ddot{u}\,\delta u + \ddot{v}\,\delta v + \ddot{w}\,\delta w)\mathrm{d}t\,\mathrm{d}V = -\int_{t_1}^{t_2}\int_V \rho\{\delta d\}^T\{\ddot{d}\}\mathrm{d}V\mathrm{d}t \tag{5.47}$$

where $\{\ddot{d}\} = \{\ddot{u}\ \ddot{v}\ \ddot{w}\}^{\mathrm{T}}$ represents the acceleration vector and $\{\delta d\} = \{\, \delta u \quad \delta v \quad \delta w \,\}^{\mathrm{T}}$ represents the vector containing the first variation of the displacements.

The expression for strain energy was derived in Section 5.2.2 for a 1D state of stress. We extend this expression for a body in a 3-D state of stress as:

$$SE = \frac{1}{2} \int_V \left( \sigma_{xx}\varepsilon_{xx} + \sigma_{yy}\varepsilon_{yy} + \sigma_{zz}\varepsilon_{zz} + \tau_{xy}\gamma_{xy} + \tau_{yz}\gamma_{yz} + \tau_{zx}\gamma_{zx} \right) dV$$

$$= \frac{1}{2} \int_V \{\varepsilon\}^T \{\sigma\} dV \tag{5.48}$$

For the linear elastic case, the stresses and strains are related through the constitutive law $\{\sigma\} = [C]\{\varepsilon\}$ [see Eq. (4.107) in Section 4.11 for more details on Hooke's law for 2D elasticity]. Hence, the strain energy becomes

$$SE = \frac{1}{2} \int_V \{\varepsilon\}^T [C]\{\varepsilon\} dV$$

Taking the first variation (which is similar to differentiating) and integrating, we have

$$\int_{t_1}^{t_2} \delta SE dt = \int_{t_1}^{t_2} \int_V \{\delta\varepsilon\}^T [C]\{\varepsilon\} dV dt \tag{5.49}$$

A body may also be subjected to other forces, such as gravity, that are expressed as force per unit volume. These are represented as $B_x$, $B_y$ and $B_z$ in the three coordinate directions. Similarly, certain regions of the structure may be subjected to traction on a surface whose vector can be represented by $\{t_s\}$, and there could also be nonconservative forces such as damping forces. Hence, the work done by the body forces, surface forces, damping elements, and the concentrated forces can be combined in $W_{nc}$. Mathematically, $W_{nc} = W_B + W_S + W_D$. The work done by the body forces is given by

$$W_B = \int_V (B_x u + B_y v + B_z w) dV = \int_V \{d\}^T \{B\} dV \tag{5.50}$$

The first variation of the body force work is given by

$$\int_{t_1}^{t_2} \delta W_B dt = \int_{t_1}^{t_2} \int_V (B_x \delta u + B_y \delta v + B_z \delta w) dV dt = \int_{t_1}^{t_2} \int_V \{\delta d\}^T \{B\} dV dt \tag{5.51}$$

The work done by the surface forces is

$$W_s = \int_S \{d\}^T \{t_s\} dS$$

The first variation of this work is given by

$$\int_{t_1}^{t_2} \delta W_s dt = \int_{t_1}^{t_2} \int_V \{\delta d\}^T \{t_s\} \, dS dt \tag{5.52}$$

Similarly, the first variation of the work done by the damping force is given by

$$\int_{t_1}^{t_2} \delta W_D dt = -\int_{t_1}^{t_2}\int_V \{\delta d\}^T \{F_D\} dV dt \tag{5.53}$$

If the damping is of viscous type, then the damping force is proportional to the velocity and is given by $\{F_D\} = \eta\{\dot{d}\}$, where $\eta$ is the damping coefficient and $\{\dot{d}\} = \{\dot{u}\ \dot{v}\ \dot{w}\}^T$ is the velocity vector in the three coordinate directions. Note that in Eq. (5.53), a negative sign is introduced to signify that the damping force always resists the motion.

Now using Eqs. (5.47) and (5.49)—(5.53) and invoking Hamilton's principle [Eq. (5.46)], we get

$$-\int_{t_1}^{t_2}\int_V \{\delta d\}^T \rho \{\ddot{d}\}\, dV\, dt - \int_{t_1}^{t_2}\int_V \{\delta\varepsilon\}^T [C]\{\varepsilon\}\, dV dt + \int_{t_1}^{t_2}\int_V \{\delta d\}^T \{B\}\, dV dt$$

$$+\int_{t_1}^{t_2}\int_V \{\delta d\}^T \{t_s\}\, dS dt - \int_{t_1}^{t_2}\int_V \{\delta d\}^T \{F_D\}\, dV dt = 0 \tag{5.54}$$

Now we substitute into this equation the assumed displacement variation in terms of shape function and nodal displacements derived earlier. Hence, we have

$$\{d(x,y,z,t)\} = [N(x,y,z)]\{u_e(t)\} \tag{5.55}$$

where $[N(x,y,z)]$ is the shape function matrix and $\{u_e\}$ is the nodal displacement vector of an element. Using this, we can write velocity, acceleration, and its first variation as

$$\{\dot{d}\} = [N]\{\dot{u}_e\}, \quad \{\ddot{d}\} = [N]\{\ddot{u}_e\} \text{ and } \{\delta d\} = [N]\{\delta u_e\} \tag{5.56}$$

Now the strains can also be written in terms of a strain-displacement relationship [1]. That is, the six strain components can be written in matrix form as

$$\begin{Bmatrix} \varepsilon_{xx} \\ \varepsilon_{yy} \\ \varepsilon_{zz} \\ \gamma_{xy} \\ \gamma_{yz} \\ \gamma_{zx} \end{Bmatrix} = \begin{bmatrix} \dfrac{\partial}{\partial x} & 0 & 0 \\ 0 & \dfrac{\partial}{\partial y} & 0 \\ 0 & 0 & \dfrac{\partial}{\partial z} \\ \dfrac{\partial}{\partial y} & \dfrac{\partial}{\partial x} & 0 \\ 0 & \dfrac{\partial}{\partial z} & \dfrac{\partial}{\partial y} \\ \dfrac{\partial}{\partial z} & 0 & \dfrac{\partial}{\partial x} \end{bmatrix} \begin{Bmatrix} u \\ v \\ w \end{Bmatrix} \tag{5.57}$$

$$\{\varepsilon\} = [B]\{d\} \tag{5.58}$$

$$\{\delta\varepsilon\} = [B]\{\delta d\} \tag{5.59}$$

Now, we consider Eq. (5.54) termwise and simplify it further. If we substitute Eq. (5.56) into the first term of the equation, which is essentially the inertial part of the governing equation, the term becomes

$$\int_V \rho\{\delta d\}^{\mathrm{T}}\{\ddot{d}\}\,\mathrm{d}V = \int_V \rho\{\delta u_e\}^{\mathrm{T}}[N]^{\mathrm{T}}[N]\{\ddot{u}_e\}\,\mathrm{d}V$$

$$= \{\delta u_e\}^{\mathrm{T}}\left[\int_V \rho[N]^{\mathrm{T}}[N]\,\mathrm{d}V\right]\{\ddot{u}_e\} = \{\delta u_e\}^{\mathrm{T}}[m]\{\ddot{u}_e\} \tag{5.60}$$

The term inside the brackets [m] is called the *element mass matrix*. The mass matrix obtained in this form is called the *consistent mass matrix*, although other forms of mass matrix exist (as explained in Section 5.4). Next, we consider the second term involving strains in Eq. (5.54). Using Eqs. (5.56) and (5.59), this second term can be written as

$$\int_V \{\delta\varepsilon\}^{\mathrm{T}}[C]\{\varepsilon\}\,\mathrm{d}V = \int_V \{\delta u_e\}^{\mathrm{T}}[B]^{\mathrm{T}}[C][B]\{u_e\}\,\mathrm{d}V$$

$$= \{\delta u_e\}^{\mathrm{T}}\left[\int_V [B]^{\mathrm{T}}[C][B]\,\mathrm{d}V\right]\{u_e\}$$

$$= \{\delta u_e\}^{\mathrm{T}}[k]\{u_e\} \tag{5.61}$$

Here the term inside the square bracket ([k]) represents the *stiffness matrix* of the formulated element. The other terms in Eq. (5.54) can be similarly written in terms of nodal displacement vector and its first variation using Eqs. (5.55) and (5.56). Now the term due to body force can be written as

$$\int_V \{\delta d\}^{\mathrm{T}}\{B\}\mathrm{d}V = \int_V \{\delta u_e\}^{\mathrm{T}}[N]^{\mathrm{T}}\{B\}\,\mathrm{d}V$$

$$= \{\delta u_e\}^{\mathrm{T}}\left[\int_V [N]^{\mathrm{T}}\{B\}\,\mathrm{d}V\right] = \{\delta u_e\}^{\mathrm{T}}\{f_B\} \tag{5.62}$$

The bracketed term represents the body force vector acting on the element. Similarly, we can write the surface forces as

$$\int_S \{\delta d\}^{\mathrm{T}}\{t_s\}\mathrm{d}S = \{\delta u\}^{\mathrm{T}}\{f_s\}, \quad \{f_s\} = \int_S [N]^{\mathrm{T}}\{t_s\}\mathrm{d}S \tag{5.63}$$

Finally, the damping force vector assuming viscous type damping can be written as

$$-\int_V \{\delta d\}^{\mathrm{T}}\eta\{\dot{d}\}\,\mathrm{d}V = -\{\delta u_e\}^{\mathrm{T}}\left[\int_V \eta[N]^{\mathrm{T}}[N]\,\mathrm{d}V\right]\{\dot{u}_e\} = -\{\delta u_e\}^{\mathrm{T}}[c]\{\dot{u}_e\} \tag{5.64}$$

The matrix $[c]$ is the *consistent damping matrix*, This form is seldom used in actual analysis. There are different ways of treating damping. In the most common damping scheme called the *proportional damping*, the damping matrix is assumed as the linear combination of the stiffness and mass matrices. That is, $[C] = \alpha[K] + \beta[M]$, where $\alpha$ and $\beta$ are the stiffness and mass proportional damping parameters, as discussed in [10].

Now, using Eqs. (5.60)—(5.64) in Eq. (5.54), we have

$$\int_{t_1}^{t_2} \{\delta u_e\}^{\mathrm{T}}([m]\{\ddot{u}_e\} + [c]\{\dot{u}_e\} + [k]\{u_e\} - \{f_B\} - \{f_s\})\, dt = 0$$

Since the first variation of the displacement vector is arbitrary, we have

$$[m]\{\ddot{u}_e\} + [c]\{\dot{u}_e\} + [k]\{u_e\} = \{R\} \tag{5.65}$$

where

$$[m] = \int_V \rho[N]^{\mathrm{T}}[N]\, dV, \quad [c] = \int_V \eta[N]^{\mathrm{T}}[N]\, dV, \quad [k] = \int_V [B]^{\mathrm{T}}[C][B]\, dV$$

$$\{R\} = \{f_B\} + \{f_s\}, \quad \{f_B\} = \int_V [N]^{\mathrm{T}}\{B\} dV, \quad \{f_s\} = \int_\tau [N]^{\mathrm{T}}\{t_s\}\, dS$$

and $V$ is the volume of the element and $\tau$ is the surface of the element containing the surface loads.

Equation (5.65) is the discretized governing equation of motion that we need to solve through the FE technique. Here $\{R\}$ is the combined force vector due to body, surface, and concentrated forces. Note that this equation is a highly coupled second-order linear differential equation. If inertial and damping forces are absent, the above equation reduces to a set of simultaneous equations that is solved to obtain the static behavior. The sizes of the matrices $[m]$, $[k]$, and $[c]$ are equal to the number of degrees of freedom an element can support. All these matrices are generated for each element and assembled to obtain the global mass matrix $[M]$, stiffness matrix $[K]$, and damping matrix $[C]$, respectively. After assembling these matrices, the displacement boundary conditions are enforced. (The flowchart in Figure 5.2 gives the procedure for analysis using FEM.) All these matrices are symmetric and banded in nature. The bandwidth BW is dictated by the node numbering of the mesh and is determined by computing the highest node difference. That is, if an element has four nodes, say with node numbers $i, j, k$, and $l$, and if the largest numerical difference among $i, j, k$, and $l$ is, say, $nl$, then the largest value of $nl$ for all elements is determined. With this value, the BW of the stiffness or mass matrix is given by BW = Largest $(nl) \times n_{dof} - 1$, where $n_{dof}$ is the number of degrees of freedom the element can support.

In static analysis, the mass and damping matrices do not enter in the picture and the FE equation can be written only in terms of assembled stiffness matrix $[K]$ and load vector $\{R\}$ as

$$[K]\{u\} = \{R\}$$

The present formulation requires modification in order to handle curved boundaries. Such a formulation is called the *isoparametric formulation,* as explained next.

### 5.4.3 Isoparametric Formulation and Numerical Integration

So far we have dealt only with FEs having straight edges. In practical structures, the edges are almost always curved, and modeling curved edges with straight-edged elements yields an enormous increase in the degrees of freedom and hence the problem size. In addition, in many practical situations, it is always not required to have uniform mesh density throughout the problem domain. Meshes are always graded from fine (in regions of high stress gradient) to coarse (in uniform stress fields). These curved elements enable us to grade the mesh effectively. With the availability of curved quadrilateral, triangular, and wedge elements, it is now possible to model 2D geometry of any complex shape.

Elements with curved boundaries are mapped to elements with straight boundaries through a coordinate transformation that involves mapping functions (i.e. functions of the mapped coordinates). This mapping is established by expressing the coordinate transformation as a polynomial of a certain order that is decided by the number of nodes involved in the mapping. Since we will be working with straight-edged elements in the mapped domain, the displacement should also be expressed as a polynomial of a certain order in the mapped coordinates. In this case, the order of the polynomial is dependent upon the number of degrees of freedom an element can support. Thus, we have two transformations, one involving the coordinates and the other involving the displacements. If the coordinate transformation is of lower order than the displacement transformation, it is called a *subparametric transformation*. That is, if an element has *n* nodes, while all the *n* nodes participate in the displacement transformation, only a few nodes participate in the coordinate transformation. If the coordinate transformation is of higher order than the displacement transformation, it is called a *superparametric transformation*. In this case, only a small set of nodes participate in the displacement transformation, whereas all the nodes participate in the coordinate transformation. In the *FE formulation*, the most important transformation is the one in which both the displacement and coordinate transformations are of same order, implying that all the nodes participate in both transformations. Such a transformation is called the *isoparametric transformation*. Figure 5.8 shows the concept of mapping for 1D and 2D elements. Next the concept of isoparametric formulation is demonstrated for 1D and 2D elements, and stiffness matrices for some simple elements are derived using this concept.

### 5.4.4 One-Dimensional Isoparametric Rod Element

Figure 5.8(a) shows the 1D rod element in the original rectangular coordinate system and the mapped coordinate system, with $\xi$ as the (1D) mapped coordinate. Note that at the two extreme ends of the rod, where the axial degrees of freedom $u_1$ and $u_2$ are defined, the mapped coordinates are $\xi = -1$ and $\xi = 1$, respectively. We now take the displacement variation of the rod in the mapped coordinates as

$$u(\xi, t) = a_0(t) + a_1(t)\xi \tag{5.66}$$

Substituting $u(\xi = -1) = u_1$ and $u(\xi = 1) = u_2$, and eliminating the constants, we can write the displacement field in the mapped coordinates as

$$u(\xi) = \left(\frac{1-\xi}{2}\right)u_1 + \left(\frac{1+\xi}{2}\right)u_2 = \left[\frac{1-\xi}{2} \quad \frac{1+\xi}{2}\right]\begin{Bmatrix} u_1 \\ u_2 \end{Bmatrix} = [N(\xi)]\{u_e\} \tag{5.67}$$

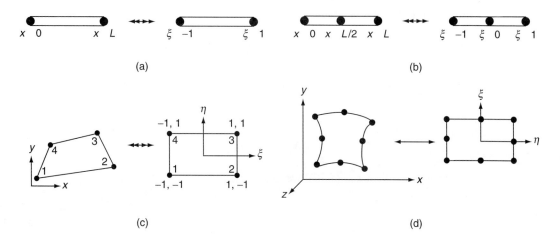

**Figure 5.8** Various isoparametric finite elements: (a) isoparametric linear rod element; (b) isoparametric quadratic rod element; (c) isoparametric quadralateral element; (d) isoparametric eight-noded element.

We also assume that the rectangular $x$-coordinate varies with respect to mapped coordinate $\xi$ in the same manner of displacement:

$$x = \begin{bmatrix} \dfrac{1-\xi}{2} & \dfrac{1+\xi}{2} \end{bmatrix} \begin{Bmatrix} x_1 \\ x_2 \end{Bmatrix} = [N(\xi)]\{x_e\} \qquad (5.68)$$

In this equation, $x_1$ and $x_2$ are the coordinates of the actual element in the rectangular $x$-coordinate system. We can see that there is one-to-one correspondence of the coordinates in the original and mapped systems.

Derivation of the stiffness matrix requires computation of the strain-displacement matrix $[B]$, and this requires the evaluation of the derivatives of the shape functions with respect to the original $x$-coordinate system. In the case of a rod, there is only axial strain, computed as

$$\varepsilon_{xx} = \frac{du}{dx} = \begin{bmatrix} \dfrac{dN_1}{dx} & \dfrac{dN_2}{dx} \end{bmatrix} \{u_1 \; u_2\}^{\mathrm{T}}$$

Hence the matrix $[B]$ becomes

$$[B] = \begin{bmatrix} \dfrac{dN_1}{dx} & \dfrac{dN_2}{dx} \end{bmatrix} \qquad (5.69)$$

However, one coordinate system can be mapped to a different coordinate system using a Jacobian $(J)$. By invoking the chain rule of differentiation, we have

$$\frac{dN_i}{dx} = \frac{dN_i}{d\xi} \frac{d\xi}{dx} \qquad (i = 1, 2) \qquad (5.70)$$

From Eq. (5.68), we have

$$x = \frac{1-\xi}{2}x_1 + \frac{1+\xi}{2}x_2$$

$$\frac{dx}{d\xi} = \frac{x_2 - x_1}{2} = \frac{L}{2} = J \tag{5.71}$$

$$\frac{d\xi}{dx} = \frac{2}{L} = \frac{1}{J}, \quad dx = Jd\xi \tag{5.72}$$

Using Eq. (5.72) in Eq. (5.70), we get

$$\frac{dN_i}{dx} = \frac{dN_i}{d\xi}\frac{1}{J} = \frac{dN_i}{d\xi}\frac{2}{L} \tag{5.73}$$

Substituting the shape functions $[N] = [N_1 N_2] = [(1-\xi)/2 \; (1+\xi)/2]$ in this equation, we can obtain the shape function derivatives with respect to mapped coordinates; hence matrix $[B]$ becomes

$$\frac{dN_1}{d\xi} = \frac{-1}{2}, \quad \frac{dN_2}{d\xi} = \frac{1}{2}, \quad [B] = \begin{bmatrix} \frac{-1}{2} & \frac{1}{2} \end{bmatrix} \tag{5.74}$$

In the case of a rod, as indicated earlier, there is only axial stress, and as a result $[C]$, the material matrix in Eq. (5.61) for evaluating the stiffness matrix, will have only $Y$, the Young's modulus of the material. The stiffness matrix for a rod is given by

$$[K] = \int_V [B]^T[C][B]dV = \int_0^L \int_A [B]^T Y[B]dA \, dx = \int_{-1}^1 [B]^T YA[B]Jd\xi \tag{5.75}$$

Finally, from Eqs. (5.69) to (5.74) for matrix $[B]$ and the Jacobian, we get the stiffness matrix for a rod as

$$[K] = \frac{YA}{L} \begin{bmatrix} 1 & -1 \\ -1 & 1 \end{bmatrix} \tag{5.76}$$

The reader can check that one can obtain the above result directly without going through isoparametric formulation by substituting the shape functions from Eq. (5.35) in Eq. (5.61) and performing a direct integration. For lower-order and straight-edged elements, the Jacobian is constant and not a function of the mapped coordinates. For complex geometries and higher-order elements, the Jacobian is always a function of the mapped coordinates. In such cases, integration of the expression for computing the stiffness matrix involves rational polynomials.

To demonstrate this, we consider a higher-order rod having three degrees of freedom, all axial, as shown in Figure 5.8(b). The displacement variation for this element in the mapped coordinate is given by

$$u(\xi, t) = a_0(t) + a_1(t)\xi + a_2(t)\xi^2 \tag{5.77}$$

Following the same procedure as in the previous case, we first substitute $u(\xi = -1) = u_1$, $u(\xi = 0) = u_2$, and $u(\xi = 1) = u_3$ in Eq. (5.77) to get the following three shape functions corresponding to the three degrees of freedom:

$$N_1 = \frac{\xi(-1 + \xi)}{2}, \quad N_2 = (1 - \xi^2), \quad N_3 = \frac{\xi(1 + \xi)}{2} \tag{5.78}$$

Next the Jacobian must be computed, for which we assume the coordinate transformation

$$x = \frac{\xi(-1 + \xi)}{2} x_1 + (1 - \xi^2) x_2 + \frac{\xi(1 + \xi)}{2} x_3 \tag{5.79}$$

In this expression, $x_1$, $x_2$ and $x_3$ are the coordinates of the three nodes of the element in the original coordinate system. Taking the derivative with respect to the mapped coordinate, we get

$$\frac{dx}{d\xi} = \frac{(2\xi - 1)}{2} x_1 - 2\xi x_2 + \frac{(2\xi + 1)}{2} x_3 = J, \quad dx = J d\xi \tag{5.80}$$

Unlike the two-noded rod case, the Jacobian in the higher-order rod case is a function of the mapped coordinate and its value changes as we move along the bar. If the coordinate $x_2$ coincides with the midpoint of the rod, the value of the Jacobian becomes $L/2$. The matrix $[B]$ in this case becomes

$$[B] = \frac{1}{J} \left[ \left( \frac{2\xi - 1}{2} \right) \quad -2\xi \quad \left( \frac{2\xi + 1}{2} \right) \right] \tag{5.81}$$

This matrix, unlike the two-noded rod, is a function of the mapped coordinate. Hence the stiffness matrix, given in Eq. (5.75), cannot be integrated in closed form. One can see that it involves integration of rational polynomials. Hence, one must resort to numerical integration. The most popular numerical integration scheme is through *Gauss quadrature*, which is explained in Section 5.4.7.

### 5.4.5 One-Dimensional Beam Element Formulation

The configuration of the beam is very similar to the rod in Figure 5.8(a). However, each node can support two degrees of freedom, namely, the transverse displacement $w(x, t)$ and rotation (slope), derived from the transverse displacement through a relation $\theta(x, t) = dw(x, t)/dx$. The procedure for formulation is same as for a rod. First, we find the isoparametric shape functions for which we need to first assume the transverse displacement variation. Since the element can support four degrees of freedom, the variation should have four independent constants; in other words, the displacement variation should be cubic with respect to the isoparametric coordinates. It is given by

$$w(\xi, t) = a_0(t) + a_1(t)\xi + a_2(t)\xi^2 + a_3(t)\xi^3 \tag{5.82}$$

Substituting the four boundary conditions, namely,

$$w(-1, t) = w_1(t), \quad dw/dx = \theta(-1, t) = \theta_1, \quad w(1, t) = w_2(t), \quad dw/dx = \theta(1, t) = \theta_2$$

and eliminating the constants $a_0(t)$ to $a_3(t)$, we write the transverse displacement field as

$$w(\xi, t) = N_1(\xi)w_1 + N_2(\xi)\theta_1 + N_3(\xi)w_2 + N_4(\xi)\theta_2 \qquad (5.83)$$

where

$$N_1(\xi) = \frac{1}{4}(2 - 3\xi + \xi^3), \quad N_2(\xi) = \frac{1}{4}(1 - \xi - \xi^2 + \xi^3),$$

$$N_3(\xi) = \frac{1}{4}(2 + 3\xi - \xi^3), \quad N_4(\xi) = \frac{1}{4}(-1 - \xi + \xi^2 + \xi^3) \qquad (5.84)$$

The coordinate transformation follows exactly that for the rod:

$$x = \frac{1 - \xi}{2}x_1 + \frac{1 + \xi}{2}x_2, \quad dx = Jd\xi, \quad J = L/2$$

where $J$ is the Jacobian and $L$ the length of the element. Note that this follows a linear transformation for spatial coordinates, and hence the formulation is not isoparametric.

The stiffness matrix can be obtained using Eq. (5.61), which requires evaluation of the strain–displacement matrix $[B]$. This requires expressions for the strains. From elementary beam theory, we obtain the stresses developed by the beam as

$$\sigma_{xx} = \left(\frac{y}{R}\right)Y$$

where the term inside the brackets represents the strain. Here, $(1/R)$ represents the radius of curvature, which can be expressed in terms of the transverse displacement as $1/R = d^2w/dx^2$. Hence, the normal strain in the beam is given by $\varepsilon_{xx} = zd^2w/dx^2$, where $z$ is the depth coordinate. Using the expression for transverse displacement in Eq. (5.83), we can write the strain displacement matrix as

$$\varepsilon_{xx} = [B]\{w\} = z[\bar{B}]\{w\} = z\left[\frac{dN_1^2}{d^2\xi} \quad \frac{dN_2^2}{d^2\xi} \quad \frac{dN_3^2}{d^2\xi} \quad \frac{dN_4^2}{d^2\xi}\right]\begin{Bmatrix} w_1 \\ \theta_1 \\ w_2 \\ \theta_2 \end{Bmatrix} \qquad (5.85)$$

Substituting $[B]$ from Eq. (5.85) in Eq. (5.61), we get

$$[K] = \int_V [B]^T[C][B]dV = \int_0^L \int_A z^2[\bar{B}]^T Y[\bar{B}]dA \, dx = \int_{-1}^1 [\bar{B}]^T YI[\bar{B}]Jd\xi \qquad (5.86)$$

Evaluating Eq. (5.86), we get the following stiffness matrix:

$$[K] = \frac{YI}{L^3}\begin{bmatrix} 12 & -6L & -12 & -6L \\ -6L & 4L^2 & 6L & 2L^2 \\ -12 & -6L & 12 & -6L \\ -6L & 2L^2 & -6L & 4L^2 \end{bmatrix} \qquad (5.87)$$

### 5.4.6 Two-Dimensional Plane Isoparametric Element Formulation

The original and mapped representations of an isoparametric quadrilateral are shown in Figure 5.8(c). Here, $x$–$y$ is the original coordinate system and $\xi$–$\eta$ is the mapped coordinate system. Each of the mapped coordinates ranges from $+1$ to $-1$. This quadrilateral element has four nodes, and each node can support two degrees of freedom. In all, the element has eight degrees of freedom and the resulting stiffness matrix is of size $8 \times 8$. The displacement variation in the two coordinate directions ($u$ along the $x$-direction and $v$ along the $y$-direction) is given in terms of mapped coordinates as

$$u(\xi, \eta) = a_0 + a_1\xi + a_2\eta + a_3\xi\eta$$
$$v(\xi, \eta) = b_0 + b_1\xi + b_2\eta + b_3\xi\eta$$

(5.88)

Substitution of the mapped coordinates at the four nodes determines the shape functions. The displacement field as well as the shape functions are given by (the reader is encouraged to work out the details):

$$\begin{Bmatrix} u \\ v \end{Bmatrix} = \begin{bmatrix} N_1 & 0 & N_2 & 0 & N_3 & 0 & N_4 & 0 \\ 0 & N_1 & 0 & N_2 & 0 & N_3 & 0 & N_4 \end{bmatrix} \{u_e\} = [N]\{u_e\}$$

(5.89)

where

$$\{u_e\} = \begin{Bmatrix} u_1 & v_1 & u_2 & v_2 & u_3 & v_3 & u_4 & v_4 \end{Bmatrix}^{\mathrm{T}}$$

$$N_1 = \frac{(1-\xi)(1-\eta)}{4}, \quad N_2 = \frac{(1+\xi)(1-\eta)}{4},$$

$$N_3 = \frac{(1+\xi)(1+\eta)}{4}, \quad N_4 = \frac{(1-\xi)(1+\eta)}{4}$$

(5.90)

The coordinate transformation between the original and mapped coordinates can be similarly written as

$$\begin{Bmatrix} x \\ y \end{Bmatrix} = \begin{bmatrix} N_1 & 0 & N_2 & 0 & N_3 & 0 & N_4 & 0 \\ 0 & N_1 & 0 & N_2 & 0 & N_3 & 0 & N_4 \end{bmatrix} \{x_e\} = [N]\{x_e\}$$

(5.91)

Here $\{x_e\} = \begin{Bmatrix} x_1 & y_1 & x_2 & y_2 & x_3 & y_3 & x_4 & y_4 \end{Bmatrix}^{\mathrm{T}}$. To compute the derivatives, we invoke the chain rule. Noting that the original coordinates are functions of both mapped coordinates $\xi$ and $\eta$, we have

$$\frac{\partial}{\partial \xi} = \frac{\partial}{\partial x}\frac{\partial x}{\partial \xi} + \frac{\partial}{\partial y}\frac{\partial y}{\partial \xi}, \quad \frac{\partial}{\partial \eta} = \frac{\partial}{\partial x}\frac{\partial x}{\partial \eta} + \frac{\partial}{\partial y}\frac{\partial y}{\partial \eta}$$

or

$$\begin{Bmatrix} \dfrac{\partial}{\partial \xi} \\[2mm] \dfrac{\partial}{\partial \eta} \end{Bmatrix} = \begin{bmatrix} \dfrac{\partial x}{\partial \xi} & \dfrac{\partial y}{\partial \xi} \\[2mm] \dfrac{\partial x}{\partial \eta} & \dfrac{\partial y}{\partial \eta} \end{bmatrix} \begin{Bmatrix} \dfrac{\partial}{\partial x} \\[2mm] \dfrac{\partial}{\partial \eta} \end{Bmatrix} = [J] \begin{Bmatrix} \dfrac{\partial}{\partial x} \\[2mm] \dfrac{\partial}{\partial \eta} \end{Bmatrix}$$

(5.92)

The numerical value of the Jacobian depends on the size, shape and orientation of the element. Also,

$$
\left\{ \begin{array}{c} \dfrac{\partial}{\partial x} \\[2mm] \dfrac{\partial}{\partial \eta} \end{array} \right\} = [J]^{-1} \left\{ \begin{array}{c} \dfrac{\partial}{\partial \xi} \\[2mm] \dfrac{\partial}{\partial \eta} \end{array} \right\} \tag{5.93}
$$

Using Eq. (5.93), we can determine the derivatives required for the computation of the matrix $[B]$. Once this is done, we can derive the stiffness matrix for a plane element as

$$
[K] = t \int\limits_{-1}^{1} \int\limits_{-1}^{1} [B]^{\mathrm{T}}[C][B] J \, \mathrm{d}\xi \, \mathrm{d}\eta \tag{5.94}
$$

where $J$ is the determinant of the Jacobian matrix and $t$ the thickness of the element. The stiffness matrix will be $8 \times 8$. Here $[C]$ is the material matrix, and assuming plane stress condition (see [5] and Chapter 4, Section 4.11 for more details on plane stress condition), we have

$$
[C] = \dfrac{Y}{1 - \nu^2} \begin{bmatrix} 1 & \nu & 0 \\ \nu & 1 & 0 \\ 0 & 0 & \dfrac{1 - \nu}{2} \end{bmatrix} \tag{5.95}
$$

Equation (5.94) cannot be integrated as such in closed form. It must be numerically integrated; for this purpose, we use Gauss quadrature, which is explained in the next subsection.

### 5.4.7 Numerical Integration and Gauss Quadrature

Evaluation of stiffness and mass matrices, specifically for isoparametric elements, involves expressions such as that in Eq. (5.94), where the elements of the matrices are necessarily rational polynomials. Evaluating these integrals in closed forms is very difficult and numerical integration techniques are necessary. Although different numerical techniques are available, Gauss quadrature (see [9] and [10] for more details) is ideally suited for isoparametric formulation, as it evaluates the value of the integral between $-1$ and $+1$, the typical range of natural coordinates in isoparametric formulation.

Consider an integral of the form

$$
I = \int\limits_{-1}^{+1} F \, \mathrm{d}\xi, \quad F = F(\xi) \tag{5.96}
$$

Let $F(\xi) = a_0 + a_1 \xi$. This function is to be integrated over a domain $-1 < \xi < 1$ with the length of the domain equal to two units. When this expression is exactly integrated, we get the value of the integral as $2a_0$. If the value of the integrand is evaluated at the midpoint (i.e. at $\xi = 0$) and multiplied by a weight 2.0, we obtain the exact value of the integral. Hence, an integral of any linear function can be evaluated. This result can be generalized for a function of any order as

$$I = \int_{-1}^{+1} F\,d\xi \approx W_1 F_1 + W_2 F_2 + \cdots + W_n F_n \tag{5.97}$$

Hence, to obtain the approximate value of the integral $I$, we evaluate $F(\xi)$ at several locations $\xi_i$, multiply the resulting $F_i$ by the appropriate weights $W_i$ and add them together.

The points at which the integrand is evaluated are called *sampling points*. In Gauss quadrature, these are the points of very high accuracy, sometimes called *Barlow points* [11]. They are located symmetrically with respect to the center of the interval and symmetrically placed points have same weights. The number of points required to integrate the integrand exactly depends on the degree of the highest polynomial involved in the expression. If $p$ is the highest degree of the polynomial in the integrand, then the minimum number of points $n$ required to integrate the integrand exactly is $n = (p+1)/2$. That is, for a polynomial of second degree ($p = 2$), the minimum number of points required is equal to 2.

Table 5.1 gives the location and weights for Gauss quadrature [9]. For 2D elements, computation of the stiffness and mass matrices involves evaluation of a double integral of the form

$$I = \int_{-1}^{1} \int_{-1}^{1} F(\xi, \eta)\,d\xi\,d\eta = \int_{-1}^{1}\left[\sum_{i=1}^{N} W_i F(\xi_i, \eta)\right]d\eta = \sum_{i=1}^{N}\sum_{j=1}^{M} W_i W_j F(\xi_i, \eta_j) \tag{5.98}$$

where $N$ and $M$ are the number of sampling points used in the $\xi$- and $\eta$-directions. Here we have considered a 2D element. A similar procedure can be extended to 3D. The sampling points of the Gauss quadrature are located so that the stresses, which are less accurate than displacements in FEM, at Gauss points are very accurate compared to other points (see [10] for more details).

**Table 5.1 Sampling points and weights for Gauss quadrature**

| Order (n) | Location ($\xi_i$) | Weight ($W_i$) |
|---|---|---|
| 1 | 0 | 2 |
| 2 | ±0.57735 02691 89626 | 1.0 |
| 3 | ±0.77459 66692 41483 | 0.55555 55555 55556 |
|   | 0.00000 00000 00000 | 0.88888 88888 88889 |
| 4 | ±0.86113 63115 94053 | 0.34785 48451 37454 |
|   | ±0.33998 20435 84856 | 0.65214 51548 62546 |
| 5 | ±0.90617 98459 38664 | 0.23692 68850 56189 |
|   | ±0.53846 93101 05683 | 0.47862 86704 99366 |
|   | 0.00000 00000 00000 | 0.56888 88888 88889 |

Numerical integration for the isoparametric triangle can also be done using Gauss quadrature. However, the Gauss points and the weights, as given in [9], are quite different. The numerical integration of the type given in Eq. (5.96) is given by

$$I = \frac{1}{2}\sum_{i=1}^{n} W_i F(\alpha_i, \beta_i, \gamma_i) \tag{5.99}$$

where $(\alpha_i, \beta_i, \gamma_i)$ is the location of the Gauss points in area coordinates.

## ▶ 5.5 NUMERICAL EXAMPLES

This section presents some numerical examples, the aim of which is to demonstrate the FE process in solving some interesting problems. All the problems treated here are essentially 1D structures consisting of rods and beams. Problems involving 2D structures are difficult to solve with hand calculations. In the following subsections, three examples are solved, two involving rod structures and the third beam structure.

### 5.5.1 Example 1: Analysis of a Stepped Bar (Rod)

A stepped bar (also called rod) of length $2L$, fixed at both ends and subjected to a central point load $P$, is shown in Figure 5.9. The aim of this example is to determine the stresses developed in this stepped bar due to the central loading.

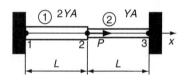

**Figure 5.9** A two-bar assembly of stepped bar with fixed ends.

To start with, the domain can be split up into two elements, one (element 1) having modulus of elasticity $2Y$ and the second having modulus of elasticity $Y$. Element 1 is made up of nodes 1 and 2, called the connectivity nodes, whereas element 2 is made up of nodes 2 and 3. Each element is of length $L$ and node 2 is loaded with a force $P$. Both segments have cross-sectional area $A$. The degrees of freedom associated with this problem are the axial deformations at nodes 1, 2, and 3, namely, $u_1$, $u_2$, and $u_3$.

We begin the analysis by considering the stiffness matrix of the rod element derived in Eq. (5.76). For element 1 having modulus of elasticity $2Y$, this can be written

$$k_1 = \frac{2YA}{L}\begin{matrix} u_1 & u_2 \\ \begin{bmatrix} 1 & -1 \\ -1 & 1 \end{bmatrix} \end{matrix} \tag{5.100}$$

Similarly, for element 2 with modulus of elasticity $Y$, the stiffness matrix is given by

$$k_1 = \frac{YA}{L}\begin{matrix} u_2 & u_3 \\ \begin{bmatrix} 1 & -1 \\ -1 & 1 \end{bmatrix} \end{matrix} \tag{5.101}$$

Now we assemble the matrices. Node 2 is common to both element 1 and element 2. Hence, its contribution from both the matrices must be added in the global stiffness matrix. Note that the middle element in the assembled matrix (Eq. (5.102)) shows the contribution of node 2 from both the elements. The assembled stiffness matrix becomes

$$[K] = \frac{YA}{L} \begin{bmatrix} 2 & -2 & 0 \\ -2 & (2+1) & -1 \\ 0 & -1 & 1 \end{bmatrix} \begin{Bmatrix} u_1 \\ u_2 \\ u_3 \end{Bmatrix} = \begin{Bmatrix} F_1 \\ F_2 \\ F_3 \end{Bmatrix} \tag{5.102}$$

The load and boundary conditions are

$$u_1 = u_3 = 0, \quad F_2 = P$$

and the FE equation becomes

$$\frac{YA}{L} \begin{bmatrix} 2 & -2 & 0 \\ -2 & 3 & -1 \\ 0 & -1 & 1 \end{bmatrix} \begin{Bmatrix} 0 \\ u_2 \\ 0 \end{Bmatrix} = \begin{Bmatrix} F_1 \\ P \\ F_3 \end{Bmatrix} \tag{5.103}$$

Deleting the first row and column and the third row and column, we obtain

$$\frac{YA}{L}[3]\{u_2\} = \{P\}, \quad u_2 = \frac{PL}{3YA}$$

The stress in element 1 is

$$\sigma_1 = Y\varepsilon_1 = Y[B]\{u\}_{\text{element}1} \tag{5.104}$$

Equation (5.74) gives the value of matrix $[B]$, and $\{u\}_{\text{element }1} = \{u_1 \quad u_2\}^{\text{T}} = \{0 \quad u_2\}^{\text{T}}$ is the nodal vector of element 1. Using these in the stress equation, we get the stress in element 1 as

$$\sigma_1 = Y\frac{u_2 - u_1}{L} = \frac{Y}{L}\left(\frac{PL}{3YA} - 0\right) = \frac{P}{3A} \tag{5.105}$$

Similarly, the stress in element 2 is given by

$$\sigma_2 = Y\varepsilon_2 = Y[B]\{u\}_{\text{element }2}, \quad \{u\}_{\text{element }2} = \{u_2 \quad u_3\}^{\text{T}} = \{u_2 \quad 0\}^{\text{T}}$$

Substituting matrix $[B]$ from Eq. (5.74) and the displacement, the stress in element 2 is given by

$$\sigma_2 = Y\frac{u_3 - u_2}{L} = \frac{Y}{L}\left(0 - \frac{PL}{3YA}\right) = -\frac{P}{3A} \tag{5.106}$$

This indicates that bar 2 is in compression.

**Observations**

1. In this case, the calculated stresses in elements 1 and 2 are exact within the linear theory for 1D bar structures. Hence, it will not help to divide element 1 or 2 further into smaller FEs.

2. For tapered bars, averaged values of the cross-sectional areas should be used for the elements.

3. We need to find the displacements first to find the stresses, since we are using the *displacement-based FEM*.

### 5.5.2 Example 2: Analysis of a Fixed Rod Subjected to Support Movement

Here again we consider a rod of length $L$ fixed at both ends and subjected to a central point load $P$. The bar configuration is shown in Figure 5.10. However, the right support moves by a small amount $\Delta$ in the horizontal direction. The aim here is to determine the support reactions at both ends. The numerical values of the parameters are $P = 6.0 \times 10^4$ N, $Y = 2.0 \times 10^4$ N/mm², $A = 250$ mm², $L = 150$ mm, and $\Delta = 1.2$ mm.

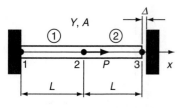

**Figure 5.10** Configuration of the bar undergoing support movement.

As before, we split the rod into two elements. This is required due to the presence of applied load in the middle of the bar. We first check to see if the bar will contact the wall on the right. To do this, we imagine that the wall on the right is removed and calculate the displacement at the right end:

$$\Delta_0 = \frac{PL}{YA} = \frac{(6.0 \times 10^4)(150)}{(2.0 \times 10^4)(250)} = 1.8 \text{ mm} > \Delta = 1.2 \text{ mm} \tag{5.107}$$

Thus, contact occurs. Next, we follow the process given in Example 1. That is, the two-element stiffness matrix is generated and assembled. As before, node 2 is common to both elements, and hence the global stiffness matrix has contributions from both the elements at the second degree of freedom. The major difference between the current example and Example 1 is the way the boundary conditions are imposed. The global FE equation is found to be

$$\frac{YA}{L} \begin{bmatrix} 1 & -1 & 0 \\ -1 & 2 & -1 \\ 0 & -1 & 1 \end{bmatrix} \begin{Bmatrix} u_1 \\ u_2 \\ u_3 \end{Bmatrix} = \begin{Bmatrix} F_1 \\ F_2 \\ F_3 \end{Bmatrix} \tag{5.108}$$

The load and boundary conditions are $F_2 = P = 6.0 \times 10^4$ N, $u_1 = 0$, and $u_3 = \Delta = 1.2$ mm. The FE equation becomes

$$\frac{YA}{L} \begin{bmatrix} 1 & -1 & 0 \\ -1 & 2 & -1 \\ 0 & -1 & 1 \end{bmatrix} \begin{Bmatrix} 0 \\ u_2 \\ \Delta \end{Bmatrix} = \begin{Bmatrix} F_1 \\ P \\ F_3 \end{Bmatrix} \tag{5.109}$$

Equation (5.109) can be expanded into three simultaneous equation. The second equation above gives

$$\frac{YA}{L}[2 \quad -1] \begin{Bmatrix} u_2 \\ \Delta \end{Bmatrix} = \{P\} \tag{5.110}$$

That is,

$$\frac{YA}{L}[2]\{u_2\} = \left\{P + \frac{YA}{L}\Delta\right\} \tag{5.111}$$

Solving this, we obtain

$$u_2 = \frac{1}{2}\left(\frac{PL}{YA} + \Delta\right) = 1.5\,\text{mm} \tag{5.112}$$

To calculate the support reaction forces, we apply the first and third equations in the global FE equation [Eq. (5.109)]. The first equation gives

$$F_1 = \frac{YA}{L}[1 - 1\,0]\begin{Bmatrix} u_1 \\ u_2 \\ u_3 \end{Bmatrix} = \frac{YA}{L}(-u_2) = -5.0 \times 10^4\,\text{N} \tag{5.113}$$

The third equation in Eq. (5.109) gives

$$F_3 = \frac{YA}{L}[0 - 1\,1]\begin{Bmatrix} u_1 \\ u_2 \\ u_3 \end{Bmatrix} = \frac{YA}{L}(-u_2 + u_3) = -1.0 \times 10^4\,\text{N} \tag{5.114}$$

### 5.5.3 Example 3: A Spring-Supported Beam Structure

This example considers a beam undergoing transverse deformation; the configuration of the beam is shown in Figure 5.11. The beam structure has a middle support and the right end is supported by a spring. Such configurations can be found in MEMS structures on a silicon substrate, where the stiffness of the substrate is modeled as a spring. Here, the properties are $P = 50.0$ kN, $k = 200$ kN/m, $L = 3.0$ m, $Y = 210$ GPa, $I = 2 \times 10^{-4}$ m$^4$.

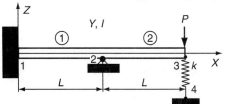

**Figure 5.11** A spring-supported beam structure subjected to transverse loads.

The beam structure is modeled with two elements, element 1 spanning nodes 1 and 2 and element 2 spanning nodes 2 and 3. The element stiffness matrix of the beam is given by Eq. (5.87). We use this to generate the elemental stiffness matrix for each element.

Unlike rods, beams support two degrees of freedom at each node, namely, the transverse displacement $w$ and the rotation $\theta$. As in the previous two examples, we generate the stiffness matrix for each element and assemble. As before, node 2 is common to both elements, and the global stiffness matrix has contributions from both elements for the transverse and rotational degrees of freedom corresponding to node 2.

Next we model the spring in the FE framework. This is similar to a rod element with stiffness contributing to nodes 3 and 4 in the transverse $z$-direction. The spring stiffness matrix is given by

$$k_3 = \begin{matrix} & w_3 & w_4 \\ & \begin{bmatrix} k & -k \\ -k & k \end{bmatrix} \end{matrix}$$

Now, adding all the stiffnesses, the global FE equation is given by

$$
\frac{YI}{L^3}
\begin{bmatrix}
12 & 6L & -12 & 6L & 0 & 0 & 0 \\
 & 4L^2 & -6L & 2L^2 & 0 & 0 & 0 \\
 & & 24 & 0 & -12 & 6L & 0 \\
 & & & 8L^2 & -6L & 2L^2 & 0 \\
 & & & & 12+k' & -6L & -k' \\
 & \text{Symmetric} & & & & 4L^2 & 0 \\
 & & & & & & k'
\end{bmatrix}
\begin{Bmatrix}
w_1 \\ \theta_1 \\ w_2 \\ \theta_2 \\ w_3 \\ \theta_3 \\ w_4
\end{Bmatrix}
=
\begin{Bmatrix}
V_1 \\ M_1 \\ V_2 \\ M_2 \\ V_3 \\ M_3 \\ V_4
\end{Bmatrix}
$$

where $k' = (L^3/EI)k$. We now apply the boundary conditions $w_1 = \theta_1 = w_2 = w_4 = 0$, $M_2 = M_3 = 0$, and $V_3 = -P$. Deleting the first three equations as well as the seventh equation (rows and columns), we have the following reduced equation:

$$
\frac{YI}{L^3}
\begin{bmatrix}
8L^2 & -6L & 2L^2 \\
-6L & 12+k' & -6L \\
2L^2 & -6L & 4L^2
\end{bmatrix}
\begin{Bmatrix}
\theta_2 \\ w_3 \\ \theta_3
\end{Bmatrix}
=
\begin{Bmatrix}
0 \\ -P \\ 0
\end{Bmatrix}
$$

Solving this equation, we obtain the deflection and rotation at nodes 2 and 3; they are given by

$$
\begin{Bmatrix}
\theta_2 \\ w_3 \\ \theta_3
\end{Bmatrix}
= -\frac{PL^2}{YI(12+7k')}
\begin{Bmatrix}
3 \\ 7L \\ 9
\end{Bmatrix}
$$

The influence of the spring $k$ is easily seen from this result. Plugging in the given numbers, we can calculate

$$
\begin{Bmatrix}
\theta_2 \\ w_3 \\ \theta_3
\end{Bmatrix}
=
\begin{Bmatrix}
-0.002492 & \text{rad} \\
-0.01744 & \text{m} \\
-0.007475 & \text{rad}
\end{Bmatrix}
$$

From the global FE equation, we obtain the nodal reaction forces as

$$
\begin{Bmatrix}
V_1 \\ M_1 \\ V_2 \\ V_4
\end{Bmatrix}
=
\begin{Bmatrix}
-69.78 & \text{kN} \\
-69.78 & \text{N} \cdot \text{m} \\
116.2 & \text{kN} \\
3.488 & \text{kN}
\end{Bmatrix}
$$

## ▶ 5.6 FINITE ELEMENT FORMULATION FOR TIME-DEPENDENT PROBLEMS

In Section (5.4.2) we derived the finite element governing equation using Hamilton's principle. The formulation resulted in three matrices: the mass matrix, the damping matrix, and the stiffness matrix. In the previous section, we dealt with numerical examples where the loads were all static; hence, we used only the stiffness matrix, the formulation of which was addressed in Sections (5.4.4)—(5.4.6). When the loads are time dependent, in addition to the elastic force provided by the stiffness matrix, two additional forces are developed, one due to the inertial force provided by the mass of the structure and the other the damping force arising due to the material property of the structure. The main effect of the damping force is to retard the motion.

Hamilton's principle (Section 5.4.2) directly yields the finite element governing equation of motion for dynamical problems given by Eq. (5.65). Unlike the static equation $[K]\{u_e\} = \{R\}$, where $\{u_e\}$ is the nodal displacement vector, Eq. (5.65) is a highly coupled second-order ordinary differential equation, the solution of which is quite difficult compared with the static case.

Two different situations arise in time-dependent problems. The first is the *free vibration* problem in which the microsystem in question is given a small disturbance and the vibration properties of the system are studied without influence of the external time-dependent loads. That is, $\{R\}$ is assumed to be zero in Eq. (5.65). If the microsystem is given a small disturbance, the inertia of the system makes it vibrate at certain frequencies due to the acceleration provided by the disturbance. (Frequency is defined as the rate at which the motion repeats itself. It is equal to inverse of the time period. In vibration terminology, it is normally referred to as the *natural frequency*. It depends on the stiffness and mass of the structure.)

Consider the MEMS cantilever structure shown in Figure 5.12, which shows the first three vibrational modes of the structure. Note that each mode deforms differently and vibrates at a different frequency.

The second class of vibration problems is the *forced vibration* problem, where the vector $\{R\}$ in Eq. (5.65) is nonzero and is a function of both space and time. The spatial function of the load is specified by indicating the degree of freedom of the finite element model on which the load acts, and its temporal variation is given as the force history, that is, the variation of the load with respect to time.

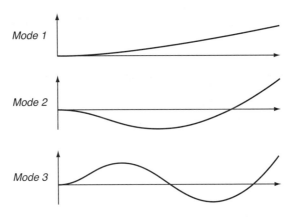

**Figure 5.12** Mode shapes of a cantilevered structure.

For a certain class of problems, it is essential to determine the free vibration parameters, the natural frequencies, and mode shapes to obtain the forced vibration response. The input forcing function has a certain frequency content that can be determined by feeding it to the Fast Fourier Transform (FFT) algorithm (see [12] and [13] for signal analysis using the FFT algorithm). All the frequencies up to the frequency bandwidth of the input signal will be excited by the forcing frequency. If these frequencies match the natural frequencies of the system, the phenomenon of *resonance* occurs. At resonance, the dynamic response of the device increases steeply, leading to the collapse of the device. Hence, resonance conditions should be avoided at all costs, and this requires that the natural frequencies and the forcing frequencies should be well separated; this aspect should be taken care of while designing microsystem devices. In addition, it is a good practice to perform a free-vibration analysis before the forced-vibration analysis.

The free vibration problems always lead to eigenvalue problems in which the eigenvalues are the natural frequencies of vibration and the eigenvectors give the deformed shape (mode shape) of the microsystem caused by the vibration. The number of natural frequencies obtained is equal to the number of degrees of freedom in the finite element model. For preliminary understanding of vibration concepts, the reader is encouraged to refer to some classic texts such as Thomson [14], Meirovich [15], and Clough and Penzin [16]. In the next section, we outline the procedure for deriving the mass and damping matrices required for finite element analysis.

### 5.6.1 Mass and Damping Matrix Formulation

As mentioned in Section 5.4.2, the mass matrix can be generated in two different ways. In the first method, the masses are generally assumed to be lumped and are derived by using appropriate laws of physics. That is, the total mass or mass moment of inertia of the system is calculated and then distributed among the degrees of freedom of the FE model in a certain manner. Such a model gives a mass matrix that is diagonal. The other way of generating the mass matrix is to use the shape functions derived in Section 5.4.1. The mass matrix generated using such a procedure is called the *consistent mass matrix*. Obviously, the lumped mass matrix is easy to operate. However, there are merits and disadvantages to using either of them (see [9]).

The expression for the *consistent mass matrix* is given by Eq. (5.65), which is

$$[M] = \int_V \rho [N]^T [N] dV$$

where $\rho$ is the density and $[N]$ is the shape function matrix. This matrix is a fully populated and banded matrix whose bandwidth is equal to that of the stiffness matrix. For a rod element of length $L$, cross-sectional area $A$, and density $\rho$, the shape function is given by Eq. (5.34). Using this shape function, the mass matrix becomes

$$[M] = \int_0^L \int_A \rho \begin{bmatrix} \dfrac{1-x}{L} \\ \dfrac{x}{L} \end{bmatrix} \begin{bmatrix} \dfrac{1-x}{L} & \dfrac{x}{L} \end{bmatrix} dx = \frac{\rho A L}{6} \begin{bmatrix} 2 & 1 \\ 1 & 2 \end{bmatrix} \tag{5.115}$$

For a beam of length $L$ and cross-sectional area $A$, the four shape functions are given by Eq. (5.37b). Substituting these in the mass matrix expression and integrating, we get

$$[M] = \frac{\rho AL}{420} \begin{bmatrix} 156 & 22L & 54 & -13L \\ & 4L^2 & 13L & -3L \\ \text{SYM} & & 156 & -22L \\ & & & 4L^2 \end{bmatrix} \tag{5.116}$$

In both these cases, we find that the matrix is symmetric and positive definite. By *positive definiteness,* it is meant that the kinetic energy $T = (1/2)\{\dot{u}\}^{\mathrm{T}}[M]\{\dot{u}\} > 0$, where $\{\dot{u}\}$ is the element velocity vector. In the lumped mass formulation, the masses can be lumped corresponding to the main degrees of freedom, which make the mass matrix diagonal. The diagonal mass matrix has very small storage requirements and hence enables faster solutions of the dynamic equations of motion.

There are certain problems such as wave propagation in which lumped mass is preferred to consistent mass. Three different methods have been reported in literature for lumping masses:

1. Adhoc Lumping.
2. HRZ Lumping.
3. Optimal Lumping.

*Adhoc lumping* is the simplest way of lumping the mass. The total mass of the structure is computed and is distributed evenly among all the translational degrees of freedom. If the element has rotational degrees of freedom, then the mass moment of inertia of the element is computed and distributed evenly among the rotational degrees of freedom.

Let us again consider a two-noded rod element of length $L$, density $\rho$, and cross-sectional area $A$. The total mass of the element is $\rho AL$. If this mass is equally distributed between the two axial degrees of freedom, the lumped mass can be written as

$$[M]_{\text{lumped}} = \frac{\rho AL}{2} \begin{bmatrix} 1 & 0 \\ 0 & 1 \end{bmatrix}$$

Now consider a three-noded quadratic bar (a bar having three nodes) having the same element properties of the two-noded bar. The total mass is again equal to $\rho AL$, which can be distributed equally among the three axial degrees of freedom. The lumped mass matrix then becomes

$$[M]_{\text{lumped}} = \frac{\rho AL}{3} \begin{bmatrix} 1 & 0 & 0 \\ 0 & 1 & 0 \\ 0 & 0 & 1 \end{bmatrix} \tag{5.117}$$

Experience has shown that this matrix gives poor results. On the other hand, if the three-noded bar is split into two halves each having mass of $\rho AL/2$, then the middle node gets the mass contribution from both halves and the mass matrix becomes

$$[M]_{\text{lumped}} = \frac{\rho AL}{4} \begin{bmatrix} 1 & 0 & 0 \\ 0 & 2 & 0 \\ 0 & 0 & 1 \end{bmatrix} \tag{5.118}$$

This mass representation gives much better results because the distribution of mass is more appropriate. Hence, in adhoc mass lumping, no fixed rules are specified for lumping procedure. It is purely left to the analyst to decide how the masses should be lumped.

The lumped mass for a beam of length $L$, density $\rho$, and cross-sectional area $A$ that has four degrees of freedom including two rotational degrees of freedom can be derived by again taking the total mass $m$ equal to $\rho AL$, which can be distributed equally between the two transverse degrees of freedom $w_1$ and $w_2$ [see Figure 5.6(b)].

The mass corresponding to rotational degrees of freedom $\theta_1$ and $\theta_2$ is derived as follows. The mass moment of inertia of a bar is given by $mL^2/3$, where $m$ is the mass of the bar. In our case, for better approximation, we split the beam into two halves of length $L/2$ and the mass moment of inertia of each half is computed and lumped onto the respective rotational degrees of freedom. The mass moment of inertia is $(1/3)(m/2)(L/2)^2 = \rho AL^3/24$. Hence, the lumped mass for the beam becomes

$$[M]_{\text{lumped}} = \frac{\rho AL}{420} \begin{bmatrix} \alpha & 0 & 0 & 0 \\ 0 & \beta L^2 & 0 & 0 \\ 0 & 0 & \alpha & 0 \\ 0 & 0 & 0 & \beta L^2 \end{bmatrix}, \quad \alpha = 210, \quad \beta = 17.5 \tag{5.119}$$

One can compare Eq. (5.119) with Eq. (5.116) and establish the correlation between the two different mass matrices.

Hinton, Rock and Zeikienwich [17] derived a new lumping scheme that uses the consistent mass matrix; it is called the *HRZ lumping* (named from the authors' initials). The diagonal coefficients are extracted from the consistent mass matrix as follows. The consistent mass matrix is first obtained. If $m$ is the total mass and $N_i$ is the shape function of the $i^{th}$ degree of freedom, then using the diagonal coefficients of the consistent mass matrix derived earlier, the HRZ lumped mass matrix is given by

$$M_{ii} = \frac{m}{S} \int_V \rho N_i^2 \, dV, \quad S = \sum_{n=1}^N (M_{ii})_{\text{consistent}} \tag{5.120}$$

Let us consider the same example of the two-node bar. The total mass of the bar is $m = \rho AL$. The consistent mass matrix for this linear bar is given by Eq. (5.115). We have $S$ in Eq. (5.120) is equal to $(2/3)\rho AL$. The shape functions $N_1$ and $N_2$ are given in Eq. (5.34). Using this and the value of $S$ and $m$, we can write the lumped mass matrix through HRZ lumping as

$$[M]_{HRZ} = \frac{\rho AL}{2} \begin{bmatrix} 1 & 0 \\ 0 & 1 \end{bmatrix} \tag{5.121}$$

This matrix is the same as that derived using adhoc lumping. The same procedure can be adapted to derive the lumped mass matrix based on the HRZ-lumping procedure for beams and 2D isoparametric elements. For beams, the consistent mass matrix is given by Eq. (5.116). For this case, we have two values of $S$, one for translational degrees of freedom and the other for rotational degrees of freedom. For translational degrees of freedom, $S = (\rho AL/420)(156 + 156) = (26/35)\rho AL$ and for rotational degrees of freedom, $S = (\rho AL/420)(4L^2 + 4L^2) = (2L^2/105)\rho AL$. Using this and the shape functions given in Eq. (5.37b), we get

$$[M]_{\text{HRZ}} = \frac{\rho AL}{78} \begin{bmatrix} 39 & 0 & 0 & 0 \\ 0 & L^2 & 0 & 0 \\ 0 & 0 & 39 & 0 \\ 0 & 0 & 0 & L^2 \end{bmatrix} \tag{5.122}$$

Comparing this with the mass matrix obtained by adhoc lumping in Eq. (5.119), we see that the translational degrees of freedom have similar mass distributions, while the rotational degrees of freedom have smaller values. Similar mass matrices for eight- and nine-noded isoparametric 2D elements can be obtained. These elements are shown in Figure 5.13.

HRZ lumping gives very good results for lower-order elements (sometimes this lumping scheme gives better results than the consistent mass formulation, especially for bending-dominated problems). It is less accurate for higher-order elements.

*Optimal lumping*, introduced by Malkus and Plesha [18], uses a numerical integration scheme to obtain a lumped mass matrix. That is, it uses the property of the shape function discussed in Section 5.4.1. The shape function takes the value of unity at the node where it is evaluated; at other nodes, its value is zero. The scheme requires an integration scheme that uses the nodes as the sampling point. This process eliminates the off-diagonal terms in the mass matrix, the *Newton–Coates method* (see [2] for more details), which is the one-third Simpson's rule, provides for integrating numerically with nodal points as the sampling points.

The 2D version of the Newton–Coates method is the *Labatto integration rule* [18]. The number of sampling points required for numerical integration is determined by the highest order of polynomial involved in computation of the mass matrix, and in general is given by

$$n = 2(p - m) \tag{5.123}$$

Here $p$ is the highest order of the polynomial, $n$ is the number of sampling points required for numerical integration, and $m$ is the highest order of the derivative appearing in the energy functional. For plane stress problem, $m = 1$, while for bending problems, $m = 2$.

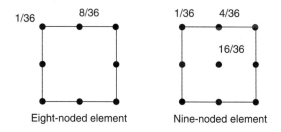

Eight-noded element        Nine-noded element

**Figure 5.13** HRZ mass lumping for 2D eight- and nine-noded elements.

Let us consider a three-noded isoparametric quadratic bar element of length $L$ having three axial degrees of freedom corresponding to the three nodes [see Figure 5.8(b)]. The isoparametric shape functions are given in Eq. (5.78). In this case, $p = 2$, $m = 1$, and hence, the minimum number of points required according to Eq. (5.123) is 2. The Newton–Coates formula for integrating the function $f(x)$ in the interval $a$ to $b$ is given by (see [2] for more details)

$$\int_a^b f(x)dx = (b-a)\left[\frac{1}{6}f(x=a) + \frac{4}{6}f\left(x = \frac{a+b}{2}\right) + \frac{1}{6}f(x=b)\right]$$

(5.124)

Now, the mass matrix of the bar in indicial notation can be written in terms of shape functions as

$$M_{ij} = \int_{-1}^{1} \rho A \, N_i N_j J \, d\xi$$

(5.125)

Here $J$ is the Jacobian and its value is equal to $L/2$ if the middle node is exactly at the center. Now using the Newton–Coates formula and noting that $b-a = 2$, we get

$$M_{ij} = \rho A L \left[\begin{array}{l} \frac{1}{6}N_i(\xi=-1)N_j(\xi=-1) + \frac{4}{6}N_i(\xi=0)N_j(\xi=0) + \\ \frac{1}{6}N_i(\xi=1)N_j(\xi=1) \end{array}\right]$$

(5.126)

Substituting the shape functions and evaluating, we get

$$[M]_{\text{optimal}} = \frac{\rho A L}{6}\begin{bmatrix} 1 & 0 & 0 \\ 0 & 1 & 0 \\ 0 & 0 & 4 \end{bmatrix}$$

(5.127)

We will get the same results if we use HRZ lumping. One can similarly obtain the optimally lumped mass matrix for 2D triangular and quadrilateral elements, as shown in Figure 5.14. From the figure, we see that some mass coefficients are zero and even have a

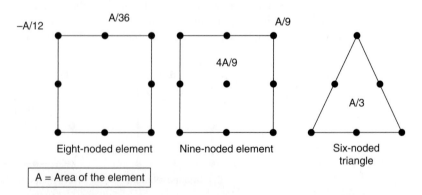

**Figure 5.14** Optimal mass lumping for six-, eight-, and nine-noded elements.

negative value. This will pose problems in the solution of dynamic equations. Some special solution schemes are required for the purpose.

Here we state the general guidelines for the choice of mass matrix. The consistent mass matrix is generally used for flexure problems such as bending of a cantilever structure, but in general, it gives poor results when the mode-shape spans more than four elements. It is generally not advised for problems requiring high-frequency input signals. The computed natural frequencies required for the solution of the dynamic equations are always upper bounds and it is expensive to store and operate as it is fully populated. However, it is ideally suited for higher-order elements.

The lumped mass matrix is used extensively in wave propagation and transient dynamics problems as it gives very few spurious oscillations. It gives good results for lower-order elements. However, for higher-order elements, one can use the optimally lumped matrix.

**Your Turn:**

We derived the optimally lumped matrix in Eq. (5.127) using the Newton–Coates formula for a three-noded bar shown in Figure 5.8(b). For this bar, derive the consistent mass matrix and the lumped matrix using the HRZ-lumping procedure.

Damping is a very complex phenomenon in structures and is difficult to ascertain exactly. It depends on a variety of factors such as material properties, environmental conditions, frequency, etc. Many times, laboratory tests do not yield repeatable results. Some of the sources of damping are material hysteresis and friction in joints. The damping matrix derived from the equation of motion [Eq. (5.65)] is seldom used, as it is difficult to measure the damping parameter $\eta$. The most common method of deriving the damping matrix is called *Rayleigh's proportional damping*, where the damping matrix is a combination of the stiffness and mass matrices given by

$$[C] = \alpha[K] + \beta[M] \tag{5.128}$$

where $\alpha$ and $\beta$ are the stiffness and mass proportional damping coefficients, respectively, to be determined experimentally. They can be measured by determining the damping ratio $\xi$ from the single-degree-of-freedom model. The damping ratio and the stiffness and mass proportional parameters are related by

$$\xi = \frac{1}{2}\left(\alpha\omega + \frac{\beta}{\omega}\right) \tag{5.129}$$

A plot of frequency $\omega$ with damping ratio $\xi$ is shown in Figure 5.15. By measuring the damping ratio at two different frequencies, one can find the stiffness- and mass-proportional damping coefficients. This damping representation has the great advantage of uncoupling the governing equation for solution of the dynamic equation in the modal domain.

## 5.6.2 Free Vibration Analysis

As mentioned earlier, free vibration analysis mainly deals with determination of natural frequencies and mode shapes. This information is required to determine the forced response of the system under investigation. Analysis of the forced response is dealt

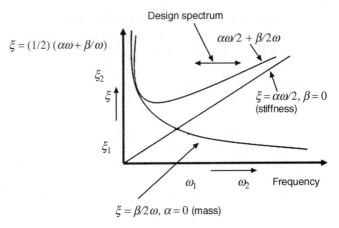

**Figure 5.15** Damping behavior as a function of frequency.

with in the next subsection. The starting point in free vibration analysis is to determine the global stiffness and mass matrix of the system under investigation and solve the associated eigenvalue problem (explained next). Note that the global stiffness and mass matrices are obtained by assembling the element stiffness and mass matrices of all the elements of the discretized system as described in Sections 5.5.2 and 5.5.3, respectively.

We begin the free vibration analysis by considering Eq. (5.65), in which the vector $\{R\} = 0$. Also, the damping matrix $[C]$ is not normally considered in the analysis. Hence, the governing dynamic equilibrium equation becomes

$$[M]\{\ddot{d}\} + [K]\{d\} = \{0\} \tag{5.130}$$

In this equation, $[M]$ and $[K]$ are the assembled mass and stiffness matrices of size $n \times n$, where $n$ is the number of active degrees of freedom of the discretized system and $\{d\}$ is the vector of nodal degrees of freedom of size $n \times 1$. In vibration analysis, we assume that the system undergoes harmonic motion due to inertia when subjected to some initial disturbance. If the system undergoes harmonic motion, the displacement is proportional and $180°$ out of phase with the acceleration. Hence the displacement, velocity, and acceleration vectors can be represented as

$$\{d(t)\} = \{\hat{d}\}\sin(\omega t), \{\dot{d}(t)\} = \omega\{\hat{d}\}\cos(\omega t), \{\ddot{d}(t)\} = -\omega^2\{\hat{d}\}\sin(\omega t) \tag{5.131}$$

In this equation, $\{\hat{d}\}$ is the frequency-domain displacement amplitude (mode shape) and $\omega$ is the natural frequency of the system that is to be determined. Substituting Eq. (5.131) in Eq. (5.130), we get

$$[K]\{\hat{d}\} = \omega^2[M]\{\hat{d}\} \tag{5.132}$$

Equation (5.132) is an eigenvalue problem with $\omega^2$ the eigenvalue or the natural frequencies and $\{\hat{d}\}$ the eigenvector or the mode shape. If the system is discretized with a FE mesh having $n$ active degrees of freedom, it will have $n$ natural frequencies and mode shapes.

### 5.6.3 Free Vibration Analysis of a Fixed Rod

The static analysis of a fixed rod subjected to a central point load was considered in Section 5.5.2. In this section, we consider the same rod and perform free vibration analysis. The main objective here is to compare the natural frequencies and mode shapes obtained from lumped mass and consistent mass formulations and draw some important conclusions from our results.

First, the rod is modeled using two elements as shown in Figure 5.16(a). The bar has length $2L$, Young's modulus $Y$, cross-sectional area $A$, and density $\rho$. Element 1 shares nodes 1 and 2 while element 2 is shared by nodes 2 and 3. The stiffness matrix of element 1 and element 2 is given by

$$k_1 = \frac{YA}{\bar{L}} \begin{array}{cc} u_1 & u_2 \\ \left[ \begin{array}{cc} 1 & -1 \\ -1 & 1 \end{array} \right] & \begin{array}{c} u_1 \\ u_2 \end{array} \end{array} \qquad k_2 = \frac{YA}{\bar{L}} \begin{array}{cc} u_2 & u_3 \\ \left[ \begin{array}{cc} 1 & -1 \\ -1 & 1 \end{array} \right] & \begin{array}{c} u_2 \\ u_3 \end{array} \end{array} \qquad (5.133)$$

where $u_1, u_2$, and $u_3$ are axial degrees of freedom at the three nodes. Assembling the element stiffness matrix of the two elements given in Eq. (5.133), we get

$$[K] = \frac{YA}{\bar{L}} \begin{array}{ccc} u_1 & u_2 & u_3 \\ \left[ \begin{array}{ccc} 1 & -1 & 0 \\ -1 & 2 & -1 \\ 0 & -1 & 1 \end{array} \right] & \begin{array}{c} u_1 \\ u_2 \\ u_3 \end{array} \end{array} \qquad (5.134)$$

Since $u_2$ is the only active degree of freedom, rows 1 and 3 are eliminated in the preceding equation, leaving the reduced stiffness matrix $[\bar{K}]$ as a single element given by

$$[\bar{K}] = \frac{YA}{\bar{L}} [2] \qquad (5.135)$$

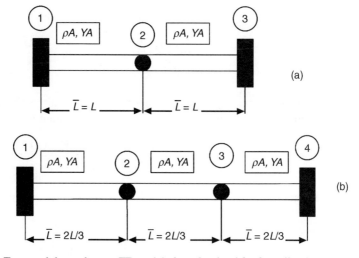

(a)

(b)

**Figure 5.16** Two- and three-element FE models for a fixed rod for free vibration analysis.

Next, we need to construct the mass matrix. We first construct the lumped mass matrix, where the total mass of the element is distributed equally across the translational degrees of freedom. The lumped mass matrix for the two elements is given by

$$
\begin{array}{cc}
u_1 \ u_2 & u_2 \ u_3
\end{array}
$$

$$
m_{l1} = \frac{\rho A \bar{L}}{2} \begin{bmatrix} 1 & 0 \\ 0 & 1 \end{bmatrix} \begin{matrix} u_1, \\ u_2 \end{matrix} \qquad m_{l2} = \frac{\rho A \bar{L}}{2} \begin{bmatrix} 1 & 0 \\ 0 & 1 \end{bmatrix} \begin{matrix} u_2 \\ u_3 \end{matrix} \tag{5.136}
$$

The mass matrix can be assembled in the same manner as the stiffness matrix, which yields the single-element reduced lumped mass matrix corresponding to the $u_2$ degree of freedom given by

$$
[\bar{M}]_{\text{lumped}} = \left[ \frac{\rho A \bar{L}}{2} [2] \right] \tag{5.137}
$$

Equations (5.136) and (5.137) are then substituted in Eq. (5.132) to get

$$
\left[ \frac{YA}{L} [2] - \omega^2 \frac{\rho A \bar{L}}{2} [2] \right] \hat{u}_2 = 0
$$

In this equation, $\omega$ is the fundamental natural frequency of this fixed rod while $\hat{u}_2$ is its mode shape. From this equation, we can write the fundamental natural frequency of the rod as

$$
\omega_{\text{lumped}} = \frac{\sqrt{2}}{\bar{L}} \sqrt{\frac{Y}{\rho}} = \frac{\sqrt{2}}{L} \sqrt{\frac{Y}{\rho}} = \frac{1.414 C_0}{L} \tag{5.138}
$$

where $C_0 = \sqrt{Y/\rho}$ is the wave speed in the rod. We next evaluate the natural frequency of this rod system using the consistent mass matrix, which, for the two-element rod system in Figure 5.16(a), is given by

$$
\begin{array}{cc}
u_1 \ u_2 & u_2 \ u_3
\end{array}
$$

$$
m_{c1} = \frac{\rho A \bar{L}}{6} \begin{bmatrix} 2 & 1 \\ 1 & 2 \end{bmatrix} \begin{matrix} u_1, \\ u_2 \end{matrix} \qquad m_{l2} = \frac{\rho A \bar{L}}{6} \begin{bmatrix} 2 & 1 \\ 1 & 2 \end{bmatrix} \begin{matrix} u_2 \\ u_3 \end{matrix} \tag{5.139}
$$

Assembling as earlier and considering $u_2$ as the only active degree of freedom, we get a single-element consistent mass matrix corresponding to $u_2$ degrees of freedom given by

$$
[\bar{M}]_{\text{consistent}} = \left[ \frac{2 \rho A \bar{L}}{3} \right] \tag{5.140}
$$

Substituting Eqs. (5.135) and (5.140) in Eq. (5.132), we get

$$
\left[ \frac{YA}{\bar{L}} [2] - \omega^2 \frac{2 \rho A \bar{L}}{3} \right] \hat{u}_2 = 0
$$

From this equation, we can obtain the natural frequency of the same rod system from the consistent mass matrix as

$$\omega_{\text{consistent}} = \frac{\sqrt{3}}{\bar{L}}\sqrt{\frac{Y}{\rho}} = \frac{\sqrt{3}}{L}\sqrt{\frac{Y}{\rho}} = \frac{1.732 C_0}{L} \tag{5.141}$$

The frequencies obtained from Eq. (5.141) and (5.138) are not the same: they differ by nearly 18% with respect to the frequency of the consistent mass value. Now the question is which of the two mass formulations gives a result closer to the exact solution given by [14]:

$$\hat{u}(x) = A\sin\left(\frac{\omega x}{C_0}\right), \omega = \frac{n\pi}{2L}\sqrt{\frac{Y}{\rho}} = \frac{n\pi C_0}{2L} \tag{5.142}$$

Here $\hat{u}(x)$ represents the mode shape and $n$ the mode number. Substituting $n = 1$ in Eq. (5.142), we get the exact fundamental natural frequency as

$$\omega_{\text{exact}} = \frac{1.57 C_0}{L} \tag{5.143}$$

From Eqs. (5.138, 5.141,) and (5.143), we find that the exact solution is between the consistent and lumped mass solutions. The frequencies predicted by the lumped mass formulation are the lower bound while those predicted by the consistent mass formulations are always the upper bound. The error in the predicted frequency in the lumped mass formulation is about 10%, while for the consistent mass formulation it is about 10.26%, Hence, it can be concluded that the error in predicted frequency is nearly the same for both the mass formulations.

However, we would like to decrease the error in the fundamental frequency. This can be done by increasing the number of elements. The three-element configuration is shown in Figure 5.16(b). Introduction of an additional element introduces an additional degree of freedom. Hence, from Figure 5.15(b), there are two active degrees of freedom, $u_2$ and $u_3$. The assembled stiffness and mass matrices are generated in the manner discussed earlier. Hence, the stiffness matrix, lumped mass matrix, and consistent mass matrix for this three-element system are given by

$$[K] = \frac{YA}{\bar{L}}\begin{matrix} & u_2 & u_3 \\ & \begin{bmatrix} 2 & -1 \\ -1 & 2 \end{bmatrix} & \begin{matrix} u_2, \\ u_3 \end{matrix} \end{matrix} \quad [M]_{\text{lumped}} = \frac{\rho A \bar{L}}{2}\begin{matrix} & u_2 & u_3 \\ & \begin{bmatrix} 2 & 0 \\ 0 & 2 \end{bmatrix} & \begin{matrix} u_2, \\ u_3 \end{matrix} \end{matrix} \quad [M]_{\text{consistent}} = \frac{\rho A \bar{L}}{6}\begin{matrix} & u_2 & u_3 \\ & \begin{bmatrix} 4 & 1 \\ 1 & 4 \end{bmatrix} & \begin{matrix} u_2 \\ u_3 \end{matrix} \end{matrix}$$

$$\bar{L} = \frac{2L}{3} \tag{5.144}$$

Note that the introduction of an additional degree of freedom yields an additional natural frequency and mode shape. Now we first compute the frequencies using the lumped mass formulation. Substituting the stiffness and mass matrices from Eq. (5.144) in Eq. (5.132), we get

$$\left[ \frac{YA}{\bar{L}} \begin{bmatrix} 2 & -1 \\ -1 & 2 \end{bmatrix} - \frac{\omega^2 \rho A \bar{L}}{2} \begin{bmatrix} 2 & 0 \\ 0 & 2 \end{bmatrix} \right] \begin{Bmatrix} \hat{u}_2 \\ \hat{u}_3 \end{Bmatrix} = \begin{Bmatrix} 0 \\ 0 \end{Bmatrix} \tag{5.145}$$

By designating $\lambda = \rho \omega^2 \bar{L}^2 / Y$ and simplifying, we get

$$\begin{bmatrix} 2 - \lambda & -1 \\ -1 & 2 - \lambda \end{bmatrix} \begin{Bmatrix} \hat{u}_2 \\ \hat{u}_3 \end{Bmatrix} = \begin{Bmatrix} 0 \\ 0 \end{Bmatrix} \tag{5.146}$$

In this equation $\lambda$ is the eigenvalue and for non-trivial solutions, we need to find the value of $\lambda$ that satisfies Eq. (5.146), which is done by setting the determinant of the matrix to zero. This process will yield a quadratic equation in $\lambda$ given by

$$\lambda^2 - 4\lambda + 3 = 0$$

The roots of this equation are $\lambda = 1$ and $\lambda = 3$, respectively. These values give the natural frequencies as

$$\omega^1_{\text{lumped}} = \frac{1.5C_0}{L}, \quad \omega^2_{\text{lumped}} = \frac{2.6C_0}{L} \tag{5.147}$$

Comparing this solution to the exact one in Eq. (5.144), we find that the error in the first mode frequency is about 4.5% and in the second mode frequency is 17.23%. Hence, it can be seen that increasing the number of elements certainly brings down the error in the fundamental frequency. However, the error introduced in the natural frequency of the higher mode, due to the increased number of elements, is significantly higher.

Next, we perform the same exercise with the consistent mass formulation. Substituting the stiffness matrix and the consistent mass matrix from Eq. (5.144) in Eq. (5.132), we get the matrix equation

$$\left[ \frac{YA}{\bar{L}} \begin{bmatrix} 2 & -1 \\ -1 & 2 \end{bmatrix} - \frac{\omega^2 \rho A \bar{L}}{6} \begin{bmatrix} 4 & 1 \\ 1 & 2 \end{bmatrix} \right] \begin{Bmatrix} \hat{u}_2 \\ \hat{u}_3 \end{Bmatrix} = \begin{Bmatrix} 0 \\ 0 \end{Bmatrix} \tag{5.148}$$

Designating $\lambda = \rho \omega^2 \bar{L}^2 / 6Y$ and simplifying, we obtain a quadratic equation in $\lambda$, solving which we get $\lambda = 1/5$ and $\lambda = 1$, which correspond to the first and second natural frequencies. These values give the first two natural frequencies as

$$\omega^1_{\text{consistent}} = \frac{1.643C_0}{L}, \quad \omega^2_{\text{consistent}} = \frac{3.674C_0}{L} \tag{5.149}$$

The percentage error introduced in the first and second natural frequencies is 4.6% and 16.94%, respectively, which are approximately the same as that of the lumped mass formulation. As in the two-element case, the lumped mass frequencies are the lower bound while the consistent mass frequencies are the upper bound, with respect to the exact values.

This study shows that the error introduced by consistent and lumped mass formulation is approximately of the same order and hence no great advantage is obtained by using the consistent mass matrix (which requires larger storage and solution time for large-scale problems) for lower-order elements such as the rod element. However, according to [9], for higher-order finite elements such as beams and plates, the consistent mass can provide better results.

### 5.6.4 Free-Vibration Analysis of Proof-Mass Accelerometer

Proof-mass accelerometers are common MEMS devices used to measure acceleration; they are used extensively in aircraft and automobiles. Typical proof-mass accelerometers have a proof mass connected to a series of beam connectors as shown in Figure 5.17. For design purposes, we can model the accelerometer as a cantilever beam connected to a heavy mass, as in Figure 5.17.

The mass of the proof-mass accelerometer is taken as $\alpha \rho AL$; $\alpha > 1$, where $\rho AL$ is the mass of the beam connector 1–2 in Figure 5.17. Using a single-beam element and assuming that node 1 is the cantilevered end, and with Eqs. (5.87) and (5.116), we can write the stiffness and mass matrices for the accelerometer as

$$[K] = \frac{YI}{L^3} \begin{array}{c} \begin{array}{cc} w_2 & \theta_2 \end{array} \\ \begin{bmatrix} 12 & -6L \\ -6L & 4L^2 \end{bmatrix} \begin{array}{c} w_2 \\ \theta_2 \end{array} \end{array}, \quad [M] = \frac{\rho AL}{420} \begin{array}{c} \begin{array}{cc} w_2 & \theta_2 \end{array} \\ \begin{bmatrix} 156 + 420\alpha & -22L \\ -2L & 4L^2 \end{bmatrix} \begin{array}{c} w_2 \\ \theta_2 \end{array} \end{array} \quad (5.150)$$

Note that the mass of the proof mass is added to the beam-connector mass corresponding to the transverse degree of freedom $w_2$. Substituting these in Eq. (5.132), we get

$$\left[ \frac{YI}{L^3} \begin{bmatrix} 12 & -6L \\ -6L & 4L^2 \end{bmatrix} - \frac{\omega^2 \rho AL}{420} \begin{bmatrix} 156 + 420\alpha & -22L \\ -2L & 4L^2 \end{bmatrix} \right] \begin{Bmatrix} \hat{w}_2 \\ \hat{\theta}_2 \end{Bmatrix} = \begin{Bmatrix} 0 \\ 0 \end{Bmatrix} \quad (5.151)$$

Again, this equation is an eigenvalue problem. In order to solve it, this matrix equation is rewritten as

$$\begin{vmatrix} 12 - \lambda(156 + 420\alpha) & -(6L - 22\lambda L) \\ -(6L - 22\lambda L) & 4L^2(1 - \lambda) \end{vmatrix}, \quad \lambda = \frac{\omega^2 \rho AL^4}{420EI} \quad (5.152)$$

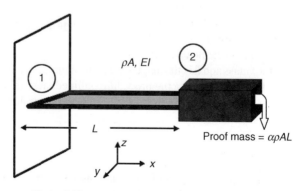

**Figure 5.17** FE model of proof-mass accelerometer.

Solving the eigenvalue problem, we obtain the two eigenvalues as

$$\lambda_1, \lambda_2 = \frac{(51 + 210\alpha) \pm \sqrt{2496 + 20160\alpha + 44100\alpha^2}}{35(1 + 12\alpha)}$$

which is further simplified as

$$\lambda_1 = \frac{0.0854}{1 + 12\alpha} \quad , \quad \lambda_2 = \frac{12\alpha + 2.8266}{1 + 12\alpha} \tag{5.153}$$

In this equation, $\alpha$ is the relative value of the mass of the proof mass in relation to the mass of the beam connectors. Taking $\alpha = 0$ simulates the condition in a simple cantilever beam, in which case $\lambda_1 = 0.0854$ and $\lambda_2 = 2.8266$. However, if the proof mass is really large, that is, $\alpha \rightarrow \infty$, then $\lambda_1 \Rightarrow 0$ and $\lambda_2 \Rightarrow 1$.

Next, we specifically determine the natural frequencies and mode shapes when the proof mass is five times heavier than the mass of the beam connectors, that is, $\alpha = 5$. For this case, we find from Eq. (5.153) the two eigenvalues as $\lambda_1 = 0.0014$ and $\lambda_2 = 1.03$. Substituting these in Eq. (5.152), we can write the natural frequencies of the accelerometer as

$$\omega_1 = \frac{0.766}{L^2} \sqrt{\frac{EI}{\rho A}} \quad , \quad \omega_2 = \frac{20.8}{L^2} \sqrt{\frac{EI}{\rho A}}$$

The corresponding eigenvectors are obtained by substituting $\lambda_1$ and $\lambda_2$ again in Eq. (5.152), which can be written as the modal matrix $[\Phi]$

$$[\Phi] = \begin{bmatrix} 0.0072L & 0.6753L \\ 1 & 1 \end{bmatrix} \tag{5.154}$$

**Your Turn:**
We derived the natural frequency and mode shapes of a proof-mass accelerometer. In this derivation we assumed that the proof mass offers very little resistance to motion. If the mass is large, however, it does offer resistance to motion. That is, we need to model the stiffness as well as the mass of the accelerometer. The stiffness of the proof mass is modeled as a spring with stiffness $k = \beta YI/L^3$, where $YI$ is the flexural rigidity

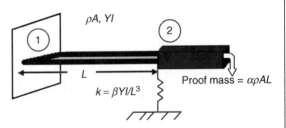

**Figure 5.18** FE model of proof-mass accelerometers with stiffness.

offered by the proof mass and $L$ is the length of the proof mass. The FE model for this case is shown in Figure 5.18. For this system, using the procedure outlined, determine the natural frequencies of the system. Comment on how the spring stiffness affects the natural frequencies of the accelerometer.

## 5.6.5 Forced Vibration Analysis

Forced vibration analysis requires solving Eq. (5.65), a highly coupled ordinary differential equation whose solution is not straightforward. There are two different methods of performing forced vibration analysis, the normal-mode method and the direct time-integration method. In the normal-mode method, we use the results of the free vibration analysis, which are the natural frequencies and the mode shapes, to determine the forced response. This is a quite popular and extensively used method, especially if the frequency content of the input forcing signal $\{R\}$ in Eq. (5.65) is small (of the order of few hundred hertz). In addition, the method is limited to linear problems. In the second method (direct time integration), the derivatives in Eq. (5.65) are replaced by the finite difference coefficients over a very small time step $\Delta t$. This converts Eq. (5.65) into an algebraic set of equations, whose solutions yield the forced response of the dependent variables at all nodes at time step $n$. To obtain the response histories at all times of interest, the procedure must be continued by updating the time step at the $(n+1)$th step as $t_{n+1} = t_n + \Delta t$ until the time of interest is reached. This method is used when the problem is nonlinear or if the forcing function has a large frequency bandwidth. In this chapter, we describe only the first method based on natural frequencies and mode shapes, the normal-mode method.

### 5.6.6 Normal Mode Method

This method requires computation of natural frequencies and mode shapes in the given system. If the system is discretized with $N$ degrees of freedom, then the FE model gives $N$ natural frequencies and mode shapes. Let these be stored in two square matrices $[\Lambda]$ and $[\Phi]$, both of which have size $N \times N$. The matrix $[\Lambda]$, which contains all the $N$ natural frequencies, is a diagonal matrix that is normally referred to as a *spectral matrix*. The matrix $[\Phi]$, normally referred to as a *modal matrix*, contains the mode shape information on all $N$ modes. It is stored columnwise. That is, the mode shape of the first mode occupies the first column of the $[\Phi]$matrix, and likewise the mode shape of the $N$th mode occupies the $N$th column of the $[\Phi]$matrix.

In order to solve Eq. (5.65) using the modal information, we use two important properties of the mode shape, called the *orthogonality conditions*, which state that *the modal matrix $[\Phi]$is orthogonal with respect to both the stiffness and mass matrices*. If $[K]$, $[M]$ and $[C]$ are the assembled stiffness, mass and damping matrix and $\{d\}$ is the assembled displacement vector, then the assembled governing equation and the orthogonality relations can be written as

$$[M]\{\ddot{d}\} + [C]\{\dot{d}\} + [K]\{d\} = \{R\}$$
$$[\Phi]^T[K][\Phi] = [\Lambda], \ [\Phi]^T[M][\Phi] = [I] \tag{5.155}$$

That is, the orthogonal conditions diagonalize the mass and stiffness matrices, which we use to uncouple the equation of motion [Eq. (5.65)]. Note that the eigenvectors in the matrix $[\Phi]$ are mass normalized.

The basic principle in solving Eq. (5.65) is to uncouple the equation for a form Eq. (A.5a), whose solution is discussed in the Appendix. In order to do this, we need to transform the dependent nodal variables $\{d\}$ in Eq (5.155), to a new set of nodal variables $\{\eta\}$ through a transformation matrix that involves the modal matrix and can be written as

$$\{d\}_{N\times1} = [\Phi]_{N\times N}\{\eta\}_{N\times1} = [\{\varphi\}_1 \quad \{\varphi\}_2 \quad \cdots \quad \cdots \quad \{\varphi\}_N]_{N\times N} \begin{Bmatrix} \{\eta\}_1 \\ \{\eta\}_2 \\ . \\ . \\ . \\ \{\eta\}_N \end{Bmatrix} \tag{5.156}$$

In the above equation, $\{\varphi\}_1, \{\varphi\}_2, \ldots \{\varphi\}_N$ are the $N$ mode shapes of system under investigation. Substituting Eq. (5.156) into the first equation in Eq. (5.155), we get

$$[M]_{N \times N}[\Phi]_{N \times N}\{\ddot{\eta}\}_{N \times 1} + [\alpha[K] + \beta[M]]_{N \times N}[\Phi]_{N \times N}\{\dot{\eta}\}_{N \times 1}$$

$$+[K]_{N \times N}[\Phi]_{N \times N}\{\eta\}_{N \times 1} = \{R\}_{N \times 1} \tag{5.157}$$

Here the damping matrix $[C]$ is replaced by the proportional damping matrix given by $[C] = \alpha[K] + \beta[M]$, as discussed in Section 5.6.1, and the subscripts indicate the sizes of the matrices. Premultiplying this equation by $[\Phi]^T$ and using the orthogonality condition in Eq. (5.155), Eq. (5.157) reduces to

$$\ddot{\eta}_r + 2\zeta_r \omega_r \dot{\eta} + \omega_r^2 \eta_r = \bar{R}_r, \quad \bar{R}_r = [\Phi]^T\{R\}, r = 1, 2, \cdots N \tag{5.158}$$

That is, the original equation [Eq. (5.155)] is uncoupled into a series of ordinary differential equations for each vibrational mode. Each of these equations represents the equation of motion of a single-degree-of-freedom system, the solution of which is given in the Appendix [see Eq. (A.9)]. In Eq. (5.158), the parameter $\zeta_r = C_r/2M_r\omega_r$ represents the damping ratio, whose definition is also given in the Appendix.

Before we proceed further in the analysis, we should note that if the number of degrees of freedom $N$ is large, then extracting all the $N$ eigenvalues and vectors is computationally prohibitive. Looking at the modal energies of each of these modes, we find that the initial modes, beginning with the fundamental mode, have higher energy than the higher-order modes. Hence, we must compute the response of all the degrees of freedom using only, say, $m$ modes, where $m \ll N$. The normal-mode method enables us to perform this analysis.

Here, we compute the eigenvalues and vectors for the first $m$ modes, where $m$ is at least two or more orders of magnitude smaller than $N$. We again store it columnwise in the modal matrix $[\Phi]$ of size $N \times m$. Hence, Eq. (5.156) can be rewritten as

$$\{d\}_{N \times 1} = [\Phi]_{N \times m}\{\eta\}_{m \times 1} = [\{\varphi\}_1 \quad \{\varphi\}_2 \quad \cdots \quad \cdots \quad \{\varphi\}_m]_{N \times m} \begin{Bmatrix} \{\eta\}_1 \\ \{\eta\}_2 \\ . \\ . \\ \{\eta\}_m \end{Bmatrix} \tag{5.159}$$

Using this transformation in the governing equation [first equation in Eq. (5.155)] and using orthogonality relations, the equation is uncoupled as before; however, to only a small set of $m$ equations, which can be easily solved.

In modal methods, we first solve for all for the generalized coordinates $\eta$. The values of $\eta$ obtained are then substituted in Eq. (5.159) to get the actual nodal displacement vector.

Satisfying the initial conditions, say $\{d(t=0)\} = \{d_0\}$, $\{\dot{d}(t=0)\} = \{v_0\}$, is not straightforward. Since we will be solving in the transformed $\eta$ domain, we need these initial conditions in terms of the generalized coordinates $\eta$. Equation (5.159) cannot be used directly to obtain initial conditions in terms of $\eta$ since $[\Phi]$ is not a square matrix and hence not invertible. Hence, we have to use an alternative procedure. Premultiplying Eq. (5.159) by $[\Phi]^T[M]$ on both sides, substituting the initial conditions, and using the orthogonality conditions [Eq. (5.155)], we get the initial conditions in terms of generalized coordinates $\eta$ as

$$[\Phi]^T[M]\{d(t=0)\} = [\Phi]^T[M][\Phi]\{\eta(t=0)] = \{\eta(t=0)\}$$

$$[\Phi]^T[M]\{\dot{d}(t=0)\} = [\Phi]^T[M][\Phi]\{\dot{\eta}(t=0)] = \{\dot{\eta}(t=0)\} \qquad (5.160)$$

## ▶ 5.7 FINITE ELEMENT MODEL FOR STRUCTURES WITH PIEZOELECTRIC SENSORS AND ACTUATORS

Modeling systems with smart-material patches is very similar to modeling conventional structures. However, additional complexities arise due to the presence of coupling terms in the constitutive law of these smart materials. Coupling introduces additional matrices in the FE formulations. This section addresses modeling of systems with piezoelectric-type smart material patches.

Piezoelectric materials have two constitutive laws, one for sensing and the other for actuation purposes. For 2D problems, the constitutive model for the piezoelectric material has the form

$$\{\sigma\}_{3\times1} = [C]^{(E)}_{3\times3}\{\varepsilon\}_{3\times1} - [e]_{3\times2}\{E\}_{2\times1} \qquad (5.161)$$

$$\{D\}_{2\times1} = [e]^T_{2\times3}\{\varepsilon\}_{3\times1} + [\mu]^{(\sigma)}_{2\times2}\{E\}_{2\times1} \qquad (5.162)$$

The first of these constitutive laws is called the *actuation law* and the second the *sensing law*. Here, $\{\sigma\}^T = \{\sigma_{xx} \quad \sigma_{yy} \quad \tau_{xy}\}$ is the stress vector, $\{\varepsilon\}^T = \{\varepsilon_{xx} \quad \varepsilon_{yy} \quad \gamma_{xy}\}$ is the strain vector, $[e]$ is the matrix of piezoelectric coefficients of size $3 \times 2$ (with units N/V-mm), and $\{E\}^T = \{E_x \quad E_y\} = \{V_x/t \quad V_y/t\}$ is the applied field in two coordinate directions, where $V_x$ and $V_y$ are the applied voltages in the respective coordinate directions and $t$ is the thickness parameter. The unit of $E_x$ (and $E_y$) is V/mm. The term $[\mu]$ is the permittivity matrix of size $2 \times 2$, measured at constant stress and having units N/V/V, and $\{D\}^T = \{D_x \quad D_y\}$ is the vector of electric displacement in the two coordinate directions with units N/V-mm (the same as the piezoelectric coefficients). $[C]$ is the mechanical constitutive matrix measure (or Hooke's law) at constant electric field. Equation (5.161) can also be written in the form

$$\{\varepsilon\} = [S]\{\sigma\} + [d]\{E\} \qquad (5.163)$$

where $[S]$ is the compliance matrix that is the inverse of the mechanical material matrix $[C]$ and $[d] = [C]^{-1}[e]$ is the electromechanical coupling matrix, where the elements of this matrix are direction-dependent and have units mm/V. In the ensuing analysis, it is assumed that the mechanical properties change very little with a change in the electric field and that as a result the actuation law [Eq. (5.161)] can be assumed to behave linearly with the electric field. Further, the sensing law [Eq. (5.162)] can be assumed to behave linearly with the stress. This assumption considerably simplifies the analysis.

The first part of Eq. (5.161) represents the stresses developed due to mechanical load and the second part gives the stresses due to voltage input. From Eqs. (5.161) and (5.162), it is clear that the structure will be stressed due to the application of the electric field even in the absence of a mechanical load. Alternatively, loading the mechanical structure

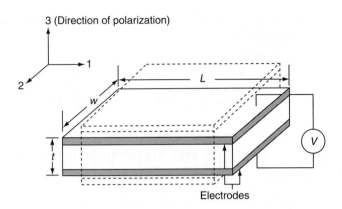

**Figure 5.19** Illustration of actuation effect in a piezoelectric plate.

generates an electric field. In other words, the constitutive law demonstrates the electro-mechanical coupling, and this can be exploited in a variety of structural applications involving sensing and actuation such as vibration control, noise control, shape control, or structural health monitoring.

The actuation using piezoelectric materials can be demonstrated using a plate of dimensions $L \times W \times t$, where $L$ and $W$ are the length and width of the plate and $t$ is its thickness. Thin piezoelectric electrodes are placed on the top and bottom surfaces of the plate as shown in Figure 5.19. Such a plate is called a *bimorph* plate. When the voltage is applied between the electrodes as shown in the figure (which is normally referred to as the poling direction), the deformations in the length, width, and thickness directions are given by

$$\delta L = d_{31}E_1L = \frac{d_{31}VL}{t}, \quad \delta W = d_{31}E_2W = \frac{d_{31}VW}{t}, \quad \delta t = d_{33}V \tag{5.164}$$

Here $d_{31}$ and $d_{33}$ are the electromechanical coupling coefficients in directions 1 and 3, respectively. Conversely, if a force $F$ is applied in any of the length, width, or thickness directions, the voltage $V$ developed across the electrodes in the thickness direction is

$$V = \frac{d_{31}F}{\mu L} \text{ or } \frac{d_{31}F}{\mu W} \text{ or } \frac{d_{33}F}{\mu LW} \tag{5.165}$$

where $\mu$ is the dielectric permittivity of the material. The reversibility between the strain and voltages makes piezoelectric materials ideal for both sensing and actuation. The FE modeling of the mechanical part is very similar to what was discussed in the previous section except that the coupling terms introduce additional energy terms in the variational statements, resulting in additional coupling matrices in the FE formulation.

Different types of piezoelectric materials are used in many structural applications. The most commonly used material is PZT, which is extensively used as bulk actuator material as it has a high electromechanical coupling coefficient. Due to their low electromechanical coupling coefficients, piezopolymers PVDF are extensively used only as sensor materials. With the advent of smart composite structures, a new brand of material called *piezofiber composite* (PFC) has been found to be a very effective actuator material in vibration/noise control applications.

Developing FE models for such structures with smart-material patches is very similar to developing them for conventional structures. Such structures, in addition to having mechanical degrees of freedom, also have additional degrees of freedom in terms of the electric field supplied by the piezoelectric patches. As explained earlier, piezoelectric patches introduce additional modeling complexities due to the presence of electromechanical coupling that manifest themselves as additional elements in the stiffness matrix due to electromechanical coupling. See [12] for an outline of a four-node FE model for a laminated composite structure with an embedded or surface-bonded piezoelectric patch. Each node of this element can support two mechanical degrees of freedom and one electrical degree of freedom. The details of this formulation are omitted here; the interested reader may refer to [12]. The FE equation for such an element is

$$
\begin{bmatrix} [K_{uu}] & [K_{uE}] \\ [K_{uE}]^{\mathrm{T}} & [K_{EE}] \end{bmatrix} \begin{Bmatrix} \{u\} \\ \{E_z\} \end{Bmatrix} = \begin{Bmatrix} \{F\} \\ \{q\} \end{Bmatrix}
\tag{5.166}
$$

where $\{u\}$ is the $8 \times 1$ vector containing eight mechanical degrees of freedom, $\{E_z\}$ is the $4 \times 1$ vector of electrical degrees of freedom, $[K_{uu}]$ is the $8 \times 8$ matrix containing elements contributed by mechanical degrees of freedom alone, $[K_{uE}]$ arises due to electromechanical coupling $[d]\{E\}$ in the constitutive model Eq. (5.163), $[K_{EE}]$ is only due to electrical field, $\{F\}$ is the $8 \times 1$ mechanical force vector, and $\{q\}$ is the $4 \times 1$ charge vector. Equation (5.166) can be expanded into two equations as

$$
[K_{uu}]\{u\} + [K_{uE}]\{E_z\} = \{F\}
\tag{5.167a}
$$

$$
[K_{uE}]^{\mathrm{T}}\{u\} + [K_{EE}]\{E_z\} = \{q\}
\tag{5.167b}
$$

From Eq. (5.167b), we get

$$
\{E_z\} = [K_{EE}]^{-1}\{q\} - [K_{EE}]^{-1}[K_{uE}]^{\mathrm{T}}\{u\}
\tag{5.168}
$$

Substituting $\{E_z\}$ in Eq. (5.167a), we get

$$
[\bar{K}]\{u\} = \{\bar{F}\}
\tag{5.169}
$$

where

$$
[\bar{K}] = [K_{uu}] - [K_{uE}][K_{EE}]^{-1}[K_{uE}]^{\mathrm{T}}, \quad \{\bar{F}\} = \{F\} - [K_{EE}]^{-1}\{q\}
$$

This process is called *static condensation*. That is, even when the additional degrees of freedom are present, they can be condensed out to yield Eq. (5.169), which is like any other FE equation.

## ► 5.8 ANALYSIS OF A PIEZOELECTRIC BIMORPH CANTILEVER BEAM

In this example, the modeling and static analysis of a piezoelectric bimorph composite beam is discussed using the beam element formulated in Section 5.4.5. The bimorph beam consists of two identical PVDF beams laminated together with opposite polarities. Figure 5.20 shows a schematic diagram of the bimorph beam. The PVDF patches are poled in such a way that strains are produced in the axial $x$-direction due to an applied electric

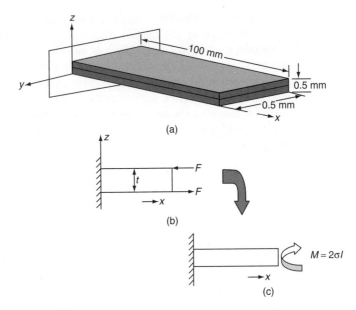

**Figure 5.20** Schematic diagram of the piezoelectric PVDF bimorph cantilever beam: (a) actual beam; (b) moment due to electrical field; and (c) cantilever beam with end moment.

field in the $z$-direction. The dimensions of the beam are taken as 100 mm × 5.0 mm × 0.5 mm. The aim of this example is to see how the cantilever deforms due to the applied voltage.

### 5.8.1 Exact Solution

The solution to this problem can be obtained from a basic strength of materials approach. This problem can be statically reduced to the problem of a cantilever beam with an end moment $M$ shown in Figure 5.20(c), where the moment $M$ must be determined from the constitutive law of the PVDF material. The beam is under a 1D state of stress with the stress acting in the $x$-direction. This stress and hence the force can be obtained as follows.

From Eq. (5.163), we have

$$\{\sigma\} = [S]^{-1}\{\varepsilon\} - [S]^{-1}[d]\{E\}$$

The inverse of the compliance matrix is the constitutive matrix $[C]$; using $[e] = [S]^{-1}[d]$, the above equation becomes

$$\{\sigma\} = [C]\{\varepsilon\} - [e]\{E_z\} \tag{5.170}$$

The first part of Eq. (5.170) is due to the mechanical load, which is zero in the present case and hence is not relevant here. We consider only the second part of Eq. (5.170). Since the beam is in a 1D state of stress, only $\sigma_{xx}$, the bending stress in the axial direction, exists. The only material property relevant here is the Young's modulus $Y$ and the relevant piezoelectric coefficient is $e_{31}$, which is the first element of the third row of the matrix $[e]$ given in Eq. (5.170). Hence, the constitutive law can be written as

$$\sigma_{xx} = -e_{31}E_z = -e_{31}V/t \tag{5.171}$$

From elementary beam theory (see Chapter 4, Section 4.2, where elementary bending theory is detailed), we have $M/I = \sigma_{xx}/z$, where $M$ is the moment acting on the cross section, $I$ the area moment of inertia of the cross section, and $z$ the coordinate in the thickness direction. Substituting for $\sigma_{xx}$ from Eq. (5.171) in the elementary beam equation, we can express the moment developed due to electrical excitation as

$$M = -\frac{2e_{31}VI}{t^2} \tag{5.172}$$

From the theory of deflection of beams, we can show that the transverse displacement $w$ $(x)$ of a cantilever beam with a tip moment $M$ is given by (see Chapter 4, Section 4.2 for details)

$$w(x) = -\frac{Mx^2}{2YI} \tag{5.173}$$

Using the value of moment $M$ from Eq. (5.172) in Eq. (5.173), we can write the displacement variation in a bimorph piezoelectric cantilever beam as

$$w(x) = \frac{e_{31}V}{Y}\left(\frac{x}{t}\right)^2 \tag{5.174}$$

The deflection at the tip can be obtained by substituting $x = L$ as

$$w(L)|_{\text{elec}} = \frac{e_{31}V}{Y}\left(\frac{L}{t}\right)^2 \tag{5.175}$$

We see from this expression that the deflection is positive, implying that the beam will deflect upward, and that the deflection is a function of the voltage: as the voltage is increased, the deflection increases.

Next, in addition to the electric field, we introduce a mechanical load $P$ applied at the tip. We have already derived the deflection for a cantilever with tip loading in Chapter 4 [Eq. (4.27)]. Hence, we write

$$w(x) = \frac{P}{YI}\left(\frac{x^3}{6} - \frac{x^2L}{2}\right) \tag{5.176}$$

The tip deflection in this case is obtained by substituting $x = L$ in Eq. (5.176):

$$w(L)|_{\text{mech}} = -PL^3/3YI \tag{5.177}$$

Note that the negative sign in Eq. (5.177) implies that the deflection is downward. Now when both mechanical and electrical loads are applied, the total deflection is the sum of the deflections due to the mechanical load [Eq. (5.177)] and the electrical load [Eq. (5.175)]:

$$w(L)|_{\text{total}} = -PL^3/3YI + \frac{e_{31}V}{Y}\left(\frac{L}{t}\right)^2 \tag{5.178}$$

Note that the area moment of inertia $I = bt^3/12$, where $b$ is the beam width and $t$ is the beam thickness (see Figure 4.8). From this expression, we see that as the voltage increases, the net downward deflection due to the mechanical load is reduced. Hence, we can set the voltage so that the total deflection $w(L)$ is zero. This is obtained by setting the left-hand side of Eq. (5.178) equal to zero. This voltage is given by

$$V = \frac{PL}{e_{31}bt} \tag{5.179}$$

It is clear that the presence of electrical load helps to eliminate the deflection of the cantilever beam due to mechanical load. This, in essence, is the main principle of actuation, and it can be exploited in a variety of applications such as vibration control, noise control, or structural shape control.

### 5.8.2 Finite Element Solution

We now solve the same problem using a FE solution. For this example, we do not need an FE with smart or electrical degrees of freedom. Here the forces generated due to electrical field must be lumped corresponding to FE degrees of freedom, which in this case is a moment given by Eq. (5.172). Since our objective here is to see how the deflection caused by pure mechanical load is negated by the electrical field from the PVDF patches, we need to retain the degrees of freedom corresponding to transverse mechanical force (shear force) at the tip of the cantilever beam.

The FE equation of the beam derived in Section 5.4 is used here:

$$\begin{Bmatrix} P_1 \\ M_1 \\ P_2 \\ M_2 \end{Bmatrix} = \frac{YI}{L^3} \begin{bmatrix} 12 & -6L & -12 & -6L \\ -6L & 4L^2 & 6L & 2L^2 \\ -12 & 6L & 12 & 6L \\ -6L & 2L^2 & 6L & 4L^2 \end{bmatrix} \begin{Bmatrix} w_1 \\ \theta_1 \\ w_2 \\ \theta_2 \end{Bmatrix} \tag{5.180}$$

We model the beam through one element only. We need first to enforce the boundary condition. The left end of the beam, designated here as node 1, has both deflection and slope equal to zero. This eliminates the first two rows and columns of the stiffness matrix in Eq. (5.180). The reduced stiffness matrix after the enforcement of boundary conditions then becomes

$$\begin{Bmatrix} P_2 \\ M_2 \end{Bmatrix} = \frac{YI}{L^3} \begin{bmatrix} 12 & 6L \\ 6L & 4L^2 \end{bmatrix} \begin{Bmatrix} w_2 \\ \theta_2 \end{Bmatrix} = \frac{YI}{L^3} [\bar{K}]\{w\} \tag{5.181}$$

The displacements and rotations can be obtained by inverting the reduced stiffness matrix given by

$$\begin{Bmatrix} w_2 \\ \theta_2 \end{Bmatrix} = \frac{L}{12YI} \begin{bmatrix} 4L^2 & -6L \\ -6L & 12 \end{bmatrix} \begin{Bmatrix} P_2 \\ M_2 \end{Bmatrix} \tag{5.182}$$

For a purely electrical load, $P_2 = 0$. As mentioned earlier, a pure electrical load causes a moment in the beam that is given by Eq. (5.172). Substituting this into Eq. (5.182), we get the tip deflection as

$$w_2 = \frac{L}{12YI}(-M_2 6L) = -\frac{M_2 L^2}{2YI} = \frac{e_{31}V}{Y}\left(\frac{L}{t}\right)^2 \tag{5.183}$$

This is the same as the exact solution derived in Eq. (5.175). If we now apply a tip vertical load $P_2 = P$ in addition to the moment $M_2$, then from Eq. (5.182), we get the tip deflection as

$$\begin{aligned} w_2 &= \frac{L}{12YI}(4L^2 P_2 - 6LM_2) \\ &= \frac{L}{12YI}\left(-4L^2 P + 6L\frac{2e_{31}VI}{t^2}\right) \\ &= -\frac{PL^3}{3YI} + \frac{e_{31}V}{Y}\left(\frac{L}{t}\right)^2 \end{aligned} \tag{5.184}$$

This result is same as the exact solution in Eq. (5.178).

## ▶ 5.9 SUMMARY

This chapter introduces readers to the powerful finite element method (FEM) as a modeling tool for microsystems and smart systems. What this chapter covers is by no means complete. The subject of FE is so vast that typically three full-length books are needed to cover all its aspects. Books dealing with the application of this method to smart systems are few, and most of the concepts must be acquired largely from the journal papers. One of the books addressing this method for smart systems is [12], which covers mostly the highlights of the method blended with few examples. Hence the present chapter alone is not sufficient for the reader to gain a good understanding in FEM. We strongly advise the reader to go through some classic texts on the subject in order to understand the finer aspects of the different topics. Some such texts are listed in the References section.

We started by describing the need for FEM and the foundation on which the method is based, namely, the energy theorems, which provide an alternative statement of equilibrium. The construction of the weak form of the governing equation was discussed next. The method of weighted residuals is first constructed as the mother of all methods; many other methods such as finite difference, method of moments, Galerkin methods, and FEMs are derived from the weighted residual method by suitable choice of the weighting functions. Next, the essentials of FEs such as derivation of shape functions, element formulation, isoparametric formulation, and numerical integration were described in detail. This was followed by some numerical examples that help in understanding the FE process. After this, the use of FEM for time-dependent problems is discussed. In this section, the concepts of mass and damping are introduced and in particular, the different methods of formulating the mass matrix are given. The method for computing the free vibration parameters is given for some simple rod structures. Next, a practical example of a MEMS accelerometer was considered, highlighting the modeling and method of obtaining free vibration response. Following this, the constitutive model for piezoceramic materials is described and the FE formulation procedure is outlined. A numerical example specifically addressing the coupling between the electric field and the mechanical field is given to indicate how these materials can be used for actuation purposes. Solution of a number of the exercises at the end of this chapter will further help in understanding this powerful method.

## ▶ REFERENCES

1. Gopalakrishnan, S., Chakraborty, A. and Mahapatra, D. R. (2008) *Spectral Finite Element Methods*, Springer-Verlag, UK.

2. Kreyszig, E. (2005) *Advanced Engineering Mathematics*, 9th ed., John Wiley & Sons, Singapore.

3. Becker, A. A. (1993) *The Boundary Element Method in Engineering*, McGraw Hill, New York.

4. Atluri, S. N. (2005) *Computer Modeling in Engineering and Sciences*, Vol. 1, Tech Science Press, USA.

5. Shames, I. H. and Dym, C. L. (1991) *Energy and Finite Element Methods in Structural Mechanics*, Wiley Eastern Limited, London.

6. Tauchert, T. R. (1974) *Energy Principles in Structural Mechanics*, McGraw Hill, Tokyo.

7. Cook, R. D., Malkus, R. D. and Plesha, M. E. (1989) *Concepts and Applications of Finite Element Analysis*, John Wiley & Sons, New York.

8. Bathe, K. J. (1996) *Finite Element Procedures*, 3rd ed., Prentice Hall, Englewood Cliffs, NJ, USA.

9. Prathap, G. (1996) Barlow points and Gauss points and the aliasing and best fit paradigms. *Computers and Structures*, 58, 321–325.

10. Varadan, V. K., Vinoy, K. J. and Gopalakrishnan, S. (2007) *Smart Material Systems and MEMS*, John Wiley & Sons, UK.

11. Doyle, J. F. (1989) *Wave Propagation in Structures*, Springer-Verlag, New York.

12. Thomson, W. T. (1998) *Theory of Vibration with Applications*, Prentice-Hall, New York.

13. Meirovich, L. (2000) *Fundamentals of Vibrations*, McGraw-Hill, New York.

14. Clough, R. W. and Penzin, J. (1975) *Dynamics of Structures*, McGraw-Hill, New York.

15. Hinton, E., Rock, T., and Zienkiewicz, O. C. (1976) A note on mass lumping and related processes in the finite element method, *Earthquake Engineering and Structural Dynamics*, 4, 245–49.

16. Malkus, D. S. and Plesha, M. E. (1986) Zero and negative masses in finite element vibration and transient analysis, *Computer Methods in Applied Mechanics and Engineering*, 59, 281–306.

## ▶ EXERCISES

5.1 The force-displacement relation for a given 1D solid is $F = ku^{3/5}$, where $F$ is the force, $u$ is the displacement, and $k$ is some known constant. Determine the work and complementary work of the solid.

5.2 Derive the strain energy expression for a tapered rod of length $L$ fixed and one end and subjected to a load P at the other end, and undergoing axial deformation $u(x)$. The rod depth varies as a function of axial coordinate $h(x) = h_0[1 + (x/L)]$, where $h_0$ is the depth of the rod at the left end where the rod is fixed and is made of a material having Young's modulus $Y$. Assume the rod width ''b'' is constant.

5.3 The governing equation for a higher-order beam having Young's modulus $Y$ and shear modulus $G$ in the $x$–$z$ plane is given by

$$u(x, z) = -z\theta, \ w(x, z) = w(x)$$

where $u(x)$ is axial motion in the $x$-direction, $w(x)$ transverse motion in the $x$-direction and $\theta$ the rotation that is not derived from transverse displacement $w(x)$.

(a) Derive the expression for strain energy.

(b) Derive the strong form of the governing differential equation and the associated essential and natural boundary conditions.

5.4 The strong form of a differential equation of a certain beam of Young's modulus $Y$ and moment of inertia $I$ undergoing transverse motion $w(x)$ is given by

$$YI\frac{d^4w}{dx^4} + \lambda w + q = 0$$

where $\lambda$ is some known constant. Derive the weak form of the above equation.

5.5 Solve the following differential equation using *method of moments* and the *Galerkin method*:

$$x^2\frac{d^2y}{dx^2} = 1$$

Assume the boundary conditions as $y(1) = y(2) = 0$. Compare your solutions with the exact solution and comment on your results.

5.6 A three-noded beam finite element is to be formulated. The degrees of freedom at the end nodes are the transverse displacement $w(x)$ and rotation $dw/dx$, and the middle node has only the transverse displacement $w(x)$ as the degree of freedom. Derive all the shape functions for this beam element.

5.7 The voltage field in a two-noded piezoelectric element is given by

$$V(x) = Ae^{-ikx} + Be^{-k(L-x)}$$

where $k$ is some known constant, $L$ the length of the piezoelectric element, and $i$ a complex number $= \sqrt{-1}$. Determine the shape functions for the voltage field.

5.8 (a) Can the Jacobian be negative? If it is negative, what are its implications?

(b) A quadrilateral element has the following coordinates for the four nodes: $(0,0)$, $(10,2)$, $(15,6)$ and $(-5,4)$. Using four-noded isoparametric shape functions, determine the value of the Jacobian.

5.9 A spring-supported beam of length $L$ and flexural rigidity $YI$ is shown in Figure 5.21. (a) State the essential and natural boundary conditions for this problem. (b) Solve this problem using a single-beam element and determine the deflection at the spring location of the beam (beam tip).

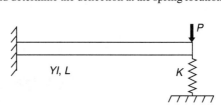

**Figure 5.21** Problem 5.9.

5.10 Numerically integrate the expression

$$\int\limits_{-1}^{1} \frac{1 + \xi + 2\xi^2}{4 + 7\xi^2}\, d\xi$$

using Gaussian integration procedure. How many integration points are required to integrate the above expression and why?

5.11 It is necessary to numerically integrate the function using Gauss Quadrature

$$\int_0^1 \frac{1}{x+1}\,dx$$

To do this, you need to transform this equation in isoparametric coordinates by assuming a suitable transformation between the actual coordinates and the isoparametric coordinate. How many Gauss points are required to integrate this integral? Compare your numerical solution with the exact results.

5.12 A beam of length $2L$ is shown in Figure 5.22.
   (a) State the essential and natural boundary conditions at the two end of the beam.
   (b) Determine the transverse deflections under the load P.

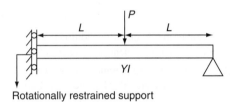

**Figure 5.22** Problem 5.12.

5.13 In certain cantilevered structures, it is necessary to provide variable rotational restraints as complete fixed boundary condition is seldom achieved. In modeling, this is accomplished by introducing a rotational spring of stiffness $K_t = 16\,YI/L$, as shown in Figure 5.23. In addition, the beam structure of length $L$, Young's modulus $Y$, area moment of inertia $I$, and density $\rho$ is also restrained in the traverse direction by two linear springs of spring stiffness $K = YI/L^3$ as shown in the figure. Using a single beam element and consistent mass matrix, determine the fundamental frequency of the system. Assume node 1 is restrained to move in the transverse direction and node 2 is rotationally restrained.

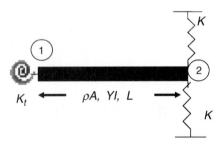

**Figure 5.23** Problem 5.13.

5.14 Figure 5.24 shows a proof-mass accelerometer attached to two springs of stiffness $K$ at a distance $\alpha a$ from the mass, as shown. Use two elements of lengths $\alpha a$ and $a$, respectively, to model the beam connector. Assume only translational degrees of freedom. Take the properties

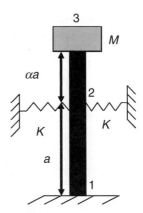

**Figure 5.24** Problem 5.14.

of beam connectors as follows: Young's modulus $Y$, area moment of inertia $I$, and density $\rho$. Determine the value of $\alpha$ for the following conditions:

(a) The relative displacement between node 2 and node 1 is limited to 0.01 units when the mass $M$ is subjected to 5 g acceleration, where g is the acceleration due to gravity.

(b) For $K = 6YI/a^3$ and $M = \rho Aa/2$, the second natural frequency is 5 times the first natural frequency.

5.15 A beam fixed at both ends is attached to a spring as shown in Figure 5.25. The stiffness of the spring is given by $K = YA/L$, where $Y$ is the Young's modulus of the beam, $A$ is its cross-sectional area, and $2L$ is its length. If the area moment of inertia of the beam is $I = \beta AL^2$, find the minimum value of $\beta$ such that the beam can still vibrate in its first mode. Use consistent mass for the beam element.

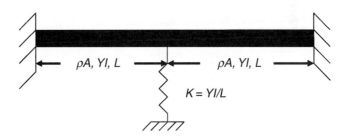

**Figure 5.25** Problem 5.15.

5.16 Two cantilevered structures are connected by a rotational spring element of stiffness $K_t = \beta YI/L$ as shown in Figure 5.26, where $Y$ is the Young's modulus of the beam, $I$ is the area moment of inertia of its cross-section, and $L$ is its length. Find the fundamental frequency of the system by using two beam elements coupled by a bending spring. Assume that the beam vibrates only in a symmetrical mode. Use lumped mass matrix for translational inertia with rotational inertia being assumed equal zero.

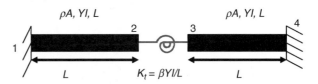

**Figure 5.26** Problem 5.16.

5.17 The equation of motion of a three-degree-of-freedom model is given by

$$
\begin{bmatrix} 1 & 0 & 0 \\ 0 & 1 & 0 \\ 0 & 0 & 1 \end{bmatrix} \begin{Bmatrix} \ddot{u}_1 \\ \ddot{u}_2 \\ \ddot{u}_3 \end{Bmatrix} + \begin{bmatrix} 3 & -1 & -1 \\ -1 & 3 & -1 \\ -1 & -1 & 3 \end{bmatrix} \begin{Bmatrix} u_1 \\ u_2 \\ u_3 \end{Bmatrix} = \begin{Bmatrix} F(t) \\ 0 \\ 0 \end{Bmatrix}
$$

Here $F(t)$ is the forcing function, which is a half-sinusoid as shown in Figure 5.27.

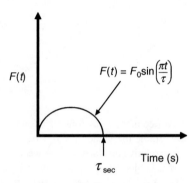

**Figure 5.27** Problem 5.17.

Determine the forced response of the system using the normal-mode method.

# CHAPTER 6

# Modeling of Coupled Electromechanical Systems

## LEARNING OBJECTIVES

After completing this chapter, you will be able to:

▶ Understand the basics of electrostatics as it relates to microsystems.

▶ Learn about how electrostatics and elastostatics problems are coupled to each other.

▶ Understand the numerical methods for solving the coupled equations of electrostatics and elastostatics, electro-thermal-elastic, and electro-magneto-elastic domains.

▶ Learn about pull-in instability and frequency pull-in.

▶ Understand how lumped inertia is calculated.

▶ Learn about Couette and squeezed film flows as they relate to microsystems.

Microsystems and smart systems do not operate in only one energy domain. For example, a device or a system that works purely in the mechanical energy domain is not considered *smart* because it cannot intelligently respond to a stimulus. There is nothing smart about a beam that bends under a transverse load or one that buckles under an axial compressive load. Likewise, an electrical or electronics circuit by itself is not considered *smart*. Smart systems or smart material systems respond intelligently to external stimuli. They always deal with more than one energy domain. Piezoelectric materials involve mechanical and electrical domains. Shape memory alloys involve mechanical and thermal domains accompanied by material phase transitions. Similarly, microsystems too work in multiple energy domains. In fact, any microsystem or smart systems device invariably involves at least two energy domains.

Since microdevices or smart devices and systems work in multiple energy domains, we almost always have to deal with coupled systems. This means that the equations governing their behavior are also coupled, which poses interesting challenges in the modeling and simulation of these devices. In this chapter, we learn about a few such challenges and how they are resolved.

Recall that we discussed only the mechanics of solids in Chapter 4. Now we move into other energy domains and, indeed, coupled domains. We begin with a short introduction to electrostatics and then see how it couples with mechanics. Following this, we consider coupling of mechanics with Couette and squeezed film flows, electro-thermal-mechanics and electro-magneto-mechanics that involve coupling among three domains.

## ▶ 6.1 ELECTROSTATICS

Electrostatics is the field of study involving electrical charges that are stationary with respect to space and time. These charges can be positive and negative. The unit of charge is the coulomb, denoted by C. The force that makes like charges repel and unlike charges attract is called the *electrostatic force*. It is one of the four fundamental forces, the other three being gravitational force and weak and strong nuclear forces. Coulomb's law quantifies the electrostatic force $\mathbf{F}_{12}$ between a pair of point charges $Q_1$ and $Q_2$ separated in vacuum by a vector $\mathbf{r}_{12}$ whose magnitude is $r$ and the direction is along the unit vector $\hat{\mathbf{r}}_{12}$ (see Figure 6.1):

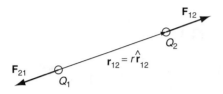

**Figure 6.1** The directions of the electrostatic forces between two point charges per Coulomb's law.

$$\mathbf{F}_{12} = \frac{1}{4\pi\varepsilon_0} \frac{Q_1 Q_2}{r^2} \hat{\mathbf{r}}_{12} \tag{6.1}$$

where $\varepsilon_0$ is the *permittivity* of free space or vacuum. We define the concept of permittivity later; for now, it can be simply understood as a constant of proportionality in the inverse square relationship in Coulomb's law. Note that $\mathbf{F}_{12}$ is the electrostatic force acting on charge $Q_2$ due to the charge $Q_1$. Since positive charges have a positive value and negative charges a negative value, we can see that Eq. (6.1) is consistent with the notion that like charges repel and unlike charges attract. Figure 6.1 makes this clear.

---

### ▶ *CURIOUS FACTS ABOUT THE INVERSE SQUARE LAWS OF ELECTROSTATICS AND GRAVITATION*

Experimental evidence shows that Coulomb's law holds for distances ranging from $10^{-3}$ m to $10^7$ m with an accuracy of $r^{2\pm10^{-16}}$ in the denominator of the expression for the electrostatic force. It may appear that it is a strange coincidence that the inverse square law (i.e., force is inversely proportional to the square of the distance) holds for the gravitational force as well. But a mathematical argument due to Laplace indicates that it must be that way, and no other, if a spherical shell of charge exerts no electrostatic force on a charged particle within it. The latter is an experimentally observed fact. A similar observation is also valid for the law of gravitational force. It is instructive to find out more about these two fundamental laws to appreciate how they were constructed from experimental facts and how Laplace's mathematical genius established that the inverse square relationship is the only relationship possible.

---

**Example 6.1**

The helium atom nucleus has two protons, each carrying a charge of $1.6022 \times 10^{-19}$ C, separated by a distance of $10^{-15}$ m. Compute the magnitude of the electrostatic force between the two protons. Use the fact that the numerical value of the reciprocal of $4\pi\varepsilon_0$ in SI units is $9 \times 10^9$.

**Solution:** Application of Coulomb's law gives

$$F = \frac{1}{4\pi\varepsilon_0} \frac{Q_1 Q_2}{r^2} = (9 \times 10^9) \frac{(1.6022 \times 10^{-19})^2}{(10^{-15})^2} \approx 231 \text{ N}$$

This is a huge force indeed and it is a repulsive force. So, we can imagine how strong the nuclear forces must be to overcome this repulsive force and hold the protons together inside the nucleus. For comparison, note that the gravitational force of attraction between the same two protons at that distance is less than $2 \times 10^{-34}$ N. So, electrostatic force is huge at small scales: this is the main reason for using the electrostatic force in microsystems.

## 6.1.1 Multiple Point Charges

When there are multiple charges, each charge experiences an electrostatic force due to every other charge. Since force is a vector quantity, we can add all such forces vectorially and get the total force. For $N$ charges, each with charge $Q_i$ and located at vectorial distance $r_i\hat{\mathbf{r}}_i$ from a charge $Q$, the electrostatic force experienced by this charge $Q$ is given by

$$\mathbf{F}_Q = \frac{1}{4\pi\varepsilon_0} \sum_{i=1}^{N} \frac{QQ_i}{r_i^2} \hat{\mathbf{r}}_i \tag{6.2}$$

The electrostatic force per unit charge is the *electric field* $\mathbf{E}$. Like any force, it is a vector with a magnitude and direction. A point charge $Q$ creates an electric field around itself. Using Eq. (6.1), we can write this field as

$$\mathbf{E}_P = \frac{1}{4\pi\varepsilon_0} \frac{Q}{r^2} \hat{\mathbf{r}} \tag{6.3}$$

where the subscript $P$ denotes that the field is given at a point $P$ located at a vectorial distance of $r\hat{\mathbf{r}}$ from charge $Q$. Because of the inverse square relationship, the field's magnitude decreases rapidly as we move away from the charge.

Electric field, being a vector, can also be added vectorially when multiple charges cause multiple overlapping fields around them. It is convenient to choose a reference point, say the origin of the coordinate system, and identify the location of each charge, $Q_i$, at vectorial distance $r_i\hat{\mathbf{r}}_i$ from the reference point. Then, per Eq. (6.3), the electric field due to all of the charges at a point $P$ located $r_P\hat{\mathbf{r}}_P$ away from the same reference point is

$$\mathbf{E}_P = \frac{1}{4\pi\varepsilon_0} \sum_{i=1}^{N} \frac{Q_i}{\|r_P\hat{\mathbf{r}}_P - r_i\hat{\mathbf{r}}_i\|^2(\hat{\mathbf{r}}_P - \hat{\mathbf{r}}_i)} \tag{6.4}$$

where $\| \cdot \|$ denotes the magnitude (or norm) of the vector; this is simply the square root of the sum of the squares of the components of the vector along the orthogonal axes of the coordinate system.

### Problem 6.1

Imagine four charges $Q_i$ $(i = 1, \ldots, 4)$ at the four corners of a square of side $10\,\text{cm}$. What must these charges be to have the following electric field at the center of the square? Assume a coordinate system at the center of the square with $x$- and $y$-axes parallel to two adjacent sides of the square.

(a) $10^{-5}\,\text{V/m}$ in the $x$-direction only

(b) $10^{-8}\,\text{V/m}$ in the $y$-direction only

(c) $10^{-6}\,\text{V/m}$ at a $45°$ angle to the $x$-direction

*Note:* One might wonder why the unit of the electric field is V/m (volt per meter) instead of the more logical N/C. The reason for this will become clear when we introduce the concept of electrical potential and its relationship to the field.

**Example 6.2**

Imagine a micromachined polysilicon cantilever beam of length $150\,\mu m$ and rectangular cross-section of $5\,\mu m$ width and $2\,\mu m$ height, with a biomolecule of charge $8.0 \times 10^{-15}\,C$ attached at its free end. If another biomolecule of charge $-1.0 \times 10^{-14}\,C$ fixed to the substrate at a distance of $1\,\mu m$ below the first, as shown in Figure 6.2, how much would the cantilever bend down at its free tip due to the electrostatic force?

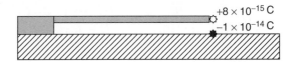

**Figure 6.2** Deflection of a cantilever due to electrostatic force between two biomolecules.

**Solution:** Since the two charges are opposite, they attract each other. Hence, the movable charge will try to move toward the fixed charge due to the electrostatic force. However, the cantilever will try to resist this. The cantilever will bend so that these two forces are equalized. In this example, it is sufficient to model the cantilever as a linear spring of spring constant, $k$, given by

$$k = \frac{3YI}{l^3} = \frac{Ywt^3}{4l^3} = \frac{3 \times (169 \times 10^9) \times (5 \times 10^{-6}) \times (2 \times 10^{-6})^3}{4 \times (150 \times 10^{-6})^3} = 1.5022\,N/m$$

Notice how small this spring constant is: you can pull this spring by about 1 m by hanging 150 g mass on it! The biological cells too have spring constants as low as this and even smaller. Hence, it is possible to manipulate cells gently with micromachined devices without causing them much mechanical damage.

If the free end of the cantilever moves down by a distance $x$, the spring force exerted by the cantilever on the charge is $kx$ in the upward direction. The downward-acting electrostatic force is then given by

$$F_e = \frac{1}{4\pi\varepsilon_0}\frac{Q_1 Q_2}{r^2} = (9 \times 10^9)\frac{(8 \times 10^{-15})(1 \times 10^{-14})}{\left[(2 \times 10^{-6}) - x\right]^2} = \frac{72 \times 10^{-20}}{\left[(2 \times 10^{-6}) - x\right]^2}$$

For equilibrium, we should have $F_e = kx$. This leads to a cubic equation in $x$:

$$\frac{72 \times 10^{-20}}{\left[(2 \times 10^{-6}) - x\right]^2} = 1.5022\,x$$

$$1.5022\,x^3 - (6.0088 \times 10^{-6})x^2 + (6.0088 \times 10^{-12})\,x - (72 \times 10^{-20}) = 0$$

Upon solving we get $x = (0.1383 \times 10^{-6})\,m$, $(1.4188 \times 10^{-6})\,m$, $(2.4429 \times 10^{-6})\,m$. The lowest of the three roots of this equation, $0.1383\,\mu m$, is the answer. In order to see why the other two roots are not the answers, we need to discuss the stability and practical feasibility of the three solutions. We do that after we introduce some more basic concepts in electrostatics, namely electrical potential and electrostatic energy.

## 6.1.2 Electric Potential

We note that a charge experiences a force everywhere in an electric field. In order to move a charge from one place to another, we need to do work against the field. An electric charge in an electric field stores energy—the *electrostatic energy*—as potential energy. This energy per unit charge is called *electric potential* (denoted by $\phi$) with volt V as its unit, which is joules per coulomb, J/C. We already begin to see how electrostatics and mechanics are being coupled to each other!

Since **E** is force per unit charge, the work done when a charge $Q$ is moved from point $A$ to $B$ in the electric field is given by

$$W_{A\to B} = -\int_{A\to B} Q\,(\mathbf{E}\cdot\mathbf{dl}) \tag{6.5}$$

where the integration is done along a path between points $A$ and $B$. The negative sign indicates that when we move along the electric field **E**, the field does the work (rather than we doing it) and when we move against the field (i.e. when **E** and **dl** have opposite signs), we must do positive work.

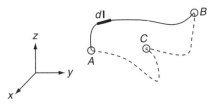

**Figure 6.3** Work done on a charge as it is moved from point $A$ to point $B$ in electric field **E**. This work per unit charge is also equal to the electric potential, $\phi$, between the two points.

In Figure 6.3, we can imagine multiple paths between points $A$ and $B$. The nature of the electrostatic field is such that the work done is independent of the path taken. Such a field is called *conservative*. Gravitation field is also conservative. We can use this property to uniquely define the electric potential difference between any two points in an electrical field. For this, we take a path between points $A$ and $B$ that passes through a third point, $C$ (the dashed curve in Figure 6.3). Now, we can rewrite Eq. (6.5) as

$$
\begin{aligned}
W_{A\to B} &= -\int_{A\to B} Q\,(\mathbf{E}\cdot\mathbf{dl}) = -\int_{A\to C} Q\,(\mathbf{E}\cdot\mathbf{dl}) - \int_{C\to B} Q\,(\mathbf{E}\cdot\mathbf{dl}) \\
&= -\int_{A\to C} Q\,(\mathbf{E}\cdot\mathbf{dl}) + \int_{B\to C} Q\,(\mathbf{E}\cdot\mathbf{dl})
\end{aligned}
\tag{6.6}
$$

If we choose point $C$ as the reference, we can denote the electric potential at any point relative to point $C$. Since we are interested in *potential difference* rather than the absolute potential, the reference point is immaterial. Hence, we define the electric potential difference as:

$$\int_{A\to C} \mathbf{E}\cdot\mathbf{dl} = \phi_A \quad \text{and} \quad \int_{B\to C} \mathbf{E}\cdot\mathbf{dl} = \phi_B$$

Therefore,

$$\frac{W_{A\to B}}{Q} = -\int_{A\to B} \mathbf{E}\cdot\mathbf{dl} = -\int_{A\to C} \mathbf{E}\cdot\mathbf{dl} + \int_{B\to C} \mathbf{E}\cdot\mathbf{dl} = -\phi_A + \phi_B = \phi_B - \phi_A \tag{6.7}$$

By using the result shown in Eq. (6.7), we now derive an important relationship between **E** and $\phi$. Consider again Figure 6.3 and let the coordinates of $A$ and a point very close to it, say $A'$, be $(x, y, z)$ and $(x + \Delta x, y + \Delta y, z + \Delta z)$. We can write the change in potential energy between $A$ and $A'$ and find that this change depends on the electric field, **E**, at point $A$. For this, we write the work done in moving a charge $Q$ from $A$ to $A'$:

$$\Delta W = W_{A \to A'} = Q\{\phi_{A'} - \phi_A\}$$
$$= Q\{\phi(x + \Delta x, y + \Delta y, z + \Delta z) - \phi(x, y, z)\} \tag{6.8a}$$
$$= Q\left\{\frac{\partial \phi}{\partial x}\Delta x + \frac{\partial \phi}{\partial y}\Delta y + \frac{\partial \phi}{\partial z}\Delta z\right\}$$

$$\Delta W = Q\left\{\left(\frac{\partial \phi}{\partial x}\hat{\mathbf{i}} + \frac{\partial \phi}{\partial y}\hat{\mathbf{j}} + \frac{\partial \phi}{\partial z}\hat{\mathbf{k}}\right) \cdot (\Delta x\,\hat{\mathbf{i}} + \Delta y\,\hat{\mathbf{j}} + \Delta z\,\hat{\mathbf{k}})\right\} \tag{6.8b}$$

$$\Delta W = Q\left\{\left(\frac{\partial \phi}{\partial x}\hat{\mathbf{i}} + \frac{\partial \phi}{\partial y}\hat{\mathbf{j}} + \frac{\partial \phi}{\partial z}\hat{\mathbf{k}}\right) \cdot (\Delta \hat{\mathbf{l}})\right\} \tag{6.8c}$$

where we have used the following mathematical concepts:

1. A function can be expressed at $(x + \Delta x)$ as a Taylor series:

$$f(x + \Delta x) = f(x) + \frac{\partial f}{\partial x}\Delta x + \frac{1}{2}\frac{\partial^2 f}{\partial x^2}(\Delta x)^2 + \cdots$$

If $\Delta x$ is sufficiently small, the approximation

$$f(x + \Delta x) \approx f(x) + \frac{\partial f}{\partial x}\Delta x$$

is valid. We apply this to $\phi(x + \Delta x, y + \Delta y, z + \Delta z)$ because the point is considered *very close* to $A$. So, we get Eq. (6.8a).

2. In Eq. (6.8b), we express the work as a dot product of two vectors.

3. In Eq. (6.8c), we recognize that $\Delta x\,\hat{\mathbf{i}} + \Delta y\,\hat{\mathbf{j}} + \Delta z\,\hat{\mathbf{k}}$ is nothing but a small vector from $A$ to $A'$, that is, $\Delta\mathbf{l}$. It is line d**l** at point $A$ instead of somewhere in the middle of the path, as shown in Figure 6.3. Now, from Eq. (6.5), we have

$$W_{A \to A'} = -\int_A^{A'} Q\,(\mathbf{E} \cdot \mathrm{d}\mathbf{l}) = -Q(\mathbf{E} \cdot \Delta\mathbf{l}) \tag{6.9}$$

From Eqs. (6.8c) and (6.9), we observe that

$$\mathbf{E} = -\left(\frac{\partial \phi}{\partial x}\hat{\mathbf{i}} + \frac{\partial \phi}{\partial y}\hat{\mathbf{j}} + \frac{\partial \phi}{\partial z}\hat{\mathbf{k}}\right) = -\nabla\phi \tag{6.10}$$

which gives the relationship between the electric field and the electric potential. We see that the electric field is the gradient of the potential. The symbol $\nabla$ is called the *del* operator and is used for denoting the *gradient*.

## ► CURIOUS FACTS ABOUT THE ∇ SYMBOL AND ITS USE IN COMPACT NOTATION

∇ is actually called the *nabla*, a Hebrew word for a stringed musical instrument of similar shape. It was apparently first used by W. R. Hamilton, a famous English mathematician and mechanician, who in fact used a sideways version of it. In vector analysis, physics, and engineering, this symbol occurs so many times, sometimes even in the same equation, that people found it difficult to say nabla and switched to the simple monosyllable *del*. Some people call it *atled*, the reverse of delta! However, del is now universally agreed upon.

- When ∇ operates on a scalar, it becomes a vector.

There are two other ways in which ∇ can operate on a scalar. Note that these two operations yield a scalar rather than a vector:

- $\nabla^2 s = \dfrac{\partial^2 s}{\partial x^2} + \dfrac{\partial^2 s}{\partial y^2} + \dfrac{\partial^2 s}{\partial z^2}$ = del-squared $s$

  This is called the *Laplacian* of a scalar $s$.

- $\nabla^4 s = \dfrac{\partial^4 s}{\partial x^4} + 2\dfrac{\partial^4 s}{\partial y^2 \partial y^2} + \dfrac{\partial^4 s}{\partial y^2}$ = del-to-the four $s$.

This is called the *biharmonic operator* on $s$. It is useful in two dimensions in certain situations such as bending of plates.

∇ can also operate on a vector to give two useful concepts: *divergence* and *curl*. The divergence operation on a vector gives a scalar and curl gives a vector.

The divergence of a vector $\mathbf{v} = v_x\hat{\mathbf{i}} + v_y\hat{\mathbf{j}} + v_z\hat{\mathbf{k}}$ is given by

$$\nabla \cdot \mathbf{v} = \frac{\partial v_x}{\partial x} + \frac{\partial v_y}{\partial y} + \frac{\partial v_z}{\partial z}$$

The curl of a vector $\mathbf{v} = v_x\hat{\mathbf{i}} + v_y\hat{\mathbf{j}} + v_z\hat{\mathbf{k}}$ is given by

$$\nabla \times \mathbf{v} = \left(\frac{\partial v_z}{\partial y} - \frac{\partial v_y}{\partial z}\right)\hat{\mathbf{i}} + \left(\frac{\partial v_x}{\partial z} - \frac{\partial v_z}{\partial x}\right)\hat{\mathbf{j}}$$
$$+ \left(\frac{\partial v_y}{\partial x} - \frac{\partial v_x}{\partial y}\right)\hat{\mathbf{k}}$$

### Problem 6.2

(a) Verify that $\nabla^2 s = \nabla \cdot (\nabla s)$.

(b) Prove that the curl of the electric field is zero, that is, $\nabla \times \mathbf{E} = \mathbf{0}$. In general, the curl of any conservative field is zero. Find out why.

Equation (6.10) can also be interpreted geometrically. If we plot equipotential curves (curves joining all points at the same potential) of an electric field, we find that they are normal to the electric field curves. Consider two charges of opposite polarity but equal magnitude as shown in Figure 6.4. The figure shows the electric field curves generated by plotting the field at every point as a short line segment that indicates the direction of the field vector and its magnitude scaled to an appropriate value. In fact, we do not need to draw the short line segments at all points to see the emerging curves; it is sufficient to draw them at a few points so that line segments meet end to end and provide continuity to the curves. Observe also that Figure 6.4 shows the equipotential curves

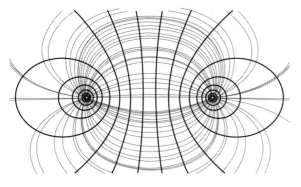

**Figure 6.4** Electric field curves (gray) and equipotential curves (black) for two point charges of opposite polarity.

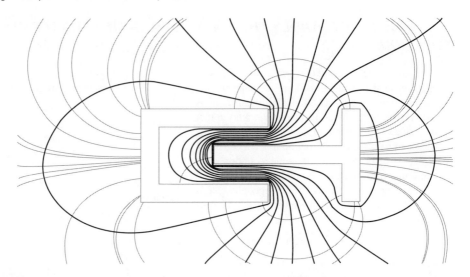

**Figure 6.5** Electric field curves (gray) and equipotential (black) curves in an electrostatic comb-drive.

and that the two sets of curves are normal to each other. That is, the tangents to the two curves are perpendicular to each other at the points of their intersection. The field curves and equipotential curves help visualize the problem, just as they helped its originator, Michael Faraday. Figure 6.5 shows the field curves and equipotential curves for a portion of a comb-drive actuator. Examine this figure to understand the pattern of its two curves, because we return to it in Section 6.2.2.

### 6.1.3 Electric Field and Potential Due to Continuous Charge

Figure 6.5 looks different from our discussion so far, because we are considering continuous charge distribution rather than point charges. But we know that charge, because of protons and electrons, can only be discrete. However, given that the basic quantum of charge is as small as $1.6022 \times 10^{-19}$ C (the magnitude of charge of an electron or proton), it is reasonable to assume that charge can be continuously distributed over a curve, surface, or volume. The real structures we deal with shortly have continuous charge distribution, not discrete charges.

When we have continuous charge along a curve, surface or volume, we can rewrite Eq. (6.4) by replacing the summation with integration. Thus, the electric field at a point $P$ due to several point charges and continuous charge along a curve $L$, surface $S$, and volume $V$ can be written as

$$\mathbf{E}_P = \frac{1}{4\pi\,\varepsilon_0} \sum_{i=1}^{N} \frac{Q_i}{\|r_P\hat{\mathbf{r}}_P - r_i\hat{\mathbf{r}}_i\|^2}(\hat{\mathbf{r}}_P - \hat{\mathbf{r}}_i) + \frac{1}{4\pi\,\varepsilon_0} \int_L \frac{\psi_L(\hat{\mathbf{r}}_P - \hat{\mathbf{r}})}{\|r_P\hat{\mathbf{r}}_P - r\hat{\mathbf{r}}\|^2}dL +$$

$$+ \frac{1}{4\pi\,\varepsilon_0} \int_S \frac{\psi_S(\hat{\mathbf{r}}_P - \hat{\mathbf{r}})}{\|r_P\hat{\mathbf{r}}_P - r\hat{\mathbf{r}}\|^2}dS + \frac{1}{4\pi\,\varepsilon_0} \int_V \frac{\psi_v(\hat{\mathbf{r}}_P - \hat{\mathbf{r}})}{\|r_P\hat{\mathbf{r}}_P - r\hat{\mathbf{r}}\|^2}dV \qquad (6.11)$$

where $\psi_L$, $\psi_S$ and $\psi_V$ are charge densities per unit length, surface, and volume, respectively. In view of Eq. (6.10), that is, since $\mathbf{E} = -\nabla\phi$, we can also write the potential as follows (the reader is urged to verify this through differentiation):

$$\phi_P = \frac{1}{4\pi\,\varepsilon_0}\sum_{i=1}^{N}\frac{Q_i}{\|r_P\hat{\mathbf{r}}_P - r_i\hat{\mathbf{r}}_i\|} + \frac{1}{4\pi\,\varepsilon_0}\int_L \frac{\psi_L}{\|r_P\hat{\mathbf{r}}_P - r\hat{\mathbf{r}}\|}\,dL +$$

$$+ \frac{1}{4\pi\,\varepsilon_0}\int_S \frac{\psi_S}{\|r_P\hat{\mathbf{r}}_P - r\hat{\mathbf{r}}\|}\,dS + \frac{1}{4\pi\,\varepsilon_0}\int_V \frac{\psi_v}{\|r_P\hat{\mathbf{r}}_P - r\hat{\mathbf{r}}\|}\,dV \qquad (6.12)$$

Note that $\psi_L$, $\psi_S$ and $\psi_V$ are functions of coordinates $x$, $y$, and $z$. That is, they can vary in space along their respective geometric entities: curve, surface and volume. In other words, the charge along the boundaries of the comb-drive (a 2D example involving only $x$ and $y$) in Figure 6.5 can be different at different points. This brings us to a point at which we need to differentiate among materials in terms of their ability to block, hold, or transmit charge.

## 6.1.4 Conductors and Dielectrics

Materials can be *conductors* (in which electrons move freely), *insulators* (where there is no movement of electrons over large distances) and *semiconductors* (where conductivity of the electrons can be changed by adding foreign species). However, in insulators we can have movement of charges over microscopic distances, thus causing polarization of the charge. Such polarized insulators are called *dielectrics*. At steady state, conductors cannot hold any charge inside them, but dielectrics can. Any charge introduced into a conductor distributes itself on its surface. This apparent movement of charges led Faraday to propose the concept of *electric displacement* **D**, a vector quantity defined in general by

$$\mathbf{D} = \varepsilon_0(1 + \chi_e)\mathbf{E} = \varepsilon_0\mathbf{E} + \varepsilon_0\chi_e\mathbf{E} = \varepsilon_0\mathbf{E} + \mathbf{P}_{ep} \qquad (6.13)$$

where $\varepsilon_0$ is the *permittivity* of free space, $\chi_e$ is the *electric susceptibility*, and $\mathbf{P}_{ep}$ is the *electric polarization* vector. While $\varepsilon_0$, equal to $8.854 \times 10^{-12}$ F/m in SI units,[1] is a universal constant, $\chi_e$ is a material property indicating how susceptible a material is to polarization. By using Eq. (6.13), we now define the permittivity $\varepsilon$, a property of materials that determines the extent to which they can be polarized and permit the creation and retention of electric dipoles inside them, as

$$\varepsilon = \varepsilon_0(1 + \chi_e) \qquad (6.14)$$

so that

$$\mathbf{D} = \varepsilon\mathbf{E} \qquad (6.15)$$

Another name for permittivity is the *dielectric constant*. Sometimes it is also defined relative to $\varepsilon_0$ and called *relative permittivity* or relative *dielectric constant* $\varepsilon_r$:

$$\varepsilon = \varepsilon_r\varepsilon_0 \qquad (6.16)$$

A vacuum, air and conductors have $\varepsilon_r = 1$ because they are not susceptible to polarization and hence $\varepsilon = \varepsilon_0$ for them. Table 6.1 shows the relative permittivities of some materials relevant to microsystems and smart systems.

---

[1] The curious unit of F/m (farad per meter) is explained later in this chapter.

**Table 6.1 Approximate relative dielectric constants (permittivities), $\varepsilon_r$, of some materials**

| Material | $\varepsilon_r$ |
| --- | --- |
| Silicon | 11.9 |
| Silicon dioxide | 3.9 |
| Silicon nitride | 7.2 |
| Glass | 4–9 |
| Distilled water | 81 |
| PZT (lead zirconium titanate) | 1300 |

*Source:* www.matweb.com

### 6.1.5 Gauss's Law

We noted in the preceding subsection that dielectric materials can hold charge inside them in the presence of an electric field, while conductors cannot. The charge resides only on the boundary of a conductor; it is on the boundary curve in 2D and on the boundary surface in 3D. Let us now consider the surface-micromachined RF switch shown in Figure 6.6. It has a polysilicon cantilever beam-like structure anchored at one end to the silicon substrate coated with a dielectric layer of silicon nitride. There is also another patterned polysilicon layer underneath the beam that acts as the ground electrode; dimensions are indicated in the figure. We want to know the charge distribution over the polysilicon beam when an electric potential (i.e. voltage), $\phi$, is applied between the beam and the polysilicon electrode beneath the beam, which is assumed to be at zero potential.

Coulomb's law alone does not make it easy to compute the distribution of charges on the polysilicon beam and we need another one known as *Gauss's law*. This is because using the general equations [Eqs. (6.11) and (6.12)] to compute the electric field and potential for real geometries, such as those in Figure 6.6, is quite tedious since a lot of integration and algebraic manipulation is necessary when there is continuous charge. A simpler way to handle such situations is to use *Gauss's law*, which can be derived from Coulomb's law and gives us a mathematical framework that is useful to write integral or differential equations; these can be solved analytically (for simple geometries) or numerically (for any geometry). Gauss's law states that

$$\oint_S \mathbf{D} \cdot d\mathbf{s} = \int_V \psi_v \, dV = Q_{\text{enclosed}} \tag{6.17}$$

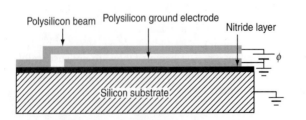

Polysilicon beam    Polysilicon ground electrode    Nitride layer

Silicon substrate

$t_b$ = Beam thickness
$t_e$ = Electrode thickness
$g_0$ = Gap between the beam and the electrode
$t_n$ = Nitride layer thickness
$t_s$ = Substrate thickness
$l_b$ = Beam length
$l_e$ = Electrode length
$l_a$ = Beam anchor length

**Figure 6.6** The mechanical structure of an RF switch.

The leftmost term in Eq. (6.17) is a surface integral taken over a closed surface $S$ that encloses volume $V$. The integrand in this term is the dot product of $\mathbf{D}$ and the differential surface element vector $\mathbf{ds}$, whose magnitude is the area of an infinitesimal (as small as you can imagine) surface patch at a point on $S$ (i.e. $\mathbf{ds}$) and direction is the unit normal at that point (i.e. $\hat{\mathbf{n}}$). The middle term in Eq. (6.17) is simply the total charge enclosed by volume $V$ because $\psi_v$ is the charge per unit volume. Therefore, the last term of Eq. (6.17) gives the total charge, $Q_{enclosed}$, within the volume enclosed by the closed surface $S$.

**Your Turn:**

Although we have not assumed anything other than Coulomb's law in this chapter, all the nuances and details of electrostatics cannot be explained merely in one section of one chapter. Refer to any book on electrostatics (two are cited at the end of the chapter) and convince yourself that Gauss's law is a direct consequence of Coulomb's law. You will then be able to appreciate also how Laplace was able to prove that the inverse square relationship is the only relationship possible in Coulomb's law.

In order to use Gauss's law to solve the problem of an RF switch, we need a mathematical theorem of vector calculus also owed to Gauss, sometimes known as the *divergence theorem* or the *Green–Gauss theorem*. It is stated as follows for any vector $\mathbf{G}$:

$$\oint_S \mathbf{G} \cdot \mathbf{ds} = \int_V (\nabla \cdot \mathbf{G})\, dV \tag{6.18}$$

By using Eqs. (6.17) and (6.18), we can write

$$\nabla \cdot \mathbf{D} = \psi_v \tag{6.19a}$$

or

$$\nabla \cdot \mathbf{D} = \frac{\partial D_x}{\partial x} + \frac{\partial D_y}{\partial y} + \frac{\partial D_z}{\partial z} = \psi_v \tag{6.19b}$$

Equation (6.19b) is the differential form of Gauss's law, while Eq. (6.17) is its integral form. In writing Eq. (6.19b), we used the fact that $\mathbf{D} = D_x \hat{\mathbf{i}} + D_y \hat{\mathbf{j}} + D_z \hat{\mathbf{k}}$ with components along the three orthogonal directions. Equation (6.19a) can be simplified further using the concepts introduced earlier in this section. We have

$$\nabla \cdot \mathbf{D} = \psi_v$$

$$\nabla \cdot (\varepsilon \mathbf{E}) = \psi_v \text{ [from Eq. (6.15)]} \tag{6.20a}$$

$$\nabla \cdot [(\varepsilon(-\nabla \phi)] = \psi_v \text{ [from Eq. (6.10)]} \tag{6.20b}$$

When $\varepsilon$ is uniform within the volume of interest, it can be taken out of the gradient operation and we write

$$\varepsilon \nabla \cdot (-\nabla \phi) = \psi_v$$

$$\nabla \cdot (\nabla \phi) = -\frac{\psi_v}{\varepsilon} \tag{6.21}$$

$$\nabla^2 \phi = -\frac{\psi_v}{\varepsilon}$$

which is known as the *Poisson equation*. For convenience, we expand the compact notation of del-squared to get

$$\nabla^2\phi = \frac{\partial^2\phi}{\partial x^2} + \frac{\partial^2\phi}{\partial y^2} + \frac{\partial^2\phi}{\partial z^2} = -\psi_v \tag{6.22}$$

If we consider a volume that does not enclose any charge, Eq. (6.22) becomes

$$\frac{\partial^2\phi}{\partial x^2} + \frac{\partial^2\phi}{\partial y^2} + \frac{\partial^2\phi}{\partial z^2} = 0 \quad \text{or} \quad \nabla^2\phi = 0 \tag{6.23}$$

This is known as the *Laplace equation*. In our problem of the RF switch, if we take a bounding box around the switch and subtract from it all the conductors, namely, the beam, the electrode, and the substrate, we get a volume that does not enclose any charge. Figure 6.7 shows a cross-section of such a volume. Thus, Eq. (6.23) is valid for the union of the hatched area of the empty space and the dark rectangle of the nitride layer. This means that Eq. (6.23) is to be solved outside the region of conductors. Since there is no limit on how much ''outside'' we need to take, we call this an ''infinite domain'' problem. For practical purposes, we restrict it to a large bounding box as shown in Figure 6.7 so that only that part needs to be discretized using a mesh or a grid and solved using either the finite difference method or FEM discussed in Chapter 5. If we solve the differential equation of Eq. (6.23), we will be able to compute the variation of $\phi$ in the hatched region of Figure 6.7 by using our knowledge of the potential on all its boundaries. For the edges of the bounding rectangular box, we assume that there is zero surface charge, that is, $\mathbf{D}\cdot\hat{\mathbf{n}} = 0$. This is true if the bounding box is sufficiently large.

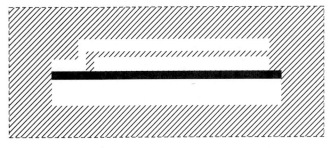

**Figure 6.7** The bounding rectangle around the structure of an RF switch with conductors subtracted. The hatched region of the empty space (air) and the dark dielectric region of the conductor satisfy the Laplace equation [Eq. (6.23)].

After determining $\phi$ using, say, the FE solution, we can compute its gradient to get the electric field $\mathbf{E} = -\nabla\phi$. Figure 6.8

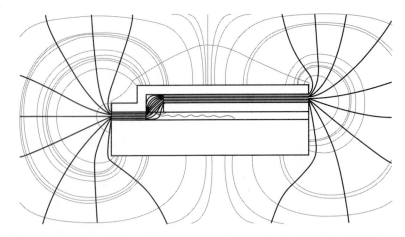

**Figure 6.8** Electric potential (black) curves and field (gray) curves for the RF switch geometry in Figure 6.7.

shows the contours of $\phi$ and the field curves of **E**, respectively, for the RF switch problem. These results were obtained using COMSOL MultiPhysics software, which uses the FE method. Notice that, per $\mathbf{E} = -\nabla\phi$, the potential and field curves are normal to each other. That is, the tangents to these curves at any point of their intersection are perpendicular to each other. We now proceed to compute the charge distribution over the surface of the polysilicon beam.

### 6.1.6 Charge Distribution on the Conductors' Surfaces

Since charge can move freely inside a conductor, at steady state (as assumed in electrostatics) there should not be any charges inside a conductor. Observe that charge inside a conductor would create an electric field, which would in turn move the charges, violating the static conditions assumed in electrostatics. Therefore, due to Eq. (6.10), the electric potential is constant in the entire conductor. And, as noted earlier, the charge resides on the boundary of the conductor. We now determine a general expression for this surface (boundary in 3D) charge $\psi_s$.

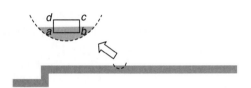

**Figure 6.9** Electric field and charge on the surface of a conductor.

Consider a small portion of a conductor such as the beam in Figure 6.6. The beam and an enlarged portion of it are shown in Figure 6.9. We consider in this portion a rectangle *abcd* that is partly within the conductor and partly outside. Let the height of this rectangle be $\Delta h$ and the width $\Delta w$. Since the work done to move a unit charge around a closed path *abcda* in a conservative field, such as the electric field **E**, is zero, we can write

$$\oint_{abcda} \mathbf{E} \cdot d\mathbf{l} = 0 \tag{6.24}$$

Now, imagine shrinking $\Delta h$ to zero so that we are left with only edges *ab* and *cd*. But the electric field on *ab* is zero because it is inside the conductor. Thus, as $\Delta h \to 0$, the line integral in Eq. (6.24) reduces to $E_t \Delta w$, where $E_t$ is the tangential component of **E**. That is,

$$\lim_{\Delta h \to 0} \oint_{abcda} \mathbf{E} \cdot d\mathbf{l} = \oint_{cd} \mathbf{E} \cdot d\mathbf{l} = E_t \Delta w = 0$$
$$E_t = 0 \tag{6.25}$$

Thus, we conclude that the electric field on the surface (i.e. boundary) of a conductor is always normal to it. Referring back to Figure 6.5, we can see that the field lines leave the conductors in the direction normal (i.e. perpendicular to the tangent) to the surface at all points and that they begin from positive charge and end on negative charge.

Next, we consider the same rectangle in Figure 6.9 and write Gauss's law [Eq. (6.17)] for the closed boundary:

$$\oint_{abcda} \mathbf{D} \cdot d\mathbf{s} = Q_{\text{enclosed}} = \psi_s \Delta w \tag{6.26}$$

where we have used the fact that the charge enclosed is only due to what is on the edge $cd$ with a surface charge density of $\psi_s$. Now, as $\Delta h \to 0$, we get

$$\lim_{\Delta h \to 0} \oint_{abcda} \mathbf{D} \cdot \mathbf{ds} = \oint_{cd} \varepsilon \mathbf{E} \cdot \mathbf{ds} = \varepsilon E_n \Delta w = \psi_s \Delta w \tag{6.27}$$

$$\psi_s = \varepsilon E_n$$

This result can be written in general for any point on the conductor as

$$\mathbf{D} \cdot \hat{\mathbf{n}} = D_n = \varepsilon E_n = \psi_s \tag{6.28}$$

Thus, once we have the potential, and hence the electric field, in the bounding box (minus the conductors), we can compute the charge distribution using Eq. (6.28).

Thus, using the concepts presented here, any electrostatic problem involving any number of conductors and dielectrics of any shape with empty spaces in between can be solved—of course, only numerically—using the FE method. For this, we need to know the geometry, the electric potential on all the conductors, and any enclosed charge. Given these, we can compute the electric potential and electric field in the space between the conductors and also the surface charge distribution on the conductors. By using Gauss' law, we can compute the charge inside the dielectrics as well. We now proceed to find the forces acting on the conductors and dielectrics.

### Problem 6.3

Use any finite element software and solve a problem involving multiple conductors and dielectrics. Verify the concepts discussed so far. Observe the effect of the size of the bounding box on the result. Explore how large it must be so that the result does not change noticeably. It will also be a good exercise for you to write your own program using either the FE method or the finite difference method.

### 6.1.7 Electrostatic Forces on the Conductors

The charges on the surface of a conductor create an electric field around it. If there are nearby conductors (creating their own fields around them), the total electric field is the combined effect of all of them. From Eq. (6.28), we can write

$$E_n = E_{1n} + E_{2n} = \frac{\psi_s}{\varepsilon} \tag{6.29}$$

where $E_{1n}$ and $E_{2n}$ are the fields interior and exterior to the conductor, respectively. But then, if we now use the argument of the rectangle $abcd$ in Figure 6.9 shrinking to edges $ab$ and $cd$ on either side of a charged sheet of zero thickness, these two components must be equal. This is because the integral in Eq. (6.26) will give twice the quantity on the right-hand side and hence each component is equal to $(\psi_s/2\varepsilon)$ per Eq. (6.28).

Only the component $E_{2n}$ equal to $(\psi_s/2\varepsilon)$ will exert a force on the charges on a conductor's surface because the other component is a field created by itself and hence, cannot exert any force. (*Note*: for the same reason, a point charge that creates a field around it cannot exert a force on itself.)

Now, if we consider an infinitesimally small surface patch of area $\Delta s$, we know that it experiences a force field of $E_{2n} = (\psi_s/2\varepsilon)$. Since it has a charge of $\psi_s \Delta s$ (remember that $\psi_s$ is the charge per unit area), the force on it is equal to $(\psi_s^2 \Delta s/2\varepsilon)$. That is,

$$\Delta F = E_{2n}\Delta q = \frac{\psi_s}{2\varepsilon}(\psi_s \Delta s) = \frac{\psi_s^2}{2\varepsilon}\Delta s \tag{6.30}$$

As $\Delta s \to 0$, we see that, in the limit, the electrostatic force at a point is

$$\mathbf{t}_e = \frac{\psi_s^2}{2\varepsilon}\hat{\mathbf{n}} \tag{6.31}$$

and is normal to the surface as indicated. Note that this is force per unit area [per Eq. (6.30)] and is like pressure (e.g. buoyancy or atmospheric pressure). In the next section, we consider the effect of this force and its coupling effects.

---

 **Your Turn:**

Deriving an expression for the electrostatic forces on dielectrics is more involved because they are polarized in an electric field and the electric field inside them is not zero. Read about this in [1] and [2]. In particular, the most general expression for electrostatic force, known as *Maxwell's stress*, is applicable to conductors and dielectrics alike. Because it is a stress, it applies to all points and not just to surfaces. But the expression reduces to surface force when there is a pure conductor or a homogeneous dielectric. To appreciate this, consider the divergence of the Maxwell's stress, $\boldsymbol{\sigma}_e$:

$$\nabla \cdot \boldsymbol{\sigma}_e = \left\{ \left( \frac{\nabla \varepsilon}{\varepsilon} - \frac{\nabla k_e}{k_e} \right) \cdot (\varepsilon \mathbf{E}) \right\} \mathbf{E} - \frac{1}{2}E^2 \nabla \varepsilon \tag{6.32}$$

where the new symbol $k_e$ stands for electrical conductivity. When there is no spatial gradient in $\varepsilon$ and $k_e$, we can see that $\nabla \cdot \boldsymbol{\sigma}_e$ is zero; this is consistent with Eq. (6.31) in the sense that only when there is a jump in material properties across an interface between two materials, that is at the boundary, $\nabla \cdot \boldsymbol{\sigma}_e$ is nonzero. The Maxwell stress can be added directly to the stress caused by mechanical forces, leading to the concept of *electroelasticity*—an interesting approach on which computational research is currently in progress.

Note also that in Eq. (6.32), we show the divergence of the Maxwell stress rather than the stress itself. This is because stress is a *tensor*, a concept not discussed here. Read in [1] about tensors and how the divergence of a tensor gives a vector, just as the divergence of a vector gives a scalar.

---

## ▶ 6.2 COUPLED ELECTROMECHANICS: STATICS

Recall Eq. (4.117) from Chapter 4, where we gave the following equations for the static equilibrium of a general elastic body:

$$\left. \begin{aligned} \frac{\partial \sigma_x}{\partial x} + \frac{\partial \tau_{xy}}{\partial y} + \frac{\partial \tau_{zx}}{\partial z} + b_x = 0 \\[2mm] \frac{\partial \tau_{xy}}{\partial x} + \frac{\partial \sigma_y}{\partial y} + \frac{\partial \tau_{yz}}{\partial z} + b_y = 0 \\[2mm] \frac{\partial \tau_{zx}}{\partial x} + \frac{\partial \tau_{yz}}{\partial y} + \frac{\partial \sigma_z}{\partial z} + b_z = 0 \end{aligned} \right\} \text{ For the interior} \tag{6.33a}$$

$$\left.\begin{array}{l} \sigma_x n_x + \tau_{xy} n_y + \tau_{zx} n_z = t_x \\ \tau_{xy} n_x + \sigma_y n_y + \tau_{yz} n_z = t_y \\ \tau_{xz} n_x + \tau_{yz} n_y + \sigma_z n_z = t_z \end{array}\right\} \text{For the boundary} \qquad (6.33b)$$

In the case of pure conductors and homogeneous dielectrics, since electrostatic force, $\mathbf{t}_e = t_{ex}\hat{\mathbf{i}} + t_{ey}\hat{\mathbf{j}} + t_{ez}\hat{\mathbf{k}}$, acts only on the boundary, it is added to the mechanical surface force, $\mathbf{t} = t_x\hat{\mathbf{i}} + t_y\hat{\mathbf{j}} + t_z\hat{\mathbf{k}}$, as follows:

$$\left.\begin{array}{l} \sigma_x n_x + \tau_{xy} n_y + \tau_{zx} n_z = t_x + t_{ex} \\ \tau_{xy} n_x + \sigma_y n_y + \tau_{yz} n_z = t_y + t_{ey} \\ \tau_{xz} n_x + \tau_{yz} n_y + \sigma_z n_z = t_z + t_{ez} \end{array}\right\} \text{For the boundary} \qquad (6.34)$$

This shows that the electrostatic problem influences the elastostatic solution. But there is coupling the other way, too: that is, the elastostatic problem influences the electrostatic problem. In order to understand this, let us revisit the RF switch problem in Figure 6.6.

The polysilicon beam in Figure 6.6 deforms due to the electrostatic force acting on it. When it does, the region (see Figure 6.7) over which the electrostatic problem was solved changes. So, we need to solve the electrostatic problem for the deformed geometry in Figure 6.10(b), not Figure 6.10(a), because the electric field, potential, and charge distribution depend on the geometry of the conductors and dielectrics. All of these quantities change as the geometry changes due to elastic deformation. Furthermore, the electrostatic force also changes and will differ from the one that caused the deformation. Now, if we solve the electrostatic problem again on the new region, we get the new electrostatic force. This force applied on the undeformed conductor(s) causes a new deformation that further changes the region over which the electrostatic force must once again be solved. This process continues until there is no change in the deformation, a situation referred to as the *convergence* of the numerical solution process. Clearly, an iterative solution procedure is necessary because of the coupling between the electrostatic and elastostatic equations. The solution that satisfies both the equations with appropriate electric potential (in the electrostatic problem) and deformations (in the elastostatic problem) is the goal here, as shown in Figure 6.11.

For simplicity, we consider only two conductors in the schematic in Figure 6.11; furthermore, the second schematic shows the electrostatic force only on the lower edge of conductor 2. In reality, it will be on all the boundaries but will be negligibly small on the boundaries that do not face the other conductor. In numerical computations, however, the

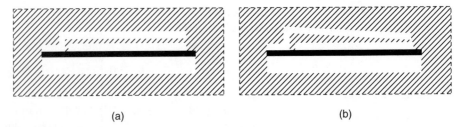

(a)                                    (b)

**Figure 6.10** (a) The region over which Eq. (6.23) must be solved to compute the electric potential and thereby the electrostatic force; (b) the changed region due to deformation of the polysilicon beam caused by the electrostatic force.

forces on the entire boundary are considered. Conductor 1 also experiences the electrostatic force, but it is not considered because it is assumed to be bulky and unable to deform much, so that it will not contribute significantly to changing the shaded region for the electrostatic calculation.

Usually, multiple iterative steps are required to make the procedure in Figure 6.11 converge. When the applied potential is increased, the entire procedure must be redone from the beginning or from the previously converged solution. The number of steps taken will be more for larger applied potentials. At some point, the solution will not converge at all because instability occurs in the problem. We analyze this problem subsequently after summarizing this method and then present an alternate method to solve the coupled problem.

***Summary of the First Method***   Consider two conductors: the first is bulky, fixed on three sides, and electrically grounded at zero potential; the second is slender, is at some applied potential $\phi_2$, and is fixed on its left edge.

**Step 1a (electrostatics)** Construct a bounding box around the two conductors in Figure 6.11(a) and subtract the two conductors to find the electric potential in the shaded region in Figure 6.11(b). That is, solve: $\nabla^2\phi = 0$ such that zero potential and $\phi_2$ potential are maintained on the boundaries of conductors 1 and 2, respectively. Compute $\mathbf{E} = -\nabla\phi$ and then $\psi_s = \varepsilon E_n$. Then compute the electrostatic force, $\hat{\mathbf{t}}_e = \{\psi_s^2/(2\varepsilon)\}\hat{\mathbf{n}}$, on the nonfixed boundaries of conductor 2.

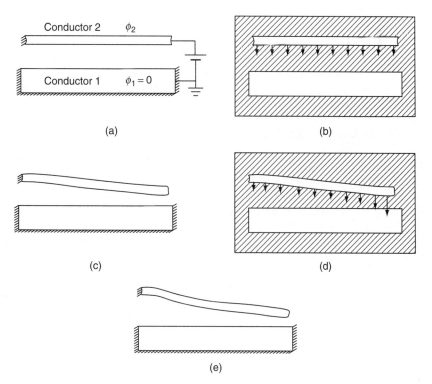

**Figure 6.11** Pictorial illustration of an iterative method of solving the coupled problem of electrostatics and elastostatics.

**Step 1b (elastostatics)** Compute the deformation of conductor 1 due to the electrostatic force. For this solve Eq. (6.33a) with the boundary equations taken from Eq. (6.34) to get the deformed geometry of conductor 1 as shown in Figure 6.11(c).

**Step 2a (electrostatics)** Subtract the deformed conductor 1 and conductor 2 and solve the electrostatic problem in the shaded region in Figure 6.11(d). Compute the changed forces on conductor 1 due to the newly computed electric potential, electric field, and surface charge.

**Step 2b (elastostatics)** Compute the deformed geometry of the conductor 1 due to the new forces applied on the undeformed configuration to get the new deformed geometry as shown in Figure 6.11(e).

**Repeat the electrostatics and elastostatics steps** until there is no more deformation of conductor 1. At this point, the electric potential will also remain unchanged. That is, we have obtained a self-consistent solution between the electrostatic and elastostatic equations.

## ▶ FINITE ELEMENT FORMULATION FOR THE ELECTROSTATICS PROBLEM

The general methodology of finite element analysis was described in Chapter 5. We now apply it to solve the partial differential equation (PDE) governing electrostatics [Eq. (6.23)], for a 3D domain of sufficiently large bounding box containing the conductors and dielectrics. Note that the discussion so far has focused on 2D domains only; the discussion that follows is more general and deals with 3D.

As is customary in FEA, we begin by writing the electrostatics PDE in the weak form by multiplying it with a trial function corresponding to the function to be solved and then integrating it over the domain. The function to be solved is $\phi(x, y, z)$. Let the weight function be $\phi_v(x, y, z)$. Then, we can write the weak form of the PDE in Eq. (6.23) as

$$\int_V (\varepsilon \nabla^2 \phi) \phi_v \, dV = 0 \qquad (6.35)$$

By using the fact that $(\varepsilon \nabla^2 \phi) \phi_v = \varepsilon \nabla \cdot (\phi_v \nabla \phi) - \varepsilon (\nabla \phi \cdot \nabla \phi_v)$, Eq. (6.35) can be rewritten

$$\int_V \varepsilon \nabla \cdot (\phi_v \nabla \phi) \, dV - \int_V \varepsilon (\nabla \phi \cdot \nabla \phi_v) \, dV = 0 \qquad (6.36)$$

By applying the divergence theorem [see Eq. (6.18)] to the first integral of Eq. (6.36), we can rewrite Eq. (6.36) as

$$\oint_S \varepsilon (\phi_v \nabla \phi \cdot \hat{\mathbf{n}}) dS - \int_V \varepsilon (\nabla \phi \cdot \nabla \phi_v) \, dV = 0 \qquad (6.37)$$

where $\hat{\mathbf{n}}$ is the outward normal to the closed surface $S$ enclosing volume $V$. From Eqs. (6.28) and (6.40), we note that $\varepsilon \nabla \phi \cdot \hat{\mathbf{n}} = \varepsilon E_n = \psi_s$. Hence, $(\varepsilon \nabla \phi \cdot \hat{\mathbf{n}})$ in the first term of Eq. (6.37) can be replaced by $\psi_s$, the surface charge density. Thus, Eq. (6.37) assumes the form

$$\int_V \varepsilon (\nabla \phi \cdot \nabla \phi_v) \, dV = \oint_S \psi_s \, \phi_v \, dS \qquad (6.38)$$

We now turn to interpolation of $\phi(x, y, z)$ and $\phi_v(x, y, z)$ using shape functions and nodal values of these quantities for a finite element (see Section 5.4.1). That is, within a finite element with $p$ nodes, we have

$$\phi_e(x, y, z) = \sum_{i=1}^p N_i \phi_i = \mathbf{N} \boldsymbol{\varphi}_e \qquad (6.39a)$$

$$\phi_{ve}(x, y, z) = \sum_{i=1}^p N_i \phi_i = \mathbf{N} \boldsymbol{\varphi}_{ve} \qquad (6.39b)$$

where $\mathbf{N}$ is a $1 \times p$ shape function matrix and $\boldsymbol{\varphi}_e$ is the $p \times 1$ nodal potential vector. Note that the weight function $\phi_{ve}$ over the element is also interpolated using the same functions and the nodal vectors, $\boldsymbol{\varphi}_{ve}$, as in the Galerkin method described in Section 5.2.3.3.

We also write the gradients of the interpolated functions $\phi_e(x, y, z)$ and $\phi_{ve}(x, y, z)$ in Eqs. (6.39a) and (6.39b):

$$\nabla \phi_e = \left( \frac{\partial \mathbf{N}}{\partial x} \hat{i} + \frac{\partial \mathbf{N}}{\partial y} \hat{j} + \frac{\partial \mathbf{N}}{\partial z} \hat{k} \right) \boldsymbol{\varphi}_e \text{ and}$$

$$\nabla \phi_{ve} = \left( \frac{\partial \mathbf{N}}{\partial x} \hat{i} + \frac{\partial \mathbf{N}}{\partial y} \hat{j} + \frac{\partial \mathbf{N}}{\partial z} \hat{k} \right) \boldsymbol{\varphi}_{ve} \qquad (6.40)$$

By using Eqs. (6.39a, 6.39b,) and (6.40), we can write Eq. (6.38) for a finite element of volume $V_e$ enclosed by the surface $S_e$ as

$$\int_{V_e} \varepsilon (\nabla \phi_e \cdot \nabla \phi_e)\, dV_e = \oint_{S_e} \psi_{se}\, \phi_{ve}\, dS_e \qquad (6.41a)$$

$$\int_{V_e} \varepsilon (\nabla \phi_e)^T (\nabla \phi_e)\, dV_e = \oint_{S_e} \psi_{se}\, \phi_{ve}\, dS_e \qquad (6.41b)$$

where we have used the fact that the dot product of two vectors can be computed using the transpose operation on one of them. Thus, Eq. (6.417b) takes the form

$$\boldsymbol{\varphi}_e^T \left[ \int_{V_e} \varepsilon \left\{ \left( \frac{\partial \mathbf{N}}{\partial x} \right)^T \left( \frac{\partial \mathbf{N}}{\partial x} \right) + \left( \frac{\partial \mathbf{N}}{\partial y} \right)^T \left( \frac{\partial \mathbf{N}}{\partial y} \right) \right. \right.$$
$$\left. \left. + \left( \frac{\partial \mathbf{N}}{\partial z} \right)^T \left( \frac{\partial \mathbf{N}}{\partial z} \right) \right\} dV_e \right] \boldsymbol{\varphi}_{ve}$$
$$= \left\{ \oint_{S_e} \psi_{se}\, \mathbf{N}\, dS_e \right\} \boldsymbol{\varphi}_{ve} \qquad (6.42)$$

Since $\boldsymbol{\varphi}_{ve}$ is arbitrary, Eq. (6.42) gives

$$\boldsymbol{\varphi}_e^T \left[ \int_{V_e} \varepsilon \left\{ \left( \frac{\partial \mathbf{N}}{\partial x} \right)^T \left( \frac{\partial \mathbf{N}}{\partial x} \right) + \left( \frac{\partial \mathbf{N}}{\partial y} \right)^T \left( \frac{\partial \mathbf{N}}{\partial y} \right) \right. \right.$$
$$\left. \left. + \left( \frac{\partial \mathbf{N}}{\partial z} \right)^T \left( \frac{\partial \mathbf{N}}{\partial z} \right) \right\} dV_e \right]$$
$$= \left\{ \oint_{S_e} \psi_{se}\, \mathbf{N}\, dS_e \right\} \qquad (6.43)$$

By taking the transpose, Eq. (6.43) can be rewritten as

$$C_e \boldsymbol{\varphi}_e = \mathbf{q}_e \qquad (6.44a)$$

where

$$C_e = \left[ \int_{V_e} \varepsilon \left\{ \left( \frac{\partial \mathbf{N}}{\partial x} \right)^T \left( \frac{\partial \mathbf{N}}{\partial x} \right) + \left( \frac{\partial \mathbf{N}}{\partial y} \right)^T \left( \frac{\partial \mathbf{N}}{\partial y} \right) \right. \right.$$
$$\left. \left. + \left( \frac{\partial \mathbf{N}}{\partial z} \right)^T \left( \frac{\partial \mathbf{N}}{\partial z} \right) \right\} dV_e \right] \qquad (6.44b)$$

$$\mathbf{q}_e = \left\{ \oint_{S_e} \psi_{se}\, \mathbf{N}^T\, dS_e \right\} \qquad (6.44c)$$

Note that $\mathbf{C}_e$ is the element capacitance matrix, $\boldsymbol{\varphi}_e$ the element potential vector, and $\mathbf{q}_e$ the element charge vector.

The next step is to assemble the element-level equations of Eqs. (6.44a)–(6.44c) into a global equation involving all the finite elements in a manner similar to the description in the last paragraph of Section 5.4.2. This yields

$$\mathbf{C}\boldsymbol{\varphi} = \mathbf{q} \qquad (6.45)$$

Application of the boundary conditions becomes relevant at this stage. At any point in the domain, we can specify either the potential or the charge but not both. This is similar to specifying either force or displacement at any point in elastic analysis. By referring to Figures. 6.12(a) and (b) of a typical electrostatic problem and its coarse mesh for the purpose of illustration, we note that potential is known at the nodes 10, 11, 18, 19, 26, 27, 14, 15, 22, 23, 30, and 31, and that specified charge is zero at all other nodes.

We then eliminate the $\phi$'s at the nodes where it is known, by deleting the corresponding rows and columns in Eq. (6.45), thus reducing it to a smaller size involving only the nodes where $\phi$ is not known. The resulting linear system is then solved. After that, the charges at the eliminated nodes are computed using Eq. (6.45).

To complete this derivation, it is pertinent to give the shape functions for the 2D problem so that the readers can implement the procedure to write their own computer program to solve the electrostatics problem in 2D. Shape functions for a three-noded [i.e. $p = 3$ in Eq. (6.39a)] triangular finite element are:

$$N_i = \frac{1}{2A} (a_i + b_i x + c_i y) \qquad (6.46)$$

where

$$a_1 = x_2 y_3 - x_3 y_2, \; b_1 = y_2 - y_3, \; c_1 = x_3 - x_2$$
$$a_2 = x_3 y_1 - x_1 y_3, \; b_2 = y_3 - y_1, \; c_2 = x_1 - x_3$$
$$a_3 = x_1 y_2 - x_2 y_1, \; b_3 = y_1 - y_2, \; c_3 = x_2 - x_1$$
$$(6.47a)$$

and

$$A = \frac{1}{2} \begin{vmatrix} 1 & x_1 & y_1 \\ 1 & x_2 & y_2 \\ 1 & x_3 & y_3 \end{vmatrix} \qquad (6.47b)$$

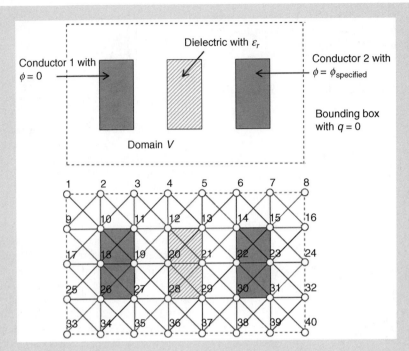

**Figure 6.12** A sample problem in electrostatics. (a) The specifications, (b) the meshed model.

and $(x_i, y_i, z_i)$ $(i = 1, 2, 3)$ are the coordinates of the triangular element. Using Eqs. (6.46) and (6.47), the element capacitance matrix of Eq. (6.44b) may be computed for the 2D triangular element of thickness $t$ (see Figure 6.12b) as

The explicitly integrated element matrix of the preceding equation makes implementation of the FE program easy without resorting to numerical quadrature, as discussed in Section 5.4.7.

$$\mathbf{C}_e = \frac{\varepsilon t}{4A} \begin{bmatrix} (b_1^2 + c_1^2) & (b_1 b_2 + c_1 c_2) & (b_1 b_3 + c_1 c_3) \\ (b_1 b_2 + c_1 c_2) & (b_2^2 + c_2^2) & (b_2 b_3 + c_2 c_3) \\ (b_1 b_3 + c_1 c_3) & (b_2 b_3 + c_2 c_3) & (b_3^2 + c_3^2) \end{bmatrix} \qquad (6.48)$$

## 6.2.1 An Alternative Method for Solving the Coupled Problem

In Figure 6.11 and the discussion so far, we referred to the use of the FE or finite difference method to solve the electrostatic problem in a large bounding box from which the conductors are removed. For this, we used the differential form of Gauss's law [Eq. (6.19b) or its simplified form in Eq. (6.23) when the region does not enclose any charge]. Alternatively, we can also solve the integral form of Gauss's law given in Eq. (6.17).

Integral equations can be solved using the boundary element method just as differential equations are solved using the FE method. The advantage of using the boundary element method is that only the boundaries, and not the interiors of the objects need to be discretized with a mesh. Thus, in the problem of RF switch [Figs. 6.10(a) and (b)], we need to work with only the boundaries of the beam, the electrode, the nitride layer, and the substrate. Furthermore, we do not need a bounding box.

Referring to Eq. (6.17), imagine that we discretize the closed surfaces of the 3D conductors and dielectrics into small triangular panels. This converts the surface integral [the left-hand side of Eq. (6.17)] into a summation. The right-hand side gives a set of charges (which need to be determined) on the panels. We can assume, for simplicity, that the charge on each panel is uniformly distributed when the triangular panel is sufficiently small. With this, Eq. (6.17) can be written in matrix form as[2]

$$
\begin{bmatrix}
c_{11} & c_{12} & \cdots & c_{1N} \\
c_{21} & c_{22} & \cdots & \vdots \\
\vdots & \vdots & \ddots & \vdots \\
c_{N1} & \cdots & \cdots & c_{NN}
\end{bmatrix}
\begin{Bmatrix}
\phi_1 \\ \phi_2 \\ \vdots \\ \phi_N
\end{Bmatrix}
=
\begin{Bmatrix}
q_1 \\ q_2 \\ \vdots \\ q_N
\end{Bmatrix}
\tag{6.49}
$$

where the number of panels is $N$. The matrix in Eq. (6.49) will consist only of numbers that depend on the geometry of the surfaces of conductors and dielectrics. This matrix is multiplied by the known potentials of all the panels. The potentials appear in Eq. (6.49) because $\mathbf{D}$ on the left-hand side of Eq. (6.17) has $\phi$ in it due to the relationship $\mathbf{D} = -\varepsilon\nabla\phi$. The right-hand side has charges corresponding to the $N$ panels. Then, we use these charges to compute the forces and solve the elastostatic problem to get the displacements.

As described in Chapter 5, FE analysis applied to elastostatics will also lead to a matrix equation

$$
\begin{bmatrix}
k_{11} & k_{12} & \cdots & k_{1M} \\
k_{21} & k_{22} & \cdots & \vdots \\
\vdots & \vdots & \ddots & \vdots \\
k_{M1} & \cdots & \cdots & k_{MM}
\end{bmatrix}
\begin{Bmatrix}
u_1 \\ u_2 \\ \vdots \\ u_M
\end{Bmatrix}
=
\begin{Bmatrix}
F_1 \\ F_2 \\ \vdots \\ F_M
\end{Bmatrix}
\tag{6.50}
$$

where $M$ is the number of entries in the displacement $\mathbf{u}$ and force $\mathbf{F}$ vectors. The force vector here is the electrostatic force. We solve for the displacement caused by this force.

The geometry of the conductors and dielectrics changes with the displacements. Hence, the matrix in Eq. (6.49) will change. Therefore, once again Eq. (6.49) and subsequently the elastostatic problem [Eq. (6.50)] needs to be solved until convergence. Some commercial microsystems simulation software use the boundary element method for solving the electrostatic problem and link it with the FE analysis algorithm for solving the elastostatic problem. In order to get a direct feel for this iterative technique, let us consider a numerical problem that has such a feature and is physically

---

[2] We need to know the boundary element method to do this. To be specific, we need the Green's function that solves the integral equation of Eq. (6.17) using which the integration is changed to summation.

similar to the coupled problem discussed here. In fact, Problem 6.8 that appears later in the chapter leads to equations that are similar to the ones used in the next example problem.

■

**Example 6.3**

Solve for $\mathbf{q} = \{\, q_1 \quad q_2 \,\}^T$ and $\mathbf{u} = \{\, u_1 \quad u_2 \,\}^T$, vectors that satisfy the following two coupled matrix equations:

$$\begin{bmatrix} 10^{-20}/(10^{-6} - u_1) & 0 \\ 0 & 10^{-20}/(10^{-6} - u_2) \end{bmatrix} \begin{Bmatrix} v \\ 2 \end{Bmatrix} = \begin{Bmatrix} q_1 \\ q_2 \end{Bmatrix}$$

$$\begin{bmatrix} 4 & -1 \\ -1 & 6 \end{bmatrix} \begin{Bmatrix} u_1 \\ u_2 \end{Bmatrix} = \begin{Bmatrix} (0.5vq_1)/(10^{-6} - u_1) \\ (q_2)/(10^{-6} - u_2) \end{Bmatrix}$$

Assume that all entries in $\mathbf{u}$ are zero when you start the iterative numerical procedure. Solve first for $v = 8$ and then for $v = 12$.

**Solution:** We begin with $\mathbf{u}^{(0)} = \{\, 0 \quad 0 \,\}^T$ and calculate $\mathbf{q}^{(1)}$ for $v = 8$ as follows, where the superscripts ''(0)'' and ''(1)'' refer to the iteration numbers:

$$\begin{bmatrix} 10^{-20}/(10^{-6} - u_1^{(0)}) & 0 \\ 0 & 10^{-20}/(10^{-6} - u_2^{(0)}) \end{bmatrix} \begin{Bmatrix} v \\ 2 \end{Bmatrix} = \begin{Bmatrix} q_1 \\ q_2 \end{Bmatrix}^{(1)}$$

$$\begin{Bmatrix} q_1 \\ q_2 \end{Bmatrix}^{(1)} = \begin{bmatrix} 10^{-20}/(10^{-6}) & 0 \\ 0 & 10^{-20}/(10^{-6}) \end{bmatrix} \begin{Bmatrix} 8 \\ 2 \end{Bmatrix} = \begin{Bmatrix} 8 \times 10^{-14} \\ 2 \times 10^{-14} \end{Bmatrix}$$

We substitute $\mathbf{q}^{(1)}$ obtained in this manner and $\mathbf{u}^{(0)}$ on the right-hand side of the following equation and then solve for $\mathbf{u}^{(1)}$:

$$\begin{bmatrix} 4 & -1 \\ -1 & 6 \end{bmatrix} \begin{Bmatrix} u_1 \\ u_2 \end{Bmatrix}^{(1)} = \begin{Bmatrix} \left(0.5vq_1^{(1)}\right)/(10^{-6} - u_1^{(0)}) \\ \left(q_2^{(1)}\right)/(10^{-6} - u_2^{(0)}) \end{Bmatrix}$$

$$\begin{Bmatrix} u_1 \\ u_2 \end{Bmatrix}^{(1)} = \begin{Bmatrix} 0.0843 \\ 0.0174 \end{Bmatrix}$$

We follow the preceding two steps to calculate $\mathbf{q}^{(2)}$ and $\mathbf{u}^{(2)}$ using $\mathbf{u}^{(1)}$. We repeat this iterative procedure until there is no further change in $\mathbf{u}^{(i)}$. The values computed in this manner are:

$$\begin{Bmatrix} u_1 \\ u_2 \end{Bmatrix}^{(2)} = \begin{Bmatrix} 0.1005 \\ 0.0202 \end{Bmatrix}, \begin{Bmatrix} u_1 \\ u_2 \end{Bmatrix}^{(3)} = \begin{Bmatrix} 0.1041 \\ 0.0208 \end{Bmatrix}, \begin{Bmatrix} u_1 \\ u_2 \end{Bmatrix}^{(4)} = \begin{Bmatrix} 0.1041 \\ 0.0208 \end{Bmatrix} \begin{Bmatrix} u_1 \\ u_2 \end{Bmatrix}^{(5)} = \begin{Bmatrix} 0.1049 \\ 0.0210 \end{Bmatrix},$$

$$\begin{Bmatrix} u_1 \\ u_2 \end{Bmatrix}^{(6)} = \begin{Bmatrix} 0.1051 \\ 0.0210 \end{Bmatrix}, \begin{Bmatrix} u_1 \\ u_2 \end{Bmatrix}^{(7)} = \begin{Bmatrix} 0.1051 \\ 0.0210 \end{Bmatrix}, \begin{Bmatrix} u_1 \\ u_2 \end{Bmatrix}^{(8)} = \begin{Bmatrix} 0.1051 \\ 0.0210 \end{Bmatrix} \begin{Bmatrix} u_1 \\ u_2 \end{Bmatrix}^{(9)} = \begin{Bmatrix} 0.1051 \\ 0.0210 \end{Bmatrix}, \cdots$$

It can be seen that the repetitive procedure converges after eight iterations. The converged value of **q** is

$$\mathbf{q}^{(8)} = \left\{ \begin{array}{c} 0.0894 \times 10^{-12} \\ 0.0204 \times 10^{-12} \end{array} \right\}$$

Thus, we have found self-consistent solution values of **q** and **u** to satisfy the two equations in the problem statement. It is important to note Eqs. (6.49) and (6.50) are solved iteratively in the same way.

Notes:

1. Interested readers may find that two coupled cubic equations in terms of $u_1$ and $u_2$ can be obtained by eliminating $q_1$ and $q_2$. In that case, iterative numerical procedure is not required. It is important to see the significance of the appearance of a cubic here. Refer back to Example 6.2 where we also obtained a cubic.

2. It may be verified that the same numerical procedure does not converge for $v = 12$. This is because for that value of $v$, there is no feasible stable solution. This point is discussed in the following subsections.

The coupled problem in Figure 6.11 in its general form cannot reveal the instability that may arise during the iterative solution and its convergence. But we did note that $v = 12$ in Example 6.3 does not converge. In order to gain insight into the nature of this issue, we need to consider simplified situations that can be solved analytically. Lumped modeling of the electrostatic problem is useful in such an analytical treatment. We used this approach in Chapter 4 where we first considered simple problems and then presented the most general equations. In contrast, here we first developed the general equations. We consider simpler models now.

## 6.2.2 Spring-Restrained Parallel-Plate Capacitor

A spring-restrained parallel-plate capacitor is the simplest model in coupled electro-mechanics. It is ubiquitous in microsystems and is seen in many capacitive sensors and electrostatic actuators. A capacitive accelerometer that measures acceleration in the direction perpendicular to a substrate wafer has a proof-mass with an electrode underneath it. The proof-mass and the ground electrode form a parallel-plate capacitor, as shown in Figure 6.13(a). An electrostatic comb-drive also has parallel-plate capacitors between every pair of moving and stationary comb-fingers that face each other [Figure 6.13(b)]. In both these cases, as in many others, one plate is fixed while the other is moving. The second plate is not completely free to move; it has an elastically deformable structure attached to it. As we learned in Chapter 4, any elastically deformable structure can be "lumped" to a spring, thus hiding all the intricacies ensuing from Eqs. (6.33a) and (6.33b). Similarly, we will show how electrostatic effects can be lumped to a *capacitor* hiding all the intricacies ensuing from Eqs. (Eqs. (6.17)–(6.23).

Figure 6.14(a) shows the schematic of a spring-restrained parallel-plate capacitor. This is a simplified model of real physical structures such as those in Figs. 6.13(a) and (b). In order to analyze this model, we first consider two infinite parallel sheets separated by a gap $g_0$, as shown in Figure 6.14(b). Let the bottom sheet be at ground (i.e. zero) potential

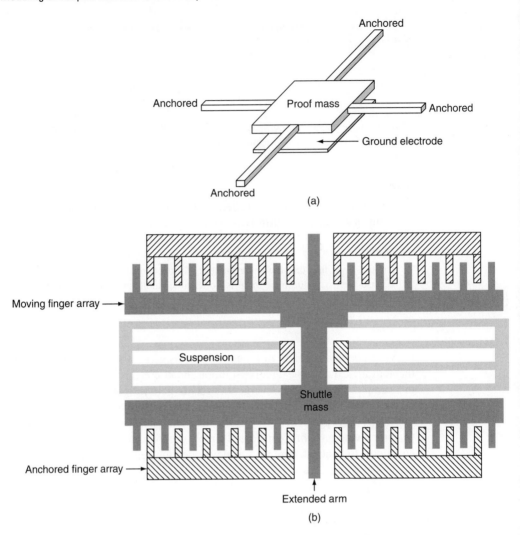

**Figure 6.13** Parallel-plate capacitors appear in many capacitive sensors and electrostatic actuators. (a) A $z$-axis capacitive accelerometer; (b) an electrostatic comb-drive.

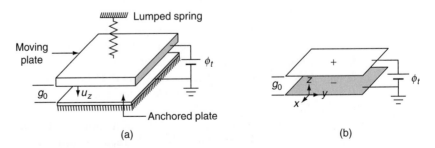

**Figure 6.14** (a) A spring-restrained parallel-plate capacitor of finite area; (b) an ideal parallel-plate capacitor formed by two infinite sheets of charge.

while the top one is at some potential $\phi_t$ V. We have to solve Eq. (6.23) (repeated here) for this configuration:

$$\nabla^2 \phi = 0$$

$$\frac{\partial^2 \phi_x}{\partial x^2} + \frac{\partial^2 \phi_y}{\partial y^2} + \frac{\partial^2 \phi_z}{\partial z^2} = 0 \tag{6.51}$$

Because of the parallelism between the plates and infinite extents of the sheets, we can see that the potential in the space between the plates does not vary with respect to $x$ and $y$. Hence, Eq. (6.51) can be reduced to

$$\frac{\partial^2 \phi_z}{\partial z^2} = 0 \tag{6.52}$$

This can be readily solved by integrating it twice,

$$\frac{\partial^2 \phi_z}{\partial z^2} = 0$$

$$\frac{\partial \phi_z}{\partial z} = C_1 \tag{6.53}$$

$$\phi_z = C_1 z + C_2$$

The two constants can be evaluated since we know the potentials at the bottom and top sheets:

$$\phi_{z=0} = C_0 = 0$$

$$\phi_{z=g_0} = C_1 g_0 + C_0 = \phi_t \quad \Rightarrow \quad C_1 = \frac{\phi_t}{g_0} \tag{6.54a}$$

Therefore,

$$\phi_z = \frac{\phi_t}{g_0} z \tag{6.54b}$$

This shows that the potential varies linearly between the plates. Hence, due to Eq. (6.10), the electric field is computed as

$$\mathbf{E} = -\nabla \phi = -\frac{\partial \phi}{\partial z} \hat{\mathbf{k}} = -\frac{\phi_t}{g_0} \hat{\mathbf{k}} \tag{6.55}$$

and we see that the electric field is constant and directed downward between the plates from the positive charge to the negative charge. Note that it is perpendicular to the conductors, as it should be [recall Eq. (6.25)]. The surface charge density, per Eq. (6.28), is

$$\psi_s = \varepsilon E_n = \frac{\varepsilon \phi_t}{g_0} \tag{6.56}$$

where $E_n$ is the normal component of the electric field $\mathbf{E}$ in Eq. (6.55) and $\varepsilon$ is the permittivity of the medium between the sheets. The electrostatic force per unit area on the top sheet can now be written using Eq. (6.31):

$$\mathbf{t}_e = \frac{\psi_s^2}{2\varepsilon}\hat{\mathbf{n}} = \frac{(\varepsilon\phi_t/g_0)^2}{2\varepsilon}(-\hat{\mathbf{k}}) = -\frac{\varepsilon\phi_t^2}{2g_0^2}\hat{\mathbf{k}} \tag{6.57}$$

where $\hat{\mathbf{k}}$ is the outward normal to the top sheet that faces the bottom sheet.

Now, by referring to Figure 6.14(a), which shows a parallel-plate capacitor of finite size of plate area (say $A$), we approximate its behavior by that of the two charged parallel sheets of infinite area in Figure 6.14(b). Therefore, for the parallel-plate capacitor of Figure 6.14(a), we can repeat Eqs. (6.56) and (6.57) as follows:

$$Q = A\psi_s = \varepsilon AE_n = \frac{\varepsilon A\phi_t}{g_0} \tag{6.58}$$

where $Q$ is the magnitude of the lumped charge on each plate and

$$\mathbf{F} = A\mathbf{t}_e = \frac{A\psi_s^2}{2\varepsilon}\hat{\mathbf{n}} = \frac{A(\varepsilon\,\phi_t/g_0)^2}{2\varepsilon}(-\hat{\mathbf{k}}) = -\frac{\varepsilon A\phi_t^2}{2g_0^2}\hat{\mathbf{k}} \tag{6.59}$$

where $\mathbf{F}$ is the lumped force on the moving plate. Thus, just as we lumped the elastic stiffness of the suspension of the proof mass of Figure 6.13(a) into a spring that restrains the moving plate in Figure 6.14(a), we have now lumped the electrical behavior. We can thus define a concept analogous to that of the spring. From Eq. (6.58) we have

$$\frac{Q}{\phi_t} = \frac{\varepsilon A}{g_0} = C \tag{6.60}$$

Note that when we increase/decrease the charge $Q$ by some factor, the potential $\phi_t$ too changes by the same factor because $C$ is a constant. This constant is called the *capacitance*. Its unit is the farad (F), named in honor of Michael Faraday, who laid the foundations of electrostatics (in addition to Coulomb and Gauss). The farad, like the coulomb, is a large unit. Most micromachined capacitors have capacitances of the order of picoF ($10^{-12}$ F, denoted pF) or smaller. From the relationship in Eq. (6.60), we can now see why permittivity has the curious unit of F/m. Indeed, permittivity is the material property that decides the capacitance (in addition to the geometry). Thus, capacitance is a concept similar to the spring constant $k$, which also depends on one material property (Young's modulus) and the geometry of a structure. Therefore, we can draw the analogy between electrostatics and elastostatics shown in Table 6.2.

The new concepts introduced in Table 6.2 are the electrostatic energy, *ESE*, electrostatic coenergy ("co" for complementary), $ESE_c$, and complementary strain energy, $SE_c$. Just as a spring stores strain energy, a capacitor stores *ESE*. Notice that this is the area under the $\phi$ versus $Q$ curve [see Figure 6.15(a)], just as *SE* is the area under the $F$ versus $u_z$ curve [see Figure 6.15(b)]. Note that $\phi = Q/C$ and $F = ku_z$ for a linear capacitor and a linear

**Table 6.2 Analogous quantities in electrostatics and elastostatics**

| Electrostatics | Elastostatics |
| --- | --- |
| Charge $Q$ | Displacement $u_z$ |
| Potential $\phi$ | Force $F$ |
| Capacitance $C$ | Spring constant $k$ |
| Electrostatic energy $ESE = \dfrac{1}{2}\dfrac{Q^2}{C}$ | Strain energy $SE = \dfrac{1}{2}ku_z^2$ |
| Electrostatic coenergy $ESE_c = \dfrac{1}{2}C\phi^2$ | Complementary strain energy $SE_c = \dfrac{1}{2}\dfrac{F^2}{k}$ |

spring, respectively. The corresponding coenergies can also be written in terms of $\phi$, in which case they are areas under the $Q$ versus $\phi$ curve and the $u_z$ versus $F$ curve, respectively. For linear relationships between $\phi$ and $Q$ (or $F$ and $u_z$), the two areas are the same. But this is not true for the nonlinear relationships shown in Figs. 6.15(a) and (b). The motivation for defining the coenergies becomes clear when we derive the mechanical force from the electrostatic energy.

The spring-restrained, parallel-plate capacitor shown in Figure 6.14(a) stores both mechanical and electrostatic energies. Let us consider these two energies together for the parallel-plate capacitor:

$$TE = SE + ESE = TE(u_z, Q) = \frac{1}{2}ku_z^2 + \frac{1}{2}\frac{Q^2}{C} \tag{6.61}$$

where the total energy, $TE(u_z, Q)$, is expressed in terms of the displacement and the charge. By taking the total derivative of $TE(u_z, Q)$ with respect to $u_z$ and $Q$, we get:

$$\mathrm{d}(TE) = \frac{\partial(TE)}{\partial u_z}\,\mathrm{d}u_z + \frac{\partial(TE)}{\partial Q}\,\mathrm{d}Q \tag{6.62}$$

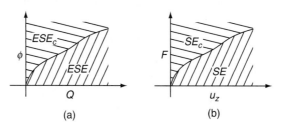

(a)　　　　　　　(b)

**Figure 6.15** Energies and complementary (co) energies in a mechanical spring and an electrical capacitor. (a) Electrostatic energy and electrostatic coenergy in a capacitor; (b) strain energy and complementary energies in a spring. For generality, a nonlinear relationship is shown between the electric potential and the charge and between the spring force and the displacement. When the relations are linear, energy and coenergy are the same.

and from Eq. (6.61) we see that

$$\frac{\partial(TE)}{\partial u_z} = \frac{\partial(SE)}{\partial u_z} + \frac{\partial(ESE)}{\partial u_z} = F_{mech} + F_{elec} \text{ and } \frac{\partial(TE)}{\partial Q} = \frac{Q}{C} = \phi \qquad (6.63)$$

Hence, we observe that $d(TE) = Fdu_z + \phi dQ$, that is, the force multiplied by infinitesimally small displacement $Fdu_z$ as well as electric potential multiplied by infinitesimally small charge $\phi\,dQ$, gives the total change in the energy $d(TE)$. This means that when we take the total energy and differentiate it with respect to the displacement and also its analogous quantity (the charge per Table 6.2), we get the force and the analogous quantity, the electric potential. This is similar to Castigliano's first theorem that we came across in Chapter 4 (see Eq. 4.33).

Similarly, if we take $SE_c$ and $ESE_c$, we get the same result as the Castigliano's second theorem (see Eq. 4.34):

$$SE_c + ESE_c = TE_c(F,\phi) = \frac{1}{2}\frac{F^2}{k} + \frac{1}{2}C\phi^2 \qquad (6.64a)$$

$$\frac{\partial(TE_c)}{\partial F} = \frac{F}{k} = u_z \text{ and } \frac{\partial(TE_c)}{\partial\phi} = C\phi = Q \qquad (6.64b)$$

Thus, we get forces and displacements and electric potential and charge in the respective energy domains. This is true when there is no coupling between the two domains.

If there is coupling, the situation is different. Consider $ESE_c$ and, by referring to Figure 6.14(a), write it as

$$ESE_c = \phi Q - ESE \qquad (6.65)$$

By taking the total derivative, we get

$$d(ESE_c) = Q\,d\phi + \phi\,dQ - d(ESE) \qquad (6.66)$$

But $ESE$ in a coupled system depends on $Q$ and $u_z$ because it is influenced by both the change in charge and the change in position. Therefore, we can write

$$d(ESE) = F_{elec}\,du_z + \phi\,dQ \qquad (6.67)$$

From Eqs. (6.67) and (6.66), we get

$$d(ESE_c) = Q\,d\phi - F_{elec}\,du_z \qquad (6.68)$$

But, since we can expand $d(ESE_c)$ as

$$d(ESE_c) = \frac{\partial(ESE_c)}{\partial\phi}d\phi + \frac{\partial(ESE_c)}{\partial u_z}du_z \qquad (6.69)$$

we get the coupled results by comparing Eqs. (6.68) and (6.69):

$$Q = \frac{\partial(ESE_c)}{\partial \phi} \tag{6.70a}$$

$$F_{elec} = -\frac{\partial(ESE_c)}{\partial u_z} \tag{6.70b}$$

The result in Eq. (6.70a) is the same as what we saw in the second equation of Eq. (6.64b). However, Eq. (6.70b) is a new result that tells us how to compute the force when $ESE_c$ is expressed in terms of $u_z$.

### Problem 6.4
Show that $\phi = -\partial(SE_c)/\partial Q$. This formula is useful in computing the change in electric potential when a mechanical force is applied to an initially charged parallel-plate capacitor.

Based on Castiglianos first theorem and Eq. (6.70b), if we want to compute the total force due to the elastostatic and electrostatic domains, we need to add $SE$ and $-ESE_c$ to get the total energy, which is then differentiated with respect to $u_z$. That is,

$$F_{\text{total}} = F_{mech} + F_{elec} = \frac{\partial(SE)}{\partial u_z} - \frac{\partial(ESE_c)}{\partial u_z} = \frac{\partial(SE - ESE_c)}{\partial u_z} = \frac{\partial TE}{\partial u_z} \tag{6.71}$$

For the parallel-plate capacitor, we have

$$TE = SE - ESE_c = \frac{1}{2}ku_z^2 - \frac{1}{2}C\phi^2 \tag{6.72}$$

By using Eq. (6.60) giving an expression for $C$ in Eq. (6.72), we get

$$TE = SE - ESE_c = \frac{1}{2}ku_z^2 - \frac{1}{2}C\phi^2 = \frac{1}{2}ku_z^2 - \frac{1}{2}\frac{\varepsilon A}{(g_0 - u_z)}\phi^2 \tag{6.73}$$

where we have substituted $(g_0 - u_z)$ for $g_0$ because we need to consider the gap as the top plate moves down by $u_z$. Note that we measure $u_z$ in the downward direction in Figure 6.14 (a). Now, from Eqs. (6.71) and (6.73) we get the total force as

$$F_{\text{total}} = \frac{\partial(SE - ESE_c)}{\partial u_z} = ku_z - \frac{1}{2}\frac{\varepsilon A}{(g_0 - u_z)^2}\phi^2 \tag{6.74}$$

For static equilibrium, this total force must be zero. Therefore,

$$ku_z - \frac{1}{2}\frac{\varepsilon A}{(g_0 - u_z)^2}\phi^2 = 0 \tag{6.75}$$

This equation is noteworthy for two reasons. First, it is a cubic equation in $u_z$, a situation similar to what we obtained in Example 6.2. There, we had three roots and indicated that taking the lowest value among the roots is the answer. The meaning of the other two roots can now be explained. Second, Eq. (6.75) shows the nonlinear coupling between the elastic force and electrostatic force. This is a simple manifestation of the coupling and iterative solution procedures discussed earlier in this section. We noted earlier that numerical solutions of problems involving complex geometries of structures fail to give insight into the problem. Now, with the help of Eqs. (6.73) and (6.75) and the numerical example that follows, we explain the phenomena that occur in coupled electromechanics.

## Example 6.4

We consider again the RF switch in Figure 6.6 (reproduced in Figure 6.16) with the numerical values given. Calculate the deflection of the free beam tip under the electrostatic force using the lumped modeling of the beam as a linear spring and the electrostatic force as a parallel-plate capacitor. The Young's modulus of the beam material is 169 GPa and the permittivity of air is $8.854 \times 10^{-12}$ F/m. Let $\phi$ vary between 0 and 1.5 V. Plot the tip deflection as a function of the applied electric potential, $\phi$. Also, plot the capacitance as a function of $\phi$. It is given that $t_b$ = beam thickness = 2 μm; $t_e$ = electrode thickness = 2 μm; $g_0$ = gap between the beam and the electrode = 1 μm; $t_n$ = nitride layer thickness = 0.2 μm; $t_s$ = substrate thickness = 200 μm; $l_b$ = beam length = 310 μm; $l_e$ = electrode length = 300 μm; $l_a$ = beam anchor length = 50 μm; $w_b$ = beam width = 5 μm; and $w_e$ = electrode width = 10 μm.

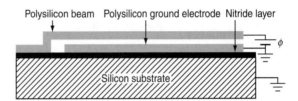

Polysilicon beam   Polysilicon ground electrode   Nitride layer

**Figure 6.16** The mechanical structure in an electrostatically actuated RF switch.

**Solution:** For lumped modeling of the beam, we approximate it as a cantilever beam of length 310 μm ($l_b$), width 5 μm ($w_b$), and thickness 2 μm ($t_b$) fixed at one end and free at the other end with a point force $F$. As per the beam theory in Chapter 4, the equivalent spring constant is given by Eq. (4.29):

$$k = \frac{3YI}{l_b^3} = \frac{3Y}{l_b^3} \cdot \frac{w_b t_b^3}{12} = \frac{Y w_b t_b^3}{4 l_b^3} = \frac{(169 \times 10^9)(5 \times 10^{-6})(2 \times 10^{-6})^3}{4(310 \times 10^{-6})^3} = 0.0567 \, \text{N/m}$$

For lumped modeling of electrostatic force, we take the area of the parallel-plate capacitor as a rectangle of length 300 μm and width 5 μm, since this is the overlapping area between the beam and the electrode. The initial gap between the moving plate of the beam model and the anchored electrode is 1 μm. Thus, we have $A = 1500 \, \text{μm}^2$ and $g_0 = 1 \, \text{μm}$. Now, Eq. (6.75) yields

$$ku_z - \frac{1}{2}\frac{\varepsilon_0 A}{(g_0 - u_z)^2}\phi^2 = 0$$

$$2ku_z(g_0 - u_z)^2 - \varepsilon_0 A\phi^2 = 0$$

$$u_z^3 - 2g_0 u_z^2 + u_z g_0^2 - \frac{\varepsilon_0 A\phi^2}{2k} = 0$$

By substituting the numerical values, we get

$$u_z^3 - 2(10^{-6})u_z^2 + (10^{-12})u_z - \frac{\varepsilon_0 A \phi^2}{2k} = 0$$
$$u_z^3 - 2(10^{-6})u_z^2 + (10^{-12})u_z - (1.1706 \times 10^{-19})\phi^2 = 0$$

By solving for the roots of the preceding equation for different values of $\phi$, we find the three values of $u_z$ shown in Table 6.3. Figure 6.17 shows the plot of the three roots of $u_z$ and Figure 6.18 the capacitance as a function of $\phi$ using the smallest root. The unusual nature of this example is worth pondering by examining Table 6.3 and Figs. 6.17 and 6.18. In the discussion below, we find that this behavior is all but natural when we analyze the coupled behavior of this system.

**Table 6.3** The three roots of the cubic equation for different values of $\phi$

| $\phi(V)$ | First Root (μm) | Second Root (μm) | Third Root (μm) |
|---|---|---|---|
| 0.0000 | 0.0000 | 1.0000 | 1.0000 |
| 0.1250 | 0.0018 | 0.9563 | 1.0419 |
| 0.2500 | 0.0074 | 0.9104 | 1.0822 |
| 0.3750 | 0.0170 | 0.8618 | 1.1212 |
| 0.5000 | 0.0312 | 0.8099 | 1.1589 |
| 0.6250 | 0.0507 | 0.7537 | 1.1956 |
| 0.7500 | 0.0733 | 0.6914 | 1.2313 |
| 0.8750 | 0.1142 | 0.6197 | 1.2661 |
| 1.0000 | 0.1699 | 0.5301 | 1.3001 |
| 1.1250 | 0.3333 | 0.3333 | 1.3333 |
| 1.2500 | Imaginary | Imaginary | 1.3659 |
| 1.5000 | Imaginary | Imaginary | 1.4293 |

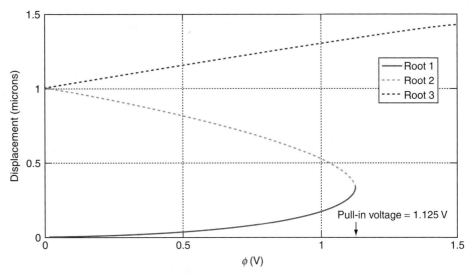

**Figure 6.17** Plot of the approximate tip deflection for different values of $\phi$ for the example in Figure 6.16 based on lumped modeling of the elastostatic and electrostatic behaviors.

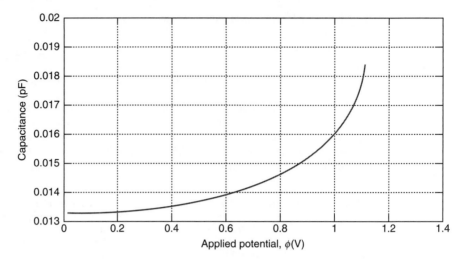

**Figure 6.18** Plot of the parallel-plate capacitance for different values of $\phi$ for the example in Figure 6.16 using the smallest root of the three roots.

We notice something unusual in Figs. 6.17 and 6.18 and Table 6.3. Initially, for low values of $\phi$, there are three real roots for $u_z$. As $\phi$ increases, the lowest root increases while the second lowest root decreases. For some value of $\phi$, the lowest and second-lowest roots have the same value, beyond which they both become imaginary. The third root (i.e. the largest root) remains real throughout but is always greater than $g_0$. This root is of no practical consequence in this problem because it corresponds to an unrealistic situation of the moving plate penetrating the fixed plate, which does not happen physically. So, of the two real roots—as long as they remain real—whichdo we take as the solution? In other words, if we had a spring-restrained parallel-plate capacitor of this type, where does the top plate stay at static equilibrium? The answer, as alluded to earlier in this chapter, depends on the stability of static equilibrium, which we discuss next.

**Problem 6.5**
If in Example 6.4 a positive charge is given to the top plate instead of applying an electric potential between the top and bottom plates, compute the displacement of the top plate. Vary $Q$ from $1.0 \times 10^{-23}$ C to $1 \times 10^{-19}$ C. At what value of $Q$ the gap between the plates becomes zero?

**Problem 6.6**
Solve Example 6.4 using a FE method and compare the lumped-model-based approximate results in Figs. 6.17 and 6.18 with the more accurate results obtained using the FE analysis. How good is the lumped model approximation?

## ▶ 6.3 COUPLED ELECTROMECHANICS: STABILITY AND PULL-IN PHENOMENON

Static force equilibrium implies that all forces are balanced perfectly and hence the system is in a state of rest. The concept of stability answers the following question: what happens if such a system is slightly perturbed from its equilibrium state? Perturbation can be due to increasing or decreasing the displacement by an infinitesimally small amount. If the system

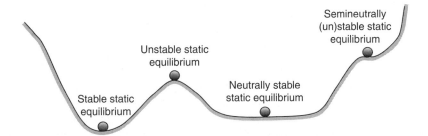

**Figure 6.19** The concept of stability of static equilibrium illustrated through a ball rolling on a curved surface.

returns to the original equilibrium state after that perturbation, we call it *stable*; if not, it is an *unstable* or *neutrally stable* system. The best way to understand this is to imagine a ball rolling on a curve under the effect of gravity (Figure 6.19). The ball can be at rest in three ways: at the bottom of the valley, at the top of the hill, or on the flat plateau. When it is at the bottom of the valley, a slight perturbation from there will surely make the ball return to the bottom again. Hence, this is a *stable static equilibrium*. This is not true of a ball carefully placed at the top of the hill: any slight perturbation makes the ball plummet down. This is *unstable static equilibrium*. In the *neutrally stable equilibrium*, a slight perturbation will not result in either returning to the original position or moving away from the perturbed position.

The analogy of a rolling ball on a curve explains two things if we take the curved landscape to be the energy landscape of the system: (a) the slope of the energy is zero for static equilibrium; (b) a local minimum-energy state corresponds to a stable static equilibrium, a local maximum-energy state corresponds to an unstable static equilibrium, and neither minimum nor maximum but a stationary energy state (due to zero slope) corresponds to a *neutral equilibrium*. As shown in Figure 6.19, a stationary energy state can occur with increasing energy on one side and decreasing energy on the other side. This state is essentially unstable but not strictly so and can be called a *semineutrally (un)stable state*.

By referring to Eq. (6.73), we can plot the energy landscapes of our coupled system that stores two energies together as a function of the displacement:

$$TE = SE - ESE_c = \frac{1}{2}ku_z^2 - \frac{1}{2}\frac{\varepsilon A}{g_0 - u_z}\phi^2 \tag{6.76}$$

By differentiating the energy in the preceding equation with respect to $u_z$ and equating it to zero, we ensure static equilibrium, which gives the force-balance equation:

$$\frac{\partial(SE - ESE_c)}{\partial u_z} = ku_z - \frac{1}{2}\frac{\varepsilon A}{(g_0 - u_z)^2}\phi^2 = 0 \tag{6.77}$$

which is a repetition of Eq. (6.75).

In order to check the stability of the static equilibrium position, we need to take the second derivative of the energy. This is because the second derivative is positive at the minimum-energy state. Furthermore, the second derivative is negative for the maximum-energy state and zero for the neutrally (un)stable energy state. Thus, we have

$$\frac{\partial^2(SE - ESE_c)}{\partial u_z^2} = k - \frac{\varepsilon A}{(g_0 - u_z)^3}\phi^2 = \lambda \tag{6.78a}$$

and

$$\lambda > 0 \Rightarrow \text{Stable}$$
$$\lambda < 0 \Rightarrow \text{Unstable} \tag{6.78b}$$
$$\lambda + 0 \Rightarrow \text{Neutrally stable (transition state)}$$

while satisfying the static equilibrium state per Eq. (6.77). These conditions explain the behavior we saw in Figure 6.17 and Table 6.3.

It is instructive to examine the total energy curve corresponding to Example 6.3. Four energy curves for $\phi = 0.25$, $0.75$, $1.125$, and $1.5$ are shown in Figs. 6.20(a) and (d). Figures 6.20(a) and (b) have all three equilibrium solutions: the first one stable (because it is a local minimum), the second one unstable (because it is a local maximum), and the third one stable but infeasible because it corresponds to $u_z > g_0$. Figure 6.20(c) is the transition state where the first two states and unstable solutions coalesce into one neutrally (un)stable solution. Figure 6.20(d) has only one solution, which, as we know, is infeasible. This shows that beyond $\phi > 1.125$, there is no feasible equilibrium state; the top plate will try to achieve the infeasible $u_z > g_0$ and in the process be pulled in toward the bottom plate. This is called the *electrostatic pull-in* phenomenon. It is a direct consequence of the coupling between the electrostatic and elastostatic domains. It is worth plotting the force curves corresponding to Figs. 6.20(b) and (d). They would show that the number of intersections between the electrostatic and spring forces reduces from three to two and then to only one as we pass the pull-in voltage.

Let us now determine this critical voltage, $\phi_{\text{pull-in}}$, called the *pull-in voltage*. The pull-in state corresponds to the transition state [see Eq. (6.78b) and Figure 6.20(c)]. Thus, we have

$$ku_z - \frac{1}{2}\frac{\varepsilon A}{(g_0 - u_z)^2}\phi^2 = 0 \tag{6.79a}$$

$$k - \frac{\varepsilon A}{(g_0 - u_z)^3}\phi^2 = 0 \tag{6.79b}$$

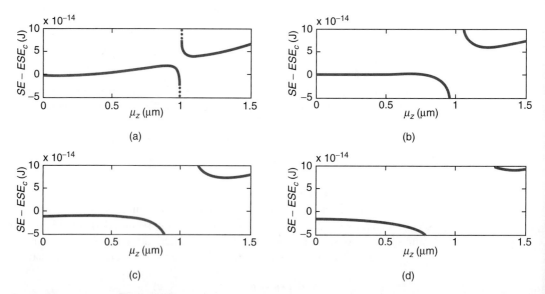

**Figure 6.20** Energy curves for a spring-restrained parallel-plate capacitor for different values of $\phi$: (a) 0.25 V; (b) 0.75 V; (c) 1.125 V; and (d) 1.5 V.

By extracting $\phi^2$ from Eq. (6.79b) and substituting for $\phi^2$ in Eq. (6.79a), we get

$$\phi^2 = \frac{k(g_0 - u_z)^3}{\varepsilon A} \tag{6.80}$$

$$ku_z - \frac{1}{2}\frac{\varepsilon A}{(g_0 - u_z)^2} \cdot \frac{k(g_0 - u_z)^3}{\varepsilon A} = 0$$

$$ku_z - \frac{k(g_0 - u_z)}{2} = 0 \tag{6.81}$$

$$2u_z - (g_0 - u_z) = 0$$

$$u_z = \frac{g_0}{3}$$

Referring to Table 6.3, we notice that the stable range of displacement is one-third the initial gap, 0.3333 μm. By substituting the final result of Eq. (6.81) into Eq. (6.80), we get the pull-in voltage as

$$\phi^2 = \frac{k[g_0 - (g_0/3)]^3}{\varepsilon A} \quad \Rightarrow \quad \phi_{\text{pull-in}} = \sqrt{\frac{8kg_0^3}{27\varepsilon A}} \tag{6.82}$$

It may be readily verified that the transition (i.e. pull-in) voltage corresponding Figure 6.20(c) is indeed 1.125 V for the lumped model of the RF switch of Example 6.3:

$$\phi_{\text{pull-in}} = \sqrt{\frac{8kg_0^3}{27\varepsilon A}} = 1.125 \text{ V}$$

We see that for an RF switch-like device, in order to close the switch we need to apply a potential greater than the pull-in voltage. But in other devices, such as accelerometers and comb-drive actuators, we want to avoid the pull-in state because it results in an electrical short circuit or mechanical damage due to sudden large, fast movement. Thus, the calculation of pull-in potential (often called the *pull-in voltage*) is an essential component of modeling of electrostatic microsystems.

There is some more subtlety in this system that we examine further. Let us say that there is a dielectric layer of thickness $t_d$ in a spring-restrained parallel-plate capacitor attached to the top of the bottom plate. Let its relative permittivity be $\varepsilon_r$. Such a layer is necessary to prevent short circuit when pull-in occurs (see Figure 6.21). Let us follow the same procedure as before to compute the pull-in voltage for this system.

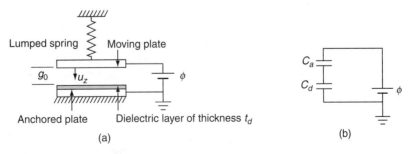

**Figure 6.21** (a) A spring-restrained parallel-plate capacitor with a dielectric layer on top of the fixed bottom plate; (b) capacitor modeling of this system.

By recasting Eq. (6.73) to account for the dielectric layer, we get the total energy, $TE$, as follows:

$$TE = SE - ESE_c = \frac{1}{2}ku_z^2 - \frac{1}{2}C_{eq}\phi^2 \tag{6.83}$$

where $C_{eq}$ is the equivalent capacitance of two lumped capacitors in Figure 6.21(a): one is the capacitor formed by the moving plate and the top of the dielectric layer, and the other is that formed by the top and bottom of the dielectric layer. These two capacitors are in series because they have the same charge and share the applied potential $\phi$. Referring to Figure 6.21(b), where $C_a$ and $C_d$ are air and dielectric parallel-plate capacitors, we can write $C_{eq}$ as follows:[3]

$$Q = C_a\phi_a = C_d\phi_d = C_{eq}\phi \Rightarrow \phi = \frac{Q}{C_{eq}} \tag{6.84a}$$

Now

$$\phi = \phi_a + \phi_d$$
$$\frac{Q}{C_{eq}} = \frac{Q}{C_a} + \frac{Q}{C_d} \Rightarrow C_{eq} = \frac{C_aC_d}{C_a + C_d} \tag{6.84b}$$

By noting that

$$C_a = \frac{\varepsilon_0 A}{g_0 - u_z} \text{ and } C_d = \frac{\varepsilon_r\varepsilon_0 A}{t_d} \tag{6.85}$$

we can find $C_{eq}$ to be

$$C_{eq} = \frac{\varepsilon_0 A}{g_0 + (t_d/\varepsilon_r) - u_z} \tag{6.86}$$

Equation (6.86) shows that a dielectric layer is equivalent to a layer of air whose thickness is equal to the thickness of the dielectric layer divided by the relative permittivity. In other words, the equivalent initial gap is $g_0 + (t_d/\varepsilon_r)$. Therefore, from Eq. (6.82), we can write the pull-in voltage for this system as

$$\phi_{\text{pull-in}} = \sqrt{\frac{8k[g_0 + (t_d/\varepsilon_r)]^3}{27\varepsilon A}} \tag{6.87}$$

An interesting question to ask now is: what happens when a potential larger than $\phi_{\text{pull-in}}$ in Eq. (6.87) is applied to the systems in Figure 6.21(a)? Unlike the spring-restrained parallel-plate capacitor in Figure 6.14(a), this one will not short electrically because of the insulating dielectric layer. Instead, the moving plate pulls down suddenly and stays put on top of the dielectric layer. More interestingly, if we now decrease the potential by a small

---

[3] This is similar to the springs in parallel discussed in Chapter 4 (see Example 4.1). Recall that springs in parallel have the same displacement and share the force but capacitors in series have the same charge and share the voltage. This difference is due to the way in which spring constant ($k$ = force/displacement) and capacitance ($C$ = charge/potential) are defined: one is the reciprocal of the other in the electromechanical analogies in Table 6.2.

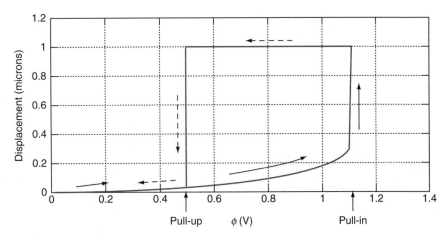

**Figure 6.22** Displacement versus voltage curve for pull-in and pull-up for a spring-restrained parallel-plate capacitor with a dielectric layer in between.

value, the moving plate does not go back up; it remains right on top of the dielectric layer. This continues to be the case until the applied potential is decreased to what can be called $\phi_{\text{pull-up}}$. At $\phi = \phi_{\text{pull-up}}$, the moving plate suddenly jumps up, as seen in Figure 6.22 when $u_z$ suddenly decreases. With further decrease in the potential, $u_z$ traverses the same curve but in the opposite direction. This curve is somewhat reminiscent of a *hysteresis-like* phenomenon, but it is actually like a *snap-through* action[4] in some mechanical structures.

In order to understand why a pulled-in moving plate does not go up until $\phi = \phi_{\text{pull-up}}$, we need to think about the forces in action: an electrostatic force acting downward (in the direction of $u_z$) and a spring force acting upward (in the direction opposite to that of $u_z$). Beyond the pull-in voltage, the electrostatic force is larger than the spring force in the range $0 \leq u_z \leq g_0$. Hence, the moving plate stays put on top of the dielectric layer until $\phi = \phi_{\text{pull-up}}$. This indicates that at $\phi = \phi_{\text{pull-up}}$, the two forces are balanced, which allows us to compute $\phi_{\text{pull-up}}$:

$$kg_0 - \frac{1}{2}\frac{\varepsilon_0 A \phi_{\text{pull-up}}^2}{(t_d/\varepsilon_r)^2} = 0$$

$$\Rightarrow \phi_{\text{pull-up}} = \sqrt{\frac{2kg_0 t_d^2}{\varepsilon_0 \varepsilon_r^2 A}}$$

(6.88)

Revisiting Example 6.4 and imagining a 0.3 μm layer of silicon dioxide ($\varepsilon_r = 3$) on top of the electrode beneath the cantilever beam of the switch, we can compute its *pull-up voltage* as:

$$\phi_{\text{pull-up}} = \sqrt{\frac{2kg_0 t_d^2}{\varepsilon_0 \varepsilon_r^2 A}} = 0.5062 \text{ V}$$

---

[4] Snap-through action can be found in many bistable structures that have two stable equilibrium states separated by an unstable state. A familiar example is the two arms of some spectacles.

This value agrees with the pull-up in Figure 6.22. What we have done so far yields only an estimate of pull-in and pull-up voltages using lumped modeling; in practice one must be more accurate by considering the full model of the system, as discussed next.

### Problem 6.7

Consider the system in Figure 6.23 in which a capacitor $C_s$ in the electrical circuit supplies the potential to the spring-restrained parallel-plate capacitor. We saw in the preceding section that the stable range of motion for the moving plate is only one-third of the gap [see Eq. (6.81)] when $C_s$ is absent. It is possible to increase the stable range of motion to $g_0$ with a suitable value of $C_s$. Find this value.

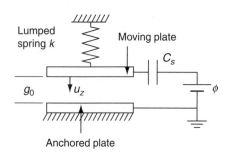

**Figure 6.23** A spring-restrained parallel-plate capacitor with an additional capacitor in the electrical circuit that suppplies the potential.

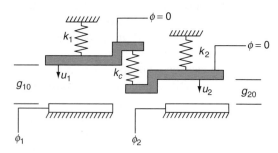

**Figure 6.24** A lumped model of a micromechanical filter.

### Problem 6.8

One type of a micromechanical bandpass filter consists of two fixed–fixed beams connected by a coupling beam. A lumped model of such a filter is shown in Figure 6.24; it has three springs connecting two moving plates. Both the moving plates are held at ground potential while the electrodes underneath them are maintained at potentials $\phi_1$ and $\phi_2$.

(a) Derive the electromechanical equilibrium equations of this system.

(b) Find all the equilibrium solutions using the numerical data $k_1 = 2\,\text{N/m}$, $k_2 = 3\,\text{N/m}$, $k_c = 0.2\,\text{N/m}$, $g_{10} = 2\,\mu\text{m}$, $g_{20} = 1\,\mu\text{m}$, $A_1 = 4 \times 10^{-10}\,\mu\text{m}^2$, $A_2 = 6 \times 10^{-10}\,\mu\text{m}^2$, $\phi_1 = 0.25\,\text{V}$, and $\phi_2 = 0.15\,\text{V}$.

(c) Study the stability of the equilibrium solutions for the given numerical data and obtain the pull-in voltage values for $\phi_1$ and $\phi_2$.

*Hint*: For (a) write the strain energy ($SE$) of the three springs and the complementary electrostatic energy ($ESE_c$) of the two capacitors, and differentiate with respect to $u_1$ and $u_2$. Equate these to zero. That is,

$$\frac{\partial(SE - ESE_c)}{\partial u_1} = 0 \text{ and } \frac{\partial(SE - ESE_c)}{\partial u_2} = 0$$

### Problem 6.9

Consider the electrostatic comb-drive shown in Figure 6.13(b). Show that the electrostatic force experienced by the shuttle mass in the direction of the length of the comb fingers is given by $(N\varepsilon_0 t V^2/2g)$ where $N$ is the number of comb-finger pairs (a pair consists of two fingers, one anchored, the other moving, that face each other), $t$ the thickness of the finger normal to the substrate, $V$ the applied voltage, and $g$ the gap between the comb-finger pairs.

## 6.3.1 Computing the Pull-In and Pull-Up Voltages for Full Models

In Section 6.2 we discussed an iterative procedure for finding the self-consistent solution between the coupled governing equations of electrostatics and elastostatics. Figure 6.11 summarized that procedure when FEA is used to solve the coupled equations. We now explain how to use the same procedure to find the pull-in voltage using the full model rather than the lumped analysis we discussed in the preceding subsection.

Unfortunately, a simple formula for the pull-in voltage such as that in Eq. (6.82) for the lumped model cannot be found for an electromechanical device of arbitrary geometry. The pull-in voltage can only be found numerically and that too by trying out various values for the applied potential. We noted that the iterative procedure of repeatedly solving electro-static and elastostatic equations until convergence gives us the equilibrium state for an applied potential. But this procedure will cease to converge when the applied potential exceeds the pull-in voltage. In fact, one would see that the number of iterations required for convergence continues to increase as we increase the applied potential. This is because increasing potential brings the conductors closer and the nonlinearity in the coupled equa-tions increases when the gap between the conductors decreases. Note that the nonlinear electrostatic force increases rapidly as the gap decreases. This can be readily seen in the lumped model where this is inversely proportional to the square of the gap [recall Eq. (6.59)].[5]

It is the nonlinearity of the system that requires an iterative solution whereas a linear system can be solved in one step. So, when nonlinearity increases, more iterations are needed for convergence. When the pull-in situation is encountered, real solutions cease to exist for the full model too, just as we saw for the lumped model. Then the iterative procedure will not converge. This means that we must solve the problem repeatedly by gradually increasing the applied potential until the procedure does not converge. Since we know that each run for a single applied potential itself needs iterations and each iteration requires numerical solution of both governing equations, we can imagine how much computation it would take to find the pull-in voltage. Indeed, it is this computational complexity that warrants simulation software customized for microsystems, even though numerous FE software programs exist.

Since it is not computationally efficient to try out several applied potential values to detect the lack of convergence, a simple bisection method can be used to find the pull-in voltage. In this, a sufficiently large voltage is used first so that the procedure does not converge. This means that we are above $\phi_{\text{pull-in}}$. Next, half that voltage is tried, and if this does not converge, the value is halved again. If this converges, we know that $\phi_{\text{pull-in}}$ lies somewhere in between the converged value of $\phi$ and the closest nonconverged value of $\phi$. Hence, their average value is tried. This process is continued until the difference between converged and nonconverged values is sufficiently small.

Alternatively, if we want the entire displacement versus voltage curve (such as the one in Figure 6.17), we can start with a small value of $\phi$ and get its converged solution. Then we can use this solution as the initial value for a run taking a slightly higher value of $\phi$. This is likely to converge in fewer iterations than that would be necessary take if we started from the original geometry (i.e. when $\phi = 0$). If we keep the increment in $\phi$ sufficiently small, we can trace the displacement-voltage curve accurately. We can then detect $\phi_{\text{pull-in}}$ when the displacement increases suddenly, as indicated by lack of convergence, or when the numerical procedure for solving the coupled equations is taking a lot of iterations to converge.

All that we discussed so far is statics! But microsystems devices are dynamic. We consider the dynamics of electrostatically actuated microsystems next.

## ▶ 6.4 COUPLED ELECTROMECHANICS: DYNAMICS

We discuss the dynamics using lumped modeling. So far, we have developed lumped modeling to arrive at a spring constant for elastic structures and capacitance for electro-statics. These two help capture the equivalent elastic strain energy and electrostatic energy, respectively. The equivalent elastic strain energy is given by

$$SE = \frac{1}{2}ku^2 \tag{6.89}$$

---

[5] Consider also that the lumped spring that captures the elastic behavior of the structure becomes nonlinear as well when the displacements are large (see Section 4.10).

where we use $u$ for generic displacement rather than $u_z$ used so far for the parallel-plate capacitor. The equivalent complementary electrostatic energy and equivalent electrostatic energy are given respectively by

$$ESE_c = \frac{1}{2}C\phi^2 \tag{6.90a}$$

$$ESE = \frac{Q^2}{2C} \tag{6.90b}$$

With dynamic mechanical systems, we need to find lumped models for the equivalent kinetic energy and dissipated energy. The lumped model for capturing equivalent kinetic energy is *mass*, or *inertia* in general. We denote this by $m$. The kinetic energy associated with it is

$$KE = \frac{1}{2}m\left(\frac{du}{dt}\right)^2 = \frac{1}{2}m\dot{u}^2 \tag{6.91}$$

The force associated with $KE$ is the inertia force given by[6]

$$F_{\text{inertia}} = \frac{d}{dt}\left(\frac{\partial KE}{\partial \dot{u}}\right) = \frac{d}{dt}(m\dot{u}) = m\frac{d^2u}{dt^2} = m\ddot{u} \tag{6.92}$$

The inertia force, like the elastic spring force, opposes the motion.

The lumped model for dissipated energy, which is primarily due to viscous damping, is a damper element with damping coefficient $b$. The force associated with this, that is, the equivalent damping force, is given by

$$F_d = b\dot{u} \tag{6.93}$$

This force too opposes the motion. Note that we purposely avoid writing the damping energy and then taking its derivative with respect to some quantity as we did with the elastic, electrostatic, and kinetic energies. This is because the fundamentals of modeling dissipation energies are far too complex and require elaborate discussion. We touch upon one specific dissipation mode in Section 6.5.

It is pertinent at this point to consider quantities analogous to $m$ and $b$ in electrical circuits., which are inductance $L$ and resistance $R$. The energies associated with these are equivalent electromagnetic energy in an inductor, given by

$$IE = \frac{1}{2}L\left(\frac{dQ}{dt}\right)^2 = \frac{1}{2}Li^2 \tag{6.94}$$

(where $i$ is electric current) and equivalent dissipated energy in a resistor, given by

$$DE = \frac{1}{2}R\left(\frac{dQ}{dt}\right)^2 = \frac{1}{2}Ri^2 \tag{6.95}$$

---

[6] This derivative may appear strange to those who do not know calculus of variations. In order to understand it, we need to know Lagrangian mechanics; see [2] and [3] in the Further Reading section. Alternatively, an intuitive way to understand this derivative is to recall that the rate of change of linear momentum (i.e. $m\dot{u}$) is equal to the applied force, which follows from Newton's second law of motion.

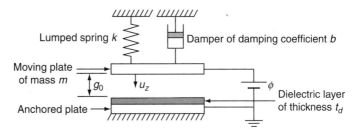

**Figure 6.25** A spring-mass-damper model for a parallel-plate capacitor with a dielectric layer.

We know that energy stored in an inductor can be recovered, just as kinetic energy can be recovered by converting it into potential energy. But dissipated energy, either in a damper or a resistor, cannot be recovered within the mechanical and electrical domains; it is converted to heat and part of that heat may be recovered by other means.

## 6.4.1 Dynamics of the Simplest Lumped Electromechanical Model

On the basis of the foregoing, we can write an equation for the dynamic behavior of the parallel-plate capacitor (see Figure 6.25) by summing all the forces acting on it:

$$m\ddot{u} + b\dot{u} + ku = \frac{\varepsilon A \phi^2}{2[g_0 + (t_d/\varepsilon_r) - u]^2} \tag{6.96}$$

Equation (6.96) is an ordinary differential equation (ODE) of second order because it involves up to the second derivative of the displacement $u$ with respect to time. Because of the term on the right-hand side (the lumped electrostatic force), this equation is nonlinear and cannot be solved analytically. If there is no force, it can be solved analytically. Furthermore, if $u$ is small, we call it a vibration equation for a single-degree-of-freedom system. Here we have only one lumped mass and thus it is a single-degree-of-freedom system. The Appendix contains solutions of this equation for no force and a periodic force. Familiarity with this Appendix is necessary for the subsequent discussion in this subsection.

### 6.4.1.1 Small Vibrations with Small Applied Potentials
In our lumped electromechanical model, the forcing function is periodic if the potential $\phi$ is sinusoidal (like an ac voltage). Let $\phi = \phi_0 \sin(\omega t)$. Then, Eq. (6.96) yields

$$m\ddot{u} + b\dot{u} + ku = \frac{\varepsilon A \phi_0^2 \sin^2(\omega t)}{2[g_0 + (t_d/\varepsilon_r) - u]^2} \tag{6.97}$$

Since $\sin^2(\omega t) = [1 - \cos(2\omega t)]/2$, for sufficiently small values of the resulting motion [i.e. when $u$ is small enough to avoid nonlinearities due to its presence in the denominator in Eq. (6.97)], resonance for it occurs when $\omega = \omega_n/2$ itself.

When we apply dc and ac voltages simultaneously, that is, when $\phi = \phi_{dc} + \phi_{ac} \sin(\omega t)$, because of $\phi^2$ in Eq. (6.96), there will be both $(\omega t)$ and $(2\omega t)$ sinusoidal terms. Hence, resonance occurs at $\omega = \omega_n/2$ and $\omega = \omega_n$.

### 6.4.1.2 Motions with Large Potentials
While small vibrations due to sinusoidal potential can readily be analyzed using the analytical solution in Eqs. (A.9), (A.11), and (A.12) of the Appendix, large-amplitude

motion needs a numerical solution because of the nonlinear forcing term: note that $u$ appears in the denominator in the electrostatic force term in Eq. (6.97). A simple method for solving Eq. (6.97) is to split it into two first-order ODEs. For this, we begin with

$$m\ddot{u} + b\dot{u} + ku = \frac{\varepsilon A \phi_0^2 \sin^2(\omega t)}{2[g_0 + (t_d/\varepsilon_r) - u]^2} = f$$

$$\ddot{u} = \frac{f - b\dot{u} - ku}{m} \tag{6.98}$$

and define $x_1 = u$ and $x_2 = \dot{u}$. Now, we can write $\dot{x}_1 = x_2$, and by using Eq. (6.98), we get

$$\dot{x}_2 = \frac{f - b\dot{u} - ku}{m}$$

Thus, we have two first-order equations that are equivalent to Eq. (6.98):

$$\dot{x}_1 = x_2 \tag{6.99a}$$

$$\dot{x}_2 = \frac{f - b\dot{u} - ku}{m} \tag{6.99b}$$

With the initial conditions given for $x_1 = u$ and $x_2 = \dot{u}$ at the starting time, we can integrate Eqs. (6.99a) and (6.99b) with respect to time and get $x_1 = u$ and $x_2 = \dot{u}$ as functions of time. Of course, the numerical integration must be used because of nonlinearity of $f$ in Eq. (6.99b). The Runge–Kutta integration scheme [3] is often used for this purpose.

### 6.4.1.3 Small Oscillations Around a Displaced State

Some electrostatic microsystems devices need to be kept at a displaced state by applying a dc voltage and then adjusting their position slightly around that displaced state; an example of such a device is discussed in Chapter 9 (see Figure 9.8). In this and similar devices, we assume that a dc voltage, $\phi_{dc}$, is applied and then a small perturbation $\delta u$ is added to $u_{dc}$. Let us assume that we have found the equilibrium displacement, $u_{dc}$, for the applied $\phi_{dc}$ by performing static analysis. Now, in order to compute the small oscillations around the displaced dc state, we linearize Eq. (6.98) as follows:

$$m(\ddot{u}_{dc} + \delta\ddot{u}) + b(\dot{u}_{dc} + \delta\dot{u}) + k(u_{dc} + \delta u) = \frac{\varepsilon A \phi_{dc}^2}{2(\bar{g}_0 - u_{dc} - \delta u)^2} \simeq \frac{\varepsilon A \phi_{dc}^2}{2(\bar{g}_0 - u_{dc})^2}\left(1 + \frac{2\delta u}{\bar{g}_0 - u_{dc}}\right)$$

$$\{m\ddot{u}_{dc} + b\dot{u}_{dc} + ku_{dc}\} + m\delta\ddot{u} + b\delta\dot{u} + k\delta u = \left\{\frac{\varepsilon A \phi_{dc}^2}{2(\bar{g}_0 - u_{dc})^2}\right\} + \frac{\varepsilon A \phi_{dc}^2}{(\bar{g}_0 - u_{dc})^3}\delta u \tag{6.100}$$

where $\bar{g}_0 = g_0 + (t_d/\varepsilon_r)$. Now, since we assumed that $u_{dc}$ is the solution for applying $\phi_{dc}$, we can cancel the terms in { } on the left- and right-hand sides of Eq. (6.100) to get

$$m\delta\ddot{u} + b\delta\dot{u} + \left(k - \frac{\varepsilon A \phi_{dc}^2}{(\bar{g}_0 - u_{dc})^3}\right)\delta u = 0 \tag{6.101}$$

The preceding equation shows that the effective stiffness, $k_{eff}$, of the displaced mass, defined as

$$k_{eff} = k - \frac{\varepsilon A \phi_{dc}^2}{(\bar{g}_0 - u_{dc})^3} \tag{6.102}$$

is smaller than $k$. This tells us that electrostatic effect decreases the stiffness for increasing values of $\phi_{dc}$. Consequently, the natural frequency of the small oscillations around a displaced state will also decrease. Indeed, we can write this changed natural frequency, $\omega_n|_{\phi_{dc}}$, as

$$\omega_n|_{\phi_{dc}} = \sqrt{\frac{k_{eff}}{m}} = \sqrt{\frac{k - [\varepsilon A \phi_{dc}^2/(\bar{g}_0 - u_{dc})^3]}{m}} \tag{6.103}$$

This shows that the natural frequency of an electrostatic device can be adjusted by applying a dc bias. One can also verify that $k_{eff}$, and hence $\omega_n|_{\phi_{dc}}$ as well, go to zero when $u_{dc} = \bar{g}_0/3$ and [see Eq. (6.87)]. This is another way of understanding the pull-in phenomenon. The overall stiffness of the coupled electrostatic-elastostatic system goes to zero at the pull-in condition and hence the system becomes unstable.

The dynamics of the one-degree-of-freedom lumped model is sufficiently rich to explain the essential features of the dynamic behavior in coupled electrostatic-elastostatic systems as done here. It is also useful to analyze other complex systems encountered in realistic situations in microsystems. We discussed in Chapter 4 how to compute the lumped spring constant of an elastic system for bars, beams, and general structures. We now show how the lumped inertia (i.e. $m$) can be obtained.

### 6.4.2 Estimating the Lumped Inertia of an Elastic System

We use a micromachined accelerometer to understand the motivation for estimating the lumped inertia of a general elastic system. Figure 6.26 shows a schematic of an in-plane micromachined accelerometer whose lumped modeling was considered in Example 4.3. We found there that its stiffness in the direction perpendicular to the beams (i.e. in the direction indicated by the dash-dot line) is 14.0626 N/m. Hence, in the lumped model of this accelerometer, we can use this value for $k$. What about $m$? As a first approximation, we can take $m$ as simply the mass of the central mass. Let the area of the central mass, $M$, be $20,000\,\mu m^2 = 2 \times 10^{-8}\,m^2$ and the density of the material be $2300\,kg/m^3$.

By using Eq. (6.97)b, we can compute its natural frequency as

$$\omega_n = \sqrt{\frac{k}{M}} = \sqrt{\frac{14.0626}{\rho_m A_{plate} t}} = \sqrt{\frac{14.0626}{2300 \times (2 \times 10^{-8}) \times (3 \times 10^{-6})}}$$

$$= \sqrt{\frac{14.0626}{1.3800 \times 10^{-10}}} = 0.3192 \times 10^6 \text{ rad/s} = 50.8 \text{ kHz}$$

This approximation does not take into account the inertia effects due to the eight beams; we can do this by referring to Eq. (6.91) to arrive at the equivalent mass of these eight beams. Note that, unlike the central mass, the entirety of the beams does not move by the same amount because the beams bend and are anchored to the substrate. So, we need to take into account this differential movement in computing the kinetic

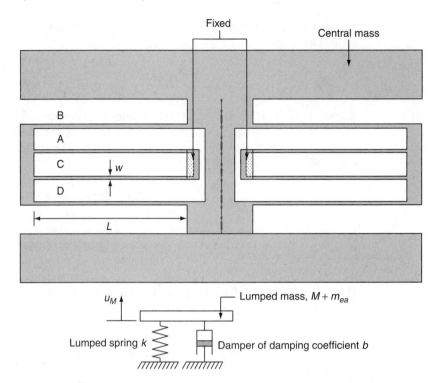

**Figure 6.26** An in-plane accelerometer and its lumped model.

energy associated with them. We then use this kinetic energy to find the equivalent mass as follows:

$$m_{eq} = \frac{KE_{\text{beams}}}{0.5\dot{u}_M^2} \tag{6.104}$$

where $\dot{u}_M$ is the velocity of the central mass.

Referring to Figure 6.26, we note that all beams obey the fixed-guided end condition (see Figure 4.13). Due to symmetry, we compute the kinetic energy of the four beams on the left and then multiply the kinetic energy by two to get $KE_{\text{beams}}$. Beams A and B are in series, as are beams C and D [see Figs. 4.17(a) and (b)]. Beams A and C are anchored at one end and guided at the other end. Their deflected shape for a transverse load $\dot{u}_M$ at the guided end is given by[7]

$$u_{A \text{ or } C}(x) = u_1(x) = \frac{Fx^2}{12YI}(3l - 2x) \tag{6.105a}$$

Since beams B and D are anchored to the moving ends of A and C, respectively, and are guided at their other ends, we can write their deflected shape as:

$$u_{B \text{ or } D}(x) = u_2(x) = \frac{Fl^3}{12YI} + \frac{Fx^2}{12YI}(3l - 2x) \tag{6.105b}$$

---

[7] This is obtained by solving Eq. (4.22) after writing the bending moment expression for this beam. See Section 4.2 for details.

where we have added the maximum deflection at the guided end of beams A and C to the deflections of beams B and D. Since the guided ends of B and D are connected to the central mass, the displacement of the central mass (i.e. $u_M$) is

$$u_M = 2 \times \frac{Fl^3}{12YI} = \frac{Fl^3}{6YI} \tag{6.106}$$

By using Eq. (6.106), we can rewrite Eqs. (6.105a) and (6.105b) as

$$u_{A \text{ or } C}(x) = u_1(x) = \frac{u_M x^2}{2l^3}(3l - 2x) \tag{6.107a}$$

$$u_{B \text{ or } D}(x) = u_2(x) = \frac{u_M}{2} + \frac{u_M x^2}{2l^3}(3l - 2x) \tag{6.107b}$$

Differentiating the displacements in Eqs. (6.107a) and (6.107b), we can compute the kinetic energy of the beams as:

$$KE_{A \text{ or } C} = \int_0^l \frac{1}{2}\rho_m A \dot{u}_{A \text{ or } C}^2 \, dx = \int_0^l \frac{1}{2}\rho_m A \frac{\dot{u}_M^2 x^4}{4l^6}(3l - 2x)^2 dx$$

$$= \frac{1}{2}(0.3428\rho_m Al)\dot{u}_M^2 = \frac{1}{2}(0.3428 m_{\text{beam}})\dot{u}_M^2 \tag{6.108}$$

where the mass of each beam is denoted by $m_{\text{beam}}$. By following the same procedure for beams B and D, we get the kinetic energy as

$$KE_{B \text{ or } D} = \int_0^l \frac{1}{2}\rho_m A \dot{u}_{B \text{ or } D}^2 dx = \int_0^l \frac{1}{2}\rho_m A \left\{\frac{\dot{u}_M}{2} + \frac{\dot{u}_M x^2}{2l^3}(3l - 2x)\right\}^2 dx$$

$$= \frac{1}{2}(0.8428\rho_m Al)\dot{u}_M^2 = \frac{1}{2}(0.8428m)\dot{u}_M^2 \tag{6.109}$$

From Eqs. (6.108) and (6.109), we can write the total kinetic energy of all the eight beams in the accelerometer as:

$$KE_{\text{beams}} = \frac{1}{2}\{4 \times (0.3428m_{\text{beam}} + 0.8428m_{\text{beam}})\}\dot{u}_M^2 = \frac{1}{2}(4.7424m_{\text{beam}})\dot{u}_M^2$$

from which we get $m_{eq}$ by comparing it with Eq. (6.104) as

$$m_{eq} = 4.7424m_{\text{beam}} = (4.7424)(\rho_m lA)$$

$$= (4.7424)\{2300 \times (200 \times 10^{-6}) \times [(5 \times 10^{-6}) \times (3 \times 10^{-6})]\}$$

$$= 0.3272 \times 10^{-10} \text{ kg}$$

Now, the corrected frequency including the inertia of the eight beams is

$$\omega_n = \sqrt{\frac{k}{M + m_{eq}}} = \sqrt{\frac{14.0626}{(1.3800 + 0.3272) \times 10^{-10}}} = 0.2870 \times 10^6 \, \text{rad/s} = 45.7 \, \text{kHz}$$

Thus, there indeed was a sizeable error (about 11%) when the inertia of the beams was not included. This is a simple analytical calculation. In order to get a more accurate result, FE analysis can be used to solve the dynamic equation of the full model.

### 6.4.3 Estimating the Lumped Damping Coefficient for the In-Plane Accelerometer

Just as we estimated the lumped mass, we can also estimate the lumped damping coefficient of the accelerometer or, for that matter, any similar structure that moves over a plane with a fluid-filled gap underneath it. This situation is illustrated in Figure 6.27 where a rigid plate is translated over a fluid film. The rigid plate is the central mass in an accelerometer. The thickness of the fluid film is equal to the gap underneath the central mass, that is, $g_0$. We assume that the area of rigid plate is very large compared with $g_0$. Therefore, we can assume that flow beneath the plate occurs purely in one dimension if we ignore the slight deviations at the entrance and exit at the plate's ends. Such a flow is called *Couette flow*.

We take a small fluid volume of length $\Delta x$, height $\Delta y$, and unit breadth (perpendicular to the paper plane) as shown in Figure 6.27 and write the equation for *conservation of mass* for this volume of fluid because the amount of fluid that goes in must come out of it, if there are no sources or sinks and if it is assumed incompressible. This gives

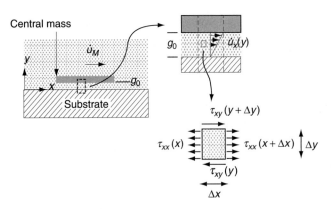

**Figure 6.27** Couette flow to model the damping on the moving central mass of an in-plane accelerometer.

$$\rho_m \left\{ \dot{u}_x(x)(\Delta y \cdot 1) + \left( \dot{u}_x(x) + \frac{\partial \dot{u}_x}{\partial x} \Delta x \right)(\Delta y \cdot 1) \right\} = 0$$

$$\frac{\partial \dot{u}_x}{\partial x} = 0 \tag{6.110}$$

where $\rho_m$ is the density of the fluid. The conclusion from the simplified form in Eq. (6.110) indicates that the velocity of the fluid beneath the plate does not vary with $x$. But it does vary with $y$ because it is zero for the fluid just over the substrate and $u_M$ just below the central mass. In order to find how it varies with $y$, we consider the *conservation of linear momentum*, which is a consequence of the Newton's second law (the rate change of linear momentum is equal to the sum of the external forces).

The contributions to the linear momentum of the fluid element come from the fluid entering and leaving. The external forces acting on the element arise from the surface forces. Figure 6.28 shows the normal stress, $\tau_{xx}$, and shear stress, $\tau_{xy}$, on the four faces of the fluid element. Note that this is a 1D flow and hence the stresses and momentum in the $y$-direction are not considered. Conservation of linear momentum in the $x$-direction gives

$$(\dot{m} \cdot \dot{u}_x)_{in} - \left[\dot{m} \cdot \left(\dot{u}_x + \frac{\partial \dot{u}_x}{\partial x}\Delta x\right)\right]_{out} - (\tau_{xx} \cdot \Delta y \cdot 1) + \left[\left(\tau_{xx} + \frac{\partial \tau_{xx}}{\partial x}\Delta x\right) \cdot \Delta y \cdot 1\right] -$$

$$-(\tau_{xy} \cdot \Delta x \cdot 1) + \left[\left(\tau_{xy} + \frac{\partial \tau_{xy}}{\partial y}\Delta y\right) \cdot \Delta x \cdot 1\right] = 0$$

$$(\rho_m \cdot \Delta x \cdot \Delta y \cdot 1 \cdot \dot{u}_x) \cdot u_x - \left[\rho_m \cdot \Delta x \cdot \Delta y \cdot 1 \cdot \left(\dot{u}_x + \frac{\partial \dot{u}_x}{\partial x}\Delta x\right)\right] \cdot \left(\dot{u}_x + \frac{\partial \dot{u}_x}{\partial x}\Delta x\right) +$$

$$+\left(\frac{\partial \tau_{xx}}{\partial x}\Delta x \Delta y\right) + \left(\frac{\partial \tau_{xy}}{\partial y}\Delta x \Delta y\right) = 0 \tag{6.111}$$

After simplifying and neglecting the term containing $(\Delta x)^2$, we get

$$-2\rho_m \dot{u}_x \frac{\partial \dot{u}_x}{\partial x} + \frac{\partial \tau_{xx}}{\partial x} + \frac{\partial \tau_{xy}}{\partial y} = 0 \tag{6.112}$$

In view of Eq. (6.110) (i.e. the *continuity equation*), Eq. (6.112) becomes

$$\frac{\partial \tau_{xx}}{\partial x} + \frac{\partial \tau_{xy}}{\partial y} = 0 \tag{6.113}$$

Now, just as we used Hooke's law for solids [see Eq. (4.3)], we use a constitutive equation for a *Newtonian fluid*—a fluid that obeys a linear relationship between the shear stress and the spatial derivative of flow velocity [4]:

$$\tau_{xx} = 2\eta \frac{\partial \dot{u}_x}{\partial x} \quad \text{and} \quad \tau_{xy} = \eta \left(\frac{\partial \dot{u}_x}{\partial y}\right) \tag{6.114}$$

where $\eta$ is the viscosity of the fluid. By substituting Eq. (6.114) into Eq. (6.113), we get

$$\frac{\partial}{\partial x}\left(\frac{2\eta \partial \dot{u}_x}{\partial x}\right) + \frac{\partial}{\partial y}\left(\frac{\eta \partial \dot{u}_x}{\partial y}\right) = 0 \tag{6.115}$$

Once again, in view of Eq. (6.110), Eq. (6.115) reduces to

$$\frac{\partial^2 \dot{u}_x}{\partial y^2} = 0 \Rightarrow \dot{u}_x = C_1 y + C_0 \tag{6.116}$$

where the two constants are found with boundary conditions at the lower and upper ends of the fluid in the gap:

$$\dot{u}_x \text{ at } y = 0 \text{ is zero} \Rightarrow C_0 = 0$$

$$\dot{u}_x \text{ at } y = g_0 \text{ is } u_M \Rightarrow C_1 = \frac{\dot{u}_M}{g_0} \tag{6.117}$$

Thus, we have

$$\dot{u}_x = \frac{\dot{u}_M}{g_0} y \tag{6.118a}$$

$$\tau_{xy} = \frac{\eta}{g_0} \dot{u}_M \tag{6.118b}$$

The shear stress, $\tau_{xy}$, creates a viscous drag on the rigid plate. This viscous force, which we call damping force, is obtained by multiplying the area of the plate with $\tau_{xy}$ as

$$F_{\text{damping}} = \frac{\eta A_{\text{plate}}}{g_0} \dot{u}_M \tag{6.119}$$

By comparison with Eq. (6.93), we can write the lumped damping coefficient for Couette flow as:

$$b_{\text{Couette}} = \frac{\eta A_{\text{plate}}}{g_0} \tag{6.120}$$

Thus, by starting from continuum analysis, we are able to arrive at the lumped damping coefficient just as we did for elastostatics, electrostatics and elastodynamics. This should convince the reader that any continuous phenomenon can be modeled in lumped form. The reader should also be convinced that the lumped parameters in Eq. (6.96) can be extracted from a full model analysis for complicated structures by equating the respective energies.

The analysis of fluids in general is not as simple as Couette flow modeling. As an example, consider an out-of-plane accelerometer where the same rigid plate (i.e. central mass) moves up and down squeezing and pulling the fluid below it. We discuss the consequences of this squeezed film in the next section.

---

**Example 6.5**

Knowing that the area of the central mass is 20,000 $\mu\text{m}^2$ with a gap of 1 $\mu$m air below it, and viscosity $2 \times 10^{-5}$ Pa s, compute the lumped damping coefficient for the accelerometer.

**Solution:** By using Eq. (6.118), we get

$$b = \frac{\eta A_{\text{plate}}}{g_0} = \frac{(2 \times 10^{-5})(20000 \times 10^{-12})}{(1 \times 10^{-6})} = 4 \times 10^{-7} \text{ N/(m/s)}$$

Note that we consider the effect of shear due only to fluid on the bottom surface of the central mass, not on the top surface. This is because the boundary condition of zero fluid velocity above the central mass extends much farther than the gap and hence the velocity derivative in Eq. (6.118) is very small.

---

**Example 6.6**

Couette flow occurs not only for the central mass but also between the interdigitated fingers of the electrostatic comb (see Figure 6.5). If the length and height of the finger are 25 μm and 2 μm, respectively, and if there are 50 pairs of parallel-plate capacitors with a gap of 2 μm, estimate the total damping coefficient due to comb fingers.

**Solution:** Once again, by using Eq. (6.120), for one pair of finger surfaces

$$b = \frac{\eta A_{\text{finger surface}}}{g_0} = \frac{(2 \times 10^{-5})(25 \times 10^{-6})(2 \times 10^{-6})}{2 \times 10^{-6}} = 5 \times 10^{-10} \text{ N/(m/s)}$$

For 50 finger pairs, the total damping coefficient is $2.5 \times 10^{-8}$ N/(m/s). Notice that this is one order of magnitude smaller than the damping coefficient of the central mass itself. This type of quick lumped analysis lets us decide if a full model analysis is necessary. It also tells us that decreasing the gap not only increases the damping force but also the capacitance. Similarly, increasing the area increases the damping coefficient and also the capacitance. Thus, there are design tradeoffs that become clear with lumped modeling.

---

**Your Turn:**

Recall from Chapter 4 that the elastostatics of solids rests on two things: equilibrium of forces and the stress-strain relationship. Elastodynamics also includes the inertia and damping forces under equilibrium, as we saw in a simple spring-mass-damper system. It is important to note that when we move from simple models to full models of realistic geometries, the concepts do not change; only the details become more mathematical. With a firm grasp of the basic concepts and with sufficient familiarity with mathematics, one can undertake the analysis of full models.

In this spirit, fluid dynamics too can be understood even in more complex situations than 1D Couette flow. All we need to know are the basic concepts used here: (a) conservation of mass, yielding the equation of continuity, (b) conservation of linear momentum, and (c) the relationship between the shear force and velocity gradient. We used all three of these in the analysis of Couette flow. Application of the same three concepts for a general 3D flow with relaxed assumptions (italicized for clarity in the sentences below) gives the following equations [4]:

$$\frac{\partial \rho_m}{\partial t} + \nabla \cdot \left( \rho \dot{\mathbf{U}} \right) = 0 \qquad (6.121)$$

This is the equation of continuity for 3D *compressible* flow where $\dot{\mathbf{U}} = (\dot{u}_x \hat{\mathbf{i}} + \dot{u}_y \hat{\mathbf{j}} + \dot{u}_z \hat{\mathbf{k}})$ is the

velocity vector with three components in the three orthogonal directions of the 3D flow. Conservation of linear momentum gives

$$\rho_m \left\{ \frac{\partial \dot{\mathbf{U}}}{\partial t} + \dot{\mathbf{U}} \cdot \nabla \dot{\mathbf{U}} \right\} = -\nabla \mathbf{p} + \eta \nabla^2 \dot{\mathbf{U}} + \mathbf{f}$$
$$+ \left( \frac{\eta}{3} + \eta_v \right) \nabla (\nabla \cdot \dot{\mathbf{U}})$$
$$(6.122)$$

where **p** is the *pressure* vector, **f** is the *body force* (e.g., gravity, centrifugal, etc.) acting in the fluid, and $\eta_v$ is the *second viscosity coefficient*. Equation (6.122) is known as the *Navier–Stokes equation* for compressible flows of Newtonian fluids in 3D. The last term on the right-hand side disappears if the flow is incompressible.

The generalized constitutive equations for a Newtonian fluid in the 3D setting are:

$$\tau_{mn} = \eta \left( \frac{\partial \dot{u}_m}{\partial n} + \frac{\partial \dot{u}_n}{\partial m} \right) - \frac{2\eta}{3} (\nabla \cdot \dot{\mathbf{U}}) \delta_{mn} \quad (6.123)$$

where $m$ and $n$ can take $x$, $y$ and $z$ values to define six shear stresses: $\tau_{xx}, \tau_{yy}, \tau_{zz}, \tau_{xy} = \tau_{yx}, \tau_{xz} = \tau_{zx}, \tau_{yz} = \tau_{zy}$.

It is a good exercise to reduce Eqs. (Eqs. (6.121)–(6.123)) to that for Couette flow and compare the result with Eqs. (6.110, 6.113) and (6.114), or to derive Eqs. (6.121) and (6.122) from first principles by following the conservation of mass and linear momentum for a small 3D volume element.

## ▶ 6.5 SQUEEZED FILM EFFECTS IN ELECTROMECHANICS

Figure 6.28 shows a solid object floating above a substrate with fluid beneath it. When the object translates up and down, rotates, or deforms, the fluid in the gap beneath is squeezed out from the sides and some of it may be compressed as well. In any case, there will be a not-so-simple pressure distribution on the bottom surface of the solid object. In fact, this pressure distribution, which we denote $p = p(x, z, t)$, will have components that are in phase with the displacement (or deformation) of the object or 90° out of phase. The in-phase component can be thought of as a spring force per unit area. This is because compressible fluid acts like a spring as it is squeezed out from the sides as the object moves down and comes back in again when the object moves up, as a consequence of the compressibility of the fluid. The out-of-phase component will be in phase with the velocity of the object. Thus this is like the damping force per unit area and is a consequence of the viscous drag on the object. Hence, the lumped model of the squeezed film effect has the two parts shown in Figure 6.28.

We can arrive at simple expressions for the lumped parameters, $k_{sqf}$ and $b_{sqf}$, in terms of the geometry and the material properties using the equations of fluid dynamics. For this, we need to use a special case of the Navier–Stokes equations that govern the fluid motion under different assumptions that are pertinent to this situation. First, note that, in the preceding paragraph, we found the pressure underneath the plate as $p = p(x, z, t)$. That is, it does not vary in the $y$-direction. This is an appropriate assumption because the gap is usually very small compared with the area of the solid object. An equation that governs such behavior is given by what is known as the *Reynolds equation* for squeezed film [5]:

$$12\eta \frac{\partial\{p(x, z, t)\,[g_0 - u_z(x, z, t)]\}}{\partial t} = \nabla \cdot \left\{(1 + 6K_n)[g_0 - u_z(x, z, t)]^3 p(x, z, t) \nabla p(x, z, t)\right\}$$

(6.124)

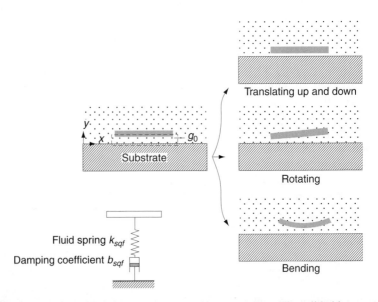

**Figure 6.28** Squeezed film effects in microsystems with narrow gaps. The solid object may translate vertically, rotate, or deform as shown. In all cases, the lumped modeling can be shown to be an additional spring and a damper.

where $Kn$ is the *Knudsen number* and is equal to the ratio of mean free path of the fluid molecules to the gap between the plate and the substrate. The nonlinearity of the preceding equation makes it difficult to analyze to extract the lumped parameters. However, it can be linearized and the lumped parameters can be derived for small motions of the solid object. We refer to a detailed discussion of the linearized analysis of this equation by Senturia [6] for a rigid oscillating plate and give only the results related to the lumped parameters:

$$b_{sqf} = \frac{96\eta l w^3}{\pi^4 g_0^3} \tag{6.125a}$$

$$k_{sqf} = \frac{8lwp_0}{\pi^2 g_0} \tag{6.125b}$$

where $l$ and $w$ are the length and width of the plate and $p_0$ the ambient pressure.

Another interesting aspect of the squeezed film effect is that damping dominates at low frequencies of motion of the plate, whereas the spring effect dominates at high frequencies. Full model analysis of the squeezed film effect requires us to solve Eq. (6.124) simultaneously with the electrostatic and elastostatic equations in an electrostatically actuated device. This indicates how involved the modeling can get in microsystems devices.

### Example 6.7

Compute the squeezed film lumped parameters, $b_{sqf}$ and $k_{sqf}$, for the central mass of the accelerometer if it moves perpendicular to the substrate. Use the following numerical data: viscosity of air $\eta = 2 \times 10^{-5}$ Pa s, plate width $w = 100\,\mu$m, plate length $l = 200\,\mu$m, gap beneath the plate $g_0 = 1\,\mu$m, ambient pressure $p_0 = 103$ kPa.

**Solution:** The squeezed film lumped parameters can be calculated as

$$b_{sqf} = \frac{96\eta l w^3}{\pi^4 g_0^3} = \frac{96(2 \times 10^{-5})(200 \times 10^{-6})(100 \times 10^{-6})^3}{\pi^4(1 \times 10^{-6})^3} = 0.0039 \text{ N/(m/s)}$$

$$k_{sqf} = \frac{8\eta l w p_0}{\pi^2 g_0} = \frac{8(200 \times 10^{-6})(100 \times 10^{-6})(103 \times 10^3)}{\pi^2(1 \times 10^{-6})} = 1669.9 \text{ N/m}$$

These values should be compared with the in-plane Couette-flow-induced damping and the suspension-related spring stiffness. This comparison will tell us if squeezed film effects are important in analyzing an electrostatically actuated microsystems device.

## ▶ 6.6 ELECTRO-THERMAL-MECHANICS

In the preceding sections, we saw how three energy domains, namely electrostatic, elastic, and fluid, are coupled to one another. We now consider another class of problems in which three domains are coupled to each other in a different way.

Figure 6.29(a) shows an electro-thermal-compliant (ETC) actuator [7] made of polysilicon using surface micromachining. The width of the narrowest beam is $2\,\mu$m and its overall height, as seen in Figure 6.29(a), is about $450\,\mu$m. The figure shows two anchors on the right-hand side. When an electric potential difference is applied between them, the

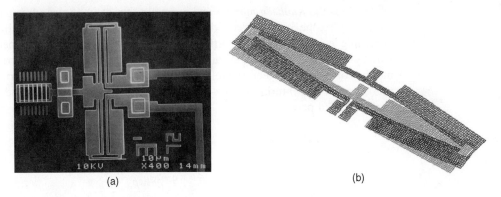

**Figure 6.29** An ETC actuator. (a) A scanning electron microscope image of the polysilicon actuator; (b) finite element simulation.

central mass on the left moves to the left. A Vernier scale (in the leftmost part of the image) is included in the device to measure the displacement of the actuator. Figure 6.29(b) shows the deformed configuration of this actuator superposed on the undeformed configuration, as determined by finite element analysis. In what follows, we answer two questions: (i) Why does the actuator deform as shown in Figure 6.29b when an electric potential is applied between the two anchors? (ii) What equations are to be solved in analyzing this?

Consider the basic building block of an electro-thermal-elastic actuator [7] in Figure 6.30. Four of these building blocks are arranged in a particular way in the ETC actuator in Figure 6.29(a). The basic ETC actuator of Figure 6.30 has four segments: a long narrow beam, a connector, a long wide beam, and a short narrow beam. The two anchors are marked with crosses. When we apply an electric potential

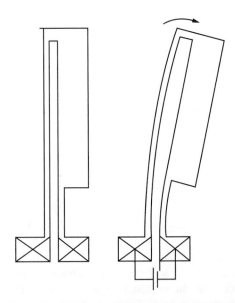

**Figure 6.30** Basic building of an ETC actuator, left: before actuation; right: after actuation.

between them, the actuator deform as shown at the right of Figure 6.30. This is because of three coupled effects:

i. First, the application of the electric potential difference between the two anchors sends an electric current through the four segments. The same current passes through the four segments. However, the current density in them is not the same because their widths are different. Thus, they have non-uniform current density.

ii. Second, the current causes the segments to heat up. Since the current density and the width are not uniform, the steady-state temperature is not uniform across the four segments.

iii. Third, the non-uniform temperature makes different segments expand to different extents. The long, narrow, and wide beams are of particular interest. In most geometries of this kind, the long narrow beam's average temperature is greater than that of the long wide beam, and thus the long narrow beam tries to expand more than the long wide beam. Hence, the actuator deflects as shown in Figure 6.30, creating create tensile bending stress on the narrow beam side and compressive bending stress on the wide beam side.

In general, any ETC actuator deflects in a particular way because of the coupling among the electric current, thermal, and elastic domains due to non-uniformity in the electric current density and in the ensuing temperature rise.

Modeling of the ETC actuators proceeds in the same way as that of the coupling effects described in the preceding paragraph. First, we have to model the electric current and then the thermal conduction, followed by elastic deflection due to the thermal load. We use the lumped analysis first and then present the general equations.

## 6.6.1 Lumped Modeling of the Coupled Electro-Thermal-Compliant Actuators

Figure 6.31 shows the basic ETC actuator and its various dimensions, all indicated as proportions of the length $L$ of the long narrow beam. The two anchored ends are held at the same temperature $T_0$. The spring of spring constant $K_s$ indicates the load on the actuator against which it produces a deflection $\Delta$. Our aim is to find $\Delta$ for applied potential $V$ between the anchors.

*Electrical Analysis*  If we assume that the width of the actuator is considerably smaller than the length, then we can assume that electric current flows only in the direction of the

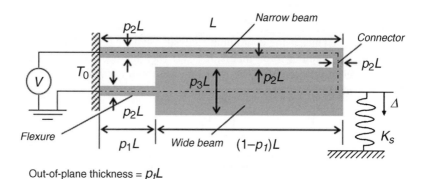

Out-of-plane thickness = $p_t L$

**Figure 6.31** Schematic of the basic ETC actuator building block.

length, as indicated by the dash–dot line in Figure 6.31. Let us consider one such slender segment, shown in Figure 6.32(a). Let $\mathbf{J}$ be the electric current density vector at every point in it. The magnitude of $\mathbf{J}$ is the charge per unit area, in other words, the electric current per unit area. The direction of $\mathbf{J}$ is along the direction of the electric field $\mathbf{E}$ at that point. $\mathbf{J}$ and $\mathbf{E}$ are related by *Ohm's law*, whose microscopic-level equation can be written as

$$\mathbf{J} = \frac{\mathbf{E}}{\rho_e} \tag{6.126}$$

where $\rho_e$ is the electrical resistivity and its reciprocal is the electrical conductivity. By integrating $\mathbf{J}$ over the cross section of the slender segment, we can determine the electric current flowing through it:

$$I = \int_A \mathbf{J} \cdot d\mathbf{A} \tag{6.127a}$$

where $d\mathbf{A} = \hat{\mathbf{n}} dA$, with $\hat{\mathbf{n}}$ denoting the unit vector normal to the cross-section area $A$. By substituting Eq. (6.126) into Eq. (6.127a), we get

$$I = \int_A \mathbf{J} \cdot d\mathbf{A} = \int_A \left( \frac{\mathbf{E}}{\rho_e} \right) \cdot d\mathbf{A} \tag{6.127b}$$

We note that current is proportional to the electric field. From Eq. (6.7), we recall that the electric field is proportional to the applied potential difference for any two points M and N:

$$- \int_{M \to N} \mathbf{E} \cdot d\mathbf{l} = -\phi_M + \phi_N = \phi_N - \phi_M \tag{6.128}$$

Therefore, we can conclude that the current is proportional to the potential difference, and the electrical resistance $R$ is given by the ratio of the potential difference to the current. Hence, $R$ is given by

$$R = \frac{- \int_{M \to N} \mathbf{E} \cdot d\mathbf{l}}{\int_A \left( \frac{\mathbf{E}}{\rho_e} \right) \cdot d\mathbf{A}} = \frac{\phi_N - \phi_M}{I} = \frac{V}{I} \tag{6.129}$$

Thus, using continuum analysis, we have derived a lumped model for electric current in a familiar form, that is, $R = V/I$ (i.e. the macroscopic Ohm's law), with the help of an electrical resistor. The next step is to express $R$ in terms of the geometry and material properties of the segment in Figure 6.32(a).

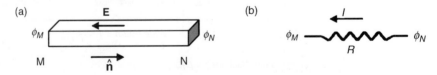

**Figure 6.32** (a) A slender segment carrying current, and (b) its lumped 'resistor' model.

Since the cross-sectional area of the segment in Figure 6.32 is uniform along its length, the electric field is constant throughout the segment. Then, according to Eq. (6.128),

$$\mathbf{E} = -\frac{(\phi_N - \phi_M)}{l}\,\hat{\mathbf{n}}.$$

Thus, Eq. (6.129) gives

$$I = \int_A \mathbf{J} \cdot d\mathbf{A} = \int_A \left(\frac{\mathbf{E}}{\rho_e}\right) \cdot \hat{\mathbf{n}}dA = -\left(\frac{\phi_N - \phi_M}{l\rho_e}\right) \int_A \hat{\mathbf{n}} \cdot \hat{\mathbf{n}}dA = -\left(\frac{\phi_N - \phi_M}{l\rho_e}\right)A \qquad (6.130)$$

Finally, Eq. (6.129) gives us the expression for $R$ for a segment of constant cross section:

$$R = \left(\frac{l\rho_e}{A}\right) \qquad (6.131)$$

### Problem 6.10

It is not uncommon to have resistor elements in microsystems in which the cross-sectional area is not uniform along the resistor length. Consider a 500 μm long, 2 μm thick resistor segment with linearly tapering width from 25 μm to 5 μm. If this segment were to be modeled as a lumped resistor with a resistance $R$, find the value of $R$. [*Hint:* use Eq. (6.129).] How does this compare with an approximate value computed using Eq. (6.131) assuming an average width of 15 μm?

On the basis of the preceding discussion, in lumped analysis, the four segments in the actuator can be treated as electrical resistors. These resistors are in series because they all have the same current passing though them (Figure 6.33):

$$R_i = \frac{\rho_e l_i}{A_i} \quad \text{for } i = 1, 2, 3, 4 \qquad (6.132)$$

By noting the relative proportions in Figure 6.31, we get the combined series-resistance as

$$R = R_1 + R_2 + R_3 + R_4 = \frac{\rho_e}{L}\left\{\frac{1}{p_t p_2} + \frac{1}{p_t} + \frac{(1 - p_1)}{p_t p_3} + \frac{p_1}{p_t p_2}\right\} = \psi_e \frac{\rho_e}{L} \qquad (6.133)$$

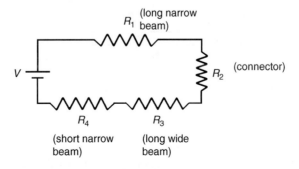

**Figure 6.33** Lumped resistor model for the ETC actuator building block in Figure 6.31.

where $\psi_e$ depends only on the proportions of the actuator as indicated in the preceding equation.

Heat is generated whenever current flows through a conductor. This is because current carriers (i.e. electrons) transfer their energy to the medium through collisions that raise the temperature of the medium. This dissipated energy per unit time is known as Joule heating and is denoted by $P_e$; it is equal to the work done by the electrons as they move to lower potential. Thus, for charge $Q$ and potential difference $\Delta\phi$, we get

$$P_e = \frac{d(\text{work})}{dt} = \frac{d(Q\Delta\phi)}{dt} = (\Delta\phi)\frac{dQ}{dt} = (\Delta\phi)I \tag{6.134}$$

Since $\Delta\phi = IR$ per Eq. (6.129), we get the expression for Joule heating, which has units J/s or watt (W):

$$P_e = I^2 R \tag{6.135}$$

The Joule heating per unit volume per unit time in each of the four segments, that is, $\hat{P}_{e_i}(i = 1 \cdots 4)$, which acts as the heat source in the thermal analysis, can be written using Eqs. (6.135), (6.132), and (6.133) as

$$\hat{P}_{e_i} = \frac{I^2 R_i}{A_i l_i} = \left(\frac{V}{R}\right)^2 \left(\frac{R_i}{A_i l_i}\right) = \left(\frac{L^2 V^2}{\psi_e^2 \rho_e^2}\right)\frac{\rho_e}{A_i^2} = \frac{L^2 V^2}{\psi_e^2 A_i^2 \rho_e} \quad \text{for } i = 1, 2, 3, 4 \tag{6.136}$$

That is,

$$\hat{P}_{e_1} = \frac{V^2}{\psi_e^2 p_i^2 p_2^2 L^2 \rho_e} \quad \hat{P}_{e_2} = \frac{V^2}{\psi_e^2 p_i^2 p_2^2 L^2 \rho_e} \quad \hat{P}_{e_3} = \frac{V^2}{\psi_e^2 p_i^2 p_3^2 L^2 \rho_e} \quad \hat{P}_{e_4} = \frac{V^2}{\psi_e^2 p_i^2 p_2^2 L^2 \rho_e} \tag{6.137}$$

***Thermal Analysis***   The slender segments are amenable to one-dimensional heat conduction analysis (see box, "Three Modes of Heat Transfer"). Consider the slender segment in Figure 6.34(a). This also can be lumped to an equivalent thermal resistor in the same way as for the electrical resistor. Here, we have Fourier's law, which is analogous to Ohm's law:

$$J_{th} = -k_{th}\frac{dT}{dx} \tag{6.138a}$$

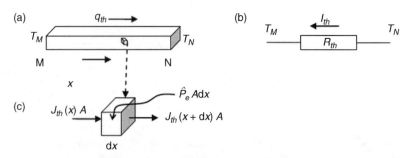

**Figure 6.34** (a) A slender segment carrying heat and (b) its lumped "resistor" model; (c) heat balance on a differential element for deriving the heat conduction equation.

or its 3D representation:

$$\mathbf{J}_{th} = -k_{th}\nabla T \tag{6.138b}$$

where $J_{th}$ is heat flux, $T$ is temperature, and $k_{th}$ is thermal conductivity. Note that Eq. (6.126) is similar to Eq. (6.138b) if we recall that $\mathbf{E} = -\nabla\phi$. Thus, temperature is analogous to electric potential. Following these analogies, the lumped thermal resistance, as indicated in Figure 6.34(b), can be written as

$$R_{th} = \frac{l}{k_t A} \tag{6.139}$$

Like electrical resistors, thermal resistors can be lumped as equivalent resistors using series and parallel arrangements. However, a lumped model for thermal part is not enough for our actuator example. We need to know the variation in temperature along each segment in order to compute the thermally induced deformation. So, we derive the thermal conduction and solve it to get the temperature profile along the segment.

In order to derive the thermal conduction equation, consider a differential (i.e. small) element of length $dx$, as shown in Figure 6.34(c), and consider the heat balance in it. Shown in the figure are the heat entering $[J_{th}(x)A]$, exiting $\left[\left(J_{th}(x) + \frac{dJ_{th}(x)}{dx}dx\right)A\right]$, and generated within it $(\hat{P}_e Adx)$. The balance of these terms, with the help of Eq. (6.138a), gives the heat conduction equation in one dimension:

$$k_t \frac{d^2 T(x)}{dx^2} + \hat{P}_{e_i} = 0 \tag{6.140}$$

---

### ▶ THREE MODES OF HEAT TRANSFER

Heat is transferred from one location to another in three distinct modes: (i) conduction, (ii) convection, and (iii) radiation. Let us understand these three terms by first noting that temperature is nothing but the vibrations of the atoms and molecules that make up substances: the higher the temperature, the greater the vibrations. Heat flows from a high-temperature location to a low-temperature location to equalize the temperatures.

Heat conduction occurs because of collisions of vibrating molecules and atoms and the ensuing diffusion effect. In diffusion, similar to the way electric current flows, there is no bulk movement of the substance. Hence, the equations governing heat conduction and electric conduction are the same, with thermal conductivity and temperature replacing the electrical conductivity and voltage, respectively. Conduction occurs in solids and fluids. Heat conduction in fluids is not as good as in solids because the molecules are much farther apart than in solids. Air is a bad conductor of heat because of reduced chance of collisions among its component molecules.

Convection is dominant in fluids. It includes diffusion and *advection*. Unlike in diffusion, in advection there is a bulk movement of molecules. Convection can occur naturally because of temperature differences in different locations of a fluid, or by force. Forced convection is what we feel when we turn on a fan. The physical property that is useful in the lumped modeling of convection is the *convective heat transfer coefficient*, $h_{conv}$. Heat flux, that is, heat current per unit area, is given by $h_{conv}(T_s - T_0)$ where $T_s$ is the temperature of a solid's surface around which there is fluid at temperature $T_0$.

Radiative heat transfer does not require a medium. It is simply electromagnetic radiation caused by a hot body. For instance, heat from the sun reaches the earth by radiation. We can feel the heat of a fire from a distance because of radiative heat transfer. All bodies above the absolute zero temperature radiate heat. Emissivity, $\varepsilon_{rad}$, is the relevant physical property to quantify a surface's radiation capability. Heat flux due to radiation is given by $\varepsilon_{rad}\sigma_{SB}(T_s - T_0)$, where $\sigma_{SB} = 5.67 \times 10^{-8}$ W/(m²K⁴) is the Stefan–Boltzmann constant.

Interested readers may consult a standard book [e.g., 8] on heat and mass transfer to understand the physical mechanisms behind these three heat transfer modes and the reasoning behind the empirical formulae describing convection and radiation heat transfer. All three modes of heat transfer are important for as simple a device as the one in Figure 6.30 when the applied potential is sufficiently large and causes a large temperature increase.

Since the connector segment is very short, the temperature distribution in it is neglected but the heat generated is taken into account in balancing the flux across its two interfaces. The temperature $T_i(x)$ in each of the three remaining segments is governed by the equations

$$k_t \frac{d^2 T_i(x)}{dx^2} + \hat{P}_{e_i} = 0 \text{ for } i = 1, 3, 4 \tag{6.141}$$

where $x$ runs from zero through the length of the corresponding segment. With convection and radiation neglected in this analysis, the only boundary conditions at the ends of each segment are due to the continuity either of temperature or of heat flux across the interface between two segments. The differential equation in Eq. (6.141) can be readily solved as

$$T_i(x) = -\frac{\hat{P}_{e_i}}{2k_t} x^2 + a_i x + b_i \tag{6.142}$$

where the constants $a_i$ and $b_i$ ($i = 1, 3, 4$) are obtained using the following six boundary conditions:

1. Temperature at the left end of the narrow beam is the ambient temperature:

$$T_1(x = 0) = T_0, \text{ that is, } b_1 = T_0 \tag{6.143a}$$

2. Temperature at the right end of the narrow beam is equal to that at the right end of the wide beam:

$$T_1(x = L_1) = T_3(x = 0), \text{ that is, } -\frac{\hat{P}_{e_1}}{2k_t} L^2 + a_1 L + T_0 = b_3 \tag{6.143b}$$

3. Continuity of heat flux across the connector along with the heat generated in it:

$$-k_t A_1 \frac{dT_1}{dx}\bigg|_{x=L_1} - k_t A_3 \frac{dT_3}{dx}\bigg|_{x=0} + \hat{P}_{e_2} A_2 L_2 = 0 \tag{6.143c}$$

4. Temperature at the left end of the wide beam is equal to that at the right end of the flexure:

$$T_3(x = L_3) = T_4(x = 0) \tag{6.143d}$$

5. Heat flux continuity across the interface between the wide beam and the flexure:

$$k_t A_3 \frac{dT_3}{dx}\bigg|_{x=L_3} - k_t A_4 \frac{dT_4}{dx}\bigg|_{x=0} = 0 \tag{6.143e}$$

**6.** Temperature at the left end of the flexure is the ambient temperature:

$$T_4(x = L_4) = T_0, \text{that is,} -\frac{\dot{Q}_{e_4}}{2k_t}p_1^2L^2 + a_4p_1L + b_4 = T_0 \qquad (6.143f)$$

Equations (6.143b)–(6.143f) can be solved to determine $\{a_1, a_3, b_3, a_4, b_4\}$. Notice that, for consistency, $b_i$s should have units of temperature and $a_i$s of temperature per unit length. The variations of temperature determined in this manner are used for elastic analysis under thermal loads.

***Elastic Analysis*** Just as in electrical and thermal analyses, the slenderness of the segments in the ETC actuator makes Euler beam theory applicable in analyzing the elastic deflections under the thermal loads. Since we are interested only in the output deflection $\Delta$ indicated in Figure 6.31, we can use energy theorems similar to Castigliano's theorems in Chapter 4 (see Eq. 4.28). We use Maizel's theorem [9] for computing $\Delta$:

$$\Delta = \sum_{i=1}^{4}\left[\int_0^{l_i}\hat{\sigma}_i(x)\,\alpha\,\{T(x)_i - T_0\}\,A_i\mathrm{d}x\right] \qquad (6.144)$$

where $\hat{\sigma}_i(x)$ is the axial normal stress induced in the $i^{th}$ segment due to a unit force applied at the output point in the desired output direction for given mechanical boundary conditions

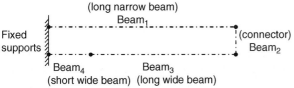

(long narrow beam)

Beam₁

Fixed supports

(connector) Beam₂

Beam₄  Beam₃
(short wide beam) (long wide beam)

**Figure 6.35** Beam model for applying Maizel's theorem to determine the desired deflection of the basic building block of the ETC actuator.

including the output spring (see Figs. 6.31 and 6.35) but in the absence of thermal loading.

The sequentially coupled analysis of the ETC actuator developed in this section is applicable to any electro-thermally actuated elastic structure that has slender segments. For general structures such as that in Figure 6.36, we need to use numerical methods (e.g. finite element analysis). Here, we give the general equations to be solved for the electrical, thermal, and elastic analysis.

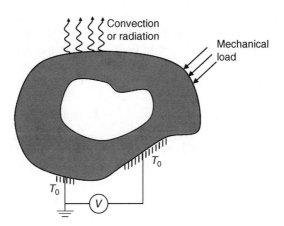

**Figure 6.36** An arbitrary electro-thermally actuated elastic structure. The general equations for modeling this are given in Section 6.6.2.

## 6.6.2 General Modeling of the Coupled ETC Actuators

Referring to Figure 6.36, we note that in order to determine the electric current distribution in the conducting continuum, we need to solve the conductive medium equation:

$$\frac{\partial^2 \phi}{\partial x^2} + \frac{\partial^2 \phi}{\partial y^2} + \frac{\partial^2 \phi}{\partial z^2} = 0$$

$$\text{or} \quad \nabla^2 \phi = 0 \tag{6.145}$$

We also note that this is exactly the same as the electrostatics equation in Eq. (6.23), with the difference that Eq. (6.145) is solved inside the body (the shaded region in Figure 6.36), whereas Eq. (6.23) is solved outside the body. That is, in the electrical conduction problem, we determine the potential everywhere inside the body for specified potentials on some portions of its boundary, as shown in Figure 6.36, by solving Eq. (6.145). Once we find the potential, the electric field is found by using the same equation as before. That is, we use Eq. (6.10): $\mathbf{E} = -\nabla \phi$; then, the microscopic form of the Ohm's law [Eq. (6.126)] to compute the electric current density, $\mathbf{J} = (\mathbf{E}/\rho_e)$. After solving for these quantities, we use Eq. (6.129) to extract a lumped quantity such as the resistance between the two electrical connection points in Figure 6.36 (or for any other pair of points).

The continuum version of the Joule heating derived in Eq. (6.135) is

$$dP_e = (d\phi)\,(dI) = (\mathbf{E} \cdot d\mathbf{l})(\|\mathbf{J}\|ds) = \left(\mathbf{E} \cdot \frac{\mathbf{J}}{\|\mathbf{J}\|}\,dl\right)(\|\mathbf{J}\|ds) = (\mathbf{E} \cdot \mathbf{J})(ds\,dl) = (\mathbf{E} \cdot \mathbf{J})dv$$

$$\Rightarrow \frac{dP_e}{dv} = p_e = \mathbf{E} \cdot \mathbf{J} \tag{6.146}$$

where $p_e$, the Joule heating per unit volume at every point inside the conductor, has units of W/m$^3$. This will appear as the heating term in the general 3D version of Eq. (6.140) of the thermal conduction:

$$k_t\left(\frac{\partial^2 T}{\partial x^2} + \frac{\partial^2 T}{\partial y^2} + \frac{\partial^2 T}{\partial z^2}\right) + p_e = 0 \tag{6.147}$$

By solving Eq. (6.147) with the Joule heating term, we can determine the temperature distribution inside the body in Figure 6.36. We then use the boundary conditions such as the points on the surface where the temperature is the ambient temperature, $T_0$, or where a heat flux $\mathbf{J}_{th} = -k_{th}\nabla T$ in Eq. (6.138b) is specified. Convection, radiation, and supplied heat flux across any portion of the boundary of the body is handled by specifying $\mathbf{J}_{th}$. For convection,

$$\mathbf{J}_{th} = h_{\text{conv}}(T - T_0) \tag{6.148}$$

where $h_{\text{conv}}$ is the heat transfer coefficient, which depends on the flow of the convecting fluid around the body. For a radiation boundary condition,

$$\mathbf{J}_{th} \cdot \hat{\mathbf{n}} = \sigma_{SB}(T + 273.13)^4 \tag{6.149}$$

where 273.13 is added to get the absolute temperature (Kelvin) of the body and $\sigma_{SB}$ is the Stefan–Boltzmann constant.

After we obtain the temperature distribution inside the body, if our interest is in computing the deflection at a point, we use the general version of the one-dimensional Maizel theorem in Eq. (6.144):

$$\Delta = \int_V \{\hat{\sigma}_x + \hat{\sigma}_y + \hat{\sigma}_z\} \, \alpha \, \{T - T_0\} \, dV \tag{6.150}$$

where $\hat{\sigma}_x$, $\hat{\sigma}_y$, and $\hat{\sigma}_z$ are the normal stresses computed by applying a unit load at the point of interest in the direction in which we want to find the deflection. On the other hand, if our interest is the deflection and stresses everywhere, we need to solve general elasticity equations [see Eq. (6.33)] with the following new equation relating stress and strain to account for the thermally induced strains:

$$\sigma_m = (\cdots) - \alpha(T - T_0) \tag{6.151}$$

where the subscript $m$ is $x$, $y$, or $z$, and $(\cdots)$ indicates the corresponding normal stress component given by Eq. (4.111). Thus, we see that coupled electro-thermal-elastic analysis requires sequential solution of three equations. A sequential solution will not suffice if we take into account the fact that some material properties (e.g. electrical resistivity, thermal conductivity, thermal expansion coefficient, Young's modulus, etc.) vary with temperature. In such a case, we need to solve the three governing equations simultaneously, which complicates matters further. We also note that we assumed that all properties (elastic, electrical, and thermal) are the same in all directions (a condition called *isotropy*) and that they do not vary within the body (a condition called *homogeneity*), but that this is not true in general. Then the equations will need to be modified slightly. The reassuring fact is that numerical methods such as the finite element, boundary element, finite difference (see Chapter 5), and others can solve almost any problem today with the help of computers.

## Example 6.8

Verify Maizel's theorem for a fixed-free bar subjected to uniform heating.

**Solution:** We simplify the general statement of Maizel's theorem in Eq. (6.150) for the one-dimensional bar:

$$\Delta = \int_V \{\hat{\sigma}_x + \hat{\sigma}_y + \hat{\sigma}_z\} \, \alpha \, \{T - T_0\} \, dV = \int_0^L \hat{\sigma}_x \, \alpha \, \{T - T_0\} \, A \, dx \tag{6.152}$$

Here, only $\hat{\sigma}_x$ is nonzero because of the one-dimensional assumption and $dV$ is replaced by $A \, dx$ since no change is assumed across the cross-sectional area. Consequently, integration is over the length of the bar. If our interest is in computing the displacement at the free end of the bar, we imagine a unit dummy load at the free end and calculate stress in the entire bar due to that to get $\hat{\sigma}_x$. From Eq. (4.1), $\hat{\sigma}_x = 1/A$ because $F = 1$. With this, and because we assumed uniform temperature rise in the bar, Eq. (6.152) becomes

$$\Delta = \int_0^L \left(\frac{1}{A}\right) \alpha \, \{T - T_0\} \, A \, dx = \alpha \, \{T - T_0\}L \quad \Rightarrow \quad \frac{\Delta}{L} = \alpha \, \{T - T_0\} \tag{6.153}$$

which is consistent with Eq. (4.5) for the thermal strain.

**Problem 6.11**

Use Maizel's theorem to solve the displacement of the midpoint of the fixed-fixed bar considered in Problem 4.1.

**Problem 6.12**

Consider again the bent-beam thermal actuator of Problem 4.5. Now, as shown in Figure 6.37, assume that there is a potential difference of $\phi$ between its two anchors. Use Maizel's theorem to compute the vertical displacement of the corner.

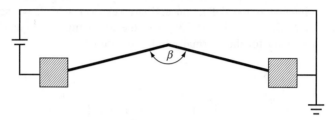

**Figure 6.37** A bent-beam thermal actuator.

## ▶ 6.7 COUPLED ELECTROMAGNET-ELASTIC PROBLEM

In order to give a feel for other types of coupled problems encountered in microsystems, we consider a specific device, the electromagnetic actuator in Figure 6.38. As can be seen, this consists of a magnetic core with winding around it. It has a cantilever-beam-like elastic *arm* that moves down due to electromagnetic force. (This magnetic actuator was developed by Ahn and Allen at Georgia Tech, Atlanta, GA.) The figure shows how they achieved the winding around the magnetic core using a sacrificial layer process. Our interest is in modeling this device so as to compute the deflection of the beam when a certain current is passed through the winding.

Without going into the derivations and detailed discussions as we did for electrostatics, we now quickly state a few important aspects of electromagnetics needed in modeling the electromagnet-based microactuator (see [10] for a detailed treatment of the electromagnetics involved; see also [6]). First, we note the there is a field, **H** in the magnetic domain that is analogous to the electric field. There are also quantities analogous to **D** (electric displacement) and **P** (polarization); they are called the magnetic flux density vector **B** and magnetization, **M**, respectively. The permeability $\mu$ is analogous to permittivity $\varepsilon$. These quantities are related to each other in much the same way as their electrical counterparts:

$$\mathbf{B} = \mu_0(\mathbf{H} + \mathbf{M}) \approx \mu\mathbf{H} \tag{6.154}$$

Compare the preceding equation with Eqs. (6.13) and (6.15).

There are laws that are fundamental to electromagnetics in terms of the quantities here that require defining a few more terms. However, for modeling the microactuator in Figure 6.38, Ampere's law is sufficient:

$$\text{Ampere's law:} \quad \int_{\text{A closed path}} \mathbf{H} \cdot d\mathbf{l} = n_{\text{turns}}I \tag{6.155}$$

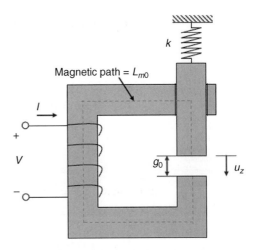

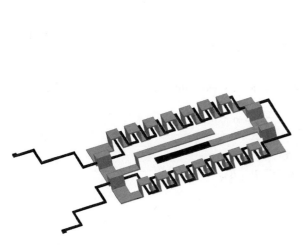

**Figure 6.38** An electromagnet-based microactuator.

**Figure 6.39** Lumped model of the electromagnet-based microactuator.

where the left-hand side is called the magnetomotive force (MMF) for a closed path and $n_{\text{turns}}$ is the number of turns in a winding carrying current $I$.

Now consider the schematic of a electromagnetic-elastic microactuator in Figure 6.39. Here, we have lumped the elastic element to a single spring of spring constant $k$ that is attached to the armature. By denoting the length of the magnetic path for zero movement of the armature by $L_{m0}$ and by referring to Figure 6.39, we can write

$$F_{MM} = H_\mu(L_{m0} + u_z) + H_g(g_0 - u_z) \qquad (6.156)$$

for any $u_z$ (armature movement), where $F_{MM}$ is the magnetomotive force, $g_0$ the initial gap between the armature and magnetic core, and $H_\mu$ and $H_g$ the magnetic fields inside the core and gap, respectively. Analogous to $\varepsilon_0$ in electrostatics, $\mu_0$ denotes the permeability of air or free space. Therefore, from Eq. (6.154) and the continuity of magnetic flux density across the interface between the core and the gap, we have

$$B = \mu_0 H_g = \mu H_\mu$$
$$\Rightarrow H_\mu = \frac{\mu_0}{\mu} H_g \qquad (6.157)$$

Now, Eq. (6.156) can be rewritten as

$$F_{MM} = \frac{\mu_0}{\mu} H_g(L_{m0} + u_z) + H_g(g_0 - u_z) \qquad (6.158)$$

This enables us to define the magnetic flux, $\varphi_M$, as

$$\varphi_M = B_g A = \mu_0 H_g A \qquad (6.159)$$

where $A$ is the cross-section area of the armature. In the microactuator in Figure 6.38, $A$ is the area of the beam face that acts like the armature. With the help of Eqs. (6.158) and (6.159), $\varphi_M$ can be obtained as

$$\varphi_M = \frac{\mu_0 A}{\frac{\mu_0}{\mu}(L_{m0} + u_z) + (g_0 - u_z)} F_{MM} \tag{6.160}$$

The ratio of the magnetomotive force to the magnetic flux is the *reluctance $R_M$*, given by

$$R_M = \frac{F_{MM}}{\varphi_M} = \frac{\frac{\mu_0}{\mu}(L_{m0} + u_z) + (g_0 - u_z)}{\mu_0 A} \tag{6.161}$$

It is interesting to note that when $\mu \gg \mu_0$, $R_M$ reduces to

$$R_M = \frac{(g_0 - u_z)}{\mu_0 A} \tag{6.162}$$

By comparing this expression with Eq. (6.60) for capacitance, it can be seen that reluctance is the magnetostatic counterpart of capacitance. This analogy can be continued further to define magnetostatic coenergy, $MSE_c$, as (see Table 6.2 for comparison)

$$MSE_c = \frac{1}{2} \frac{F_{MM}^2}{R_M} \tag{6.163}$$

whose derivative with respect to $u_z$ gives the negative of the force on the vertically moving armature. By balancing this force with the mechanical spring force, we get the equilibrium equation for the actuator:

$$-\frac{\partial(MSE_c)}{\partial u_z} + ku_z = 0 \tag{6.164}$$

Since the form of Eq. (6.164) is similar to that of the electrostatic equations, the behavior of this system is the same as before. Therefore, pull-in occurs here also.

Note also that $F_{MM} = nI$ due to Ampere's law. If we know the electrical circuit's resistance, we can compute the current $I$ to be put into Eq. (6.163) in terms of the given voltage, $V$. This completes the coupled modeling of an electro-magneto-elastic device.

**Your Turn:**
The preceding section presented only lumped modeling of electromagnetics. It is important to see the connection between continuum modeling and lumped modeling. Consult a standard book on electromagnetics [10] and convince yourself that you know which equations to solve when considering the 3D geometry of a real electro-magneto-elastic microactuator, and draw analogies to the coupled electromechanical-elastic problem.

## ▶ 6.8 SUMMARY

Just as Chapter 4 began with the basics of deformation of slender elastic bodies and ended with the equations for the general deformation of elastic solids, this chapter too started with the basics of electrostatics. Starting from Coulomb's law for the force between two charges, we derived the general equations for the electric field for a general continuum and thereby computed the electrostatic forces. We then explained the coupling between electrostatic and elastic field equations and the numerical methods for solving them. After developing the lumped analysis of electrostatics, we explained the inherent instability in the coupled electrostatic-elastic problem—the so-called pull-in phenomenon.

Coupled electrostatic-elasto-dynamics was also discussed. In this context, we showed how lumped inertia can be computed from continuum models. This was followed by a brief discussion of the relevant fluidic effects necessary to estimate lumped damping coefficients. Thus, a three-domain coupling problem emerged in electrostatically actuated micromechanical structures.

The three-domain coupled problem involving electrical, thermal, and elastic domains was also discussed in this chapter. Here also, we considered both continuum and lumped analysis in order to give the requisite understanding for solving any electrothermal actuator problem. We introduced Maizel's theorem, an energy theorem from thermoelasticity, to perform quick hand calculations for the electro thermal actuator problems.

The electro-magneto-elastic problem was also considered in the context of an electromagnet-based microactuator. The details of the continuum modeling of electromagnetics (magnetostatics, to be specific) were not considered but its analogy to electrostatics was explained by lumped modeling. It is worth noting that the pull-in phenomenon exists in this case too.

As seen in this chapter, coupling in micro and smart systems can be quite complicated. Here we discussed in detail only a few important and widely applicable coupling problems of microsystems. Several others exist, and coupling among the physical, chemical, and biological domains is not uncommon. All of these can be modeled both in general full models as well as lumped analysis. This, as is clear from this chapter, requires understanding of the basic physics (and chemistry if we consider chemical and biological phenomena) of the respective domains and their mathematical ramifications. We mention here that coupled modeling of microsystems gives us an opportunity to learn other subjects with an interdisciplinary focus. It also reinforces the understanding of our own specific disciplinary subject (mechanical, electrical, etc.) and what aspects of it are useful in other areas.

## ▶ REFERENCES

1. Tamm, I. E. (1976) *Fundamentals of the Theory of Electricity*, Mir Publishers, Moscow.
2. Alwan, A. and Ananthasuresh, G. K. (2006) Coupled electrostatic-elastic analysis for topology optimization using material interpolation, *Journal of Physics: Conference Series*, **34**, 264–70.
3. Conte, S. D. (1980) *Elementary Numerical Analysis: An Algorithmic Approach*, McGraw-Hill, New York.
4. Batchelor, G. K. (1993) *An Introduction to Fluid Dynamics*, Cambridge University Press, Cambridge, UK.
5. Pratap, R., Mohite, S., and Pandey, A. K. (2007) Squeeze film effects in MEMS devices, *Journal of the Indian Institute of Science*, **87**(1), 75–94.
6. Senturia, S. D. (2001) *Microsystems Design*, Kluwer Academic Publishers, Boston, USA.
7. Mankame, N. and Ananthasuresh, G. K. (2001) "Comprehensive thermal modeling and characterization of an electro-thermal-compliant microactuator," *Journal of Micromechanics and Microengineering*, **11**(5), pp. 452–462.

8. Incropera, F. P. and DeWitt, D. P. (1996) *Fundamentals of Heat and Mass Transfer*, John Wiley & Sons, New York.

9. Kovalenko, A. (1969) *Thermoelasticity—Basic Theory and Applications*, Walters-Noordhoff Publishing, Groningen, The Netherlands.

10. Inan, U. S. and Inan, A. S. (1998) *Engineering Electromagnetics*, Addison-Wesley, Reading, Massachusetts, USA.

## ▶ FURTHER READING

1. Greenwood, D. T. (1977) *Classical Dynamics*, Prentice Hall, Englewood Cliffs, NJ.

2. Meirovitch, L. (1986) *Elements of Vibration Analysis*, McGraw-Hill, New York.

3. Thompson, W. T. (1990) *Theory of Vibrations with Applications*, CBS Publishers, Delhi, India.

4. Guckel, H., Klein, J., Christenson, T., Skrobis, K., Laudon, M., and Lovell, E. G., "Thermo-magnetic metal flexure actuators," *Technical Digest of the Solid State Sensors and Actuators Workshop*, Hilton Head Island, SC, 1992, pp. 73–75.

5. Comtois, J. and Bright, V., "Surface micromachined polysilicon thermal actuator arrays and applications," *Technical Digest of the Solid-State Sensors and Actuators Workshop*, Hilton Head Island, SC, 1996, p. 174.

## ▶ EXERCISES

6.1 Imagine three parallel rectangular plates of area $A$ stacked one over the other and separated by an air gap of $g$. That is, there is a gap $g$ above the middle plate and the top plate, and there is a gap $g$ below the middle plate and the bottom plate. If the middle plate moves by a distance $x$, show that the change in capacitance is given by $(\varepsilon_0 Ax/g^2)$.

6.2 Figure 6.40 shows two polysilicon beams of length 200 μm, width 10 μm and thickness 2 μm separated by an in-plane gap of 4 μm. The beams overlap, as shown in the figure, over a length of 120 μm. Assume that the Young's modulus is 155 GPa. Use suitable lumped modeling and obtain the pull-in voltage for this pair of beams.

**Figure 6.40** Two in-plane beams with electrostatic actuation.

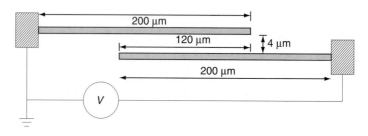

6.3 The schematic of an in-plane accelerometer with a crab-leg suspension is shown in Figure 6.41. Assume it is made of silicon whose Young's modulus is 169 GPa. The thickness everywhere is 25 μm.
   (a) Estimate the change in capacitance of this accelerometer per unit $g$, the acceleration due to gravity, using lumped modeling.
   (b) What are the free and damped-free natural frequencies of the accelerometer? Take into account damping on the proof mass and the combs and the effective mass of the suspension.

6.4 In order to perform a self-test on the accelerometer in Exercise 6.3, how much voltage needs to be applied to get a displacement of 0.5 μm?

**Figure 6.41** An accelerometer with a crab-leg suspension. Each square in the grid has size $5\,\mu m \times 5\,\mu m$.

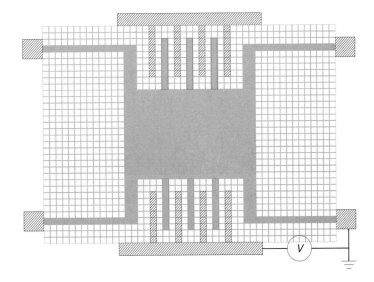

6.5 If there is a ground electrode underneath the proof mass with a gap of $2\,\mu m$, then what is the pull-in voltage for out-of-plane motion of the accelerometer in Figure 6.41?

6.6 What is the change in capacitance per unit $g$, the acceleration due to gravity, if the schematic in Figure 6.41 is used as an out-of-plane accelerometer?

6.7 In dynamic analysis of a spring-restrained parallel-plate capacitor, pull-in occurs for a slightly smaller voltage than the static pull-in voltage computed in Eq. 6.82. This is called the dynamic pull-in. Obtain a formula for the *dynamic pull-in voltage*.

6.8 Consider a cantilever beam with an electrode underneath as in Figure 6.6. For this, instead of the translating spring-restrained parallel-plate capacitor, a more appropriate model in electrostatics is a torsional spring-restrained rotating plate, as shown in Figure 6.42. Derive an equation that determines the pull-in voltage for this model in terms of the torsional spring constant $\kappa$, the length $l$ of the plate, and the width $w$ of the plate. (*Note:* Area of the plate $= lw$.)

**Figure 6.42** A torsional spring-restrained rotating-plate capacitor.

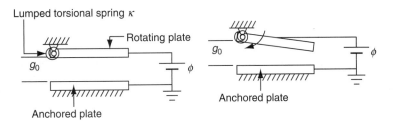

# CHAPTER 7

# Electronics Circuits and Control for Micro and Smart Systems

## LEARNING OBJECTIVES

After completing this chapter, you will be able to:

▶ Understand various aspects of electronics devices and circuits.

▶ Learn essentials of signal conditioning circuits.

▶ Get an overview of control theory and various implementations.

As discussed in earlier chapters, electronics forms the brain and nerves of microsystems. While in sensors electronics is required for signal conditioning and data acquisition, in actuators they are required for control and as driver circuits. Of course, much of data processing these days is performed by digital electronics circuits. Many of these building blocks are discussed in various contexts in the electrical and electronics engineering disciplines [1]–[3]. However, a reasonable background in some of these topics is essential for a complete understanding of microsystems. In this chapter we present aspects of electronics and control systems that are essential to understand the operation of micro-systems and smart systems.

Section 7.1 introduces the basic physics of electronic devices and delves into various circuits that are essential for microsystems applications. Electronics amplifiers form the basic electronics building blocks in most sensing systems. In Section 7.2, the basic ideas of these amplifiers are discussed, followed by a brief description of widely used IC-based operational amplifiers (op-amp). The basic working principles of several useful op-amp circuits are also shown. Further, more complex, yet practical, electronics circuits in typical sensing applications are introduced in Section 7.3. Capacitance- and frequency-sensing circuits are discussed in Section 7.4.

While electronics amplifiers and other circuit aspects are relevant in the context of sensors, design and implementation of control systems are important in the deployment of smart systems involving actuators. Basic control theory is overviewed in Section 7.5 and some simple practical circuits for application in smart systems are explained in Section 7.6.

## ▶ 7.1 SEMICONDUCTOR DEVICES

### 7.1.1 The Semiconductor Diode

#### 7.1.1.1 A p-n Junction Diode

A semiconductor diode is the most fundamental semiconductor device and is fabricated by forming a p-type region in an n-type semiconductor. Figure 7.1 shows the schematic structure and other details of such a diode. Ideally, this device may be thought of as one in which a p-type layer is in close contact with an n-type layer. This may be considered an ideal approach that results in formation of a p-n junction. A qualitative theory of p-n junctions relevant to a proper understanding of all semiconductor devices is presented in the following paragraphs.

Assuming that the p- and n-layers have doping concentrations of $N_A$ and $N_D$, respectively, on formation of the junction, mobile electrons and holes are depleted from the immediate vicinity of the junction. This leaves a positively charged (charge density $= q_e N_D$) region near the n-side of the junction and a negatively charged (charge density $= -q_e N_A$) region near the p-side, where $q_e$ is the *charge of an electron*. This causes an electrostatic field across the junction and the potential changes continuously through the junction region. This built-in potential $V_{bi}$ across the junction is high enough to make the net mobile charge flow across the junction zero under thermal equilibrium conditions (i.e. when the external applied voltage across the junction is zero). The polarity of $V_{bi}$ is positive on the n-side and negative on the p-side of the junction. When an external voltage $V$ having a polarity opposite to $V_{bi}$ (i.e. p-side positive and the n-side negative with $V_{bi} \approx 0.8$–$0.9$ V in a silicon p-n junction diode) is applied, the barrier for positive mobile carriers from p- to n-region and the barrier for electron transfer from n- to p-region are lowered below $V_{bi}$ by a value $V$. This results in a net current flow from p- to n-region. This bias condition is referred to as the *forward-bias* condition and the diode is said to be *forward biased by a voltage V*. The current $I_D$ is related to $V$ by:

$$I_D = I_0(e^{q_e V/k_B T} - 1) \tag{7.1}$$

The quantity $k_B T/q_e$ is the thermal voltage, equal to about 25 mV at room temperature. $I_0$ is called the *reverse saturation current* because when $V$ is made negative, the minimum value of $I_D$ is limited to $-I_0$, showing that there is a net current from n-region to p-region.

**Your Turn:**

Explain how a forward-biased diode can be used as a temperature sensor.

For moderate reverse voltages, the current through the diode is small and hence this is called the *reverse blocking region*. On the other hand, for sufficiently large negative voltages (above a reverse breakdown voltage $V_{Br}$) the reverse blocking behavior breaks

(a)         (b)

**Figure 7.1** A p-n junction diode: (a) simplified geometry; (b) standard circuit symbol.

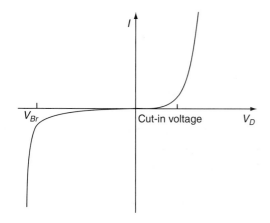

**Figure 7.2** *I–V* characteristics of a typical p-n junction diode showing forward- and reverse-biased conditions. The forward-biased current becomes significant above the cut-in voltage. For silicon, this value is 0.6 V.

down and the diode conducts in the opposite direction. There are several possible mechanisms for this reverse breakdown. One of the dominant mechanisms is *avalanche breakdown*, caused by the release of a large number of electrons and holes generated by the collision of the mobile charges with the lattice atoms in the space charge layer. The forward- and reverse-biased *I–V* characteristics of a typical p-n junction diode are shown in Figure 7.2. One can easily design rectifier circuits on the basis of these p-n junction diode characteristics.

Half-wave rectification can be achieved with a single diode connected in series with the load, as shown in Figure 7.3(a). The input to this circuit is a sine wave voltage [Figure 7.3(c)]. The diode allows the signal to pass through it only during the positive half of the input voltage cycle; during the other half cycle, the diode does not allow the signal to pass through it. As a result, the voltage across the load is as shown in Figure 7.3(d), and hence the name *half-wave rectifier*. It can be shown that the average dc value of this output is given by $V_{m1}/\pi$, where $V_{m1}$ is the peak value of the voltage, which is 0.6 V less than the peak value $V_{ms}$ of the signal voltage.

### Problem 7.1
Design a half-wave rectifier circuit and analyze its operation. Show that the average DC value is $V_m/\pi$, where $V_m$ is the peak value of the voltage.

Full-wave rectification can be achieved with the Wheatstone bridge-type connection shown in Figure 7.3(b). In this case, diodes D1 and D2 become forward biased and conduct in the positive half of the cycle (terminal A positive and terminal B negative). The current flow path is shown by the solid arrows in the figure. During the negative half of the cycle (terminal A negative and terminal B positive), diodes D3 and D4 become forward biased and the other two diodes become reverse biased. The current flow path is shown by the dotted arrows. It may be noted that for both input-signal polarities, the direction of current flow through the load is the same and the voltage developed has the polarity marked with the wave shape of the voltage [Figure 7.3(e)]; hence, the name *full-wave rectifier*. In this case, it can be shown that the average value of this dc voltage is $2V_{m2}/\pi$, where $V_{m2}$ is the peak value of the voltage, which is approximately 1.2 V less than the peak value $V_{ms}$ of the signal voltage.

### Problem 7.2
Design a full-wave rectifier circuit and analyze its operation. Show that the average DC value is double that for half wave rectification.

#### 7.1.1.2 Schottky Diode
The Schottky diode is a semiconductor diode with a low forward voltage drop and a very fast switching action. This is a metal-semiconductor (MS) junction, and, hence, is operated with only majority carriers of the semiconductor region. As this junction has only one semiconductor type (e.g. n-type) with the other being a metal, the space charge layer is present only on the semiconductor region. Hence, the built-in voltage $V_{bi}$ in this type of MS junction is lower than that of a p-n junction. Typically in silicon MS contact, $V_{bi}$ is approximately 0.6 V and allows current flow with lower forward-bias voltage.

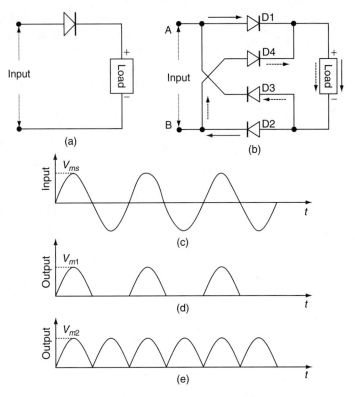

**Figure 7.3** (a) Half-wave and (b) full-wave rectifier bridge circuits; (c) input voltage waveform to both circuits; (d) and (e) output voltage (voltage across the load) waveforms, respectively, in circuits of parts (a) and (b).

The operation of this device is determined by the *work function* of the metal electrode. The work function is a characteristic property for any interface of a material with a conduction band; it is the energy needed to move an electron from the Fermi level into a vacuum. The Fermi level of a metal is inside the conduction band. In metals such as Pt and Pd, significant changes in MS work function are observed when the metal electrode is exposed to certain gases. Hence, microsensors for sensing gases such as $H_2$ and $NO_x$ can be built using these metal electrodes by forming a Schottky junction with semiconductors (e.g. Si, GaAs, GaN) [4].

**Your Turn:**

Learn the operating principle of a gas sensor using a Schottky diode and list typical gases that can be sensed using such a device. Find out the electrode material used in each case.

### 7.1.1.3 Tunnel Diode

This is a special type of p-n junction in which both p-region and n-region are doped very heavily. As a result, the space charge layer (barrier) width around the junction is very narrow, of the order of a few (1–5) nanometers. Quantum-mechanical analysis shows that

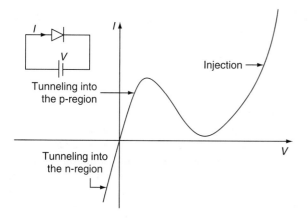

**Figure 7.4** Typical *I–V* characteristics of a tunnel diode [3].

when this barrier width is very small, electrons and holes can cross the barrier by a mechanism known as *tunneling*, without having to go over the barrier. In the conventional p-n junction diodes where the space charge layer is relatively wide (in the range 100 nm—several μm), electron transport takes place over the barrier and hence the current flow is negligible up to about 0.5 V. The tunneling mechanism in tunnel diodes allows the current to flow well below 0.6 V. Typical *I–V* characteristics of tunnel diode are shown in Figure 7.4. The region of *I–V* characteristics marked "injection" corresponds to the current due to injection over the barrier, as in the usual mode of p-n junction diode operation. The figure also shows the current due to tunneling of carriers from p-region to n-region and vice versa. The negative resistance in the *I–V* characteristics occurs where the current transport mechanism changes from the tunneling mechanism to normal injection over the barrier of a conventional p-n junction. This negative resistance region can be used in realizing oscillators. Tunnel diodes find applications in high-frequency operations.

### 7.1.2 The Bipolar Junction Transistor

The bipolar junction transistor (BJT; Figure 7.5) is a three-terminal semiconductor device *(observe the two junctions)* that is widely used for amplifying or switching applications in microelectronics. BJTs are so named because their operation involves both electrons and holes. The BJTs can be of two types, npn and pnp. The schematic structures of both types and their circuit symbols are shown in Figure 7.5. An n-type substrate is used as collector in an npn transistor. Various layers are diffused into this n-type collector region to create the p-base, the n-emitter, and $n^{++}$ collector contact of the npn BJT. The "+" signs in the figure indicate the doping levels. Typically, the doping concentration of the base region is greater than the n-collector and less than the $n^+$ emitter regions. A heavily doped $n^{++}$ region is provided close to the collector electrode

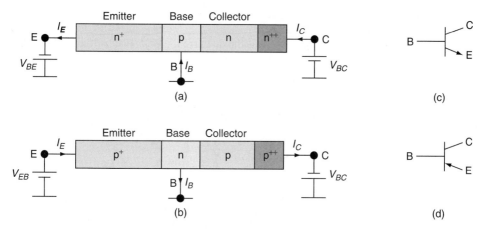

**Figure 7.5** Basic structure of (a) npn and (b) pnp bipolar junction transistor. Parts (c) and (d) show the circuit symbols for parts (a) and (b), respectively.

to interface the metal layer to the semiconductor [see Figure 7.5(a)] and to reduce the collector region series resistance. A pnp transistor has opposite doping in each of these regions [see Figure 7.5(b)].

The basic principle of operation and structure of npn and pnp transistors is the same, the difference being only in the type of charge carriers involved in the transistor action. Both may be thought of as devices in which two p-n junctions are arranged back to back with a common middle region (base region) whose width is very small, so that carriers injected from the forward-biased emitter region into the base region flow through it, and almost all the injected carriers are collected by the reverse-biased collector base junction to give rise to the collector current. Figure 7.5 shows the situation when the emitter-base junction is forward biased and the collector-base junction is reverse biased. It may be noted that the power supply polarities and directions of the currents $I_E$, $I_B$ and $I_C$ in the emitter, base, and collector terminals are exactly opposite in the pnp case and npn case.

In a pnp transistor biased as shown in Figure 7.5(b), the emitter current $I_E$ consists mainly of holes injected from the emitter to the base, and a small fraction of them is lost by recombination with electrons in the base, giving rise to the base current $I_B$. The remaining holes are collected by the collector, giving rise to collector current $\alpha I_E$ ($\alpha < 1$).

The total collector current is the sum of the reverse saturation current $I_{C0}$ due to the reverse bias $V_{BC}$ and the current $\alpha I_E$ due to the collection of the fraction of holes injected from emitter to base. Thus,

$$I_C = \alpha I_E + I_{C0} \tag{7.2}$$

where $I_{C0}$ is the reverse saturation current $I_0$ of the reverse-biased collector-base junction. In general, $I_{C0}$ is negligible compared to the hole current collected. The same concepts hold true for npn transistors, except that, instead of the hole current, $I_E$ and $I_C$ consist mainly of electron flow. The current flow directions in this case are opposite to those in pnp transistors because the actual direction of current flow is opposite to that of electron flow, whereas in the pnp transistor, the direction of current flow is the same as that of the flow of holes (which are positive mobile charges). In both npn and pnp transistors, by applying Kirchhoff's law we can write

$$I_E = I_C + I_B \tag{7.3}$$

Defining the ratio $\beta = I_C/I_B$ as the dc current gain and substituting this in Eq. (7.3), we can easily show the following:

$$I_C = \beta I_B + (\beta + 1)I_{C0} \tag{7.4}$$

and that

$$\beta = \frac{\alpha}{1 - \alpha} \tag{7.5}$$

Since $\alpha$ is very close to unity, the dc current gain $\beta$ is large. For instance, if $\alpha = 0.99$, then $\beta = 99$. We can see that in most BJTs a small base current causes a large collector current and that the ratio of the collector current to the base current, known as the current gain ($\beta$ or $h_{FE}$), is usually very large.

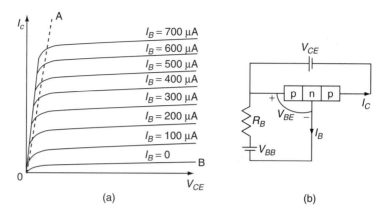

**Figure 7.6** (a) Output characteristic of BJT connected as shown in (b).

Typical output characteristics, $I_C$ versus $V_{CE}$, of a BJT are shown in Figure 7.6(a) for input values of base current varying from $I_B = 0$ to $I_B = 700\,\mu A$ in steps of $100\,\mu A$. These characteristics can be divided into three different regions as follows:

1. In the region between OA and the *y*-axis, the voltage drop across the transistor is very low. However, the current flow through the transistor will be high depending upon the base drive current $I_B$. It can be easily seen that when $V_{CE}$ is small and less than the voltage $V_{BE}$ across the emitter-base junction, the collector-base junction becomes forward biased. Thus, in this region of operation both emitter-base and collector-base junctions are forward biased. This region is called the *saturation region* because in this region the collector current does not increase linearly with the base current and Eq. (7.4) does not hold good. In digital circuits this mode corresponds to a logical ON or a closed switch.

2. The characteristic OB corresponds to $I_B = 0$. Below this region, the collector current $I_C$ is very small and both junctions are reverse biased. This region is called the *cut-off region* and the transistor is in the OFF state. In digital circuits this mode corresponds to a logical OFF or an open switch.

3. In the region between OA and OB the transistor is operating in the *forward active* region, with the emitter-base junction forward biased and the collector-base junction reverse biased. The collector current is linearly related to the base current through Eq. (7.4). Transistors operating in this region can be used as amplifiers.

When used as a switching device, the transistor should, ideally, appear as a short circuit (between the output terminals) when turned ON and an open circuit when turned OFF. The transistor is turned ON by applying a sufficiently large base current so that the transistor operates in the saturation region. This causes a large value of $I_C$ and a small value of $V_{CE}$. When the base current is made negative, the device is driven toward cut-off, causing the output current $I_C$ to be small. The BJT is in the cut-off region when both emitter-base and collector-base junctions are in the reverse-biased condition. The switching time in a real device, however, may be limited by carrier lifetime and the base currents applied.

When used in an amplifier, the voltages required at various terminals are arrived at using suitable biasing circuits. Furthermore, an amplifier has an input and an output terminal. Therefore, one of the terminals of the transistor is kept in common. Consequently, various configurations are possible for a transistor amplifier. Among the three possibilities for this three-terminal device, the common-emitter configuration is used often because the switch in this configuration can be driven with a very small input (base) current to achieve high collector current.

**Problem 7.3**

Derive Eqs. (7.4) and (7.5). Find the current gain for a transistor with $\alpha = 0.998$.

**Problem 7.4**

The emitter current of a pnp BJT is 0.1 mA, $\alpha = 0.98$, and the reverse saturation current $I_{C0} = 10\,\mu A$. Determine the collector current $I_C$ and the base current $I_B$ and mark their directions.

## 7.1.3 MOSFET

The metal-oxide-semiconductor field-effect transistor (MOSFET) is different from BJTs in both operation and structure. This three-terminal device consists of source, drain and gate regions. In addition, the substrate (body) can also be biased separately when required. Usually this is connected with the source. Generally, MOSFETs are classified into two categories, n-channel MOSFETs (nMOS) and p-channel MOSFETs (pMOS). Each of them can be of enhancement or depletion type. Both types of MOSFETs are useful in signal conditioning circuits for sensors discussed in Section 7.4.

Figure 7.7(a) shows a typical schematic structure of an nMOS that operates in enhancement mode. The gate electrode is made up of heavily doped polycrystalline silicon over a thin gate oxide grown on a p-type silicon substrate. In current very large-scale integration (VLSI) circuits, the gate oxide thickness is in the range 2–20 nm depending on the technology used. The $n^+$ regions formed in the substrate p-region serve as the source and the drain for electrons. The field oxide of thickness approximately 1 μm serves as a passivating layer for the $n^+ p$ source-drain junctions, and also serves as an insulating layer over which interconnecting metal can be laid out in the integration circuits.

When the gate voltage is zero, the source and drain regions are isolated from each other and act as two back-to-back connected diodes. Hence the drain current is only the reverse leakage current of the drain region. When the drain voltage $V_{DS}$ is applied, a drain current $I_D$ can flow if the path (called the *channel*) between the source and drain is formed. This is achieved by two approaches: (a) by inducing a field-induced charge layer of electrons or (b) by physically introducing an n-layer between the source and the drain.

The first approach is achieved by applying a positive voltage $V_{GS}$ to the gate with respect to the substrate and the source as shown. The minimum voltage required for achieving this channel chargesheet layer (also called the *inversion layer*) of electrons is defined as the threshold voltage $V_{th}$ of the enhancement MOSFET. The drain current can be increased by increasing the channel charge using higher gate voltages. This transistor is designated as an nMOS operating in the enhancement mode.

In the second approach, an n-type layer is present between the source and the drain regions below the gate oxide; thus, $I_D$ flows even in the absence of $V_{GS}$. A negative voltage $(-V_{th})$ must be applied to the gate with respect to the source/substrate to deplete this region and remove the electrons so that the electrons are not available for drain current flow. This MOSFET is designated as a depletion-mode nMOS. In current VLSIs, depletion-mode MOSFETs are not used because of excess power dissipation caused when the device is in the ON state but the gate voltage is absent.

The device structure of p-channel enhancement MOSFETs is similar to that of the n-channel enhancement MOSFETs in Figure 7.7(a). Here the substrate is n-type and the source-drain regions are $p^+$ regions. The gate region is a heavily doped p-type polysilicon.

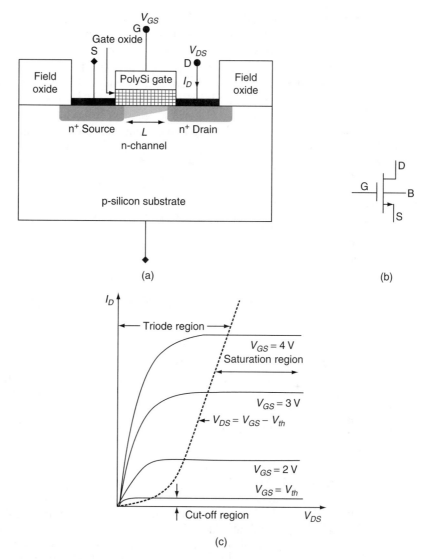

**Figure 7.7** (a) Schematic of an n-channel enhancement MOSFET; (b) symbol; and (c) output characteristics.

The drain voltage $V_{DS}$ is negative with respect to the source and the threshold gate voltage $V_{th}$ is negative, so that positively charged mobile holes are induced by this negative gate voltage. The drain current is achieved by the flow of these field-induced positive charges (inversion layer of holes) from source to drain, causing drain current flow from the source toward the drain. Hence, the symbol for enhancement pMOS is same as that of nMOS [Figure 7.7(b)] except that the direction of the current flow arrow (near S) in the symbol is reversed.

The typical output characteristics of n-channel enhancement MOSFETs are shown in Figure 7.7(c). The output characteristics of p-channel enhancement MOSFETs are also similar to those of nMOSs, but with $I_D$ and $V_{DS}$ negative and the running parameter $V_{GS}$ varying from 0 to larger negative values.

To summarize, an enhancement-mode MOSFET has three different modes, depending on the voltages at the terminals. The operation of an nMOS in these three modes is discussed on the following page.

### 7.1.3.1 Cut-Off Mode

Cut-off mode, also called subthreshold or weak inversion mode, occurs when $V_{GS} < V_{th}$. The threshold voltage for this mode is usually defined as the gate voltage at which a selected value of current $I_{D0}$ occurs (e.g. $I_{D0} = 1\,\mu A$). In this mode, the transistor is expected to be turned OFF and there would be no conduction between drain and source. However, in a practical device, the Boltzmann distribution of electron energies allows some of the more energetic electrons at the source to enter the channel and flow to the drain, resulting in a current that is an exponential function of the gate-source voltage:

$$I_D \approx I_{D0} e^{[(V_{GS} - V_{th})/nV_T]} \tag{7.6}$$

where $V_{th}$ is the threshold voltage and $n$ the slope factor given by

$$n = 1 + \frac{C_D}{C_{OX}} \tag{7.7}$$

Here $C_D$ is the capacitance per unit area of the depletion layer and $C_{OX}$ the capacitance per unit area of the oxide layer. The depletion layer capacitance is across the silicon-SiO$_2$ interface.

Since current levels are low in subthreshold operation of MOSFETs, very low-power analog circuits are designed to take advantage of this subthreshold conduction. But this exponential subthreshold $I$–$V$ relation depends on the threshold voltage of the device, and hence is strongly affected by any manufacturing variations in oxide thickness, junction depth, and body doping. In a long-channel device, the subthreshold current is practically independent of the magnitude of the drain voltage $V_{DS}$. For shorter channel lengths [region $L$ in Figure 7.7(a)], the current depends on device various parameters (channel doping, junction doping, etc.).

### 7.1.3.2 Triode Mode

The triode mode, also called linear region or ohmic mode, occurs when $V_{GS} > V_{th}$ and $V_{DS} < (V_{GS} - V_{th})$. Under these conditions the transistor is turned ON and a channel is created to allow current to flow between the drain and the source. The transition from cut-off region to linear region is abrupt. The MOSFET in this mode operates like a resistor controlled by the gate voltage relative to both source voltage and drain voltage $V_{DS}$. The current from drain to source is

$$I_D = \mu_n C_{OX} \frac{W}{L} \left[ (V_{GS} - V_{th}) V_{DS} - \frac{V_{DS}^2}{2} \right] \tag{7.8}$$

where $\mu_n$ is the effective mobility of charge carriers, $W$ the gate width, $L$ the channel length, and $C_{OX}$ the gate oxide capacitance per unit area.

### 7.1.3.3 Saturation Mode (Active Mode)

Saturation mode occurs when $V_{GS} > V_{th}$ and $V_{DS} > (V_{GS} - V_{th})$. The switch is turned ON, and a channel would be formed to allow current to flow between the drain and the source. Since the drain voltage is higher than the gate voltage, the electrons spread out and conduction is through a broader cross-section extending away from the interface and deeper in the substrate. The onset of saturation mode is also known as *pinch-off*, indicating the lack of a channel region near the drain. The drain current, which is weakly dependent upon the drain voltage and controlled primarily by the gate–source voltage, is given by the relation

$$I_D = \mu_n C_{OX} \frac{W}{2L} (V_{GS} - V_{th})^2 (1 + \lambda V_{DS}) \tag{7.9}$$

This is obtained by substituting $V_{DS} = (V_{GS} - V_{th})$ in Eq. (7.8), with an additional factor $(1 + \lambda V_{DS})$ involving the channel-length modulation parameter $\lambda$ that incorporates the dependence of the current on the drain voltage. This effect is also referred to as *channel-length modulation.*

In the preceding discussions a simplified algebraic model that works well for conventional technology has been used. Based on this model, the design parameters of a MOSFET are transconductance $g_m$ and output resistance $r_0$ given by

$$g_m = \frac{dI_D}{dV_{GS}} = \frac{2I_D}{V_{GS} - V_{th}} \tag{7.10}$$

$$r_0 = \frac{dV_{DS}}{dI_{DS}} = \frac{1 + \lambda V_{DS}}{\lambda I_D} = \frac{(1/\lambda) + V_{DS}}{I_D} \tag{7.11}$$

As pointed out already, these equations become inaccurate for short channel lengths.

MOSFET is commonly used to amplify or switch electronics signals, and is by far the most common field-effect transistor in both digital and analog circuits. We now describe a modification of MOSFETs to reduce their power consumption, which is crucial for reducing the overall size of digital circuits.

**Your Turn:**

A MOSFET can have an air gap instead of an oxide layer between the metal gate and the silicon substrate [5]. This can be used as a capacitive pressure sensor. From your knowledge of the MOSFET, explain the principle of operation of this pressure sensor.

## 7.1.4 CMOS Circuits

CMOS is a major class of ICs used in microprocessors, microcontrollers, and several other digital logic circuits. CMOS technology is also used in a variety of analog circuits including image sensors, data converters, and integrated transceivers, some of which may be integrated with microsystems. CMOS uses complementary and symmetrical pairs of pMOS and nMOS. Because of this reason, CMOS is also called complementary-symmetry metal-oxide semiconductor.

Important characteristics of CMOS devices are high noise immunity and low static power consumption. Power is drawn only when the transistors in the CMOS device are switching between ON and OFF states; in the static condition (in both ON and OFF states), power drawn is negligible. Because of these advantages, CMOS also facilitates integration of multiple logic functions on a chip. Accordingly, CMOS processes and their variants dominate the industry, and a majority of modern IC manufacturing is done by CMOS processes. Furthermore, since the basic processes in CMOS and microsystems fabrication are similar, the emphasis in microfabrication has been toward approaches that are totally compatible with CMOS technologies in order to facilitate their integration.

CMOS technology uses a combination of pMOS and nMOS to implement logic gates and other digital circuits. Typical commercial CMOS circuits have millions of transistors

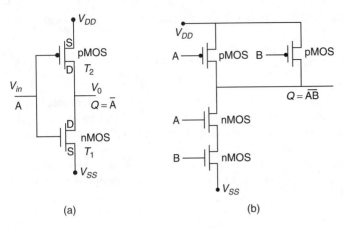

(a)                                                          (b)

**Figure 7.8** Examples of CMOS logic circuits: (a) inverter;
(b) NAND gate. Here pMOS and nMOS refer to p-channel
and n-channel MOSFETs.

of both types on a silicon chip. In CMOS logic, a set of nMOSs (pull-down network) are connected between the output and the ground. Another set of pMOSs (pull-up network) connects the output and the supply high ($V_{DD}$). As individual transistors along these networks conduct, the output node connects to one of the supply levels. Individual inputs are connected to both these networks complementarily, so that when an n-type transistor on the pull-down path is OFF, the p-type on the pull-up path is ON, and vice versa. These operations are clear from the examples of CMOS circuits in Figure 7.8. The transistor with a bubble at the gate represents p-channel devices.

In Figure 7.8(a), if the input $V_{in}$ of the inverter is high, the nMOS $T_1$ is ON and the pMOS $T_2$ is OFF because the gate to source voltage of $T_2$ is zero. Hence, the output voltage $V_o$ is low. When $V_{in}$ is low, $T_1$ is OFF and the $V_{GS}$ of $T_2$ negative. Therefore, $T_2$ is ON and hence $V_o$ is at $V_{DD}$ (high). In both static conditions, since either $T_1$ or $T_2$ is OFF, the current drawn from $V_{DD}$ is small and is equal to the leakage current of one of them. As a result, the static power loss is negligibly small in CMOS inverters and other CMOS logic circuits.

Referring to the NAND gate shown in Figure 7.8(b), if both A and B inputs are high, both the nMOSs will conduct, neither of the pMOSs will conduct, and a conductive path is established between the output and $V_{SS}$, bringing the output to low. If either A or B inputs is low, one of the nMOS transistors will not conduct, one of the pMOS transistors will be ON, and a conductive path is established between the output and $V_{DD}$, bringing the output to high. A tabular summary of output states of a digital device with respect to various possibilities at its input gates is often called a *truth table*.

**Problem 7.5**
Make the truth table for the NAND gate of Figure 7.8(b) and identify which transistors would be ON and which OFF in each case.

CMOS circuits are basic building blocks in digital electronics because this technology offers high input resistance, which translates to simple, low-power logic gates. These circuits typically have low current consumption. However, the new technology of bipolar CMOS (BiCMOS) integrates BJTs and CMOS technology into a single device.

Bipolar transistors offer high speed, high gain, and low output resistance. Furthermore, bipolar devices are preferred for realizing accurate reference voltages and when very low noise is required. BiCMOS technology is handy in developing power FETs for switching or regulating high currents. The reason is that digital control circuits for the purpose can be designed with low complexity and integrated with the power transistors. This technology has also found applications in amplifiers and analog power management circuits. BiCMOS circuits use the characteristics of each type of transistor most appropriately. Examples of BiCMOS circuits include RF oscillators, bandgap-based references, and low-noise circuits.

## ▶ 7.2 ELECTRONICS AMPLIFIERS

An amplifier is a circuit that increases the amplitude of voltage, current, or both. In general, it is a multiport circuit with ports to connect the energy sources, one or more inputs, and one or more outputs. Amplifiers have several applications: some common examples are audio amplifiers in a home stereo or a public address system, and RF and microwave applications such as radio transmitters and receivers. Conventional circuit analysis of amplifiers focuses only on the signal path between input port(s) and output port(s). This is quite different from the approach followed for other devices in which total energy is tracked. The steady-state relationship of the input to the output of an amplifier expressed as a function of the input frequency is called its *transfer function*:

$$G(f) = \frac{P_{out}(f)}{P_{in}(f)} \tag{7.12}$$

The magnitude of the transfer function is called the *amplifier gain.*

Amplification occurs due to the nonlinear resistive behavior of the transistor. In the simplest case, we can consider an amplifier as a three-port device, one port for a dc power supply, another for an input signal, and the third for the output. The inclusion of the power supply causes this circuit to have a nonlinear transfer characteristic with an incremental gain. The nonlinear behavior of the transistor ensures that the output can supply more power than is available from the input signal, the additional power being drawn from the dc power supply.

It is common practice to model the transistor with dependent sources, which do not conserve energy, and set either the voltage or the current depending on some other variable in the network. These models are based on the behavior of active devices such as BJTs and MOSFETs in these amplifiers.

Transistor-based amplifiers can be realized in various configurations. For example, with a BJT we can realize common-base, common-collector, or common-emitter amplifiers. Using a MOSFET we can realize common-gate, common-source, or common-drain amplifiers. Each configuration has different characteristic features such as gain, input impedance, output impedance, etc. As one can imagine, amplifier circuits can be built by assembling these devices with proper biases and then appropriately connecting the input and output ports. However, this approach is rarely pursued, except for some specialized requirements. In other situations one uses ICs such as operational amplifiers (popularly known as *op-amps*).

**Problem 7.6**

Design the common-emitter amplifier shown in the circuit below using a transistor having $\alpha = 0.99$. Assuming a supply voltage of 5 V, choose appropriate values for resistors so that the collector current is 1mA. Determine

(a) The dc operating point.
(b) The ac gain.
(c) The maximum input signal level possible without being clipped at the output.

### 7.2.1 Operational Amplifiers

The op-amp is a direct-coupled high-gain amplifier to which an external feedback is added to control its overall response characteristics. It can therefore be used to perform a variety of linear (and some nonlinear) functions and is often referred to as the *basic linear IC*.

This voltage-amplifying circuit has two (differential) input ports and an output port. A high input impedance and low output impedance are important characteristics of a typical circuit. Owing to these advantages, the integrated op-amp is considered a versatile, predictable, and economical system building block. They have all the advantages typical of monolithic ICs: small size, high reliability, low cost, good temperature tracking, and low offset voltage and current.

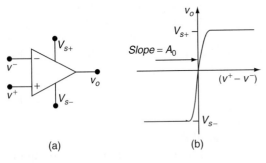

(a)                    (b)

**Figure 7.9** (a) The standard symbol for an operational amplifier; (b) the input-output relation of an ideal op-amp.

The standard op-amp symbol is shown in Figure 7.9(a). Op-amps have two power supply ports [one positive $(V_{s+})$ and the other negative $(V_{s-})$ with respect to system ground], in addition to two input ports and one output port. However, the power supply connections are often omitted in circuit diagrams. Internally, an op-amp IC contains a differential amplifier as the first stage that produces the difference of these two inputs and amplifies this signal; a high-gain stage that provides significant voltage gain; and finally, an output stage that can provide reasonable current drive to an external load without loading the high-gain stage. The transfer function of op-amps usually shows a large gain at low frequencies.

One of the input terminals is called the *noninverting input* (denoted "+") and the other the *inverting input* (denoted "−"). It should be reiterated that these symbols (+ and −) are not associated with the actual polarity of the applied voltages. The voltages at these inputs with respect to a system ground are indicated by $v^+$ and $v^-$, respectively. A typical transfer characteristic of an ideal op-amp is plotted in Figure 7.9(b), where the output voltage ($v_o$ is plotted against the difference of the inputs ($v^+ - v^-$), and the voltage gain has a slope of $A_0$ around the origin. The output saturates at a voltage that is 1 or 2 V below the power supply voltages. $A_0$ is called the *dc open-loop gain*. As suggested by this graph, the amplifier produces no output if the input terminals are at the same voltage (i.e. $v^+ = v^-$). In other words, this amplifier has a difference-mode gain of $A_0$ but its common-mode gain is zero.

It may, however, be noted that real amplifiers may show an output even when both inputs are equal. Thus, the common-mode gain ($A_C$) of a real op-amp is not zero, even though it is typically much less than the difference-mode gain ($A_0$). The common-mode rejection ratio (CMRR) is defined as the ratio between $A_0$ and $A_C$. In real op-amps, the CMRR is a large but finite number (>1,000, often close to 100,000). Obviously, the CMRR of an ideal op-amp is infinite.

We can define the linear region of op-amp operation as that part of the transfer characteristic well away from the saturation levels. In this region, the linearized transfer characteristics can be obtained as

$$v_o = A_0(v^+ - v^-) + A_C \frac{v^+ + v^-}{2} \tag{7.13}$$

In this expression, the factor of 2 in the common-mode term is based on the conventional definition of the common-mode signal as the average of the two inputs. In general, the output of a typical op-amp is also dependent on the actual power supply voltages. Mismatch between transistors inside the IC may cause the transfer characteristics of real op-amps not to pass through the origin. Usually this effect is taken into account by including an equivalent offset voltage ($v_{off}$) along with the differential input ($v^+ - v^-$). The corrected linear transfer characteristics is

$$v_o = A_0(v^+ - v^- - v_{off}) + A_C \frac{v^+ + v^-}{2} \tag{7.14}$$

**Table 7.1 Ideal values and typically reported values of various parameters of op-amps**

| Parameter | Ideal value | Typical value |
|---|---|---|
| Open-loop gain | Infinite | $2 \times 10^5$ |
| Bandwidth | Infinite | 1 MHz |
| Input impedances | Infinite | 2 MΩ |
| Offset voltage | Zero | 5 mV |
| Slew rate | Infinite | 0.5 V/µs |
| CMRR | Infinite | 90 dB |
| Noise | Zero | $30\,\text{nV}/\sqrt{\text{Hz}}$ |

*Note:* Modified after [6].

The characteristics of an ideal op-amp differ from those of a practical IC op-amp, as seen from Table 7.1. As one can notice, the parameters for a typical op-amp approach those of an ideal device. In addition, real op-amp parameters may drift over time and with changes in temperature or input conditions. However, integrated FET or MOSFET op-amps are closer to these ideal conditions than are bipolar ICs. The latter have higher input impedance, although they usually have lower input offset drift and noise characteristics. As a result, practical op-amp ICs are moderately complex ICs.

Op-amps built using bipolar transistors require nonzero input currents for their operation. However, in MOS-based op-amps, the input currents are extremely small because of the capacitive nature of the input impedance. These small input currents charge and discharge the input capacitances of the MOS transistors. In applications where these effects are significant, the input capacitances of the first stage of amplification are shown outside of the op-amp symbol. In all other cases the op-amp can be assumed to have zero input current. As a common practice, whenever limitations of real devices can be ignored, an op-amp is replaced in the circuit model with a black box having a gain. Operation of the overall circuit and its performance parameters are analyzed based on the nature of the feedback circuit, an approach demonstrated in the next section.

## 7.2.2 Basic Op-Amp Circuits

Common op-amp circuits are analyzed here in order to demonstrate their versatility. The inverting amplifier in Figure 7.10 has an op-amp with a resistor connected as a feedback between its output and inverting input. The feedback resistor has the role of stabilizing the circuit in the linear region. As the output increases, $v^-$ increases, which in turn lowers the output. For example, if $A_0 = 10^5$ and the supply voltage is 10 V, then the value of $|v^+ - v^-|$ that ensures linear amplification is less than 100 µV. Since this value is much smaller than

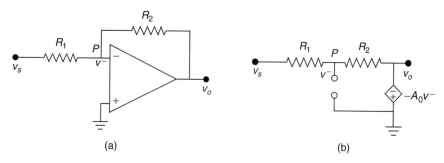

**Figure 7.10** (a) The inverting amplifier; (b) its equivalent circuit.

the supply and the typical output voltages, the circuit in Figure 7.10(a) can be made to operate safely in the linear region.

In the linearized equivalent circuit of this amplifier shown in Figure 7.10(b), the op-amp is replaced by a linear voltage source with gain $A_0$. The input is represented by an open circuit. This circuit can be analyzed by noting that $v_o = -A_0 v^-$ and writing Kirchhoff's current law at node P in the figure:

$$\frac{v_s - v^-}{R_1} = \frac{v^- - v_o}{R_2} = \frac{v^- + A_0 v^-}{R_2} \tag{7.15}$$

Equation (7.15) may be rearranged to yield $v^-$:

$$\frac{v^-}{v_s} = \frac{R_2}{R_2 + R_1 + A_0 R_1} = \frac{R_2}{A_0 R_1} \left[ \frac{1}{1 + (1/A_0)[1 + (R_2/R_1)]} \right] \tag{7.16}$$

Substituting $v_o = -A_0 v^-$ in Eq. (7.16) and rearranging, we obtain

$$\frac{v_o}{v_s} = -\frac{R_2}{R_1} \left[ \frac{1}{1 + (1/A_0)[1 + (R_2/R_1)]} \right] \tag{7.17}$$

**Problem 7.7**
Obtain Eqs. (7.16) and (7.17) from Eq. (7.15).

**Problem 7.8**
For the inverting amplifier shown in Figure 7.10, show using first principles that

$$\frac{v_o}{v_s} = -\frac{R_2}{R_1}$$

when the input impedance is very high. Note that when input impedance is very high, the inverting terminal is at virtual ground (i.e. $v^- = 0$ in this case).

Thus, for a sufficiently high-gain op-amp ($A_0 \to \infty$) we have $v_o/v_s \to (-R_2/R_1)$ for a small differential signal. The voltage gain of this op-amp circuit is the ratio of the values of these two resistors and is independent of $A_0$. Therefore, one can design amplifiers with the desired gain just by changing these resistor values. This is called an inverting amplifier to indicate the presence of the negative sign that gives rise to the 180° phase shift between $v_o$ and $v_s$.

The open-loop transfer function of the op-amp has a pole at $s_0$. This transfer function can be written as follows: $A(s) = A_0 s_0/(s_0 + s)$, where in steady-state conditions $s_0 = j\omega_0$, $s = j\omega$ and $\omega = 2\pi f$ ($f$ being the frequency of operation). When $s = s_0$, the open-loop gain $A_0$ falls to a magnitude $A_0/\sqrt{2}$ (3 dB value). $A_0 s_0$ is the gain bandwidth product of the op-amp. This open-loop gain versus $\omega$ is shown by the dotted line in Figure 7.11. By replacing $A_0$ with $A(s) = A_0 s_0/(s_0 + s)$ in Eq. (7.17), it can easily be shown that the magnitude of the sinusoidal- steady-state gain of this circuit (Figure 7.10) versus $\omega$ is as shown by the solid line in Figure 7.11.

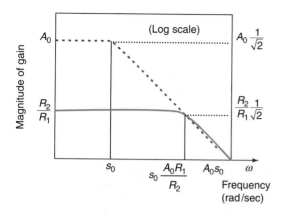

**Figure 7.11** Frequency response for an inverting amplifier. The dashed line is the transfer function of an op-amp without the feedback [7].

The closed-loop transfer function not only has this pole shifted to higher frequency $s_c = s_0 A_0 R_1 / R_2$ but also has the low-frequency gain scaled down from $A_0$ to $A = R_2 / R_1$. It may be noted that the gain-bandwidth product of the amplifier, $A_0 s_0$, remains the same as the gain-bandwidth product $A_0 s_0$ of the op-amp and is not affected by the specific values of $R_2$ and $R_1$ (assuming that $R_2 > R_1$). This fundamental fact suggests that the gain-bandwidth product is an important figure of merit for an op-amp.

As indicated earlier, op-amps are one of the most widely used analog ICs and can be used in several configurations. Some useful circuits using op-amps are indicated in Table 7.2 for ready reference.

**Table 7.2 Some examples of op-amp-based circuits**

| Type | Circuit | Applications |
|---|---|---|
| Noninverting amplifier<br>$\dfrac{v_o}{v_s} = \dfrac{R_1 + R_2}{R_1} = 1 + \dfrac{R_2}{R_1}$ | | Weinbridge oscillator circuits (to realize the amplifier of gain $= 3$). |
| Voltage follower | | Buffer stage to prevent loading of a circuit by the next stage. |
| Transimpedance amplifier<br>$v_o = -i_{in} R_2$ | | Sensing current from photodetector. Digital to analog converters. |
| Transconductance amplifier<br>$i_o = v_{in}/R_1$ | | Low-voltage dc and ac voltmeters; LED and Zener diode tester circuits. |

*(Continued)*

**Table 7.2** (*Continued*)

| Type | Circuit | Applications |
|------|---------|--------------|
| Integrator $v_o = \dfrac{-1}{RC}\displaystyle\int v_{in}(t)\,\mathrm{d}t$ | | To generate sawtooth voltage from a square wave input. |
| Differentiator $v_o = R_2 C \dfrac{\mathrm{d}v_{in}}{\mathrm{d}t}$ | | To generate a square wave voltage from a sawtooth input voltage. |

### Problem 7.9

Draw the circuit diagram of a precision half-wave rectifier using an op-amp and sketch the input and output waveforms one below the other, taking the input to be a sine wave. Explain the working of this rectifier.

### Problem 7.10

In the circuit shown, $Z_1$ is a parallel combination of $R_1$ and $C_1$ and $Z_2$ is a parallel combination of $R_2$ and $C_2$. Determine the condition and the frequency for which this circuit will work as **(a)** an integrator and **(b)** a differentiator.

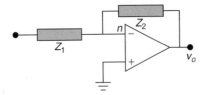

## ▶ 7.3 SIGNAL CONDITIONING CIRCUITS

Electronics circuits such as amplifiers mentioned in earlier sections are widely used with sensing devices (components). The output of the sensor may be a change in charge, voltage, resistance, capacitance, resonant frequency, or a combination of these. The circuit connecting this sensor should be capable of measuring these fundamental sensor outputs, if necessary, by converting them to voltage and then amplifying the voltage.

In order to improve the sensitivity of the sensing device, it is usually necessary to avoid electrical loading of sensors by the electronics circuits. This requires devices with high input impedance. Op-amps and FETs (when integrated with the sensor) assume significance in this context. When electronics devices are integrated with the sensor, the noise added can be minimized, which in turn improves the sensitivity. In most cases the relative value is more important than the absolute value because the necessary correction can be accommodated during sensor calibration. The instrumentation amplifier discussed in this section meets the above requirements because it amplifies the difference of voltage at its input terminals.

## 7.3.1 Difference Amplifier

An ideal instrumentation amplifier amplifies only the difference between two signals and rejects any signals common to the two input terminals. Even though the basic op-amp amplifies the difference between the two input terminals, it would be better to have a difference amplifier whose gain depends upon the ratio of two resistors so that the gain is independent of the op-amp open-loop gain, as in the case of the inverting and noninverting amplifiers. This is achieved with the circuit shown in Figure 7.12. The input voltages are $v_1$ and $v_2$.

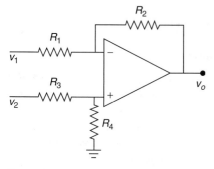

**Figure 7.12** Op-amp difference amplifier.

This circuit can be analyzed using the principle of superposition. In this approach, first we determine the output voltage $v_{o1}$ with $v_2 = 0$ (i.e. the input port 2 connected to ground). We next consider $v_1 = 0$ (terminal 1 connected to ground) and determine the output voltage $v_{o2}$. It can be easily shown, using the virtual short concept, that the two output voltages $v_{o1}$ and $v_{o2}$ are given by the following two equations:

$$v_{o1} = -\frac{R_2}{R_1}v_1 \tag{7.18}$$

$$v_{o2} = \left(1 + \frac{R_2}{R_1}\right)\frac{R_4}{R_3 + R_4}v_2 \tag{7.19}$$

Since the net output voltage is the sum of the individual terms, we have $v_o = v_{o1} + v_{o2}$. Thus we obtain

$$v_o = \left(1 + \frac{R_2}{R_1}\right)\frac{R_4/R_3}{1 + (R_4/R_3)}v_2 - \frac{R_2}{R_1}v_1 \tag{7.20}$$

For an ideal difference amplifier, $v_o$ should be equal to zero when $v_1 = v_2$. Examining Eq. (7.20), we see that this condition can be met only if we have

$$\frac{R_2}{R_1} = \frac{R_4}{R_3} \tag{7.21}$$

Then the output voltage is given by

$$v_o = \frac{R_2}{R_1}(v_2 - v_1) \tag{7.22}$$

This indicates that the amplifier has a differential gain of $A_d = R_2/R_1$.

Another important requirement of difference circuits is high input resistance. The differential input resistance can be determined using the circuit shown in Figure 7.13 by imposing the condition in Eq. (7.21) and taking $R_1 = R_3$ and $R_2 = R_4$. The input resistance is then defined as $R_i = V_I/I$. Using the virtual short concept, we can write $V_I = 2IR_i$. The input resistance therefore is $R_i = V_I/I = 2R_1$.

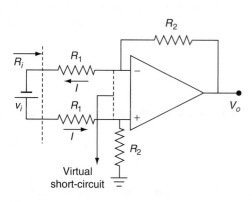

**Figure 7.13** Circuit for determining the input resistance of op-amp difference amplifier.

The disadvantage of the above difference amplifier design is that both high gain and high input resistance cannot be achieved without using extremely large resistance values.

*Note:* If the condition in Eq. (7.21) is not met, then $v_o \neq 0$ when $v_1 = v_2$. The common-mode voltage $v_c$ is given by $v_c = (v_1 + v_2)/2$ and the common-mode gain $A_c$ is given by $A_c = v_o/v_c$. The CMRR is expressed in decibels as

$$\text{CMRR [dB]} = 20\log_{10}\left|\frac{A_d}{A_c}\right| \tag{7.23}$$

Ideally, CMRR should be very high, tending to infinite.

## 7.3.2 Instrumentation Amplifier as a Differential Voltage Amplifier

We have shown in the previous section that it is difficult to obtain high input resistance and high gain in a difference amplifier with reasonable resistor values. To achieve this we need to change two resistance values and still maintain equal ratios between $R_2/R_1$ and $R_4/R_3$. This difficulty can be circumvented by using the circuit shown in Figure 7.14, which is referred to as the *instrumentation amplifier*.

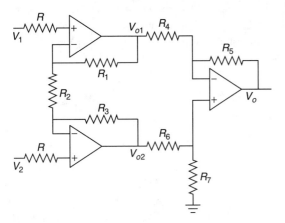

**Figure 7.14** Schematic of the instrumentation amplifier.

An instrumentation amplifier is a widely used op-amp-based circuit with low dc offset, low drift, low noise, high open-loop gain, high CMRR, and high input impedances. These are achieved by connecting a noninverting buffer to each input of the differential amplifier to increase the input impedance. With these features, such an amplifier can be used in test and measurement equipment where great accuracy, low noise, and stability are essential. Popular examples of instrumentation amplifiers include AD620, MAX4194, LT1167, and INA128.

These amplifiers are very useful in amplifying the output of several types of sensors. The output of a sensor is usually measured across its two terminals, neither of which may happen to be at the ground potential due to biasing provided to the sensor and/or inherent noise. Hence, a differential amplifier is usually required with such a sensor. The instrumentation amplifier, usually an IC, is used for such requirements. Figure 7.14 shows such a typical amplifier, which contains three op-amps. These are arranged so that there is one op-amp each to buffer the two inputs and a third to produce the desired level of output with good impedance matching. The small signal gain of the circuit is

$$\frac{V_o}{V_2 - V_1} = \left(1 + \frac{2R_1}{R_2}\right)\frac{R_5}{R_4} \tag{7.24}$$

where $R_1 = R_3$ and $R_5/R_4 = R_7/R_6$. In this circuit, since the input signal voltage is applied directly to the noninverting terminals of the two input amplifiers, the input resistance is very large (ideally infinite), which is one desirable characteristic of the instrumentation amplifier. Also, since the difference gain is a function of the resistor $R_2$, the amplifier gain can easily be varied with the adjustment of only one resistor.

Thus, the buffer amplifier in the circuit provides gain and prevents the output resistance of the connected sensors from affecting the input impedance of the op-amp circuit. In turn, this prevents the input resistance at the op-amp input circuit from loading the connected sensor.

Ideally, the common-mode gain of an instrumentation amplifier should be zero. However, mismatches between components cause some common-mode gain.

**Example 7.1**

(a) For the circuit shown in Figure 7.15, derive an expression for $V_{o1}$ and $V_{o2}$ and hence determine the expression for $V_o$.

(b) (i) Determine $V_o$ if the circuit in Figure 7.15 is modified by removing the resistor $R_2$ and shorting resistors $R_1$ and $R_3$. (ii) What is the difference in performance between this circuit and the one in Figure 7.15?

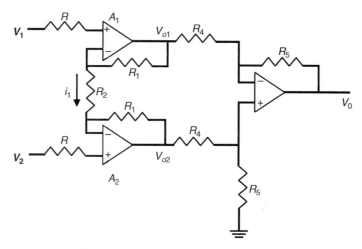

**Figure 7.15** Circuit diagram for Example 7.1.

**Solution:** (a) As suggested in the text, we choose $R_7 = R_5, R_6 = R_4$, and $R_3 = R_1$. The circuit is redrawn with these values.

The current through the resistor $R_2$ is $i_1 = \dfrac{v_1 - v_2}{R_2}$. The same current $i_1$ flows through both the resistors $R_1$ because the current entering the op-amp is negligibly small.

$$v_{o1} = v_1 + i_1 R_1 = v_1 + \frac{v_1 - v_2}{R_2} R_1 = \left(1 + \frac{R_1}{R_2}\right) v_1 - \frac{R_1}{R_2} v_2$$

$$v_{o2} = v_2 - i_1 R_1 = v_2 - \frac{v_1 - v_2}{R_2} R_1 = \left(1 + \frac{R_1}{R_2}\right) v_2 - \frac{R_1}{R_2} v_1$$

Therefore,

$$(v_{o2} - v_{o1}) = (v_2 - v_1)\left(1 + \frac{2R_1}{R_2}\right)$$

The voltage $(v_{o2} - v_{o1})$ is the differential input voltage to the differential stage amplifier. This gives

$$v_o = \frac{R_5}{R_4}(v_{o2} - v_{o1}) = \frac{R_5}{R_4}\left(1 + \frac{2R_1}{R_2}\right)(v_2 - v_1)$$

The gain of the instrumentation amplifier is

$$\frac{v_o}{v_2 - v_1} = \frac{R_5}{R_4}\left(1 + \frac{2R_1}{R_2}\right)$$

**(b)** When the resistor $R_2$ is removed (opened) and the feedback resistors $R_1$ are replaced by a short circuit, the two amplifiers $A_1$ and $A_2$ serve as unity gain amplifiers. Hence, we get

$$v_{o2} - v_{o1} = v_2 - v_1$$

Therefore,

$$v_o = \frac{R_5}{R_4}(v_2 - v_1)$$

Thus the gain of this instrumentation amplifier is fixed by the ratio of the resistors $R_5$ and $R_4$. For case (a), the gain of the instrumentation amplifier can be adjusted by varying a single resistor $R_2$, which can be easily achieved using a potentiometer. On the other hand, in case (b) the gain adjustment involves the ratio of the two sets of resistors $R_5$ and $R_4$, and hence becomes more involved.

**Example 7.2**

In the instrumentation amplifier in Example 7.1(a) the resistance values are $R_1 = 50\,\text{k}\Omega$, $R_5 = 20\,\text{k}\Omega$, and $R_4 = 10\,\text{k}\Omega$. The resistance $R_2$ is a series combination of a fixed resistance $R_f$ and a potentiometer having maximum value $R_p$. Determine $R_f$ and the range of $R_p$ so that the differential gain $A_d$ can be varied from 6 to 202.

**Solution:** From Example 7.1(a),

$$A_d = \frac{R_5}{R_4}\left(1 + \frac{2R_1}{R_2}\right)$$

The lower gain limit is 6. We first determine $R_{2(\text{max})}$ required for this purpose.

Therefore, $6 = \dfrac{20}{10}\left(1 + \dfrac{2 \times 50\,\text{k}}{R_{2(\text{max})}}\right)$. This gives $R_{2(\text{max})} = 50\,\text{k} = R_f + R_p$ (1)

The upper limit for $A_d = 202$. This can be achieved when the resistance of the potentiometer is zero so that $R_{2(\text{min})} = R_f$. Thus we have

$$A_{d(\text{max})} = 202 = \frac{20}{10}\left(1 + \frac{100\,\text{k}}{R_f}\right)$$

This gives, $R_f = 1\,\text{k}$ (2). From (1) and (2), we see that the potentiometer range should be from 0 to $49\,\text{k}\Omega$.

### 7.3.3 Wheatstone Bridge for Measurement of Change in Resistance

One of the simplest circuits used widely for improving signals from sensors is the Wheatstone bridge shown in Figure 7.16. When only a bias voltage $V_b$ is applied and all resistances are equal, voltages $V_{02}$ and $V_{01}$ at the output terminals of the bridge are both equal to $V_b/2$, so that $V_o = 0$. The resistor locations on the membrane of the sensor are

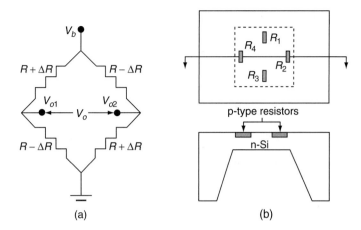

**Figure 7.16** Wheatstone bridge circuit: (a) layout; (b) schematic on a piezoresistive pressure sensor [8].

designed so that the resistors in one pair of opposite arms of the bridge circuit experience an increase in resistance $(R + \Delta R)$ and the resistors of the other opposite arms of the bridge decrease to $(R - \Delta R)$, as shown in the figure. This causes the voltages $V_{o1}$ and $V_{o2}$ to be different from $V_b/2$, resulting in an output signal $V_o$. In a measurement system, this differential voltage, $V_o = V_{o2} - V_{o1}$, is the electrical signal indicating the force or pressure acting on the sensor. With four active elements, the output voltage is

$$V_o = V_b \frac{\Delta R}{R} \tag{7.25}$$

Since $\Delta R$ is proportional to the sensing variable (e.g. force or pressure), this can be rewritten as

$$V_o = V_b \times F \times S \tag{7.26}$$

where $F$ is the force or pressure and $S$ is the sensitivity, or output voltage, of the sensor in mV/V of excitation with full-scale input. The resistor arrangement on a diaphragm of a pressure sensor is shown in Figure 7.16(b).

Normally full-scale output voltages are in the 10–100 mV range and need to be amplified by a data acquisition system. The amplifier required for this stage should amplify the differential input voltage and reject the common-mode input voltage. Therefore, the instrumentation amplifier discussed in the previous section is well suited for this requirement.

**Example 7.3**

The piezoresistive pressure sensor bridge circuit in Figure 7.16(a) is operating at a power supply voltage of 10 V. The sensitivity of this sensor was found to be 50 mV/bar. The Wheatstone bridge is balanced and hence its output voltage is zero when the input pressure is zero. The output of the bridge is connected to a differential amplifier having CMRR = 200 and differential gain $A_d = 5$. When the sensing pressure is 3 bar, determine **(a)** the output voltage $V_o$ of the bridge circuit and **(b)** the output voltage $V_A$ of the amplified signal.

**Solution:** **(a)** The output voltage of the Wheatstone bridge when the sensing pressure is 3 bar is $V_d = 50 \times 3 = 150$ mV.

**(b)** The amplifier differential gain $A_d = 5$. Hence the differential voltage $V_d$ is amplified to 750 mV. Referring to Figure 7.16(a), when the supply voltage is 10 V, even when the pressure is zero, we obtain $V_{o1} = V_{o2} = 5$ volts. The signal voltage is superimposed on this common mode voltage of the differential amplifier. As CMMR = 200 and the differential gain $A_d = 5$, the common mode gain

$$A_c = \frac{A_d}{\text{CMRR}} = \frac{5}{200} = 0.025.$$

The common mode voltage gets amplified by this $A_c$ and gives an output $= 0.025 \times 5 = 0.125$ V. Therefore, when the pressure is 3 bar the amplified output is $V_A = A_d V_o + A_c V_c = (5 \times 0.150 + 0.025 \times 5)$ V $= 0.875$ V.

◼

## 7.3.4 Phase-Locked Loop

The phase-locked loop (PLL) is one of the basic building blocks of electronics circuits and is used in several applications such as motor-speed controllers, filters, frequency-synthesized transmitters, receivers, and microsystems. As illustrated in Figure 7.17, the PLL consists of a phase detector, a low-pass filter (LPF), a dc amplifier, and a voltage-controlled oscillator (VCO).

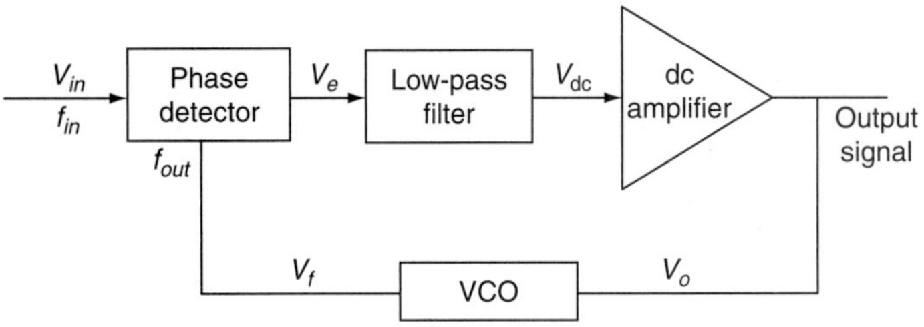

A phase detector is basically a comparator that compares the phase of the input signal (frequency $f_{in}$) with that of the feedback signal (frequency $f_{out}$). The phase detector receives two digital signals, one from the input and the second as feedback from the output. The loop is locked when these two signals are of the

**Figure 7.17** Block diagram showing the operating principle of a PLL.

same frequency and have a fixed phase difference. The output of a phase detector, called the *error voltage*, $V_e$, is applied to the LPF, which removes any high-frequency components and produces a smooth dc voltage. This dc voltage is amplified and then applied to the input of the VCO, whose output frequency is proportional to the value of the input dc voltage. If the frequency of the input signal shifts slightly, the phase difference between the input signal and the VCO output signal will begin to increase. This changes the control voltage to the VCO in such a way as to bring the VCO frequency back to the same value as the input signal.

The PLL has three modes of operation. In the *free-running mode*, there is no input frequency (or voltage) and the VCO runs at a fixed frequency called the center frequency or the free-running frequency, $f_o$. *Capture mode* requires an input frequency to be applied. In this mode, the VCO frequency changes continuously to match the input frequency. The PLL is said to be in *phase-lock mode* when the VCO output frequency becomes equal to the input frequency $f_{in}$. The feedback loop maintains the lock when the frequency of the input signal changes. If the two inputs to the phase detector are of exactly the same frequency and phase, the output of the phase detector is zero; otherwise there is an output proportional to their phase difference.

Owing to the limited operating range of the VCO and the feedback connection of the PLL circuit, two important frequency bands are specified for a PLL. The *capture range* $f_c$ of a PLL is the range of input frequencies centered around the VCO free-running frequency $f_o$, over which the output signal frequency of the VCO can lock with the input signal frequency from an unlocked condition. Once the PLL has achieved "capture," it can maintain lock with the input signal over a somewhat wider frequency range (centered around $f_o$) called the *lock range* $f_L$.

Phase detectors can be broadly classified into two types: digital detectors and analog detectors. The digital detectors are simple to implement with digital devices. However, they are sensitive to the harmonic content of the input signal and to changes in the duty cycle of the input signal and VCO output voltage. Analog detectors are of monolithic type such as the CMOS MC 4344/4044. They respond only to the transitions in the input signals, thus making them insensitive to harmonic content and duty cycle. The output voltage is independent of variations in amplitude and the duty cycle of the input waveform. Analog detectors are generally preferred to digital detectors, especially in applications where accuracy is critical.

### 7.3.4.1 Phase Detector

A phase detector takes two input voltages and produces a dc voltage proportional to the phase difference. To understand the principle of phase detectors, consider two input voltages $V_{in}$ and $V_f$ whose waveforms are shown in Figure 7.18(a), with a phase difference $\phi$. A phase detector can be implemented using an exclusive OR gate, which gives output $V_o$ only when one of the two inputs is present, as shown in Figure 7.18(a). Integrating the output voltage gives an average output voltage that is a linear function of the phase difference $\phi$, as shown in Figure 7.18(b). The average output voltage $V_{o(dc)}$ can be expressed as

$$V_{o(dc)} = \frac{V_{DD}}{\pi}\phi, \quad \text{for } 0 < \phi < \pi$$

$$V_{o(dc)} = \frac{V_{DD}}{\pi}(2\pi - \phi), \quad \text{for } \pi < \phi < 2\pi$$

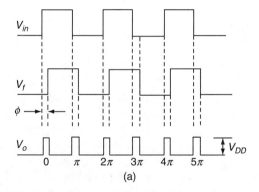
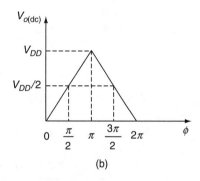

(a)                                         (b)

**Figure 7.18** (a) Phase detector output voltage $V_o$ resulting from phase difference between two input signals $V_{in}$ and $V_f$ to exclusive OR gate; (b) the dc output voltage versus phase difference $\phi$.

## 7.3.5 Analog-to-Digital Converter

The output signals of most physical systems (such as temperature and pressure gauges, flow transducers) are analog or continuous functions of time. However, these signals must be converted into binary form to make processing possible in the digital domain and to take advantage of their efficiency and reliability. The circuit that performs this conversion is called an *analog-to-digital converter* (ADC or A/D converter). This is a mixed-signal device because it has both analog and digital functions. ADC provides an output that digitally represents the input analog voltage or current level. Most ADCs convert an input voltage to a digital word, but the true definition of an ADC includes the possibility of an input current. There are many types of ADCs depending on the type of conversion technique used. Among them, the successive-approximation technique is the most commonly used, mainly because it offers excellent tradeoffs in resolution, speed, accuracy, and cost.

ADC has an analog reference voltage (or current) against which the analog input is compared. The digital output word tells us what fraction of the reference voltage (or current) is the input voltage (or current). Thus the ADC is basically a divider. Considering an analog input voltage $V_{in}$, voltage output $V_o$, and reference voltage $V_{ref}$, the input/output transfer function is given by the relation

$$V_o = 2^n G \frac{V_{in}}{V_{ref}} \tag{7.27}$$

Here $G$ is the gain factor, generally considered to be unity, and $n$ is the number of bits (for an $n$-bit ADC). We illustrate the function of an ADC considering a three-bit ADC with a binary output.

In a three-bit ADC the number of bits is $n = 3$ and therefore there are $2^3 = 8$ possible output codes. The difference between each output code is $V_{ref}/2^3$. Thus if the $V_{ref}$ is 8 V, every time we increase the input voltage by 1 V, the output code will increase by 1 bit. This means that the least significant bit (LSB) represents 1 V, which is the smallest increment this converter can resolve. Thus we can say that the resolution of this ADC is 1.0 V because voltages as small as 1 V can be resolved.

Resolution can also be expressed in bits. In this example, the LSB given by (000) represents 1 V, and if the input voltage is 5.5 V and $V_{ref} = 8$ V, then the binary output will be 101. This is further illustrated in Figure 7.19.

The larger the number of bits, the better the resolution. Alternately, smaller $V_{ref}$ gives smaller steps but possibly at the expense of noise. In the above three-bit ADC, if $V_{ref} = 0.8$ V, the resolution of the ADC is $0.8/8 = 0.1$ V and the LSB is 0.1 V, allowing us to measure smaller voltage ranges (0−0.8 V) with greater accuracy. The problem with higher resolution (more bits) is the cost. Also, the smaller LSB means it is difficult to find a really small signal because it becomes lost in the noise, thus reducing the signal-to-noise ratio (SNR) performance of the converter. When the reference voltage is reduced, the input dynamic range may be compromised. There can also be a loss of signal due to the noise in this approach, reducing SNR performance.

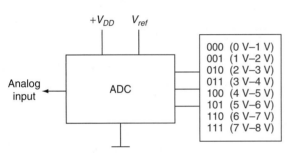

**Figure 7.19** Three-bit ADC input/output scheme.

### 7.3.5.1 Quantization Error

The quantization error can be best explained by an example. For this purpose we consider the three-bit ADC. This ADC produces an output code of zero (000) when the input analog voltage is zero. As the input voltage increases toward $V_{ref}/8$, the error also increases because the input is no longer zero; the output code remains at zero till the input reaches $V_{ref}/8$. This happens because the input voltage in this range is represented by a single code. When the input reaches $V_{ref}/8$, the output code changes from 000 to 001, where the output exactly represents the input voltage and the error is reduced to zero. As the input voltage increases past $V_{ref}/8$, the error again increases until the input voltage reaches $V_{ref}/4$, where the error again drops to zero. This process continues through the entire input range, so that the error plot is a sawtooth, as in Figure 7.20. It may be noted that the magnitude of the error ranges from zero to the LSB.

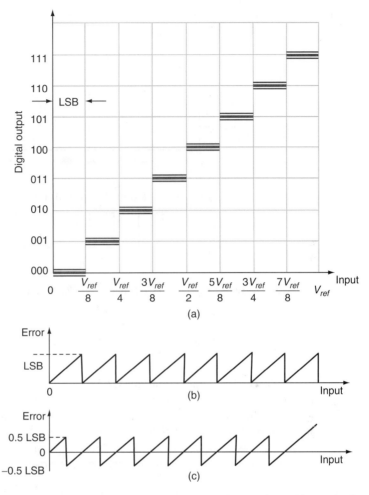

**Figure 7.20** (a) Digital output; (b) quantization error versus input voltage; (c) quantization error versus input voltage with 0.5 LSB offset in the input.

The maximum error here is 1 LSB. This zero-to-LSB range is known as *the quantization uncertainty* because there is a range of analog input values that could have caused any given code and we are uncertain exactly what input voltage caused a given code. The maximum quantization uncertainty is known as the *quantization error*. But an error of 0 to 1 LSB is not as desirable as an error of ±0.5 LSB. Hence, an offset is generally introduced into the ADC to force an error range of ±0.5 LSB. As shown in Figure 7.20(c), the output changes from 000 to 001 with an input value of 0.5 LSB rather than 1 LSB.

We illustrate the analog-to-digital conversion in a three-bit ADC by considering the analog input signal voltage in Figure 7.21(a). The ADC samples this input voltage at periodic intervals determined by the sampling time $T_s$ and a three-bit binary number $(b_1 b_2 b_3)$ is assigned to each sample, as shown in Figure 7.21(b) for the case of three-bit ADC. Notice that the output bits are sequential. This eliminates the need for separate pins for each bit of the output and hence reduces the cost and size of ADC.

In the above example, the $n$-bit binary number is a binary fraction that represents the ratio between the unknown signal input $V_s$ and the reference voltage $V_{ref}$ of the ADC. As

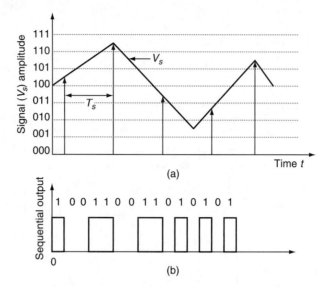

**Figure 7.21** (a) Analog signal sampled at intervals of $T_s$; (b) sequential output of the digital signal.

already pointed out for $n = 3$, each binary fraction is $V_{ref}/2^3 = V_{ref}/8$; the output voltage of the three-bit ADC is as shown in Figure 7.21(b).

### 7.3.5.2 Successive-Approximation ADC

As pointed out earlier, the basic principle of operation of an ADC is to use the comparator to determine whether or not to turn on a particular bit of the binary number in the output. It is typical for an ADC to use a digital-to-analog converter (DAC or D/A converter) to determine one of the inputs to the comparator. Successive-approximation ADC is the most popular type due to its excellent tradeoffs in resolution, speed, accuracy, and cost. Figure 7.22 illustrates a

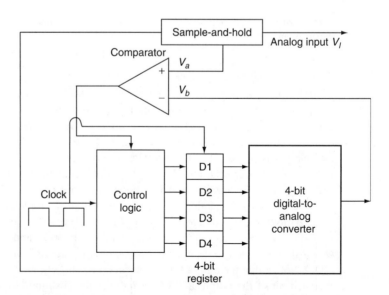

**Figure 7.22** Working principle of a four-bit SAC. The control logic increments the bits, starting with MSB.

four-bit successive-approximation converter (SAC) with 1 V step size. This ADC consists of an analog comparator, a four-bit DAC, along with a logic control and a four-bit register. The logic control synchronizes the operation of the converter with a timing (clock) signal. The comparator converts the analog voltages to digital signals. It has two inputs, $V_a$ and $V_b$, and gives a binary output voltage. Referring to Figure 7.22, if $V_a > V_b$, the output is high (logic 1) and if $V_a < V_b$, the output is low (logic 0). A sample-and-hold circuit is commonly used to hold the input voltage constant during the conversion process. The algorithm for the operation of a successive-approximation ADC is described in Example 7.4 below.

---

**Example 7.4**   Operation of a Successive-Approximation ADC

Using a successive approximation ADC, convert an analog voltage $V_a = 7.5$ V.

**Solution:**

**Step 1:** Clear all the bits in the register. The first pulse from the ring counter sets the DAC four-bit register and a ring counter so that MSB = 1 and all others are 0: $B_3 = 1$, $B_2 = B_1 = B_0 = 0$. Hence for $B_3B_2B_1B_0 = 1000$, the output voltage $V_b$ of the DAC is 8 V, which is compared by the comparator. Since $V_a < 8$ V, the MSB in the register is set to 0. (Note that if $V_a \geq 8$ V, the MSB in the register would have been maintained at 1.)

**Step 2:** The second pulse from the ring counter sets $B_2 = 1$, maintaining $B_1 = B_0 = 0$. Also, $B_3$ remains at 1 or 0 depending on the condition set by step 1 (here $B_3$ is at 0). That is, $B_3 = 0$, $B_2 = 1$, and $B_1 = B_0 = 0$. Thus, we note that for $B_3B_2B_1B_0 = 0100$, the output $V_b$ of the DAC is 4 V. This voltage is compared with $V_a$, which is 7.5 V. As this $V_a > 4$ V, $B_2$ is set to 1 at the end of step 2.

**Step 3:** The third pulse from the ring counter sets $B_1 = 1$; $B_0$ remains at 0. $B_3$ and $B_2$ remain as at the end of step 2. That is, $B_3 = 0$, $B_2 = 1$, $B_1 = 1$, and $B_0 = 0$. Thus, for $B_3B_2B_1B_0 = 0110$, the output voltage $V_b$ of the DAC is 6 V, which is compared by the comparator. Since this is less than the voltage $V_a$ (= 7.5 V), $B_1$ is maintained at 1.

**Step 4:** The fourth pulse from the ring counter sets $B_0 = 1$. $B_3$, $B_2$, $B_1$ remain as they were at the end of step 3. Thus, for $B_3B_2B_1B_0 = 0111$, the output $V_b$ of DAC is 7 V, which is lower than $V_a$ (= 7.5 V). Therefore, $B_0$ in the register is maintained at 1.

At the end of final step, the desired number at the counter gives the ''Read'' output. The results of the conversion steps are shown in Figure 7.23. For an $n$-bit ADC; the conversion process takes $n$ clock periods. Thus for an 8-bit ADC and 10 MHz clock, the conversion takes $8 \times 10^{-7}$ s = 800 ns.

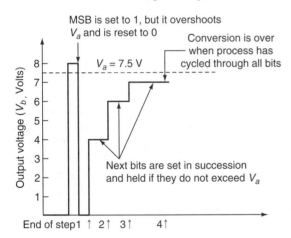

**Figure 7.23** Successive-approximation process for $V_a = 0.75$ V.

---

## ▶ 7.4 PRACTICAL SIGNAL CONDITIONING CIRCUITS FOR MICROSYSTEMS

### 7.4.1 Differential Charge Measurement

Many of the practical sensors discussed in Chapter 2 rely on measuring capacitance (directly) or charge (indirectly) electrically to convert the measured variable (displacement, pressure, etc.) to electrical form. Because of the capacitive nature of the input

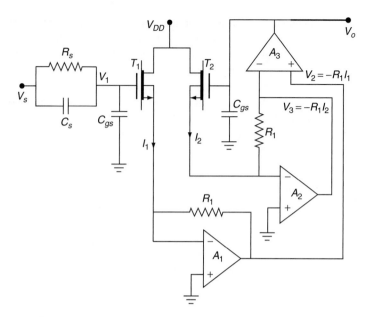

**Figure 7.24** Differential charge-measurement analog circuit [7].

impedance, MOS transistors are very useful for measuring charge. Figure 7.24 shows a differential charge-measurement circuit with two depletion-mode MOSFETs $T_1$ and $T_2$ and three op-amps $A_1$, $A_2$ and $A_3$ in a feedback configuration. Depletion-mode FETs have a conducting channel even in the absence of any dc gate bias, and hence drain currents $I_1$ and $I_2$ cannot be assumed to be zero, even if the gate-to-source voltage $V_S$ is zero [7].

The elements $R_S$ and $C_S$ represent some general impedance that couples the signal voltage $V_s$ to the gate of the FET $T_1$. Both FETs ($T_1$ and $T_2$) are identical and have the same bias voltage $V_{DD}$. Their input capacitances $C_{gs}$ are shown separately in the circuit. The transimpedance amplifiers ($A_1$ and $A_2$ in Figure 7.24) convert the two currents into voltages that are applied to the op-amp $A_3$ driving the gate of an FET $T_2$. As the op-amp $A_3$ does not have feedback, it has a very high gain. An increase in $I_2$ causes a decrease in $V_o$. The virtual ground approximation at the input of this op-amp ensures $I_1 = I_2$. In other words, the op-amp circuit ensures that the two FETs carry the same current. This requires that the output voltage $V_o$ be equal to the voltage $V_1$, and this voltage is directly related to the charge on its input capacitance $C_{gs}$. This leads to the relation between $V_o$ and $V_s$

$$\frac{V_o}{V_s} = \frac{1 + sR_sC_s}{1 + sR_s(C_s + C_{gs})} \tag{7.28}$$

The charging of $C_{gs}$ through the parallel combination of $R_s$ and $C_s$ has unity transfer function at low frequencies. At higher frequencies (well above the reciprocal of the charging time constant), the charge on $C_{gs}$ is determined by a capacitive divider:

$$\frac{V_o}{V_s} = \frac{C_s}{C_s + C_{gs}} \tag{7.29}$$

Circuits of this type are widely used for measuring charges at high impedance nodes without drawing any dc current from these nodes.

## 7.4.2 Switched-Capacitor Circuits for Capacitance Measurement

Another circuit for measuring charge involves a combination of FET switches, capacitors, and MOSFET op-amps. Enhancement-mode FETs are used in the example circuit in Figure 7.25. They do not carry any current in the absence of gate voltage but their channel begins to conduct as $V_{GS}$ exceeds $V_{th}$. The operation of this circuit depends on two independent signal streams $S_1$ and $S_2$. The waveforms are illustrated in Figure 7.26. It may be noted that signals $S_1$ and $S_2$ do not overlap.

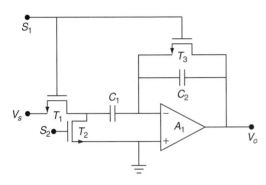

**Figure 7.25** Schematic of a switched-capacitor circuit.

When $S_1$ is high and $S_2$ is low, transistors $T_1$ and $T_3$ are turned ON and transistor $T_2$ is turned OFF. Because of the conduction path through $T_3$, the op-amp behaves as a noninverting voltage follower with its noninverting input connected to the virtual ground. Therefore, $V_0$ is zero; however, capacitor $C_1$ charges to a value $Q_1 = C_1 V_S$ because its right-hand terminal is at virtual ground and $T_1$ is ON. This charge remains on $C_1$ even when $S_1$ returns to a low value, because $T_2$ is OFF and there is no conduction path through $T_1$, $T_3$, or the op-amp input. Next, when $S_2$ is high, $T_2$ is turned ON and the voltage on the left-hand node of $C_1$ is pulled to zero. (Note that $S_1$ is low whenever $S_2$ is high. This ensures that when $T_2$ is ON both $T_1$ and $T_3$ are OFF.) However, since the capacitor cannot be discharged instantaneously as $T_2$ turns ON, the inverting input voltage becomes negative, making the output be strongly positive. The capacitive coupling to the inverting input through $C_2$ tends to pull the inverting input back toward zero. The circuit settles when the inverting node reaches virtual ground by the discharge of $C_1$. The positive charge $C_1 V_S$, which was on the left-hand plate of $C_1$, has flowed to ground while the negative charge $-C_1 V_S$ on the right-hand plate of $C_1$ has shifted to the left-hand plate of $C_2$. This requires that the right-hand plate of $C_2$ have a charge $+C_1 V_S$.

However, this charge must equal $C_2 V_o$. (Note that the inverting voltage swings to ON.) Hence, we conclude that the output voltage swings to a value given by

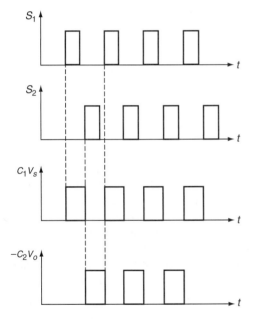

**Figure 7.26** Schematic waveforms in the switched-capacitor inverter. Charging and discharging transients are neglected [7].

$$V_o = (C_1/C_2)V_S \qquad (7.30)$$

Typically, the rate of charge transport thorough this circuit is $C_1 V_S$ coulombs (of charge) per clock period. This amount of charge is placed on $C_1$ when $S_1$ is high and $S_2$ is low, and is then transferred to $C_2$, hence to the output, when $S_2$ is high and $S_1$ is low. Hence, this circuit may be used for measuring the unknown capacitance $C_1$ in microsystems, using a known capacitance $C_2$.

## 7.4.3 Circuits for Measuring Frequency Shift

Cantilever beams fabricated using silicon, silicon nitride (SiN), or polymers are widely used in microsystems. One of the most popular applications is as cantilever arrays in

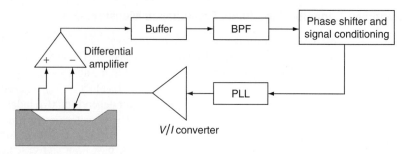

**Figure 7.27** A circuit for measuring frequency shift in microsystems [9].

biosensors for medical diagnostic applications. Cantilevers are easy to fabricate in large numbers and hence these sensors are inexpensive. However, higher-order dependence on dimensions makes them very sensitive to process parameter variations. Another factor that affects their performance is residual stress. The resonant frequency of such systems using cantilevers is given by

$$\omega_0 = \sqrt{\frac{k}{m}} \tag{7.31}$$

where $k = F/\delta = Ywt^3/4l^3$, where $Y$ is Young's modulus, $\delta$ is the deflection of the cantilever tip, $w$ and $t$ are the width and thickness of the beam, and $L$ is the length of the cantilever structure. A change in the force applied to a cantilever can shift the resonance frequency, and frequency shift can be measured using the circuit in Figure 7.27. The working of this system is based mainly on the frequency-stabilizing capability of the PLL, as discussed in Section 7.3.4 above.

## ▶ 7.5 INTRODUCTION TO CONTROL THEORY

The concept of control originated in the early 1900s and has since advanced to play a major role in almost all branches of engineering. Control systems have become integral parts of space vehicles, missiles, robotic systems, etc. In microsystems and smart systems, these techniques are employed for controlling displacements (velocities or accelerations), pressure (force or stress), temperature, humidity, viscosity, etc. A control system regulates the output variable by providing an actuating signal to the process under its supervision. The reader is advised to go through control systems textbooks [10]–[12] for a complete understanding of modern control theory. In this section some topics directly relevant for the actuation and control of smart systems are discussed.

### 7.5.1 Simplified Mathematical Description

A proper mathematical model is necessary for the analysis and design of control systems. In a single-input–single-output (SISO) system, a transfer-function-based approach is most suitable for analysis. This is usually done either in the Laplace (complex frequency) or Fourier (frequency) domain. (Remember that both Fourier and Laplace transforms convert a differential equation to an algebraic equation.) The Laplace transform method is a convenient tool for the analysis and design of linear control systems with continuous data [10]–[12].

Dynamic systems can be modeled through differential equations. Generally, many practical systems/subsystems can be represented by second-order differential equations.

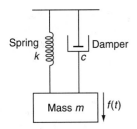

**Figure 7.28** A typical spring-mass-damper system.

As an example, let us consider the simple mechanical system in Figure 7.28. The vibrational motion of the spring-mass-damper system is governed by the differential equation

$$m\frac{d^2x}{dt^2} + c\frac{dx}{dt} + kx = f(t) \tag{7.32}$$

Here, $m$ denotes the mass of the system, $c$ the viscous damping coefficient, and $k$ the effective stiffness of the system. The terms $x$, $dx/dt$, $d^2x/dt^2$ on the left-hand side of Eq. (7.32) represent displacement, velocity, and acceleration of elements of this system, respectively, when a point force $f(t)$ is applied. Note that this is a second-order system with one degree of freedom. When a Laplace transformation is applied to both sides of Eq. (7.32), assuming that the initial conditions are zero, we get

$$(ms^2 + cs + k)X(s) = F(s) \tag{7.33}$$

The transfer function of the system, defined as the ratio of output to input in the Laplace domain, is

$$G(s) = \frac{X(s)}{F(s)} = \frac{1}{ms^2 + cs + k} \tag{7.34}$$

Normally for analysis of a system, we consider that it is excited by an impulse function (whose Laplace transform can be considered to be unity for all practical purposes) and then measure the output or response. It can be seen that the transfer function of a system is the Laplace transform of the impulse response. The evaluation of the ''impulse response'' is, thus, very important in all control systems, including adaptive control systems. If we are interested in time or frequency response, we can consider step or sinusoidal inputs, respectively.

For convenience, we may factorize the denominator in Eq. (7.34) as

$$G(s) = \frac{1}{(s - \alpha_1)(s - \alpha_2)} \tag{7.35}$$

where

$$\alpha_1 = -\frac{c}{2m} + \sqrt{\left(\frac{c}{2m}\right)^2 - \frac{k}{m}}$$

$$\alpha_2 = -\frac{c}{2m} - \sqrt{\left(\frac{c}{2m}\right)^2 - \frac{k}{m}}$$

From Eq. (7.35), it is clear that the system in Figure 7.28 has no zeros but has two poles at $\alpha_1$ and $\alpha_2$. Note that $\alpha_1$ and $\alpha_2$ can be real or complex depending on the radical under the square roots. For the design of controller, it is necessary that the real part of $\alpha_1$ and $\alpha_2$ should be negative. This aspect is dealt in more detail in Section 7.5.4 when we discuss the stability of a control system.

### 7.5.2 Representation of Control Systems

A block diagram is a simplified schematic used to represent a complex system. Block diagrams along with the transfer functions of the constituents provide a good cause–effect

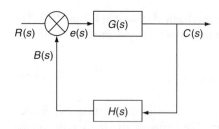

**Figure 7.29** Block diagram of a feedback control system. $G(s)$ is the system transfer function and $H(s)$ is the feedback element transfer function.

relation for the system under consideration. In principle, one can determine the transfer functions of individual elements of a complex system and then determine the overall transfer function of the system by constructing its overall block diagram. In general, any control system with appropriate sensing inputs may be represented by the block diagram.

Both open-loop and closed-loop approaches exist for control. In closed-loop (or feedback) control systems, the controller uses the output of the system in determining the corrective action required to put the system within bounds. Figure 7.29 shows an example of a closed-loop system. The closed-loop transfer function of this system $M(s)$ is given by

$$M(s) = \frac{C(s)}{R(s)} \tag{7.36}$$

where $C(s) = G(s)e(s)$ and $B(s) = H(s)C(s)$. The actuating signal, commonly referred to as the error signal, is $e(s) = R(s) - B(s)$. Therefore,

$$M(s) = \frac{G(s)}{1 + G(s)H(s)} \tag{7.37}$$

The polynomial in the denominator of Eq. (7.37) is known as the *characteristic polynomial* of the system. It may be mentioned that this representation may be extended to multivariable systems by replacing $C(s)$ and $R(s)$ with vectors $\mathbf{C}(s)$ and $\mathbf{R}(s)$ with as many dimensions as the number of variables. This would make $\mathbf{M}(s)$ a matrix.

Since the system poles are determined as those values at which the transfer function becomes infinity, it follows that the poles of the closed-loop system are obtained by solving the equation

$$F(s) = 1 + H(s)G(s) = 0 \tag{7.38}$$

This equation, in fact, is the *characteristic equation* of the system. In other words, the roots of the characteristic polynomial in Eq. (7.38) are the poles and the roots of the numerator polynomial $G(s)$ are the zeros of the transfer function. Poles and zeros are complex frequencies at which the transfer function becomes infinity and zero, respectively. We observe that the poles of $F(s)$ are the open-loop control system poles since they are contributed by $H(s)G(s)$.

The block diagram approach can become confusing for real-life systems with several branches. For complicated systems, the signal flow graphs introduced by Mason may be used for a more detailed description and indeed to simplify complex block diagrams for further analysis. Most references cited give further details of this process [10]–[12]. These approaches are useful in analyzing the stability of the system as discussed in Section 7.5.4. An alternative approach to representing the behavior of a system is by state-space modeling, as discussed in the next section.

## 7.5.3 State-Space Modeling

A state-space representation is another mathematical model of a physical system and shows the relationship among the input, the output and a set of variables called the state variables in a set of first-order differential equations. The state-space representation is clearly a time-domain approach. Unlike the Laplace or frequency-domain approaches, this representation is not limited to systems with linear components and zero initial conditions. In general, for a multi-input, multi-output system, the state-space model turns out to be a matrix differential equation with state, input, and output vectors.

States of a system are defined as those variables whose knowledge allows us to determine the response, for a given input, completely for all future times with minimum amount of information. Mathematically, a dynamic system is defined by a differential equation. In this equation all the $n$ derivatives are defined and it requires $n$ initial conditions for its solution. We may choose to call each of the variables $y$ and each of first $(n - 1)$ derivatives *state variables.* The number of state variables required to model a differential equation is equal to the order of the differential equation. For example, an $n$th-order dynamic system is given by

$$\frac{d^n y}{dt^n} + a_{n-1}\frac{d^{n-1} y}{dt^{n-1}} + \cdots + a_1\frac{dy}{dt} + a_0 y = R(t) \tag{7.39}$$

In state-space modeling, the characteristics of this system are analyzed by reducing this $n$th-order differential equation to a set of coupled $n$ first-order differential equations, with each equation defining one state. This set of equations is called the *state equations.*

In Eq. (7.39), $R(t)$ represents the forcing function. Let us first assume that $R(t)$ does not involve any dynamics, that is, no derivative of $R(t)$ exists and it is represented by $f(t)$. We can now define $y(t)$, $dy/dt, \ldots$, $d^{n-1}y/dt^{n-1}$ as a set of $n$ state variables and represent them by [13]:

$$x_1 = y$$

$$x_2 = \frac{dy}{dt} = \frac{dx_1}{dt}$$

$$x_3 = \frac{d^2 y}{dt^2} = \frac{dx_2}{dt} \tag{7.40}$$

$$\vdots$$

$$x_n = \frac{d^{n-1} y}{dt^{n-1}} = \frac{dx_{n-1}}{dt}$$

The $n$th state equation is obtained by using the definition in Eq. (7.40) and is given by

$$\frac{dx_n}{dt} = f(t) - a_0 x_1 - a_1 x_2 - \cdots - a_{n-2}x_{n-1} - a_{n-1}x_n \tag{7.41}$$

Equations (7.40) and (7.41) can now be put in the matrix form as

$$\{\dot{x}\} = [A]\{x\} + [B]f \tag{7.42}$$

where we have

$$\{x\} = \begin{Bmatrix} x_1 \\ x_2 \\ \vdots \\ x_n \end{Bmatrix}, \quad [A] = \begin{bmatrix} 0 & 1 & 0 & \cdots & 0 \\ 0 & 0 & 1 & \cdots & 0 \\ \vdots & \vdots & \vdots & \vdots & \vdots \\ -a_0 & -a_1 & -a_2 & & -a_{n-1} \end{bmatrix}, \quad [B] = \begin{Bmatrix} 0 \\ 0 \\ \vdots \\ 1 \end{Bmatrix}$$

The output can be written in the form

$$y = [1 \quad 0 \quad \cdots \quad 0] \begin{Bmatrix} x_1 \\ x_2 \\ \vdots \\ x_n \end{Bmatrix}$$

(7.43)

$$y = [C]\{x\}$$

Here $[A]$ is called the *state matrix*, $[B]$ is called the *input matrix*, and $[C]$ is called the *output matrix*. [This is often called the *companion form* of state equation for the dynamic system defined by Eq. (7.39).]

Now let us consider the generalized case where the right-hand side of Eq. (7.39) contains its time derivatives along with the forcing function $f(t)$ as well. That is,

$$R(t) = b_0 f(t) + b_1 \frac{df}{dt} + \cdots + b_{n-1} \frac{d^{n-1}f}{dt^{n-1}} + b_n \frac{d^n f}{dt^n}$$

(7.44)

In this case, defining the state variables is not straightforward. The earlier definition of state variables in Eq. (7.40) will not yield a unique solution. In this case, the state variables must be so chosen such that they will eliminate the derivatives of input $f(t)$. This can be accomplished if we define the $n$ state variables as

$$x_1 = y - c_0 f$$

$$x_2 = \frac{dy}{dt} - c_0 \frac{df}{dt} - c_1 f = \frac{dx_1}{dt} - c_1 f$$

$$x_3 = \frac{d^2 y}{dt^2} - c_0 \frac{d^2 f}{dt^2} c_1 \frac{df}{dt} - c_2 f = \frac{dx_2}{dt} - c_2 f$$

(7.45)

$$\vdots$$

$$x_n = \frac{dx_{n-1}}{dt} - c_{n-1} f$$

where we have

$$c_0 = b_0, \quad c_1 = b_1 - a_1 c_0, \quad c_2 = b_2 - a_1 c_1 - a_2 c_0, \ldots,$$
$$c_n = b_n - a_1 c_{n-1} - \cdots - a_{n-1} c_1 - a_n c_0$$

The choice of the above state variables ensures the uniqueness of the solution of the state equation. Now the reduced first-order state equation can be written as

$$\frac{dx_1}{dt} = \dot{x}_1 = x_2 + c_1 f$$

$$\frac{dx_2}{dt} = \dot{x}_2 = x_3 + c_2 f$$

$$\vdots$$

(7.46)

$$\frac{dx_{n-1}}{dt} = \dot{x}_{n-1} = x_n + c_{n-1} f$$

$$\frac{dx_n}{dt} = \dot{x}_n = -a_n x_1 - a_{n-1} x_2 - \cdots - a_1 x_n + c_n f$$

The above set of equations can be written in matrix form as

$$
\left\{ \begin{array}{c} \dot{x}_1 \\ \dot{x}_2 \\ \cdots \\ \dot{x}_{n-1} \\ \dot{x}_n \end{array} \right\} = \begin{bmatrix} 0 & 1 & 0 & \cdots & 0 \\ 0 & 0 & 1 & \cdots & 0 \\ \vdots & \vdots & \vdots & & \vdots \\ 0 & 0 & 0 & \cdots & 1 \\ -a_n & -a_{n-1} & -a_{n-2} & \cdots & -a_1 \end{bmatrix} \left\{ \begin{array}{c} x_1 \\ x_2 \\ \vdots \\ x_{n-1} \\ x_n \end{array} \right\} + \begin{bmatrix} c_1 \\ c_2 \\ \vdots \\ c_{n-1} \\ c_n \end{bmatrix} f \tag{7.47}
$$

$$
y = \begin{bmatrix} 1 & 0 & \cdots & 0 & 0 \end{bmatrix} \left\{ \begin{array}{c} x_1 \\ x_2 \\ \vdots \\ x_{n-1} \\ x_n \end{array} \right\} + c_0 f \tag{7.48}
$$

Furthermore, these equations can be written in the form

$$
\{\dot{x}\} = [A]\{x\} + [B]f
$$
$$
y = [C]\{x\} + Df \tag{7.49}
$$

Equation (7.49) is the state-space representation of Eq. (7.39) and includes derivatives of the forcing function.

One can now obtain the transfer function of the system from the state equation (7.49) by taking its Laplace transform:

$$
s\{\hat{x}(s)\} - \{x(0)\} = [A]\{\hat{x}(s)\} + [B]\hat{f}(s)
$$
$$
\hat{y}(s) = [C]\{\hat{x}(s)\} + D\hat{f}(s) \tag{7.50}
$$

Here $\{\hat{x}(s)\}$ and $\hat{f}(s)$ are the Laplace transforms of the state vector $\{x(t)\}$ and the forcing function $f(t)$, respectively. Transfer functions are normally derived assuming zero initial conditions. From the first of Eq. (7.50), we have

$$
s\{\hat{x}(s)\} = \{s[I] - [A]\}^{-1} [B] \hat{f}(s) \tag{7.51}
$$

Using this in the second equation of Eq. (7.50), we can rewrite the transfer function as

$$
\frac{\hat{y}(s)}{\hat{f}(s)} = G(s) = [C]\{s[I] - [A]\}^{-1}[B] + D \tag{7.52}
$$

It may be noticed that the transfer function computation involves computation of $\{s[I] - [A]\}^{-1}$. The determinant of the matrix $[s[I] - [A]]$ will give the characteristic polynomial of the transfer function and the eigenvalues of $[A]$ will give the poles of the system.

■

**Example 7.5**    State-space representation of a spring-mass system.

**Solution:**    Let us now consider a simple single-degree-of-freedom spring-mass vibratory system, the governing differential equation of which is $m\ddot{x} + c\dot{x} + kx = f(t)$, where $m$ is the mass of the system, $c$ the viscously damped damper coefficient, and $k$ the stiffness of the system. For state-space representation of the system, we define the state variables $x_1(t) = x(t)$, $x_2(t) = \dot{x}(t)$. Using these state variables, the governing equation reduces to two first-order equations (state equations) written in matrix form as

$$\begin{Bmatrix} \dot{x}_1 \\ \dot{x}_2 \end{Bmatrix} = \begin{bmatrix} 0 & 1 \\ -\dfrac{k}{m} & -\dfrac{c}{m} \end{bmatrix} \begin{Bmatrix} x_1 \\ x_2 \end{Bmatrix} + \begin{Bmatrix} 0 \\ \dfrac{1}{m} \end{Bmatrix} f$$

$$y = \begin{bmatrix} 1 & 0 \end{bmatrix} \begin{Bmatrix} x_1 \\ x_2 \end{Bmatrix}$$

The above equation is in the conventional form of state equations given by Eq. (7.43). Substituting the matrices $[A]$, $[B]$, $[C]$ and $D$ derived from the above equation in Eq. (7.46), we can write the transfer function as

$$G(s) = \frac{1}{ms^2 + cs + k}$$

This is same as what was obtained in Eq. (7.33) by taking Laplace transformation on the governing equation.

■

In designing controllers for practical multi-input-multi-output systems, one must depend extensively on discretized mathematical models derived using FE techniques. The discretized FE governing equation of such complex structures can be written in the form [13]

$$[M]\{\ddot{x}\} + [C]\{\dot{x}\} + [K]\{x\} = \{f\} \tag{7.53}$$

Here $[M]$, $[C]$, and $[K]$ are the mass, damping, and stiffness matrices, respectively, all of dimension $n \times n$. $\{x\}$ is the degree of freedom vector and $\{f\}$ is the force vector, both of which are $n$-dimensional vectors. This equation is similar to the single-degree-of-freedom equation in Example 7.5 and hence the state-space equation is also of the similar form. The reduced state-space form of Eq. (7.53) and its corresponding output vector are given by

$$\begin{Bmatrix} \dot{x}_1 \\ \dot{x}_2 \end{Bmatrix} = \begin{pmatrix} [0] & [I] \\ -[M]^{-1}[K] & -[M]^{-1}[C] \end{pmatrix} \begin{Bmatrix} x_1 \\ x_2 \end{Bmatrix} + \begin{Bmatrix} 0 \\ [M]^{-1} \end{Bmatrix} \{f\} \tag{7.54}$$

$$\{y\} = \begin{bmatrix} [I] & [0] \end{bmatrix} \begin{Bmatrix} x_1 \\ x_2 \end{Bmatrix} \tag{7.55}$$

This is similar to the standard form given in Eq. (7.49), where we can clearly identify matrices $[A]$, $[B]$, and $[C]$, respectively. Notice that Eq. (7.54) represents a $2n \times 2n$ system. This shows that an $n \times n$ second-order system [Eq. (7.53)], when reduced to state-space form, becomes $2n \times 2n$ of the first-order system.

Using Eq. (7.55), we can write the Laplace-domain *output–input* relation in terms of the state vector as

$$\{\hat{f}\} = [G]^{-1}[C]\{\hat{x}\} \tag{7.56}$$

where $[G]$ is the gain matrix. If $r$ states are chosen for input feedback to reduce the overall order of the system, $[G]$ will be of dimension $n \times r$. Once we represent the governing equation in state-space form, we can determine the transfer function using Eq. (7.51). However, normally, the order of the FE system is quite high, especially for dynamic systems. Because of this, it is impossible in practice to design controllers by considering the entire FE system. In most control applications for structural problems such as vibration or noise control, only the first few modes are targeted based on their energy content. In such situation, one must reduce the order of the system using suitable modal-order reduction techniques.

### 7.5.4 Stability of Control Systems

A control system design should adhere to some basic concepts that ensure system stability. It is intuitive that a system will be stable if it has enough damping to remove the transients and provide a bounded steady-state response. That is, a system is said to be *stable* if a finite-duration bounded input causes a finite-duration bounded response. Conversely, a system is said to be *unstable* if a finite-duration input causes the response to diverge from its initial value.

Restricting ourselves to linear systems with system parameters that do not change with time, we know that the system can be described by a linear constant coefficient differential equation. The solution for such an equation takes the form

$$y(t) = Ae^{r_1 t} + Be^{r_2 t} + Ce^{r_3 t} + \cdots \tag{7.57}$$

In this equation, constants $A, B, C, \ldots$ are determined using the initial conditions. The terms $r_1, r_2$, etc., are the roots or eigenvalues of the characteristic polynomial. The stability of Eq. (7.57) depends on the values of $r_i$ $(i = 1, 2, \ldots)$. If they are negative and real, the output tends to zero (value) as $t \to \infty$. Such a system, where all $r_i$ are real and negative, is said to be stable. If the roots of the characteristic polynomial are real and positive, then the output $y(t)$ in Eq. (7.57) grows without bound as $t \to \infty$. Such a system is unstable. If all $r_i$ are purely imaginary, then the system exhibits oscillatory response (continuous oscillations) due to the presence of sine and cosine terms in the output equation. Finally, if all $r_i$ are complex having negative real parts, the system exhibits an exponentially decaying response with time.

Hence, the stability of the system can be determined based on roots of its characteristic polynomial. In other words, the stability of a linear system can be determined by checking whether any root of its characteristic polynomial is in the right half of the $s$-plane or on the imaginary axis. It may be noted that the characteristic polynomial of a system does not depend on the input. Hence, the stability of linear systems is independent of the input excitations.

It may be recalled that a system is said to be stable if the output is bounded for any bounded input. Obviously in such cases, all roots of the characteristic equation will have a negative real part. Although this condition appears simple enough, determination of stability is not a trivial matter. It is not possible in many cases to take the inverse Laplace transformation of the transfer function. Furthermore, even if one can, it may not be easy to find roots of higher-order polynomials. Therefore, one has to resort to approaches that do not require us to factor the polynomial and find its roots. In what follows, we present such approaches to determining the stability of a linear control system directly from its transfer function.

Analysis and design of control systems may be carried out in either the time domain or the frequency domain. The frequency-domain analysis describes the system in terms of its

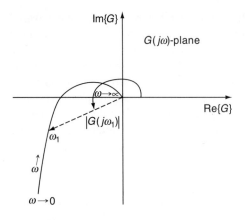

**Figure 7.30** Polar plot of the function $G(j\omega)$.

response to sinusoidal input signal. For a linear system, the input and output frequencies are the same. The transfer function for a sinusoidal input may be obtained by replacing $s$ with $j\omega$. The sinusoidal response, which represents the steady-state behavior of the system, is usually a function of a complex variable and can be represented in terms of magnitude and phase, which are functions of frequency. Graphical plots of the magnitude and the phase as the frequency is varied from zero to infinity are important representations used in analyzing and understanding linear control systems. These graphical representations may be made as polar plots, Bode plots, or magnitude-versus-angle plots.

A *polar plot* is a plot of the magnitude versus phase of the transfer function in polar coordinates, as $\omega$ is varied from zero to infinity. Mathematically, this maps the positive imaginary axis of the $s$-plane to the phase of the curve $G(j\omega)$, as shown in Figure 7.30. The polar plot of the open-loop transfer function, $[G(s)\,H(s)]$, of a feedback system leads to the Nyquist plot, which is useful in analyzing the stability of the complete system; this is discussed in Section 7.5.4.2.

Although in most cases a rough sketch of the polar plot may be sufficient, even this is difficult for complex transfer functions. In such cases one may use Bode plots, which consist of two separate plots, one for magnitude (in decibels) and the other for phase angle, both plotted versus log $\omega$. References [10]–[12] provide details on Bode plots, polar plots, and the Nyquist criterion.

### 7.5.4.1 Routh–Hurwitz Criterion

The characteristic equation of a linear system may be written in the form

$$F(s) = a_0 s^n + a_1 s^{n-1} + a_2 s^{n-2} + \cdots + a_{n-1}s + a_n = 0 \qquad (7.58)$$

In this expression, all the coefficients are real numbers. In order to ensure stability, one needs to ensure that all the roots have negative real parts. The necessary conditions for this to happen are that all coefficients of the polynomial have the same sign and none of the coefficients vanish. The sufficient condition for stability of a system is that its Hurwitz determinants, as obtained below, must be all positive. For convenience, the Routh–Hurwitz array for the characteristic equation of Eq. (7.58) is created as

$$
\begin{array}{c|cccc}
s^n & a_0 & a_2 & a_4 & . & . \\
s^{n-1} & a_1 & a_3 & a_5 & . & . \\
s^{n-2} & b_1 & b_2 & b_3 & . & . \\
s^{n-3} & c_1 & c_2 & & . & \\
\vdots & & & & \\
s^1 & f_1 & \\
s^0 & a_n &
\end{array}
$$

where $a_i$ are coefficients from the polynomial while the other coefficients (Hurtwitz determinants) are given by

$$b_1 = (a_1 a_2 - a_0 a_3)/a_1; \quad b_2 = (a_1 a_4 - a_0 a_5)/a_1; \quad \ldots$$

The process is continued till we get zero as the last coefficient of the third row. In a similar way, the coefficients at this other rows are also computed as

$$c_1 = (b_1 a_3 - a_1 b_2)/b_1; \quad \ldots$$

The necessary condition so that all roots have negative real parts is that all the elements in the first column of the array have the same sign. The number of changes of sign equals the number of roots with positive real parts. This test fails if the first element of any one row is zero while others are not, or if elements in one row are all zero. In both cases, special approaches are available to work through such situations [10].

**Example 7.6** Evaluation of Stability of a System
The characteristic polynomial of a system is given by $F(s) = s^3 + 4s^2 + 2s + 1$. Determine its stability.

**Solution:** The first two rows of the Routh–Hurwitz array for this polynomial are given by

$$1 \quad 2 \quad 0$$
$$4 \quad 1 \quad 0$$

The first element in the next row can be calculated using expressions for $b_n$:

$$b_1 = 7/4$$

The next element in this row is zero. Further, the last row is

$$c_1 = 3/7$$

Now, we can construct the complete Routh–Hurwitz array as

$$
\begin{array}{cccc}
s^3 & 1 & 2 & 0 \\
s^2 & 4 & 1 & 0 \\
s^1 & 7/4 & 0 & 0 \\
s^0 & 3/7 & 0 & 0
\end{array}
$$

Note that there is no sign change in the first column; the system will not have a root in the right half of the $s$-plane and hence is stable.

Note that this test gives us the number of roots, if any, that exist to the right half of the $s$-plane. This algebraic approach, however, does not give the location of these roots on the $s$-plane and hence gives no guidance for design. In other words, this approach is useful for testing if the system is absolutely stable or not. However, this approach can also be conveniently used for lower-order systems and it is relatively simple to implement. Furthermore, it may be noted that this criterion is applicable only if the characteristic equation is algebraic with real coefficients. If there are exponential functions of $s$, this approach fails.

### 7.5.4.2 Nyquist Criterion
An approach that sheds additional light on the degree of stability of a system is based on the Nyquist criterion for identifying the poles located on the right half of the $s$-plane. This is a

frequency-domain technique based on conformal mapping and complex variable theory. The method involves plotting the open-loop frequency response function and looking at the frequency amplitude at the resonant frequencies. From this, one can make inferences on the stability of the system. The main advantage of this criterion is that one can modify the control design by reshaping the frequency response plots.

If the open-loop transfer function $G(s)H(s)$ is a stable function and if the number of poles of the characteristic equation in the right-half plane is zero, the closed-loop system is stable if the Nyquist plot of $G(s)H(s)$ does not enclose $(-1, j0)$ in the $GH$ plane. If poles exist on the right half of the $s$-plane, the Nyquist plot should encircle the critical point $(-1, j0)$ clockwise as many times as the number of poles of $G(s)H(s)$ on the right side of the $s$-plane. The $GH$ plane is the complex plane where $G(s)H(s)$ is plotted. It may be noted that the critical point $(-1, j0)$ in this case is the origin of the $F(s)$ plane where $F(s) = 1 + G(s)H(s)$.

A point is said to be enclosed by the path of $G(s)H(s)$ if it lies on the left of the path as $s$ is traversed from $+j\infty$ to 0 along the imaginary axis. For systems that have poles, this may have to be extended to $-j\infty$ along the imaginary axis, and further from $-j\infty$ to 0 to $+j\infty$ along a semicircle of infinite radius in the right half of the complex plane, as shown in Figure 7.31.

The Nyquist plot in Figure 7.31 indicates that the curve does not encircle the critical point $(-1, j0)$ and hence indicates a stable system. This figure shows the curve only for the negative real part of $s$. The complete Nyquist plot is required only if we need to determine the number of points on the right side of the $s$-plane.

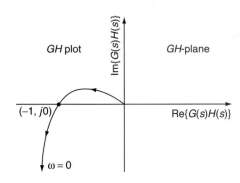

**Figure 7.31** The Nyquist plot for a system with no poles for the open-loop gain.

## ▶ 7.6 IMPLEMENTATION OF CONTROLLERS

### 7.6.1 Design Methodology

The design of a control system involves locating the poles at appropriate positions so that the system response satisfies specified requirements and at the same time ensures the stability of the system. The design of control system thus involves the design of a suitable compensation to modify the response of a plant in order that the stability of the system is ensured. Stability analysis is a very important preliminary step that determines how stable or unstable the system is and indicates to the designer the kind and amount of compensation necessary to ensure stability. The second step in the design process is to analytically determine the parameters for the compensator. As mentioned earlier, an unstable system will have roots in the right half of the $s$-plane, and to stabilize the same we need to move these roots to the left half of the $s$-plane. In general, one can move the roots by

1. Changing gain.
2. Changing plant.
3. Placing a dynamic element in the forward transmission path.
4. Placing a dynamic element in the feedback path.
5. Feed back all or some of the states.

However, in practice the first two cases are seldom permissible. Hence, the design of a good control system involves addition of an element with a gain in the control loop. The gain of a control system can be a constant or a variable depending upon the control system design. The basic or minimum system is determined by having a closed loop with unity feedback. Normally, sensors are assumed ideal and only an amplifier is added between the error signal and the plant. The gain is then set accordingly to meet the steady-state response and bandwidth requirements, and is dictated by the stability analysis.

The gain is an important parameter governing the controller design. Any increase in gain proportionately increases the output. Further, this increase in gain increases the bandwidth and makes the response faster and more accurate. An increase in gain can also decrease damping and may cause offset. Damping can be improved by introducing a forward path with a derivative of the error signal. On the other hand, offset caused by the proportional controller can be eliminated by using an integrator. In general, controllers are designed with forward paths with PD, PI, and PID responses.

$$PD \ Controller = Proportional + Derivative$$

$$G(s) = K_p + K_d s = K_d\left(s + \frac{K_p}{K_d}\right) \tag{7.59}$$

$$PI \ Controller = Proportional + Integral$$

$$G(s) = K_p + \frac{K_i}{s} = K_p\left(\frac{s + (K_i/K_p)}{s}\right) \tag{7.60}$$

$$PID \ Controller = Proportional + Integral + Derivative$$

$$G(s) = K_p + \frac{K_i}{s} + K_d s = \frac{K_d s^2 + K_p s + K_i}{s} \tag{7.61}$$

PID controllers can be thought of as combinations of PI and PD control actions. These are extremely successful and have been used in many applications. The transfer function of a PID controller is schematically represented in Figure 7.32. In Eqs. (7.59)–(7.61), $K_p$, $K_d$, $K_i$ are the gain parameters, which are adjustable.

As seen from the transfer function, the PD controller introduces an additional zero. The design requires that the zero be placed at an appropriate location and the gain adjusted accordingly. In the time domain, this translates to

$$c(t) = K_p e(t) + K_d \frac{de(t)}{dt} + C \tag{7.62}$$

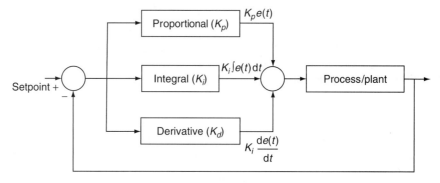

**Figure 7.32** Block diagram of a PID controller.

The time tuning parameter can be obtained from

$$K_d = K_p T_d \tag{7.63}$$

The derivative term can predict the future error based on current values, thus improving the stability. In order to get faster response time, we can introduce PI controllers with the time-domain behavior:

$$c(t) = K_p e(t) + K_i \int e(t)dt + C \tag{7.64}$$

The proportional part of control is effective when the system is not at the set point. If the system is at the set point, the error term is zero, and hence proportional action does not produce an output. In such cases, the integral action makes possible suitable control. In the PI controller, the proportional term considers the current value of $e(t)$ only at the time of the controller calculation, and the integral term (integration of error over time) carries the error history summed from the initiation of the controller to the present time. The design involves determining the coefficients $K_p$ and $K_i$. It is possible to define the reset time tuning parameter $T_i$ using

$$K_i = \frac{K_p}{T_i} \tag{7.65}$$

This parameter gives a separate weight to the integral term so the influence of integral action can be independently adjusted. As in PD control, $T_i$ has units of time and hence is always positive. Smaller values of $T_i$ result in larger contribution of the integral term in Eq. (7.65).

PI controllers are usually tuned based on the Ziegler–Nichols technique [14], which is as follows. First $K_i$ is set to zero and, using proportional action only, $K_p$ is increased from 0 to a critical value $K_{cr}$ where the output first exhibits sustained oscillations. If the period of such oscillations is $P_{cr}$, the controller coefficients are obtained from the experimental values using

$$K_p = 0.45K_{cr}$$
$$K_i = 1.2\frac{K_p}{P_{cr}} \tag{7.66}$$

The design of a PID controller requires a pole to be placed at the origin and two zeros at the desired location for adjustment of the dynamic response. The two zeros may be real or complex depending on the gains used but they will always be in the left half of the $s$-plane. Using the Ziegler–Nichols tuning technique mentioned above, the coefficients in the PID controller are obtained from the experimental values of $K_{cr}$ and $P_{cr}$ using

$$K_p = 0.6K_{cr}$$
$$K_i = 2\frac{K_p}{P_{cr}} \tag{7.67}$$
$$K_d = \frac{K_p P_{cr}}{8}$$

## 7.6.2 Circuit Implementation

Controllers designed by the procedure discussed in the above paragraphs may be implemented in several ways. A simple op-amp circuit for implementing PID controller is shown in Figure 7.33. The transfer function for this circuit is given by

$$\frac{V_{out}}{V_{in}} = -\left[\frac{R_2}{R_1} + \frac{C_1}{C_2} + sR_2C_1 + \frac{1}{sR_1C_2}\right] \tag{7.68}$$

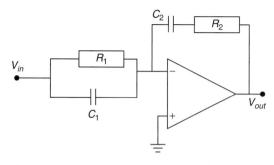

**Figure 7.33** Implementation of a PID controller using op-amp circuit.

The corresponding control coefficients for this circuit are

$$K_p = \frac{R_2}{R_1} + \frac{C_1}{C_2}; \;\; K_d = R_2 C_1; \;\; K_i = \frac{1}{R_1 C_2} \quad (7.69)$$

### 7.6.3 Digital Controllers

Control systems using various forms of digital devices as the controller may be classified as digital control systems. The controller used in these may range from a microcontroller to an ASIC to a personal computer. These digital systems operate on discrete signals that are samples of the sensed signal, rather than on continuous signals. Since most parameters in a plant operating continuously would vary continuously, DACs and ADCs are required in the implementation of such digital controllers. However, several additional blocks, as shown in Figure 7.34, are required to replace a continuous controller by a digital controller.

In Figure 7.34, the clock connected to the DACs and ADCs sends out pulses at a periodic interval $T$, which in turn makes these converters send their output signals to ensure that the controller output is synchronized with the feedback from the plant. In this digital system design, a discrete function $H_z(z)$ must be found so that, for a piecewise constant input to the continuous system $H(s)$, the sampled output of the continuous system equals the discrete output. This is illustrated in Figure 7.35. Suppose the signal $u(k)$ represents a sample of the input signal $u(t)$. These samples $u(k)$ are held constant over the interval $kT$ to $(k+1)T$ to produce a continuous-looking signal $\widehat{u}(t)$. The output of the plant $y(k)$ will be the same piecewise signal as if the continuous $u(t)$ had gone through $H(s)$ to produce the continuous output $y(t)$.

In such discrete systems, $z$-transforms replace the Laplace transforms used in continuous control systems. Another difference is quantization error due to their finite precision. It may be noted that special care is required in setting the error coefficients to account for the lag effects of the sample-and-hold circuits. Yet digital controllers are often preferred as they are inexpensive, less prone to noise effects, and are easy to scale or reconfigure. A stable analog controller may become unstable when converted as a digital system, especially if large

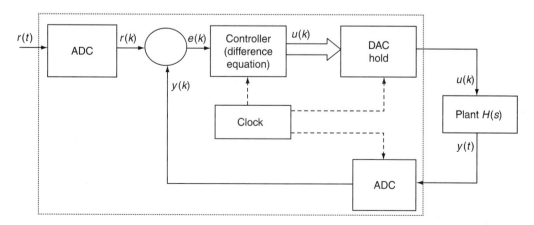

**Figure 7.34** Block schematic of a digital control system.

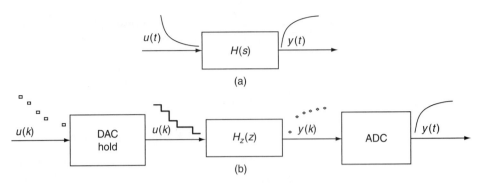

**Figure 7.35** (a) A continuous system and (b) its digital replacement (shown with signals at each stage).

sampling intervals are used. Furthermore, aliasing while sampling can modify the choice of values for various cut-off parameters. Hence the sampling rate is crucial for the stability of a digital control system and fast sampling may be required. Incidentally, approaches used to determine stability in the Laplace domain such as Bode and Nyquist criteria apply to $z$-domain transfer functions as well.

When a large number of identical systems is required, it is economical to use parts programmed and tested at the time of manufacture. In such applications microcontrollers are preferred. On the other hand, the field-programmable memory available on many microcontrollers allows easy revision and even updates in the field. Hence, programmable memory reduces product lead time in some applications.

### 7.6.3.1 Microcontroller

As mentioned earlier, one of the approaches to implement digital control is to use a microcontroller. A microcontroller-based design would be appropriate in large-volume production scenarios and where the end user does not need to alter the control. As is evident from the block schematic in Figure 7.36, a microcontroller is a small computer on a single chip consisting of CPU, clock, timers, I/O ports, and memory. In general, microcontrollers are simple computers with low power consumption, usually operating at low clock speeds,

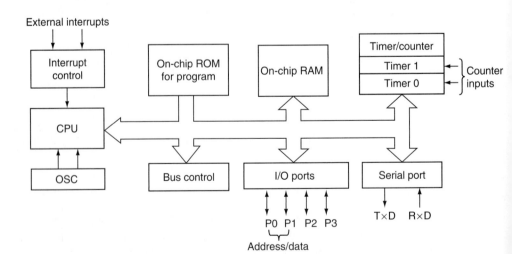

**Figure 7.36** Simplified internal block diagram of 8051 microcontroller.

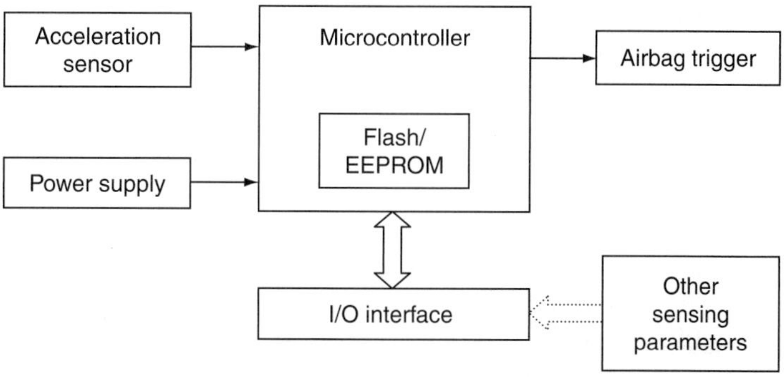

**Figure 7.37** Block schematic of an airbag trigger system in automobiles using a microcontroller.

typically designed for dedicated but low-computation-intensity applications. Such a level of integration helps in drastically reducing the number of chips and printed circuit board space. Applications include control of implantable medical devices and airbag deployment systems in automobiles.

In contrast to general-purpose microprocessors, microcontrollers do not have an external address or data bus since they integrate memory within the chip. In the absence of the bus, fewer pins are required for the interface and thus the chip can be placed in a small (hence inexpensive) package. However, integrating the memory and other peripherals on a single chip and testing them as a unit adds to the overall production cost per chip, although this is still less than the total cost of the complete system without such integration levels.

Microcontrollers provide real-time response to events in the system they are controlling. When certain events occur, an interrupt system signals the processor to suspend the current activity and begin a specific interrupt service routine (ISR). Unlike general-purpose microprocessors, microcontrollers typically do not have a math coprocessor, and hence floating-point arithmetic is performed by software. These programs must fit in the available on-chip program memory. On-chip memory of microcontrollers ranges from electrically erasable programmable read-only memory (EEPROM) to flash memory.

Microcontrollers usually contain a number of general-purpose I/O pins that are software configurable to either an input or an output state. Many applications need to read sensors that produce analog signals requiring ADCs. Mixed-signal microcontrollers integrate analog components to control nondigital electronics systems. I/O ports make it possible to receive and transmit data without CPU intervention.

The application of microcontrollers in the triggering mechanism of airbags in automobiles is illustrated by the simplified schematic in Figure 7.37. When a crash happens, the inputs from the acceleration (deceleration) sensor are compared with a predefined set of parameters (based on other sensor data available from the onboard network), and the microcontroller evaluates the situation and takes the decision to activate the airbag triggers.

**Your Turn:**

Use the datasheet for a commercial accelerometer and microcontroller to write a program to deploy airbags. Improvise this by incorporating infor mation from other sensors, for example, temperature to compensate in the accelerometer output, speed to avoid unnecessary ignition, etc.

### 7.6.3.2 Programmable Logic Controller

Another variant of the digital controller is a programmable logic controller (PLC) used in automation of a plant or a smart system operating at severe environmental and/or process conditions. Like the microcontroller, the PLC is also designed for multiple input and output arrangements, but with extended operating temperature ranges and immunity to electrical noise. Unlike microcontrollers that operate at 5 V or less, PLCs typically have a supply voltage requirement of 24 V. Control through them is typically simple and linear and does not usually depend on history of variation in the control variable. PLCs usually have logic for single-variable feedback control loop with a proportional, integral derivative or PID controller.

PLCs are specifically suited for severe operating conditions such as dust, moisture, temperature, heat, or vibration. Its I/O interface may be either built in or via external I/O modules of a network including the PLC. PLCs have built-in communications ports (e.g. RS-232) and can be configured to communicate over a network to other systems. In general, discrete signals are sent using either voltage or current; the latter are less sensitive to electrical noise than voltage inputs in an industrial environment. In a PLC using 24 V dc I/O, values above 22 V dc may represent ON, values below 2 V dc may represent OFF, and intermediate values are undefined. Analog signals are typically interpreted as integer values by the PLC, with accuracy depending on the device and the number of bits that store the data.

## ▶ 7.7 SUMMARY

In sensors and actuators, electronic circuits are required for signal conditioning and implementation of controls. This chapter presents various aspects of electronic circuits pertinent to micro and smart systems although most students are expected to have studied some related topics already.

The discussion begins with the simplest of electronic devices, a p-n junction diode. These are useful in many practical electronic circuits as sensing elements. Various configurations of a bipolar junction transistor (BJT) are discussed next, which can be used in signal amplifiers. Field effect transistors (FETs) have an edge over BJTs and are more popular in digital and analog circuits. Metal-oxide-semiconductor FETs (MOS-FETs), can be used in active or triode or cut-off modes. Most ICs are fabricated by a variant of this technology known as CMOS, which has p-MOS and n-MOS transistors as its building blocks.

Although transistors can be used to build amplifying circuits, this approach is seldom followed. Rather, a class of ICs—operational amplifiers, popularly known as op-amps—is used in most situations. Most op-amps have nearly infinite input impedance, very high bandwidth, low offset voltage, and high slew rate. Apart from their use as inverting and non-inverting amplifiers with required gain, they can also be used in integrating and differentiating circuit configurations by changing external components.

Op-amp-based differential amplifiers and their more popular extensions, known as instrumentation amplifiers, are widely used in sensing circuits. Another simple circuit used at low signal levels is the Wheatstone bridge. Phase-locked loops using phase detectors are useful in sensing circuits for measuring changes in the resonant frequency of the element. Analog-to-digital converters are required in interfacing between digital and analog circuits. Circuits for measuring charge, capacitance, and frequency shift are also discussed in this chapter.

After a quick introduction to control theory, including state-space modeling and stability analysis, this chapter also provides implementation approaches. Op-amp and digital implementations of control schemes are discussed. Digital controllers are pervasive in applications of microsystems, such as an airbag trigger system.

In the course of discussing these devices and circuits, some interesting applications have been alluded to. You will see some of these circuits being used in practical systems in the case studies of micro and smart systems in the next chapter.

► **REFERENCES**

1. Millman, J. and Halkias, C. C. (2003) *Integrated Electronics: Analog and Digital Circuits and Systems*, McGraw-Hill, Electrical and Electronic Engineering Series, Tata McGraw-Hill Publishing Company Ltd., New Delhi.

2. Neamen, D. A. (2003) *Electronic Circuit Analysis and Design*, 2nd ed., Tata McGraw-Hill Publishing Company Ltd., New Delhi.

3. Achuthan, M. K. and Bhat, K. N. (2007) *Fundamentals of Semiconductor Devices*, Tata McGraw-Hill Publishing Company Ltd., New Delhi.

4. Morrison, S. R. (1981–1982) Semiconductor gas sensors, *Sensors and Actuators*, **2**, 329–41.

5. Bolañosa, M. F., Abeléa, N., Potta, V., Bouveta, D., Racinea, G-A., Querob, J. M. and Ionescua, A. M. (2006) Polyimide sacrificial layer for SOI SG-MOSFET pressure sensor, *Microelectronic Engineering*, **83**(4–9), 1185–188.

6. http://www.national.com/profile/snip.cgi/openDS=LM741 (accessed March 31, 2010).

7. Senturia, S. D. (2001) *Microsystem design*, Kluwer Academic Publishers, Boston, MA, USA.

8. Bhat, K. N. (2007) Silicon micromachined pressure sensors, *Journal of Indian Institute of Science*, **87**(1), 115–31.

9. Xia, X., Zhang, Z., and Li, X. (2008) A Latin-cross-shaped integrated resonant cantilever with second torsion-mode resonance for ultra-resoluble bio-mass sensing, *Journal of Micromechanics and Microengineering*, **18**(3) (Available online: stacks.iop.org/JMM/18/035028).

10. Ogata, K. (2001) *Modern Control Engineering*, 4th ed., Prentice Hall, Englewood Cliffs, NY.

11. Nagrath, I. J. and Gopal, M. (2008) *Control Systems Engineering*, 5th ed., Anshan Publishers, New Delhi.

12. Kuo, B. C. (1994) *Automatic Control Systems*, 9th ed., Prentice Hall, Englewood Cliffs, NJ.

13. Varadan, V. K., Vinoy, K. J. and Gopalakrishnan, S. (2006) *Smart Material Systems and MEMS: Design and Development Methodologies*, John Wiley & Sons, Chichester, London, UK.

14. Ziegler, J. G. and Nichols, N. B. (1942) Optimum Settings for Automatic Controllers, *ASME Transactions*, **64**, 759–68.

► **EXERCISES**

7.1 A noninverting amplifier was designed to have a gain of 100 with an ideal op-amp and suitable feedback resistors. The power supply voltages are $V_{CC} = 10\,V$ and $V_{EE} = -10\,V$. The gain bandwidth product of the op-amp is $10^6\,Hz$. Determine the output voltage $V_0$ of this amplifier when the input voltage is
(a) $150\,mV$.
(b) $0.01\sin(10^5 t)$.

7.2 Using the figure in Table 7.2 for the circuit diagram of a noninverting op-amp, design the resistor $R_2$ if $R_1 = 1\,k\Omega$ such that the voltage gain of the amplifier is 100. Assume that the open-loop gain $A_0$ of the op-amp is infinite.

7.3 Determine the input voltage $V_s$ when the output voltage of the amplifier in Problem 7.2 is 1 V. Assume that $V_{CC} = 10$ V and $V_{EE} = -10$ V. Also, determine the input voltage $V_s$ when the output voltage is 1 V and the open-loop gain of the op-amp is finite with a value $A_0 = 10^5$.

7.4 (a) Draw the circuit diagram of a precision half-wave rectifier using an op-amp and sketch the input and output waveforms one below the other, taking the input as a sine wave.
    (b) Explain the working of the above rectifier.

7.5 The sensitivity of a piezoresistive pressure sensor operating at 5 V power supply was found to be 100 mv/bar (1 bar = $10^5$ N/m$^2$). It gives linear output up to 10 bar pressure.
    (a) Design a differential amplifier circuit with differential gain $A_d = 5$ using an op-amp to amplify the signal output. Choose appropriate power supply voltage for the op-amp.
    (b) The Wheatstone bridge of the sensor is perfectly balanced with zero offset voltage. However, the amplified output gives an offset voltage when its input terminals are connected to the output of the pressure sensor. This is attributed to the finite CMRR of the amplifier whose differential gain is 5. Determine the above output offset voltage if the CMRR of the amplifier is 500.

7.6 The open-loop transfer function of a unity feedback system is given by $G(s) = K/[s(Ts+1)]$ where $K$ and $T$ are positive constants. The unit step response of this system has overshoot, which can be controlled with amplifier gain. By what factor should the gain be changed to reduce the overshoot from 75% to 25%?

7.7 Check the stability of a system whose characteristic equation is given by
    (a) $s^4 + 6s^3 + 12s^2 + 18s + 3 = 0$;
    (b) $3s^4 + 5s^3 + 2s^2 + 10s + 3 = 0$.

7.8 Consider a system with open-loop transfer function

$$G(s)H(s) = \frac{4s + 1}{s^2(s+1)(2s+1)}$$

Sketch the Nyquist contour for this and determine its closed-loop stability.

7.9 For a third-order process with open-loop transfer function as $G(s) = 250/[(s+10)(s+5)^2]$, use a Routh array to determine the critical gain.

# CHAPTER 8

# Integration of Micro and Smart Systems

## LEARNING OBJECTIVES

After completing this chapter, you will be able to:

▶ Understand challenges in the integration of electronics and microsystems.

▶ Learn technologies for packaging of microsystems, highlighting special issues.

▶ Understand case studies of microsystems integrated with electronics.

▶ Get an overview of a smart material used in vibration control.

As mentioned in previous chapters, the success of microsystems and smart systems lies in the ability to integrate them with the required electronics. Several possibilities exist for such integration in the case of microsystems. Some electronics is often integrated with these microsystems and are packaged together. Therefore the integration, packaging, and reliability considerations described in this chapter are particularly relevant to microsystems.

Integration and packaging of microsystems are important in their own right, as in ICs and VLSIs. Packaging of ICs, although not explored as widely as the rest of their development processes, is very crucial, and is often the most expensive step. In microelectronics, packaging provides protection from the environment and electrical interfaces. However, in the context of microsystems, the scope and requirements of packaging delve into several additional aspects such as microfluidic interfaces without compromising the fundamental aspect of protection.

If the microsystem is fabricated on silicon, it may even be possible to build the electronics circuits on the same wafer. Approaches for integration of electronics and microsystems are discussed briefly in Section 8.1. Since packaging is a crucial step in developing microsystems, a review of various technologies used in microelectronics and microsystems packaging are discussed. Some special considerations for the latter are discussed in Section 8.2. In several other smart structures where the sensors and/or actuators are comparatively large, such integration is not feasible and therefore other building blocks of the system, including electronics, may have to be packaged separately and then interfaced together.

Two examples of microsensors integrated with electronics are described in detail in Section 8.3. The design, fabrication, and characterization of a piezoresistive pressure sensor are discussed there along with schematics and results. A force-balanced feedback approach for capacitive sensing useful for bulk and surface-micromachined accelerometers is also described there. Control of structural noise and vibrations is a major challenge in aerospace and civil engineering. In this context the finite element modeling of a simple beam was introduced in Chapter 5. The use of piezoelectric actuators to dampen vibrations is illustrated through another case study in Section 8.4.

## ▶ 8.1 INTEGRATION OF MICROSYSTEMS AND MICROELECTRONICS

The integration of microsensors with electronics has several advantages when dealing with small signals. The function of electronics is to ensure that microsystem components operate correctly. The state of the art in microsystems is to combine them with ICs, utilizing advanced packaging techniques to create system-on-a-package (SOP) or system-on-a-chip (SIP). However, in such cases it is important that the process used for microsystem fabrication not adversely affect the electronics. Microsystems can be fabricated as pre- or postprocessing modules that are integrated by standard processing steps. The choice of integration depends on the application and different aspects of implementation technology. Various approaches for this integration with microelectronics are considered in this section.

In general, three possibilities exist for monolithic integration of CMOS and microsystems: CMOS first, MEMS first, and MEMS in the middle. In addition, a hybrid approach called multichip module (MCM) is also used often for such integration. Each of these methods has its own advantages and disadvantages.

It may be recalled from Chapter 3 that a number of materials like ceramics are used in the fabrication of various microsystems, unlike in CMOS. Annealing of polysilicon or sintering of most ceramics generally requires high processing temperatures, often exceeding that allowed in CMOS. It may be noted that, aluminum layers may need to be diffused at temperatures above 800°C and thus cause performance degradation. Hence, if ceramic processing at a higher temperature is involved, it may be preferable to fabricate the microsystem first. In contrast, if the microsystems involve delicate surface micromachined structures, several common CMOS processes such as lift-off may degrade microsystem performance. Hence, the choice of process sequence is highly dependent on the particular microsystem at hand.

### 8.1.1 CMOS First

In the CMOS-first approach (Figure 8.1), initially developed at UC Berkeley [1], CMOS circuits are first fabricated using conventional processes, and polysilicon microstructures are then fabricated after passivating the CMOS portion with a nitride layer. To prevent the rapid thermal annealing (RTA) of polysilicon (done at 1000°C for stress relief) from affecting CMOS performance, tungsten metallization is used as the conducting layer instead of aluminum.

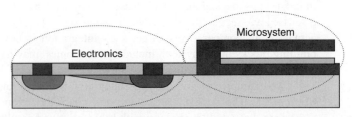

**Figure 8.1** CMOS-first approach.

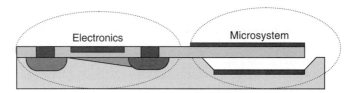

**Figure 8.2** MEMS-first approach.

## 8.1.2 MEMS First

In the MEMS-first approach [2] and [3], on the other hand, microstructures that require higher processing temperatures are first fabricated on a silicon wafer (Figure 8.2). In one such process developed at Sandia National Laboratories, shallow trenches are first anisotropically etched on the wafer and the microsystem is built within these trenches. Silicon nitride and a sacrificial oxide may be deposited within these trenches for the microstructures. A polysilicon layer on top of these layers helps establish contacts with the subsequent CMOS processing. Chemical–mechanical planarization (CMP) and high-temperature annealing are done to optimize this polysilicon layer. The sacrificial oxide covering the microstructure is removed after fabrication of the CMOS device.

## 8.1.3 Other Approaches of Integration

Ideally, in fabricating microstructures along with electronics, it might be possible to use the same material layer to perform different roles in electronics and in mechanical structures. It may be recalled that most microelectronics processes require oxide, polysilicon, metal, and nitride. Utilizing these layers in microsystems requires only a few additional steps of strengthening, masking, and etching.

However, incompatibilities in fabrication processes for microsystems and electronics devices often make such monolithic integration difficult. MCM packaging is an efficient way to integrate microsystem with microelectronics circuits, since it supports a variety of die types on a common substrate without significant changes in the fabrication process of either component. Several sensors, actuators, or a combination can be combined in a single chip using the MCM technique. In this approach, both surface and bulk micromachined components may be integrated with electronics. However, extended procedures are required in this approach to release and assemble the microstructures without degrading the package or other dies in the module. It may be recalled from Chapter 3 that the release etch step is often deferred until the die is put on the package base. Therefore care should be taken to prevent HF, commonly used for this step, from reacting with parts of the package.

In one of the variants of MCM called MCM-D (multichip module—deposited), the dies are embedded in cavities milled on the base substrate and then a thin-film interconnect layer is deposited on top of the components. Holes for interconnecting vias and the windows in the dielectric overlay above the microsystem are usually etched. In the MCM-D approach shown in Figure 8.3, the interconnect layers are

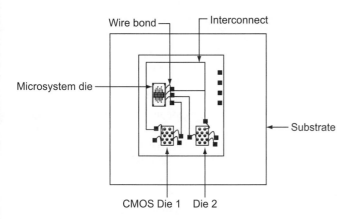

**Figure 8.3** Microsystem and CMOS interfaced through MCM-D package approach.

deposited on the substrate and the dies are mounted above these. Interconnecting the die and the packaging is done by wire bonding. Wet etching using HF can, however, be used for releasing surface micromachined structures part of this chip after shielding the rest with a photo resist [4].

## ▶ 8.2 MICROSYSTEMS PACKAGING

Packaging is the technique of establishing interconnections among the various subsystems and providing an appropriate operating environment for the microsystem components to process the information they gather. Challenges in packaging design depend on the overall complexity of the ultimate device application. Package size, choice of its shape and material, alignment of the device, mounting to provide isolation from shock and vibration, and sealing are some of the many concerns in microsystem packaging. Furthermore, special considerations such as biocompatibility may have to be examined in designing the packaging of a system for biomedical applications. Many important lessons learnt in microelectronics can be adapted to the packaging of microsystems. Packaging is an expensive process since it seeks to protect relatively fragile structures integrated into the device. For a standard IC, the packaging process may take up to 95% of the total manufacturing cost. Issues in microsystems packaging are even more difficult to solve due to stringent requirements for processing and handling and the diversity and fragility of the microstructures providing windows for sensing and actuation. Packaging of microsystems is by no means trivial.

The process technologies for microsystems were developed so closely with microelectronics processing that most early packaging approaches for microsystems were directly adapted from those of microelectronics. In state-of-the-art microelectronics, the device normally accesses the outside world via electrical connections alone and the systems are totally sealed and isolated. In contrast, most microsystems need physical access to the outside world, either to react mechanically to an external parameter or to sense a physical variable. Therefore, unlike electronics packaging where a standard package can be used for a variety of applications, microsystem packages tend to be customized.

In microsystems, mechanical structures and electrical components are combined to form a functional system. In packaging, these electrical and mechanical components are interconnected and the electrical inputs are interfaced with external circuits. Microsystem components can be extremely fragile and must be protected from mechanical damage and hostile environments.

### 8.2.1 Objectives of Packaging

The main objective of packaging is to integrate all components of a system in such a way that cost, mass, and complexity are minimized. The microsystem package should protect the device while still letting it perform its intended functions, without attenuating the signal in a given environment [5,6].

In general, packages provide mechanical support, electrical interfaces to the other system components, and protection from the environment, as shown in Figure 8.4. In the context of microsystems, packages additionally provide an interface between the system and the physical world. Sensing components of microsystems require such an interface to the outside environment. For example, a

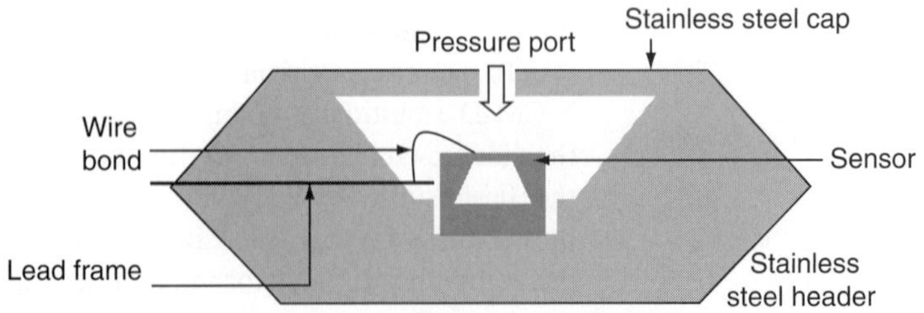

**Figure 8.4** Schematic of a pressure sensor packaged inside a manifold. The electrical and pressure ports, environmental protection, and mechanical support are apparent.

pressure sensor packaging requires the incorporation of a pressure port (Figure 8.4) to transmit fluid pressure to the sensor. Hence, major differences are expected between standard semiconductor device packages and microsystem packages.

Since most microsystems are miniaturized mechanical systems, protection and isolation from thermal and mechanical shock, vibration, acceleration and other physical damage during operation is critical to their performance. The coefficient of thermal expansion (CTE) of the package material should be equal to that of silicon for reliability because thermal cycling may cause cracking or delamination if they are unmatched.

The interface between the microsystem and signal lines is usually made with wire bonds or flip-chip die attachments and multilayer interconnections. Wire bonds and other electrical connections to the device should be made with care taken to protect the device from scratches and other physical damage. Power supply and ac signals to the microsystems are given through these connections and interfaces. These packages should also be able to distribute signals to all components within the package.

Many microsystems are designed to measure variables from the surrounding environment. Microsystem packages must protect the micromachined parts from environments; at the same time they must provide interconnections for electrical signals as well as access to and interaction with external environments. The hermetic packaging generally useful in microelectronic devices is not suitable for such microsystems. These devices might be integrated with the circuits or mounted on a circuit board. Special attention in packaging can protect a micromachined device from hostile surroundings and mechanical damages. Elements that cause corrosion or physical damage to the metal lines as well as other components such as moisture remain a concern for many microsystems.

Most present-day microsystems do not dissipate substantial high power, and their temperatures usually do not increase substantially during operation. Hence, unlike in many ICs, thermal dissipation from microsystems is not a serious problem. However, as the integration of microsystems with other high-power devices such as amplifiers in a single package increases, the need for heat dissipation arises to ensure proper operation. Thus, thermal management is an important consideration in package design. This may become even more important in power-generating microsystems such as microturbines when they become practical. In general, the primary motives in heat-transfer design are to diffuse heat as rapidly as possible and to maximize heat dissipation from the system shell to the environment. Heat spreading in thin space is one of the most important modes of heat transfer in compact electronics equipment and microsystems [7].

## 8.2.2 Special Issues in Microsystem Packaging

The design of microsystem packages, although it follows a path similar to that of microelectronics packaging, must address several unique challenges. Some of these, as well as their typical solutions, are described in the following paragraphs [7].

### 8.2.2.1 Release of Structures

During the fabrication of polysilicon structures by surface micromachining techniques, these structures must be protected by silicon dioxide layers against damage or contamination. In order to release these polysilicon structures, the oxide layers are etched out, often by HF acid solution. An associated risk is *stiction*, a phenomenon by which microstructures tend to stick to one another after release. This is caused by capillary action of droplets of rinse solutions used after etching and may be reduced by incorporating dimples into the structures. Other solutions such as freeze drying or critical $CO_2$ drying are also useful to reduce stiction during release. To further reduce the possibility of stiction over the lifetime of the device, nonstick dielectric films may be coated on the released structures. The issue here is the timing of this release etch vis-à-vis packaging. If done prior to the start of packaging, it may weaken the structure; but if done during or after packaging, contamination is possible, and hence there are incompatibility issues.

### 8.2.2.2 Die Separation

Dicing is a common process in microelectronics fabrication for separating mass-produced devices. The standard die separation method adopted for silicon ICs is to saw through the wafer using a diamond-impregnated blade. The blade and the wafer are flooded with high-purity water while the blade spins at 45,000 rpm. This creates no problem for standard ICs because the surface is essentially sealed from the effects of water and silicon dust. However, if a released microsystem is exposed to water and debris, the structures may break off or get clogged and the moisture may adversely affect performance. Therefore in microsystems, *release of structures is usually done only after dicing*. To further avoid damaging these structures, laser dicing is preferred for microsystems [8].

### 8.2.2.3 Die Handling and Die Attach

During automated processing, vacuum pick-up heads are commonly used for handling microelectronics dies. As these may not be used for microsystems due to their delicate structures, additional clamp attachments are required to handle microsystem dies, possibly by their edges. However, the requirement for this special equipment may be eliminated by *wafer-level encapsulation*. In this approach, a capping wafer is used during dicing in such a way that a protective chip is attached to each microsystem chip. These wafers are bonded using direct bonding with silicon or anodic bonding of silicon with glass.

The process of attaching the diced silicon chip to the pad in the package is known as *die attach*. Two common approaches for this purpose are *adhesive die attach* and *eutectic die attach*. In the former, adhesives such as polyimide, epoxy, or silver-filled glass epoxy are used as die attach materials. However, many die attach materials outgas during curing and these vapors and moisture may be deposited on structures and cause stiction or corrosion, resulting in possible degradation of performance. Possible remedies for this include using low-outgassing materials and/or the removal of vapors during die attach curing. In eutectic die attach, alloys with low melting points, such as Au–Si, are used to provide hermetic sealed packages. For example, atomic percent (80%–20%) Au–Si has a melting point of 363°C.

Hermetic packages provide good barriers to liquids and gases. In hermetic packages, the electrical interconnections through a package must confirm hermetic sealing. Wire bonding is the popular technique to electrically connect the die to the package. However, there is a high chance of structural failure as microstructures are subjected to ultrasonic vibrations during the wire bonding process. Even so, in many applications, parts are hermetically sealed as this increases the reliability and minimizes the possibility of outgassing.

### 8.2.2.4 Interfacial Stress

Recall that thermal annealing, required for strengthening microstructures fabricated with polysilicon, is a concern in integrating microsystems with electronics. Similarly, high-temperature processes during device packaging (such as use of hard solders for die attach or package lid sealing) may introduce additional thermal stress. Microsystems involve several materials with varying CTEs. Subjecting these complex devices to high temperatures may cause device deformation, misalignment of parts, change in resonant frequency of structures, and buckling in long beam elements. Lower-modulus die attach materials is one of the limited solutions to solveing these problems.

### 8.2.2.5 Microsystem Packaging: Issues and Strategies

On the basis of the discussion above, one can adapt a three-level packaging strategy for microsystem packaging: at the die level, device level, and system level. Die-level

**Table 8.1 Issues in packaging: challenges and some solutions**

| Packaging Parameters | Challenges | Possible Solutions |
|---|---|---|
| Release etch and dry | Stiction | Freeze drying, supercritical $CO_2$ drying, roughening of contact surfaces such as dimples, nonstick coatings |
| Dicing/cleaving | Contamination risks, elimination of particles generated | Release dice after dicing, cleaving of wafers, laser sawing, wafer-level encapsulation |
| Die handling | Device failure, top die face is very contact sensitive | Fixtures that hold microsystem die by sides rather than top face |
| Stress | Performance degradation and resonant frequency shifts | Low-modulus die attach, annealing, compatible thermal expansion coefficients |
| Outgassing | Stiction, corrosion | Low-outgassing epoxies, cyanate esters, low-modulus solders, new die attach materials, removal of outgassing vapors |
| Testing | Applying nonelectric stimuli to devices | Test using wafer-scale probing, and finish with cost-effective and specially modified test systems |

*Note:* Modified after [9].

packaging involves passivation and isolation of delicate and fragile devices. These devices must be diced and wire-bonded. Device-level packaging involves connecting the power supply, signal, and interconnection lines. System-level packaging integrates microsystems with signal-conditioning circuitry or ASICs for custom applications. The major barriers in the microsystems packaging technology can be attributed to lack of information and standards for materials as well as the shortage of cross-disciplinary knowledge. Table 8.1 presents solutions to some of the challenges in microsystems packaging.

Another point to be noted is that design standards for packaging of microsystems are yet to be created. Apart from certain types of pressure and inertial sensors used in the automotive industry, most microsystems are custom-built. Hence, standardized design and packaging methodologies are virtually impossible at this time due to the lack of sufficient data in these areas.

## 8.2.3 Types of Microsystem Packages

Although microsystem design is a relatively new field, methods of packaging of very small mechanical device are not new. For example, the aerospace and watch industries have been performing such tasks for a very long time. However, microsystem applications usually require design and optimization of specialized packages depending on the application. In general, the possible packages for microsystems can be categorized as follows:

1. **Metal packages:** These are often used in monolithic microwave integrated circuits (MMICs) and hybrid circuits due to their effective thermal dissipation and ability to provide shielding. In addition, these packages are sufficiently rugged, especially for larger devices. Hence, metal packages are often preferred in microsystem applications also. Materials like CuW (10/90), Silver$^{TM}$ (Ni–Fe alloy), and CuMo (15/85) are good thermal conductors and have higher CTE than silicon.

2. **Ceramic packages:** This is one of the most common types of packaging in the microelectronics industry due to its low mass, low cost, and ease of mass production. Cceramic packages can be made hermetic and adapted to multilayer designs, and can be integrated easily for signal feed through lines. Multilayer packages reduce the size and cost of integrating multiple microsystems into a single package. The electrical performance of the packages can be tailored by incorporating multilayer ceramics and interconnect lines.

3. **Low–temperature cofired ceramic (LTCC) packages:** These are constructed from individual pieces of thick film ceramic sheets called *green sheets*. Metal lines are deposited in each film by thick-film processing such as screen printing. Vertical interconnects are formed by punching via holes and filling them with a suitable metal paste. The unfired layers are then stacked, aligned, laminated together, and fired at a high temperature. The microsystems and the necessary components are then attached using epoxy or solders and wire bonds are made. The processing steps involved in LTCC have already been discussed in Chapter 3 in the context of microsystem fabrication (see Section 3.9.3.5).

4. **Plastic packages:** These are common in electronics industry because of their low manufacturing cost. However, the hermetic seals generally required for high reliability are not possible using plastic packages. Plastic packages are also susceptible to cracking during temperature cycling.

## 8.2.4 Packaging Technologies

Due to the wide variety of interfaces required for functioning, each microsystem may need its own packaging methods. The type of packaging should be decided at the very beginning of device development.

### 8.2.4.1 Wire Bonding

Wire bonding is one of the oldest and still the most flexible method of providing electrical connection between the chip and the external leads of the package. This uses a thin (e.g. 25 μm diameter) wire, usually Al or Au. Gold wires are severed with hydrogen flame and melted to form a ball that is then brought into contact with the bond pad on the chip. Adequate pressure, heat, and ultrasonic force are applied to the ball to effect a metallurgical weld. The wire is then run to the required tip of the lead frame in a gradual arc; pressure and ultrasonic forces bond the wire on to the metal lead post. This is shown in Figure 8.5. On the other hand, aluminum wire bonding uses primarily ultrasonic forces and hence can withstand higher temperatures during further processing.

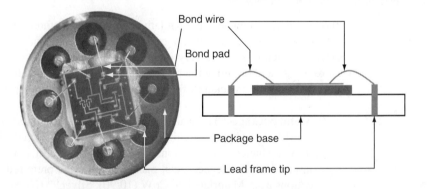

**Figure 8.5** Photograph and schematic of a wire-bonded chip.

#### 8.2.4.2 Flip-Chip Assembly

Flip-chip assembly is the most favored technology for high-frequency applications because the short bump interconnect (between the die and the package) can reduce parasitics. In flip-chips, an IC die is placed on a circuit board with bond pads facing down and directly joining the bare die to the substrate. Bumps form electrical contacts as well as mechanical joints to the die. This reduces the electrical path length and the associated capacitance and inductance, making them particularly suited for high-density RF applications. The minimization of parasitic capacitance and inductance can also reduce the signal delay in high-speed circuits. The technology was developed by IBM during the 1960s and was called *controlled collapse chip connection* (C4) [7].

Flip-chip bonding involves bonding the die face down with a package substrate. Electrical connections are made by plated solder bumps between bond pads on the die and metal pads on the substrate. The attachment is intimate with relatively small spacing ($\sim$100 μm) between die and substrate. In flip-chip assemblies the bumps serve as (a) electrical contacts to the substrate and (b) mechanical support.

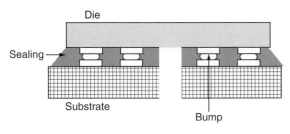

**Figure 8.6** Schematic of flip-chip bonding.

Figure 8.6 shows the flip-chip design of a microsystem package [8]. Since the active surface of microsystem is placed towards the substrate, the cavity will protect the movable elements of microsystem. The standoff distance between the die and the package can be accurately controlled by the bump height. Flip-chip technology is, therefore, a very flexible assembly method suitable for several applications. Figure 8.6 presents the flip-chip bonding process on a ceramic substrate such as alumina. The bump with an acute tail makes it easy to deform, making the bonding area more stable under thermal conditions. However, another consideration in deciding bump height is that taller bumps introduce additional series inductances, thus degrading high-frequency performance.

Flip-chip bonding is attractive in the microsystems industry because of its ability to package a number of dice closely on a single package with multiple levels of electrical traces. A similar system can be built with wire bonding, but may require a larger area and raise reliability issues due to the number of gold wires within the package. The process is self-aligning, because the wetting action of the solder aligns the chip's bump to the substrate pads and compensates for slight misalignment between them. Another feature of this process is that it allows removal or replacement without scrapping the components. However, flip-chip packaging may not be compatible with microsystems with micro-structures that should be exposed to open environment.

The primary advantages of flip-chip assembly are [10]:

1. Size and weight reduction.
2. Applicability to existing chip designs.
3. Performance enhancement and increased production.
4. Feasibility of chip replacement.
5. Increased input/output (I/O) capability, especially for RF and optical interfaces.

The reliability of this scheme depends on the difference between the CTE of the substrate and that of the chip material, which may introduce thermal and mechanical stresses on the bumps.

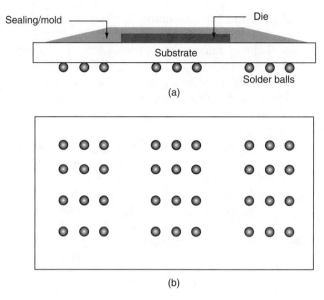

**Figure 8.7** Schematic of a PBGA package. (a) Cross-sectional view; (b) bottom view of package.

### 8.2.4.3 Ball-Grid Array

A ball-grid array (BGA) is a surface-mount chip package containing a grid of solder balls for interconnections. This approach leads to small size, high lead count (since it is a surface mount), and low parasitic inductance. In this scheme, a flexible circuit tape is used as the substrate and a low-stress elastomer is used for the die attachment. The die is mounted face down and the electrical connections are made by bonding. The leads are encapsulated with epoxy for protection. Both ceramic and plastic variants are available. A miniaturized version called a microball-grid array ($\mu$BGA) yields package sizes very close to die size [7].

The plastic-ball-grid array (PBGA) in Figure 8.7 has become one of the most popular packaging alternatives in the semiconductor industry. Its advantages include high lead count (typically >208 leads), reduced coplanarity problems as there are no leads to bend, and minimized handling issues. Because of the large ball pitch of a PBGA, the overall package and board assembly yields are good. Furthermore, PBGA has an improved design-to-production cycle time and can also be used in few-chip-packages? (FCPs) and MCMs configurations.

### 8.2.4.4 Embedded Overlay

The embedded overlay concept for microsystem packaging is derived from the chip-on-flex (COF) process widely used for microelectronics packaging [11]. COF is a high-performance multichip packaging technology in which the dies are encased in a molded plastic substrate and interconnections are made using a thin-film structure formed over these components. The electrical interconnections are made through a patterned overlay while the die is embedded in a plastic substrate. Chips are attached face down on the COF overlay using polyimide or thermoplastic adhesives. The substrate is molded after bonding the chips around the components using a plastic mold-forming process such as transfer, compression, or injection molding. The electrical connections are made by laser drilling via holes.

### 8.2.4.5 Wafer-Level Packaging

A cost-effective method for microsystem packaging is wafer-level packaging [12]. Designing the packaging schemes and incorporating them into the device manufacturing process itself can reduce the overall cost. Versatile packaging may be needed for many devices in which

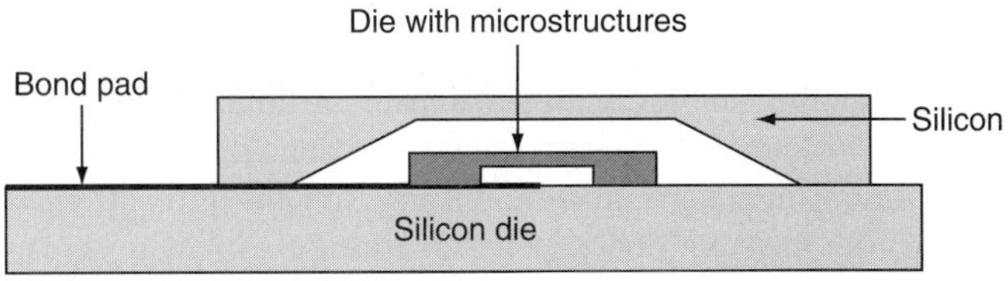

**Figure 8.8** Silicon wafer-level microsystem packaging.

microsystems and microelectronics are on a single chip. Since microsystems have movable structures built on the surface of the wafer, the addition of a cap wafer on the silicon substrate makes them useful in many applications. The cap provides protection against handling damage and also avoids atmospheric damping. The substrate with the active device is bonded to a second wafer, which need not be of the same material as the substrate. The bonding is done by glass frit or by anodic bonding in which the bond is created by an electrical potential. Precision-aligned wafer bonding is key in high-volume, low-cost packaging of microsystems. Examples include micromachined inkjet heads and piezoresistive pressure sensors.

The wafer-level package, which protects the device at the wafer stage (before dicing), is clearly a choice to be made at the product design stage. This involves extra fabrication processes in which a micromachined wafer is bonded to a second wafer with appropriate cavities etched on it. Cavities are formed by the silicon etching methods described in Chapter 3. Figure 8.8 shows a schematic diagram of a device after wafer-level packaging. This approach enables the microsystem to move freely in vacuum or inert atmosphere with hermetic bonding, which prevents any contamination of the structure. Etching cavities in a blank silicon wafer and bonding it with the substrate by placing over the device can make a hermetic seal.

*Microscale riveting* or *eutectic bonding* [13] can be performed by directional electro-plating for rivet formation through cavities formed on the wafer. The wafer joining can be done at room temperature and low voltages. The protected devices after microriveting can be treated in the same way as IC wafers during the dicing process. Once the joining is complete, the resulting chips can be handled the same way as IC chips during the remaining packaging steps such as wire bonding and molding for plastic packages.

In eutectic bonding, a gold layer is sputtered and used for heating; this eventually melts and causes the bonding. The temperature of this layer rises with the flow of current through a microheater, further activating the bonding process. The effectiveness of the microheater depends on the materials and design of the geometrical shape of the structure. For example, a high temperature of 1000°C can be created using microheaters. This setup can be used for silicon–gold as well as silicon–glass eutectic bonding. Conventional bonding takes one hour to reach the temperature, while localized eutectic heating takes less than five minutes. Phosphorus-doped polysilicon and gold resistive heaters are used in the silicon-to-glass fusion and silicon-to-gold eutectic bonding process, respectively. Some of the common methods of wafer bonding are compared in Table 8.2.

**Table 8.2 Comparison of wafer-bonding schemes**

| Type | Bonding Temperature | Sensitivity to Roughness | Hermeticity |
|---|---|---|---|
| Fusion bonding | 1000–1100°C | Less than 1 nm | Yes |
| Anodic bonding | 430–470°C | Less than 1 μm | Yes |
| Epoxy bonding | <300°C | Not critical | No |
| Eutectic bonding | 360–370°C | Typically 1 μm | Yes |

## 8.2.5 Reliability and Key Failure Mechanisms

Reliability studies and mechanisms of failure are as essential in microsystems as in any other systems. Extensive studies are done on the reliability and mechanisms of failure of various microsystem components. Reliability requirements for microsystems are significantly different and depend on specific applications, especially in systems with unique microsystems. Hence, standard reliability testing is not possible until a common set of reliability requirements is developed. The chain of events that establishes the reliability of a microsystem package is described in Table 8.3.

Understanding the reliability of systems comes from a knowledge of their failure behavior and failure mechanisms. The common failure mechanisms of microsystems are summarized as follows.

### 8.2.5.1 Stiction

In most cases, microsystem failure is caused by stiction and wear. Stiction occurs due to microscopic adhesion when two surfaces come into contact, for example, during electrostatic actuation. This adhesion is caused by van der Waals forces resulting from the interaction of instantaneous dipole moments of atom. To reduce stiction, various dielectric or polymer-thin-film materials are often coated onto the surface of one of these moving structures.

In-use stiction can also be caused by mechanical shock during packaging or operation. For example, as the sensitivity of the micromechanical accelerometer is increased, the stiffness of the proof-mass suspension must be decreased, also decreasing the mechanical shock required to bring the mechanical elements into contact with each other. In-use stiction can occur when released microstructures are exposed to humid environments. Water vapor can condense and the water droplets formed in the narrow gap between the surfaces exert an attractive force that pulls the microstructure toward the substrate. This causes eventual collapse and permanent adhesion of microstructures.

In-use stiction of contacting structures can be reduced by reducing the adhesion energy by providing hydrophobic surface coating. This can eliminate water-mediated adhesion. Weakly adhesive materials such as fluorinated polymer (e.g. polytetra fluoro ethylene PTFE) still has a surface tension of about $20 \, \text{mJ/m}^2$. (Note: An adhesion energy of $100 \, \text{J/m}^2$ is high enough to counterbalance the restoring elastic force). Reducing the geometrical area of contact using bumps and side-wall spacers, adding side-wall-spacing supports on

**Table 8.3 Typical steps required to establish reliability of microsystems packaging**

| Procedure | Details |
|---|---|
| Microsopic examination | Optical microscopy at all stages of packaging |
| Electrical tests | Failure identification by electrical functional tests before and after a bake cycle |
| X-ray analysis | Analyses of internal connections and alignments at decapsulation hotspots using x-ray |
| Hermeticity | Verify sealing environment for hermeticity of the cavity by residual gas analysis |
| Decapsulation test | Visual and electrical verification of internal aspects of the component after opening the package cap |
| Destructive and nondestructive analyses | Bond pull and die shear tests and IR imaging and verification material properties on the die |

substrate, and increasing surface roughness of microstructures in the processing stage reduces in-use stiction and enhances reliability.

### 8.2.5.2 Wear

Wear occurs due to the movement of one surface over another, and is caused by the removal of material from a solid surface by some kind of mechanical action. The primary mechanisms of wear in microsystems are adhesion, abrasion, corrosion, and surface fatigue. Corrosion is due to one or more of the following [7]:

1. Moisture ingress.
2. Loss of hermetic contact.
3. Galvanic corrosion.
4. Crevice corrosion.
5. Pitting corrosion.
6. Surface oxidation.
7. Stress corrosion.
8. Corrosion resulting from contaminants or micro-organisms.

### 8.2.5.3 Delamination

Microsystems may fail due to delamination of bonded thin-film materials. Failure of bonding between dissimilar materials or wafer-to-wafer bonding can cause delamination.

### 8.2.5.4 Dampening

Microsystems may have mechanical parts whose resonant frequency is critical to their operation. Dampening is caused by many variables including atmospheric gases. Good sealing is, therefore, critical for microsystems. Those microsystems with moving parts are therefore more susceptible to environmental failure than other packaged systems.

### 8.2.5.5 Fatigue

The effect of continuous or cyclic variation of loading characteristics, even if the load is below some threshold, produces fatigue in a structure and may cause it to fail. Such cyclic loading causes the formation of microcracks that weaken the material over time and create localized deformations. The time until a structure ultimately fails depends on various factors including material properties, geometry, and variations in load characteristics. It is believed that in microsystems, for example, structures made of polysilicon formed by deposition techniques, such as CVD, have rough surfaces that result from the plasma etching used in the final stages of microsystem processing. Under compressive loading, these surfaces come into contact and their wedging action produces microcracks that grow during subsequent tension and compression cycles. This fatigue strength is strongly influenced by the ratio of compressive to tensile stresses experienced during each cycle.

### 8.2.5.6 Mechanical Failure

Changes in elastic properties of materials due to exposure to the atmosphere or temperature may affect the resonant and damping characteristics of microstructures such as beams and will cause a change in device performance.

## ▶ 8.3 CASE STUDIES OF INTEGRATED MICROSYSTEMS

In this section we present a few case studies of sensing devices integrated with electronics as illustrations of microsystem integration.

## 8.3.1 Pressure Sensor

One of the first sensors fabricated using micromachining techniques was the pressure sensor. These are widely used in aerospace, automobile, biomedical, and defense applications. In general, pressure sensors are based on measuring the mechanical deformation caused in a membrane when it experiences stress due to pressure of a fluid. It may be noted that silicon is brittle and breaks at a maximum strain of 2%. However, its Young's modulus is very high (190 GPa), and below the elastic limit strain is linearly related to stress. The resulting strain or displacement (which is a measure of the pressure on the membrane) is converted to electrical signals. Several principles are used to sense strain: piezoelectricity, piezoresistivity, changes in capacitance, changes in resonant frequency of vibrating elements in the structure, and changes in optical resonance. A vast majority of commercial pressure sensors today use piezoresistive sensing, which measures changes in the electrical resistance caused primarily by geometrical changes.

Most conventional pressure sensors use *strain gauges*. These sensors use the change in the resistance of conductors due to the change in their physical dimensions (deformations) when subjected to strain. These are attached to the membrane and give information about the strain experienced by the membrane due to the applied pressure. However, if these sensors are made of semiconductors, the change in the resistance is predominantly due to a change in resistivity rather than to deformations. This effect, known as *piezoresistivity*, is a phenomenon by which the electrical resistance of a material changes in response to mechanical stress. In piezoresistive pressure sensors, semiconductor resistors are laid out on a membrane to sense the strain from the membrane that is subjected to external pressure. Incidentally, the discovery of piezoresistive effect in silicon and germanium in 1954 is commonly cited as the stimulus for silicon-based sensors and micromachining. The first piezoresistive pressure sensor based on diffused resistors in a thin silicon diaphragm was demonstrated in 1969.

Piezoresistive pressure sensors are simple to fabricate and provide a directly readable electrical output. Further, they can be easily integrated with microelectronics circuits, since the fabrication processes are compatible with IC fabrication techniques. Therefore, we focus our discussion on piezoresistive pressure sensors.

### 8.3.1.1 Design Considerations of a Piezoresistive Pressure Sensor

The piezoresistive effect can be quantified using the *gauge factor*. The resistance of a rectangular conductor is expressed by

$$R = \frac{\rho_e l}{wt} \tag{8.1}$$

where $\rho_e$ is the resistivity and $l$, $w$, and $t$ are the length, width, and thickness of the conductor, respectively. When the resistor is subjected to strain, the relative change in resistance is given by

$$\frac{\Delta R}{R} = \frac{\Delta \rho_e}{\rho_e} + \frac{\Delta l}{l} - \frac{\Delta w}{w} - \frac{\Delta t}{t} \tag{8.2}$$

Here $\Delta l$, $\Delta w$, $\Delta t$, and $\Delta \rho_e$ are the changes in the respective parameters due to the strain. If the resistors experience tensile stress along the length, the resistor's thickness and width decrease whereas the length increases. Poisson's ratio, $v$, relates the change in length to the change in width and thickness of the piezoresistors by the following equation:

$$\frac{\Delta w}{w} = \frac{\Delta t}{t} = -v\frac{\Delta l}{l} \tag{8.3}$$

**Table 8.4 Gauge factors of different types of strain gauges**

| Type of Strain Gauge | Gauge Factor |
|---|---|
| Metal foil | 1–5 |
| Thin-film metal | ≈2 |
| Diffused semiconductor | 80–200 |
| Polycrystalline silicon | ≈30 |

The gauge factor $G$ (strain sensitivity) is given by

$$G = \frac{\Delta R/R}{\varepsilon} = 1 + 2v + \frac{\Delta \rho_e/\rho_e}{\varepsilon} \tag{8.4}$$

where $\varepsilon = \Delta l/l$ is the strain. The first two terms in Eq. (8.4) represent the change in resistance due to dimensional changes and are dominant in metal gauges, while the last term is due to the change in resistivity. In semiconductor gauges, the resistivity change is larger than the dimensional change by a factor of about 50, and hence the dimensional change is generally neglected.

As shown in Table 8.4, the gauge factors of different types of strain gauges can differ dramatically. For metals, $\rho$ does not vary significantly with strain and $v$ is typically in the range of 0.3 to 0.5, leading to a gauge factor of only about 2. In semiconductor strain gauges, the piezoresistive effect is very large and hence the gauge factor is high. Gauge factors up to 200 for p-type silicon and up to 140 for n-type silicon have been reported.

The basic structure of a piezoresistive pressure sensor consists of four sensing elements connected in a Wheatstone bridge configuration to measure stress in a thin membrane (see Figure 8.9). When fabricated using silicon bulk micromachining by anisotropic etching, the membrane is invariably rectangular or square and is realized on (100) plane with the edges in the <110> directions. The stress is a direct consequence of the membrane deflection in response to an applied pressure. Therefore, the design of a

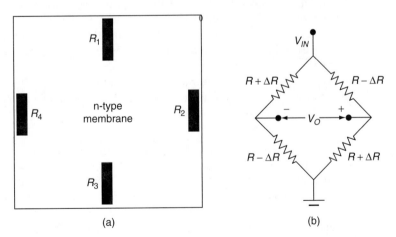

(a)  (b)

**Figure 8.9** (a) Schematic of a single-crystal piezoresistive pressure sensor showing arrangement of p-type piezoresistors on an n-type membrane; (b) Electrical connection of resistors in the Wheatstone bridge arrangement.

pressure sensor involves (a) determination of the thickness and geometrical dimensions of the membrane and (b) location of the piezoresistors on the membrane to achieve maximum sensitivity.

From the theory of plates, the maximum deflection $w_0$ of a square membrane of thickness $h$ and side $2a$ takes place at the center of the plate and is related to the applied pressure $p$ by

$$p = Y \frac{h^4}{a^4} \left[ g_1 \frac{w_0}{h} + g_1 \left( \frac{w_0}{h} \right)^3 \right] \tag{8.5}$$

Here, $Y$ is the Young's modulus and $g_1$ and $g_2$ are related to Poisson's ratio and are given by

$$g_1 = \frac{4.13}{1 - v^2} \quad \text{and} \quad g_2 = \frac{1.98(1 - 0.585v)}{1 - v}$$

The first term on the right-hand side of Eq. (8.5) is due to the stress distribution caused by pure bending. In this case the central plane of the diaphragm is not stretched or compressed. This is true if the deflection of the diaphragm is small compared to its thickness. Otherwise the central plane of the diaphragm is stretched like a balloon, as represented by the second term in the equation. As can be seen from Eq. (8.5), the second term is a source of nonlinearity in the piezoresistive pressure sensors. Therefore, even though a reduction in the thickness $h$ would increase the deflection $w_0$ and hence the sensitivity of the pressure sensor, it is necessary to keep $h$ sufficiently thick so that $w_0$ at the maximum operational pressure is small compared to $h$.

The burst pressure, defined as the maximum pressure at which the diaphragm ruptures, is another criterion for choosing the thickness of a diaphragm. The maximum stress in a square diaphragm is at the midpoints of its edges and is given by the relation

$$\sigma_{\max} = p \left( \frac{a}{h} \right)^2 \tag{8.6}$$

Thus, for a given value of $2a$, the thickness $h$ should be chosen such that at the maximum operating pressure $P_{\max}$ of the sensor, $\sigma_{\max}$ estimated using Eq. (8.6) is less than the burst pressure $\sigma_{\text{burst}}$. For single-crystal silicon, an ideal value of $\sigma_{\text{burst}}$ is 7 GPa (1 Pa = 1 N/m²).

The next design criterion is to assign proper locations for the piezoresistors on the diaphragm so that the sensitivity is maximum. It may be noted that the resistance change can be calculated as a function of stress using the concept of the piezoresistive coefficient. Stresses in the longitudinal ($\sigma_l$) and transverse ($\sigma_t$) directions (with respect to the direction of current flow in the resistor) contribute to the change in resistance. The fractional change in the resistance has a linear dependence on the stress in both these directions. Assuming that mechanical stresses are constant over the footprint of a resistor, the relative resistance change is given by

$$\frac{\Delta R}{R} = \sigma_l \pi_l + \sigma_t \pi_t \tag{8.7}$$

where $\pi_l$ and $\pi_t$ are the longitudinal and transverse piezoresistive coefficients, respectively.

It has been observed that piezoresistive coefficients depend greatly on crystal orientation as well as on dopant type and concentration. It may be recalled that the separation of valence and conduction band levels is very different for n-type and p-type in silicon and is a strong function of doping concentration. Table 8.5 shows piezoresistive coefficients for n-type and p-type {100} wafers with doping levels below $10^{18}$ cm$^{-3}$. These values decrease at higher doping concentrations. For {100} wafers, the piezoresistive

**Table 8.5  Piezoresistive coefficients for p-type and n-type {100} wafers and doping levels below $10^{18}$ cm$^{-3}$**

| Si Wafer Type | $\pi_l \, (10^{-11} \, \text{Pa}^{-1})$ | $\pi_t \, (10^{-11} \, \text{Pa}^{-1})$ | Direction |
|---|---|---|---|
| p-type | 0 | 0 | <100> |
| p-type | 72 | −65 | <110> |
| n-type | −102 | 53 | <100> |
| n-type | −32 | 0 | <110> |

coefficients for p-type elements are maximal in the <110> directions and vanish along the <100> directions. Therefore, p-type piezoresistors must be oriented along the <110> directions to measure stress.

In piezoresistive pressure sensor design layout on single-crystal {100} substrates, the four diffused p-type piezoresistive sense elements are located near the edges of the n-type silicon diaphragm, as shown in Figure 8.9, where the stress is maximum, as given by Eq. (8.6). Here, the two piezoresistors $R_1$ and $R_3$ are placed perpendicular to two opposite edges of the membrane, and the other two ($R_2$ and $R_4$) are placed parallel to the other two edges. When the membrane deflects downward (into the plane of the paper), causing tensile stress at the edges of the membrane surface, both resistors $R_1$ and $R_3$ experience longitudinal tensile and transverse stress given by $\sigma_l = p(a/h)^2$ and $\sigma_t = vp(a/h)^2$, respectively. Hence they show an increase in resistance given by

$$\frac{\Delta R}{R} = \pi_l p \frac{a^2}{h^2}(1 - v) \tag{8.8}$$

Similarly, resistors $R_2$ and $R_4$ laid out parallel to the edge of the diaphragm experience transverse and longitudinal stresses given by $\sigma_t = p(a/h)^2$ and $\sigma_l = vp(a/h)^2$, respectively. Therefore, these show a decrease in resistance given by

$$\frac{\Delta R}{R} = \pi_t p \frac{a^2}{h^2}(1 - v) \tag{8.9}$$

In single-crystal (100) silicon, $\pi_l$ and $\pi_t$ are equal in magnitude. Therefore, the absolute value of the four resistance changes can be made equal when the resistors are arranged in the locations shown in Figure 8.9 to achieve maximum sensitivity.

The resistors are chosen to be equal in magnitude so that the bridge in Figure 8.9(b) is balanced when the pressure is zero. When the diaphragm is subjected to pressure, the imbalance created in the bridge results in an electrical output $V_O$ of polarity shown in Figure 8.9(b) and given by

$$V_O = \frac{\Delta R}{R} V_{IN} = p \frac{a^2}{h^2}(1 - v)\pi_l V_{IN} \tag{8.10}$$

where $R$ is the zero-stress resistance of each resistor and $V_{IN}$ is the bridge supply voltage.

In the low-pressure range, flat-diaphragm pressure sensors are not suitable because the sensitivity would have to be increased considerably by making the $(a/h)$ ratio extremely large, resulting in a high degree of nonlinearity. To improve the sensitivity and linearity simultaneously, specialized geometries such as diaphragms with a rigid center or boss have been introduced to increase the stiffness limiting the maximum deflection of the diaphragm.

### 8.3.1.2 Performance Parameters of Pressure Sensors

At this stage let us look at some of the important performance parameters of a pressure sensor. A typical response (output voltage of the Wheatstone bridge versus the pressure) of a pressure sensor in the literature [14] is shown in Figure 8.10. The *sensitivity S* is defined as the relative change of output voltage per unit change of applied pressure (slope of the linear portion of the curve in Figure 8.10). This is expressed in mV/bar (1 bar = $10^5$ Pascal) for a given input voltage. To be specific, sensitivity can also be expressed in mV/V/bar as

$$S = \frac{\Delta V_O}{\Delta p} \frac{1}{V_{IN}} = \frac{\Delta R}{\Delta p} \frac{1}{R} \tag{8.11}$$

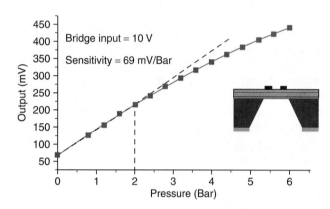

**Figure 8.10** Typical characteristics of a pressure sensor.

The *offset voltage* ($V_{OFF}$) is the output voltage of the pressure sensor when the applied pressure is zero. An offset voltage of about 65 mV can be seen in the performance curve in Figure 8.10. This is mainly due to some residual stress on the membrane or variability in the four resistors. It can be compensated by using external resistors along with signal conditioning electronics.

Now we come to the concept of *non-linearity*. The output voltage does not vary linearly with pressure when the applied pressure exceeds a certain value. Nonlinearity is defined with reference to the full-scale output corresponding to the maximum pressure at which the sensor can be used. This can be better appreciated by referring to the output voltage $V_O$ versus applied pressure $p$ and the endpoint straight line in Figure 8.11; the nonlinearity $NL_i$ of a pressure sensor at a specific pressure $p_i$ is defined as follows:

$$NL_i = \frac{V_{OI} - (V_{OM}/p_m)p_i}{V_{OM}} \times 100\% \tag{8.12}$$

where $V_{OI}$ and $V_{OM}$ are respectively the output voltages at a pressure $p_i$ and the full-scale output voltage at the maximum pressure input $p_m$. Thus, the nonlinearity can be either positive or negative depending upon the calibration point. The nonlinearity of the pressure sensor is the maximum value of $NL_i$.

Nonlinearity in the piezoresistive pressure sensors is caused mainly by the following:

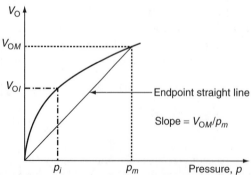
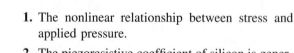

**Figure 8.11** Output voltage $V_O$ versus pressure $p$ and the corresponding endpoint straight line.

1. The nonlinear relationship between stress and applied pressure.

2. The piezoresistive coefficient of silicon is generally considered to be independent of stress. In practice, however, this is not really true at high accuracy.

3. A difference in piezoresistive sensitivity between the resistors of the Wheatstone bridge.

**Example 8.1**

Determine **(a)** the burst pressure and **(b)** the maximum pressure up to which a pressure sensor works in the linear range of operation in the following three cases for square silicon diaphragms with thickness $h$ and diaphragm side length $2a$: (i) $h = 5\,\mu m$, $a = 250\,\mu m$; (ii) $h = 10\,\mu m$, $a = 250\,\mu m$.

 **(c)** Determine the sensitivity of piezoresistive pressure sensors fabricated using the three membranes, with the four resistors laid out as in Figure 8.9. Assume that the gauge factors of the piezoresistor are $G_L = 100$ and $G_T = -100$. The yield strength (maximum stress) of silicon is $\sigma_{max} = 7\,GPa$.

**Solution:** The maximum stress is experienced by the membrane in the middle of the membrane edge and is given by $\sigma_{max} = p(a/h)^2$. The silicon membrane will burst at a pressure $P_b$ at which $\sigma_{max} = 7\,GPa = 7 \times 10^4\,bar = 7 \times 10^9\,Pa$ (1 Pa $= 1\,N/m^2$). The linear range of operation exists till the deflection $\omega_o \ll h$ [see Eq. (8.5)].

Case (i) $h = 5\,\mu m$, $a = 250\,\mu m$

**(a)** $p_b = \dfrac{\sigma_{max}}{(q/h)^2} = 7 \times 10^9 / 2500 = 2.8 \times 10^6\,Pa = 28\,bar$ (1 bar $= 10^5\,Pa$)

**(b)** From Eq. (8.5), in the linear range the cubic term in $\omega_o/h$ is negligible and we can write

$$p = Yg_l \frac{h^4}{a^4}\frac{\omega_0}{h} \quad \text{where } g_1 = \frac{4.13}{1 - v} = 5.9 \text{ for silicon.}$$

$$p = 170 \times 10^9 \times 5.9 \times \frac{\omega_o}{h}\frac{h^4}{a^4} = 10^{12}\frac{\omega_o}{h}\frac{h^4}{a^4}\,Pa$$

Considering $\omega_o = h/10$, we obtain the maximum pressure $p_{max}$ up to which linearity can be maintained as

$$p_{max} = 10^{12} \times 0.1 \times 1.6 \times 10^{-7} = 1.6 \times 10^4\,Pa = 0.16\,bar$$

Case (ii) $h = 10\,\mu m$, $a = 250\,\mu m$  Following the same procedure as in case (i), we obtain

**(a)** $p_b = 7 \times 10^9 / 625 = 11.2 \times 10^6\,Pa = 112\,bar$

**(b)** $p_{max} = 10^{11} \times 2.56 \times 10^{-6} = 2.56 \times 10^5\,Pa = 2.56\,bar$

**(c)** Sensitivity for 1 volt input is obtained by using eq. (8.11) $S = \dfrac{\Delta R}{R\Delta p} = \dfrac{G_L(1 - v)\varepsilon}{\Delta p}$ The strain is related to the stress through Young's modulus, $Y = \dfrac{Stress}{strain} = \dfrac{\sigma}{\varepsilon}$. The stress on the resistors at the center of the membrane edges is given by Eq. (8.6), $\sigma = P(a/h)^2$. Substituting these in the equation for sensitivity, we get $S = G_L(1 - v)\dfrac{1}{\Delta p}\dfrac{\sigma}{Y} = \dfrac{G_L}{Y}(1 - v)\left(\dfrac{a}{h}\right)^2$

Thus we have sensitivity $S = \dfrac{100}{170 \times 10^9}0.7\left(\dfrac{a}{h}\right)^2 = 4.1 \times 10^{-10}\left(\dfrac{a}{h}\right)^2$

Case (i) $h = 5\,\mu m$, $a = 250\,\mu m$

$$S = 4.1 \times 10^{-10}\left(\frac{250}{5}\right)^2 = 1.025 \times 10^{-3} = mV/Pa = 102.5\,mV/bar.$$

Case (ii) $h = 10\,\mu m$, $a = 250\,\mu m$. In this case $h$ is twice that in the case (i). Therefore the sensitivity is four times lower. Thus $S = \dfrac{102.5}{4} = 25.63\,mV/bar.$

 Table 8.6 compares the burst pressure, sensitivity, and $p_{max}$ for square membrane with side $2a = 500\,\mu m$, Poisson ratio $v = 0.3$, and $\sigma_{max} = 7\,GPa$.

 In summary, it can be seen that for a given value of $a$, when $a/h$ is decreased by increasing $h$, the burst pressure $p_b$ goes up, the sensitivity $S$ goes down, and the operating maximum pressure to which linearity is maintained goes up.

**Table 8.6 Effect of membrane thickness on $p_b$, maximum operating range, and sensitivity $S$**

| Thickness, $h$ (μm) | $a/h$ | Burst Pressure $p_b$ (bar) | Max Operating Pressure with Linear Output (bar) | Sensitivity, $S$ (mV/bar) |
|---|---|---|---|---|
| 5 | 50 | 28 | 0.16 | 102.5 |
| 10 | 25 | 112 | 2.56 | 25.63 |

## Example 8.2

An n-type square membrane having side $2a = 500$ μm and thickness $h = 10$ μm is used in a piezor-esistive pressure sensor. The p-type piezoresistors with $R = 1$ kΩ are arranged as shown in Figure 8.9. Determine **(a)** the change $\Delta R$ in the resistors and **(b)** the sensitivity when the membrane is subjected to a pressure of 1 bar (= 100 kPa), noting that the gauge factor of the resistors is $G = 100$ in both the longitudinal and transverse directions. Assume that Poisson's ratio is 0.3 and Young's modulus is 170 GPa for silicon. The input voltage to the bridge is 1 V.

**Solution:** **(a)** From Eq. (8.4) and Eq. (8.8), we obtain

$$\varepsilon = p\left(\frac{a}{h}\right)^2 (1 - v)\frac{1}{Y},$$

where $a = 250$ μm and $h = 10$ μm. Substituting all the values, we obtain $\varepsilon = 2.57 \times 10^{-4}$ for $p = 100$ kPa and $\dfrac{\Delta R}{R} = G\varepsilon = 100 \times 2.57 \times 10^{-4} = 2.57 \times 10^{-2}$. Since $R = 1$ kΩ, we obtain $\Delta R = 25.7$ Ω.

**(b)** Substituting in Eq. (8.11) for the sensitivity of the pressure sensor $S$ is equal to $\dfrac{\Delta R}{R}\dfrac{1}{P} = \dfrac{2.57 \times 10^{-2}}{100}$ volts/kPa $= 0.257$ mV/kPa $= 25.7$ mV/bar (since 1 bar $= 100$ kPa)

## Example 8.3

The output voltage $V_o$ as a function of the pressure for a piezoresistive gauge pressure sensor is shown in Table 8.7 for pressure ranging from 0 to 300 kPa ($3 \times 10^5$ N/m$^2$). Determine the maximum nonlinearity of the pressure sensor **(a)** if the maximum operating pressure (full scale) is chosen to be 300 kPa; **(b)** if the maximum pressure is chosen to be 200 kPa.

**Table 8.7 Output voltage versus pressure**

| Pressure (kPa) | $V_o$ (mV) |
|---|---|
| 0 | 0 |
| 50 | 51.5 |
| 100 | 106.6 |
| 150 | 159 |
| 200 | 210 |
| 250 | 255 |
| 300 | 300 |

**Solution:** Case (a): <u>Full scale is at 300 kPa</u>. The nonlinearity at any pressure is obtained as follows: Assume that $V_o$ versus $P$ is linear from $V_o = 0$ to $V_o = V_m = 300\,\text{mV}$ as $P$ is varied from 0 to 300 kPa. At each pressure $P$, determine the difference between the experimental value of $V_o$ and the value obtained from the linear plot. If this value is $\Delta V_o$ then the nonlinearity is given by dividing $\Delta V_o$ by the output voltage at 300 kPa. Multiply this by 100 to obtain the percentage nonlinearity. Repeat the above calculation for each value of $P$. The maximum nonlinearity is the maximum value among all nonlinearity magnitudes. The results are shown in Tables 8.8 and 8.9 for the cases of full scale (a) 300 kPa and (b) 200 kPa, respectively.

**Table 8.8  Case (a): full scale is at 300 kPa**

| Pressure, $P$ (kPa) | $V_o$ Experimental Value (mV) | $V_o$ for the Linear Case = $V_L = \frac{V_m}{300}P$ (mV) | $\Delta V_o = (V_o - V_L)$ (mV) | Percentage Nonlinearity $\frac{\Delta V_o}{V_m} \times 100$ |
|---|---|---|---|---|
| 0 | 0 | 0 | 0.0 | 0 |
| 50 | 51.5 | 50 | 1.5 | 0.5 |
| 100 | 106.6 | 100 | 6.6 | 2.2 |
| 150 | 159 | 150 | 9.0 | 3.0 |
| 200 | 210 | 200 | 10.0 | 3.33 |
| 250 | 255 | 250 | 5.0 | 1.67 |
| $300 = P_m$ | $300 = V_m$ | 300 | 0.0 | 0.0 |

**Table 8.9  Case (b): full scale is at 200 kPa**

| Pressure, $P$ (kPa) | $V_o$ Experimental Value (mV) | $V_o$ for the Linear Case = $V_L = \frac{V_m}{200}P$ (mV) | $\Delta V_o = (V_o - V_L)$ (mV) | Percentage Nonlinearity $\frac{\Delta V_o}{V_m} \times 100$ |
|---|---|---|---|---|
| 0 | 0 | 0 | 0.0 | 0.0 |
| 50 | 51.5 | 52.5 | 1.0 | 0.476 |
| 100 | 106.6 | 105 | 1.6 | 0.762 |
| 150 | 159 | 157.5 | 1.5 | 0.714 |
| $200 = P_m$ | $210 = V_m$ | 210 | 0.0 | 0.0 |

From the last column it can be seen that the maximum nonlinearity is 3.33% when the pressure is 200 kPa for case (a).

It can be seen from the last column that the maximum nonlinearity is 0.762% when the pressure is 100 kPa for case (b).

*Important Note*: The percentage nonlinearity referred to the maximum output voltage is less if the maximum operating range of a given pressure sensor is limited to lower values.

### 8.3.1.3 Some Practical Implementation Aspects

The isolation between piezoresistors in Figure 8.9 is provided by the p-n junction formed by the p-type resistors with the n-type membrane layer. The leakage current of a p-n junction increases exponentially with temperature. Therefore, isolation between the resistors of these sensors becomes poor at temperatures above about 100°C. For such situations, piezoresistors are deposited on an oxide layer grown on the diaphragm region so that individual resistors are isolated from each other. Isolation is maintained even at temperatures exceeding 300°C.

It should be noted that silicon, when deposited over the oxide layer, is polycrystalline. Hence, the total resistance is determined by the resistance of the silicon grains and that of the grain boundaries, the latter being more significant. Within the grains, current transport is by carrier drift and the resistivity of the grain region behaves essentially like that of the single-crystal material. Therefore as mobility decreases at higher temperature, the resistivity of the grain region increases. On the other hand, depletion regions developed at the grain boundaries due to the charge trapping become potential barriers for carrier transport. At higher temperatures, more carriers can overcome these barriers, and hence the resistance at the grain boundaries decreases with increasing temperature. The barrier height at the grain boundary is a function of doping concentration. Therefore, the temperature coefficient of resistivity (TCR) of polysilicon can be tailored and almost made nearly zero by adjusting the boron doping concentration so that the positive TCR of the grains is balanced by the negative TCR of the grain boundary regions.

It has been observed that the TCR of boron-doped LPCVD polysilicon resistors can be negative, approach zero, or be positive depending on the doping concentration. For doping concentration greater than $3 \times 10^{19}$ cm$^{-3}$, polysilicon has nearly zero TCR. The piezoresistive effect in polysilicon film originates from each grain, and hence, this effect is nearly isotropic and is less than in single-crystal silicon. The longitudinal gauge factor ($G_L$) of boron-implanted polysilicon is larger than the transverse gauge factor ($G_T$) by a factor of about 3. Therefore, piezoresistors on polysilicon are arranged as shown in Figures 8.12 and 8.13 so as to experience maximum longitudinal stress for better sensitivity. Recall that this arrangement is different from that on single-crystal silicon.

Furthermore, in a rectangular membrane, the deformation along the narrower direction is greater than that along the wider direction and hence the polysilicon gauges should be arranged along the narrower direction to increase sensor sensitivity. The diaphragm is conveniently realized by bulk micromachining of silicon-on-insulator (SOI) wafers. The SOI wafer is fabricated by silicon fusion bonding (SFB) the oxidized silicon wafers and back etching the top wafer to reduce the thickness of the SOI layer so that it forms the diaphragm. The SOI approach makes possible integration of electronics with the pressure sensor [14].

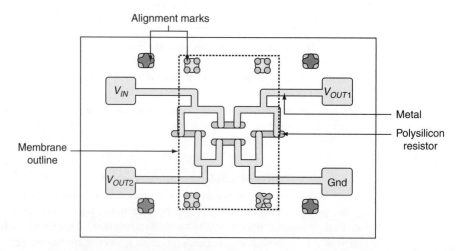

**Figure 8.12** Schematic of the top view of a piezoresistive pressure sensor [14].

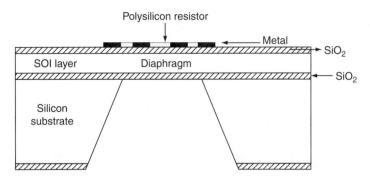

**Figure 8.13** Schematic cross-section of polysilicon piezoresistive pressure sensor.

The diaphragm thickness (SOI layer thickness) is chosen based on the burst pressure considerations. The burst pressures can be estimated using the ANSYS for membrane thickness and the aspect ratio of the diaphragm. Considering a safety factor of 5, the membrane thickness is estimated to be 15 μm for a maximum pressure ($P_{MAX}$) of 10 bar and burst pressure of $5P_{MAX}$.

### 8.3.1.4 Design of Electronics Circuits

The recent trend in industry is to provide offset and temperature compensations with specifically designed ASICs where the active circuits also amplify the output voltage of piezoresistive bridge to standard CMOS voltage levels (0–5 V). Error coefficients particular to a sensor are permanently stored in on-board electrically programmable memory such as EEPROM. Most sensor manufacturers have developed their own proprietary circuit designs. However, a few general-purpose signal-conditioning ICs, such as the MAX 1457 from Maxim Integrated Products of Sunnyvale, California, are commercially available.

There are several benefits of using the SOI approach for integrating micromachined silicon sensors with electronics. Design and process integration of source-follower amplifier circuits with piezoresistive polysilicon pressure sensor is used in illustrating the merits of SOI. Here, the SOI layer is used for the diaphragm and its thickness and lateral geometry decide the mechanical properties such as burst pressure and sensitivity. The doping concentration controls the electrical characteristics such as threshold voltage, mobility and transconductance of the MOSFET in the electronics circuit.

A source-coupled differential amplifier configuration (Figure 8.14) can be used for this integrated approach. For process compatibility, the resistors of the amplifier circuits are fabricated along with piezoresistors of the sensor. The resistors for the amplifier in the figure are realized using boron-doped polysilicon. The gate electrode for the MOSFET is realized using a heavily phosphorus-doped polysilicon. The MOSFETs for the amplifier are fabricated using an SOI wafer with layer thickness of 11 μm. For the circuit in Figure 8.14, drain resistors ($R_D = R_{D1} = R_{D2}$) of 10 kΩ with $R_D/R_S$ equal to 2.5 and threshold voltage around 500 mV gives a differential voltage gain of 5. The differential-mode and common-mode characteristics of this differential amplifier are presented in Figure 8.15 ([14] and [15]). A CMRR of 230 has been achieved with a differential gain of 4.6.

### 8.3.1.5 Integration of Pressure Sensor

A schematic cross-section of the MOSFET integrated piezoresistive pressure sensor [15] fabricated on SOI wafers (layer thickness of 11 μm) is shown in Figure 8.16. A cross-sectional view of this device is in Figure 8.16(a) and the composite mask layout of the MOS integrated pressure sensor and its photograph are in Figure 8.16(b). Photographs of the

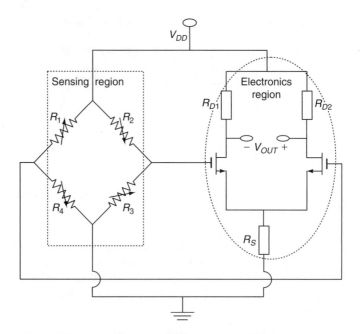

**Figure 8.14** Circuit diagram of MOSFET integrated pressure sensor.

device [15] taken at the wafer level after dicing the individual chips appear in Figure 8.16(c) and (d), respectively. The bond pads of size $150\,\mu m \times 150\,\mu m$ and the Al interconnect of $30\,\mu m$ width can be seen very clearly. The resistors connected in the form of Wheatstone bridge as well as the resistors forming part of the electronics and the two MOSFETs with a common source can also be seen. The $V_{DD}$ and $V_{in}$ in Figure 8.16(d) are power supply voltages and GND is the ground pad. $S_{OUT}^-$ and $S_{OUT}^+$ are the output pads of the pressure sensor, and it can be seen that they are connected to the two gate terminals

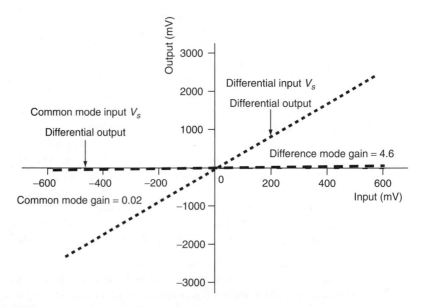

**Figure 8.15** Common-mode and differential-mode output of the differential amplifier fabricated on SOI wafer [14,15].

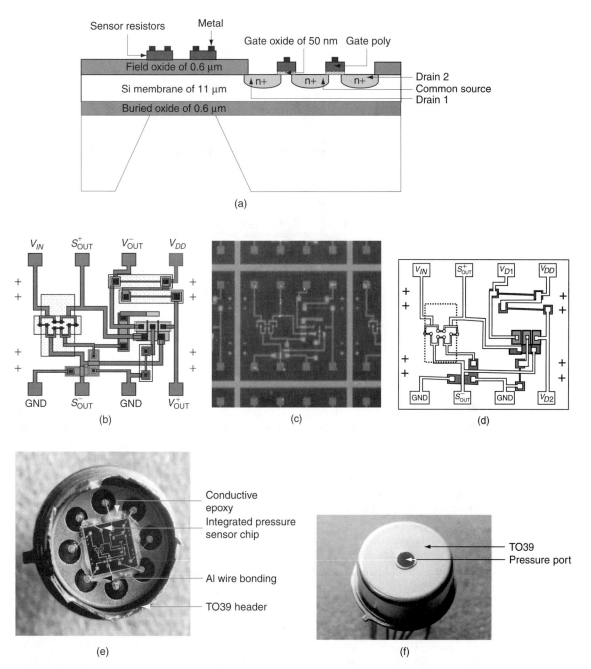

**Figure 8.16** Various stages of integration and packaging of the pressure sensor: (a) Schematic cross-section of MOSFET integrated piezoresistive pressure sensor; (b) composite mask layout of the pressure sensor; (c) photograph at the wafer level. Scribe lines between the individual IC chips, the interconnecting aluminum, and the bond pads can be seen [15]; (d) photograph of one of the 3 mm × 3 mm integrated chips after dicing from the wafer. The dotted line shows the location of the diaphragm of the pressure sensor [15]; (e) a close-up photograph of an individual die mounted on a TO39 metal header. The aluminum wires connecting the bond pads of the pressure sensor chip and the leads of the header are seen. The individual leads of the header are isolated from each other by insulating material [16]; (f) packaged MOS integrated pressure sensor ready for testing [16]; and (g) schematic of packaged pressure sensor mounted in a test jig (see page 26).

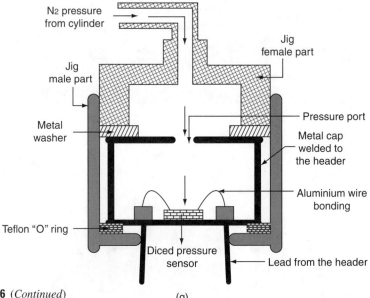

**Figure 8.16** (*Continued*)                    (g)

(input) of the amplifier. $V_{D1}$ and $V_{D2}$ are the pads connected to the drain regions of the two MOSFETs that have a common source region.

A close-up photograph [16] of an individual die mounted on a TO39 metal header can be seen in Figure 8.16(e). The full view [16] of the packaged device showing the pressure port and the leads coming out of the TO39 header is in Figure 8.16(f). These devices are tested with a 10 V supply for the monolithic chip containing the pressure sensor and the amplifier; the special jig arrangement shown in Figure 8.16(g) is used for this purpose. Apart from other conveniences, the integrated sensor shows excellent linearity. The output voltage versus pressure characteristics of the final packaged device is shown in Figure 8.17. An output voltage of 0.956 V was obtained with the packaged MOS integrated sensor at 3 bar gauge pressure with 10 V input to the integrated chip. The integrated sensor showed excellent linearity. For a full-scale output of 0.956 V at 3 bar pressure, the maximum nonlinearity of 0.2% has been achieved. Even for a full-scale output of 1.59 V at 5 bar pressure, the maximum nonlinearity is found to be within 0.3%. The sensitivity at the sensor output is 60 mV/bar at $V_{in} = 10$ V. The sensitivity of the integrated sensor output is 270 mV/bar/10 V input. The differential gain of the amplifier portion of this integrated device is 4.5.

### 8.3.2 Micromachined Accelerometer

The basic structure of an accelerometer consists of an inertial mass suspended from a spring as shown in Figure 8.18. In this system, the relative position of the inertial mass is disturbed from its equilibrium position due to an externally applied acceleration or force, and this in effect measures the latter.

Under dynamic conditions, the spring–mass system obeys the second-order equation

$$M\frac{d^2u}{dt^2} + b\frac{du}{dt} + ku = F_{ext}(t) = Ma \qquad (8.13)$$

Here $M$ is the inertial mass, $k$ is the spring constant (N/m), $b$ is the damping coefficient (Ns/m) or the damping factor, and $a$ is the acceleration due to the external force $F_{ext}$.

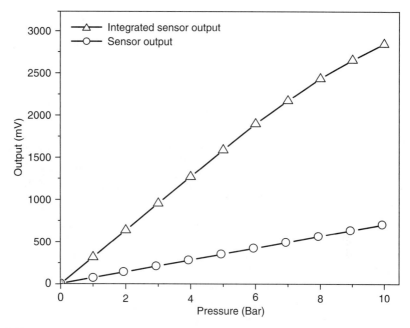

**Figure 8.17** Output voltage versus pressure of an integrated pressure sensor with supply voltage 10 V. The results obtained at the terminals [$S_{OUT}^+$ and $S_{OUT}^-$ in Figure 8.16(d)] of the pressure sensor output are also shown. Sensor membrane thickness 11 μm; membrane size $500 \times 1000$ μm$^2$; input 10 V.

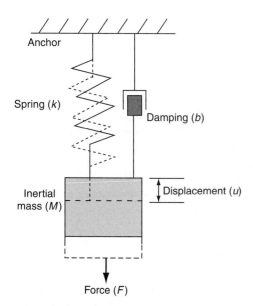

**Figure 8.18** Spring–mass system illustrating the concept of an accelerometer.

Solutions of this equation for various cases of damping coefficient are derived in the Appendix. The frequency response of the system in response to a periodic force is also shown there in detail. However, if the force is of unit step nature, the response of the system is somewhat different. This is a common scenario in the context of typical measurement scenarios for accelerometers.

Note that the initial (at $t = 0$) equilibrium position of the mass is at $u = 0$. The steady-state equilibrium position (for large values of time $t$) can be obtained as $x_0 = F_{ext}/k = F_0/k$. When unit step force $F_0$ is applied at $t = 0$, the mass moves from $u = 0$ toward $u = F_0/k$. If the damping ratio $\xi$ is negligible, the system will possess total energy $F_0 u_0$ when the mass arrives at $u_0$ due to the work done by the force $F_0$. As the potential energy in this position is $F_0 u_0/2$, the mass has the same value for kinetic energy at this position. This means that the mass will pass through the point $u_0$ until it reaches $u_1$ such that $F_0 u_1 = k u_1^2/2$. Thus at $u_1 = 2u_0$ (which is not an equilibrium position) the mass moves back towards $u = 0$. In the same way, the mass will oscillate between 0 and $2u_0$. When the effect of damping is considered, the oscillation dies down and the mass finally settles at $u = u_0$, when the existing energy is completely consumed by the damping effect. Therefore, the response of the system to a unity step force strongly depends on the damping ratio $\xi$ as shown in Figure 8.19.

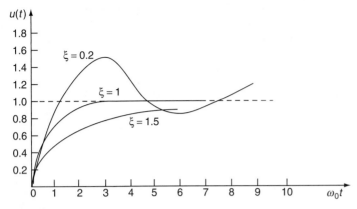

**Figure 8.19** Displacement versus time of the mass for a unity step input force.

**Problem 8.1**

See the solution in the Appendix (Section A.3) of Eq. (8.13) for the case when the spring mass system is set to continuous oscillations by a periodic driving force $F_{ext} = F_0 \sin \omega t$. Using the solution found there, show that the amplitude will give a resonance peak for damping ratio $\xi < 0.7$. (Note that $\xi = 0.7$ is called the critical damping condition.) Plot the amplitude response for the following three cases:

**(a)** $\xi = 0.2$
**(b)** $\xi = 0.7$
**(c)** $\xi = 1.2$

### 8.3.2.1 Performance Parameters

In the real world, acceleration cannot remain constant over a long period of time. Hence the output of an acceleration sensing system is either a short-lived, fast-varying signal or an alternating signal. Therefore, one of the most important characteristics of an accelerometer is its dynamic performance or frequency response. From Eq. (8.13) we can see that under steady-state conditions, the external force is exactly balanced by the restoring force exerted by the spring, and the resulting displacement $u$ can be expressed in terms of the force $F_{ext}$ as follows:

$$F_{ext} = Ma = ku \quad \text{(steady state)} \tag{8.14}$$

Therefore, the sensitivity $S$ of the accelerometer is expressed by

$$S = \frac{u}{a} = \frac{M}{k} \tag{8.15}$$

An ideal accelerometer should have uniform sensitivity over a large frequency band. The dynamic behavior of the system can be analyzed using the simple mass–spring model described by the differential equation Eq. (8.13). As discussed in Chapters 6 and 7, the damping coefficient $b$ is due to losses experienced by the system (e.g. friction) as it moves. It can also be due to the aerodynamic drag of the mass or squeeze film compression due to the trapped air around the mass. Gravity, electrostatic, and thermal forces contribute to $F_{ext}$. If $F_{ext}$ has the form of a sine wave, the movement of the mass is also sinusoidal at the same frequency since this is a linear system. The magnitude and relative phase of this sine wave are functions of frequency. From Eq. (8.13), we can write the mechanical transfer function from an acceleration to a displacement of the mass as

$$H(s) = \frac{u(s)}{a(s)} = \frac{1}{s^2 + (k/M)s + (k/M)} = \frac{1}{s^2 + (\omega_0/Q)s + \omega_0^2} \tag{8.16}$$

where $\omega_0 = 2\pi f_0$ (with $f_0$ the resonance frequency) and $Q$ is the quality factor. From Eq. (8.16), it can be seen that the resonance frequency of this system is given by the relation

$$\omega_0 = \sqrt{\frac{k}{M}} \tag{8.17}$$

From Eqs. (8.15) and (8.17) it can be seen that the sensitivity can be increased by reducing the spring constant or increasing the mass $M$ and that this will reduce the resonance frequency.

Yet another important parameter for accelerometers is the total noise equivalent acceleration ($a_{\text{noise}}$), defined as

$$a_{\text{noise}} = \sqrt{(8\pi k_B T f_0)/QM} \qquad (8.18)$$

where $k_B$ and $T$ are the Boltzmann constant and temperature, respectively. Therefore, the main specifications for an accelerometer are the range, sensitivity, resolution (in terms of $g$), bandwidth (Hz), cross-axis sensitivity, and immunity to shock. The range, usually expressed in terms of Earth's gravitation $g$, is the maximum $g$ value that can be sensed using the accelerometer. The sensitivity has already been defined in Eq. (8.15). It may be recalled from Eqs. (8.15) and (8.17) that the sensitivity is related to the resonant frequency by $S = 1/\omega_0^2$. The bandwidth is the frequency range for which the sensed signal is linearly proportional to $g$. This is usually taken as one-fifth the resonant frequency. Thus the resonance frequency and hence the bandwidth are lower when the sensitivity is higher.

Generally, the resonance frequency will be low for accelerometers useful in inertial navigation systems. The range and bandwidth requirements vary depending upon the application. The full range of acceleration for airbag crash-sensing accelerometers is $\pm 50g$ with bandwidth about 1 kHz. On the other hand, for measuring engine knock or vibration the range is only about $1g$ with the ability to resolve very small accelerations below $100\,\mu g$ and bandwidth greater than 10 kHz.

Acceleration is a vector quantity. Hence, in most accelerometers, the proof mass is aligned with respect to coordinate planes to resolve the direction aspect. When accelerometers with three different directions are required, the purity of the direction information depends on constraining the mass to move in one of the coordinate axis directions. In these cases cross-axis sensitivity is also specified. In certain applications such as those implanted in human body, the sensor should operate with very low power to save battery life.

### 8.3.2.2 Design of an Accelerometer

It is evident from Eq. (8.15) that displacement $u$ can be used as a measure of the acceleration $a$. Piezoelectric-, piezoresistive- and capacitive-sensing methods have been used for sensing the displacement in the spring–mass system of the accelerometer. In the piezoelectric approach, either the spring itself is piezoelectric or it contains a piezoelectric thin film to provide a voltage directly proportional to the displacement. In the second approach, piezoresistors are used to sense the inertial stress induced in the spring during displacement of the mass. In the capacitive-sensing approach, the mass forms one side of a parallel-plate capacitor and the change in the capacitance of this moving plate with respect to a fixed electrode is used as a measure of displacement and hence of external force. Even though the piezoresistive-sensing approach is used in many sensors because of its simplicity, the sensitivity in accelerometers using them is highly temperature-dependent. Therefore, capacitive-sensing systems are popular and have been widely used by industry. In this section the focus is on capacitive-type acceleration sensors.

The structure of an SOI-based accelerometer [17] consists of a seismic mass (released from the SOI wafer) supported by four folded beams that act as the spring anchored to the substrate (Figure 8.20). The effect of location of the folded beam with respect to the square mass is

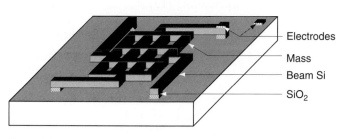

Electrodes

Mass

Beam Si

SiO$_2$

**Figure 8.20** Structure of the SOI-based capacitive-sensing accelerometer.

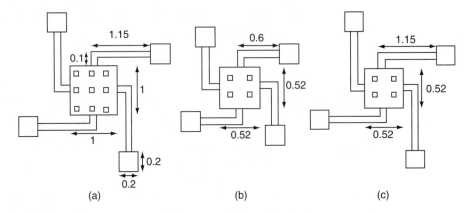

**Figure 8.21** Top view and dimensions of the seismic mass and supporting beams analyzed using ANSYS: (a) model UZ1; (b) model UZ2; and (c) model UZ3. All dimensions are in millimeters.

studied using the commercially available FEA software ANSYS to achieve best $z$-axis sensitivity and minimum cross-axis sensitivity. The analysis has shown that best results can be achieved if the beams are located midway on the square side of the mass. Three models, UZ1, UZ2, and UZ3 (Figure 8.21), have been studied in this context. UZ1 has a seismic proof-mass area of $1 \times 1 \, mm^2$, while UZ2 and UZ3 have $0.52 \times 0.52 \, mm^2$. In micromachined accelerometers, the mass is proportional to the area of the proof-mass structure. The length of the support arm in UZ1 and UZ3 is 1.15 mm while in UZ2 it is 0.6 mm.

Finite element harmonic analysis can be done using ANSYS to study the effects of acceleration and frequency on the deflection of the mass. These results [18] are presented in Figures 8.22 and 8.23. It may be noted that structure UZ1, which has a larger mass and the long supporting beams, gives the best results in terms of sensitivity. The stress analysis has shown that the maximum stress is located in the region where the beam is attached to the mass. The maximum stress estimated using ANSYS is presented in Figures 8.24. It may be noted that the maximum stress is greatest with structure UZ1 and that it is well within the yield strength of silicon, which is 7 GPa.

### 8.3.2.3 Fabrication and Characterization

The SOI wafer is oxidized and this $SiO_2$ layer is then patterned by photolithography. Using the patterned $SiO_2$ as a mask, the top silicon layer is etched at 65°C for 30 minutes in 28% KOH solution containing 30% tert-butanol solution to avoid convex corner undercutting [19].

Owing to the anisotropy of the KOH etching, the walls of the holes in the seismic mass are not vertical but are at an angle of 54.74°. This does not affect the performance of the accelerometer because the purpose of the holes is to enhance the release process and control the damping. The oxide at the top of the SOI wafer (the etch mask layer) is next etched away. To increase the conductivity of the silicon layer, blanket boron diffusion is

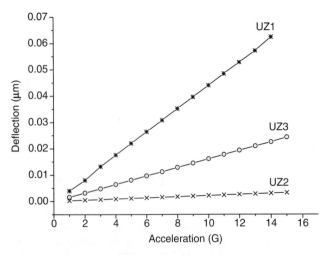

**Figure 8.22** Deflection versus acceleration (dynamic analysis) of devices UZ1, UZ2, and UZ3.

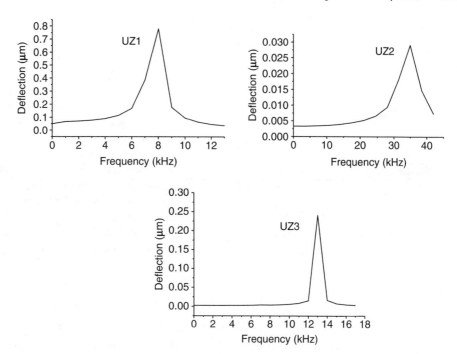

**Figure 8.23** Deflection versus frequency (dynamic analysis) of devices UZ1, UZ2, and UZ3.

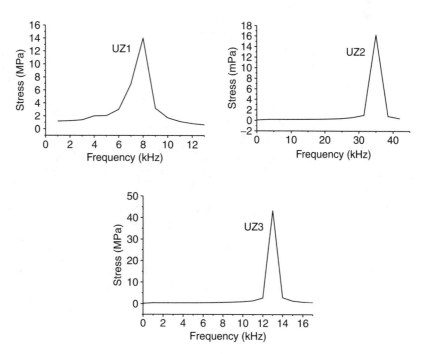

**Figure 8.24** Maximum stress versus frequency (dynamic analysis) in devices UZ1, UZ2 and UZ3.

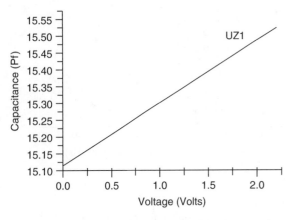

**Figure 8.25** Typical result obtained by electrostatic actuation of the accelerometer structure UZ1.

carried out at 1050°C for 30 minutes. Next, the seismic mass and the folded beams are released by etching in buffered HF solution for over 2 hours. Chrome gold metallization is used to provide electrical contacts to the mass and substrate forming the two parallel plates of the capacitive accelerometers.

An approach to testing a capacitive accelerometer is by measuring the capacitance variation under electrostatic actuation. The results obtained as a function of dc bias applied are presented in Figure 8.25. The capacitance increases linearly with the applied dc voltage because the deflection and hence the change in capacitance are small compared to the initial value at zero bias. A capacitance change of 0.3 pF is observed when the voltage is varied from 0 to 2 V, which corresponds to about 0.019 μm deflection of the mass in an air gap of 1 μm. This value corresponds to a predicted acceleration of about 3$g$ (Figure 8.22) for UZ1.

A schematic of an accelerometer fabricated using SOI wafer is shown in Figure 8.26 and the frequency response of a typical accelerometer obtained using a laser doppler vibrometer is shown in Figure 8.27. The measured frequency for this structure is 35.5 kHz [20]. On the basis of the measured dimensions, the mass of the structure is 6.45 μg. The spring constant of the structure can be evaluated using Eq. (4.43) as 304 N/m. The calculated resonant frequency is therefore 34.54 kHz.

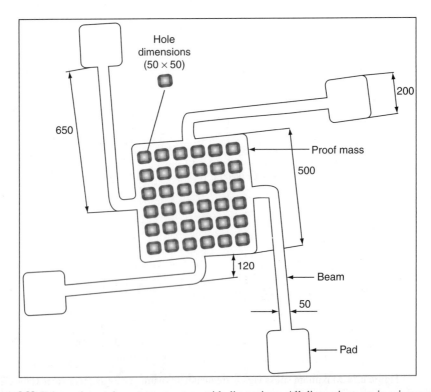

**Figure 8.26** Schematic accelerometer structure with dimensions. All dimensions are in micrometers.

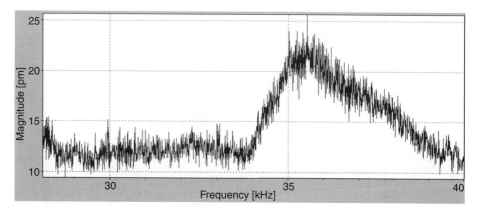

**Figure 8.27** Typical frequency response of an accelerometer.

### 8.3.2.4 Bulk Micromachined Accelerometer

In an accelerometer with a capacitive-sensing system it is possible to use electrostatic actuation to force the mass back into its equilibrium position, and the electrostatic force required for this can be used as a measure of the inertial force $F_{ext}$. This approach, called the "force balance" method, can be implemented conveniently when accelerometer has a differential capacitor for sensing.

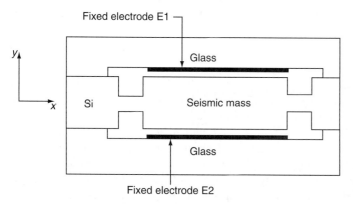

**Figure 8.28** Differential capacitive-type accelerometer.

A schematic diagram of a bulk silicon micromachined accelerometer with differential capacitor sensing is shown in Figure 8.28. The two fixed electrodes are placed on the glass plates as shown. When the system is subjected to acceleration in the negative y-direction, the seismic mass experiences the inertial force and moves in the positive y-direction, resulting in change in the capacitance between the mass and the fixed plates. The capacitance difference $\Delta C$ between the two capacitances can be used as a measure of acceleration. The symmetric design and differential sensing reduces the effect of thermal mismatch, if any, and linearizes the $\Delta C$ versus capacitance relationship.

### 8.3.2.5 Commercial Capacitive Microaccelerometer

The well-known commercial ADXL series of accelerometers from Analog devices are capacitive-type devices [21]. They make use of the force-balance technique and are fabricated using the polysilicon surface micromachining approach described below.

Figure 8.29 shows the schematic of such an accelerometer. In this approach, as discussed in Chapter 3, the structural material is doped polycrystalline silicon of about $2\,\mu m$ thickness deposited on $SiO_2$. The polysilicon is patterned to realize the structure, which consists of a mass plate having several fingers on both sides as shown. This structure is released by etching the sacrificial oxide layer, at the same time retaining the oxide underneath the four anchor regions so that the structure is suspended over the substrate by four thin beam flexures. Thus, the rectangular central mass $M$ is electrically isolated from

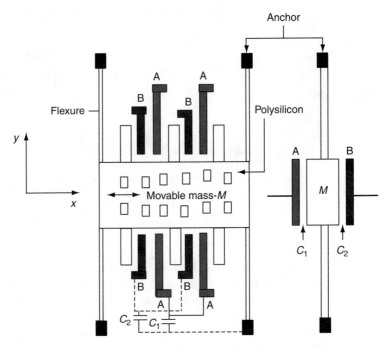

**Figure 8.29** Capacitive accelerometer using surface micromachining: schematic top view (left) and the equivalent lumped model (right). A and B are stationary.

the substrate to which it is anchored, can be moved along the $x$-direction by an inertial force, and serves as the seismic mass of the beam–mass structure. The oxide is not etched from underneath the doped polysilicon fingers A and B, which serve as fixed fingers that are electrically isolated from the substrate. The small holes etched in the mass plate assist in etching the sacrificial layer and also adjust the damping parameter.

It can be seen from the schematic top view in Figure 8.29 that there are two stationary electrodes A and B on both sides of any finger of the seismic mass plate at distances of the order of 1 to 1.5 μm. All the fingers marked A are connected together into one group to form a variable capacitance $C_1$ with respect to the fingers of the movable mass $M$. Similarly, the fingers marked B are connected together forming a capacitor $C_2$, as shown by the equivalent representation on the right in Figure 8.29. When the external force (acceleration) is absent, the two capacitances are identical ($C_1 = C_2$) and hence the output signal proportional to the difference in capacitance is zero. Owing to an applied acceleration, the suspended structure is displace, resulting in an imbalance between the capacitances' half bridge. Referring to Figure 8.29, if the displacement is to the left, capacitance $C_1$ increases and capacitance $C_2$ decreases. The capacitances of the two capacitors are formed by the sidewalls of the fingers. These capacitors are very small because the polysilicon layer is usually only about 2 μm thick. The overall capacitance in the ADXL05 rated at ±5$g$ is only of the order of 0.1 pF even with several tens of fingers, and the change in capacitance for a change of 1$g$ is about 0.1 fF (=100 aF) [21]. As this is a very minute change in capacitance, it must be measured with on-chip integrated electronics to ensure that the effect of parasitic capacitances is minimized.

### 8.3.2.6 Force-Balanced Feedback in Capacitive-Sensing Accelerometer
The force balance for measuring the acceleration is achieved using a closed-loop feedback circuit to electromechanically force the movable seismic mass to its equilibrium position

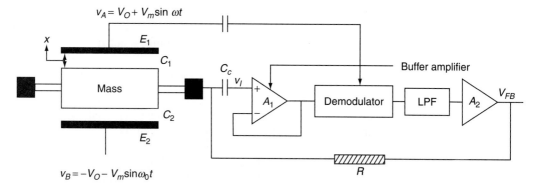

**Figure 8.30** Accelerometer with schematic of the force-balanced feedback circuit.

when it is subjected to acceleration [22]. Figure 8.30 shows a simplified schematic diagram of the capacitive accelerometer with a force-balanced feedback arrangement. In this arrangement, dc voltages $V_O$ of opposite polarity and equal value, along with sinusoidal voltages that are out of phase but of equal amplitude and frequency so that $v_B = -v_A$, are applied to the electrodes $E_1$ and $E_2$. The ac signal output $v_i$ is taken from the movable mass plate through a coupling capacitor $C_c$ and fed to a buffer stage $A_1$. This is followed by a synchronous demodulator. The output of the demodulator is filtered by a low-pass filter and then the dc voltage is amplified by the output amplifier $A_2$. The output dc voltage $V_{FB}$ is fed back to the seismic mass plate through a high resistance. The feedback path usually has a feedback buffer amplifier (not shown).

As the voltages $v_A$ and $v_B$ are out of phase, the total voltage appearing across the capacitors $C_1$ and $C_2$ is $2v$ where $v = V_O + V_m \sin \omega t$. The net voltage appearing at the mass plate is

$$v_C = 2v \frac{C_1}{C_1 + C_2} - v = \frac{C_1 - C_2}{C_1 + C_2} v \tag{8.19}$$

When the acceleration is zero, the movable mass plate is in a balanced position at an equal distance $d$ between the two fixed plates $E_1$ and $E_2$. Hence, when the acceleration is zero the two capacitors $C_1$ and $C_2 = C = \varepsilon_0 A/d$, where $\varepsilon_0$ is the permittivity of free space and $A$ is the area of the capacitor plates ($E_1$ and $E_2$ are fixed electrodes). Since $C_1 = C_2 = C_0$, the dc voltage and signal voltage on the mass for this situation are zero. Hence the input ac voltage $v_i = 0$ and the feedback voltage $V_{FB}$ is zero.

When a downward acceleration $a$ acts on the frame supporting the accelerometer, the mass moves toward the upper fixed plate $E_1$ by a distance $x$ (say). As a result of this upward movement of the mass, the capacitance $C_1$ will increase and $C_2$ will decrease to new values given by

$$C_1 = \frac{\varepsilon_0 A}{d - x} \quad \text{and} \quad C_2 = \frac{\varepsilon_0 A}{d + x}$$

Applying these values in Eq. (8.19), when $x \ll d$, we obtain the voltage appearing on the mass plate as

$$v_C = \frac{x}{d}(V_O + V_m \sin \omega t) \tag{8.20}$$

The dc component of $v_C$ is not transferred to the electronics circuit due to the presence of the coupling capacitor. However, the ac signal voltage can pass through the capacitor and appears as the signal voltage, $v_i$. From Eq. (8.20) this is noted to be

$$v_i = \frac{x}{d} V_m \sin \omega t \tag{8.21}$$

If the total open-loop gain from the buffer, demodulator, and operational amplifier is $A_0$, the feedback dc voltage through resistor $R$ is given by

$$V_{FB} = A_0 V_m \frac{x}{d} = A_0 V_m \bar{x} \tag{8.22}$$

Here $\bar{x} = x/d$. This dc voltage appears at the central plate, and thus the dc voltage between the fixed top electrode $E_1$ and the moving plate is reduced to $V_O - A_0 V_m \bar{x}$. At the same time the dc voltage between the fixed bottom electrode $E_2$ and the movable middle plate is increased by the same extent. As a result, the net electrostatic force on the movable plate caused by $V_{FB}$ is in the downward direction. The displacement of the central plate thus decreases by the electromechanical feedback, and this decrease is significant if the open-loop gain $A_0$ is large enough.

The electrostatic force $F_e$ acting on the movable mass plate having area $A$ can be expressed as [using Eq. (6.59 in Chapter 6)]

$$F_e = \frac{A\varepsilon_0}{2d^2} \left[ \frac{(V_O - A_0 V_m \bar{x})^2}{(1 - \bar{x})^2} - \frac{(V_O + A_0 V_m \bar{x})^2}{(1 + \bar{x})^2} \right] \tag{8.23}$$

Using the condition that $\bar{x} \ll 1$, we can easily show that the electrostatic force $F_e$ applied on the movable plate can be simplified to

$$F_e = \frac{2A\varepsilon_0 V_O V_m A_0 \bar{x}}{d^2} \tag{8.24}$$

The net force acting on the movable plate in steady-state condition is given by

$$Ma - kx - \frac{2A\varepsilon_0 V_O V_m A_0 \bar{x}}{d^2} = 0 \tag{8.25}$$

Solving for $x$ in Eq. (8.25), assuming that $A_0$ is very large and that the second term in the left-hand side is small, we obtain

$$\bar{x} = \frac{Mad^2}{2A\varepsilon_0 V_O V_m A_0} \tag{8.26}$$

By using Eq. (8.26) in Eq. (8.22), it is seen that the feedback voltage $V_{FB}$ is related to the acceleration $a$ by the expression

$$V_{FB} = \frac{d^2}{2A\varepsilon_0 V_O} Ma \tag{8.27}$$

It is thus evident from Eq. (8.27) that the feedback voltage can be used as a measure of the inertial force $F = Ma$ and hence the acceleration.

**Example 8.4**

Figure 8.31 shows the cross-section of a capacitive-sensing accelerometer. The seismic mass is made of silicon and has the following dimensions: thickness 200 μm, the top view is a square having edge length 1000 μm. This mass is suspended by two silicon beams of thickness $t_b = 10$ μm, length $L = 200$ μm, and breadth $b = 16$ μm and anchored onto the rigid silicon frame. The density of silicon is 2330 kg/m$^3$ and the Young's modulus of silicon is $Y = 170$ GPa. Assume that the spring constant $k_b$ of each of the beams is given by the relation $k_b = \frac{Ybt_b^3}{L^3}$. The gap between the fixed electrode and the seismic mass is $d = 2$ μm so that $C_{01} = C_{02}$ when the force is zero, where $C_{01}$ is the capacitance of the top electrode with respect to the top surface of the mass and $C_{02}$ is the capacitance of the bottom fixed electrode with respect to the bottom surface of the mass. The mass is heavily doped silicon so that its conductivity is very high. The frame is subjected to a force $F$ in the negative $z$-direction as shown in the figure. As a result, the mass is deflected a vertical distance of 0.1 μm in the positive $z$-direction. Determine the following:

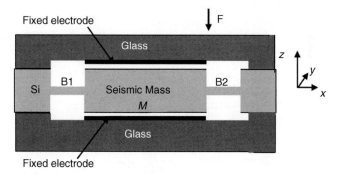

**Figure 8.31** Cross-section of a capacitive-sensing accelerometer.

**(a)** The steady-state capacitances $C_1$ and $C_2$ of the mass with respect to the upper electrode and lower fixed electrode, respectively.

**(b)** The steady-state force $F$ and the acceleration experienced by the mass.

**(c)** The sensitivity and resonance frequency of the accelerometer.

**Solution:** **(a)** When the frame is accelerated in the negative $z$-direction, the mass will be deflected in the positive $z$-direction by an extent $z$. In this example, $z = 0.1$ μm. As a result the capacitance $C_{01}$ increases above its equilibrium value $C_{01}$ to a value $C_1$, and $C_{02}$ decreases to $C_2$. In this example the area of the capacitor is $A = 1000$ μm $\times 1000$ μm $= 10^{-6}$ m$^2$. Using the standard formula for the parallel-plate capacitor, the capacitances $C_1$ and $C_2$ are obtained as follows:

$$C_1 = \frac{\varepsilon_0 A}{(d - z)} = \frac{(8.854 \times 10^{-12}) \times 10^{-6}}{1.9 \times 10^{-6}} = 4.66 \text{ pF}$$

$$C_2 = \frac{\varepsilon_0 A}{(d + z)} = \frac{(8.854 \times 10^{-12}) \times 10^{-6}}{2.1 \times 10^{-6}} = 4.216 \text{ pF}$$

**(b)** In steady-state conditions, referring to Eq. (8.14), $F = Ma = k_{\text{eff}}\, z$. In this example, $M = \text{volume} \times \text{density} = (10^{-6} \times 200 \times 10^{-6}) \times 2300 = 4.6 \times 10^{-7}$ kg

From Chapter 4, the spring constant $k_b$ per beam is

$$k_b = \frac{Eb\, t_b^3}{L^3} = \frac{1.7 \times 10^{11} \times 16 \times 10^{-6} \times (10)^3}{(200)^3} = 340 \text{ N/m}$$

As the mass is supported by two beams, the effective spring constant is $k_{\text{eff}} = 2k_b = 680$ N/m:

$$F = k_{\text{eff}} z = 680 \times 10^{-7} = 6.8 \times 10^{-5} \text{ N}$$

The acceleration is $= \frac{F}{M} = \frac{6.8 \times 10^{-5}}{4.6 \times 10^{-7}} = 147.8$ m/sec$^2$. Taking the acceleration due to gravity as $g = 9.8$ m/sec$^2$, the above acceleration is 15.08g.

(c) The sensitivity is defined as $S = \frac{\text{deflection, } z}{\text{acceleration}}$. In this example, the steady-state displacement, $z = 0.1\,\mu$m $= 10^{-7}$ m and the acceleration $= 15.08$ Therefore, the sensitivity $= \frac{10^{-7}}{15.08} = 6.63$ nm/g. Referring to Eq. (8.17), the resonance frequency of the accelerometer

$$f_0 = \frac{\omega_0}{2\pi} = \frac{1}{2\pi}\sqrt{\frac{k_{\text{eff}}}{M}} = 6.119 \text{ kHz}.$$

## Example 8.5

In Example 8.4, a force-feedback circuit is used with the circuit in Figure 8.32. Assuming $V_0 = 10$ V and $V_m = 1$ V, determine the feedback voltage $V_{\text{FB}}$ necessary to achieve the force-balance condition.

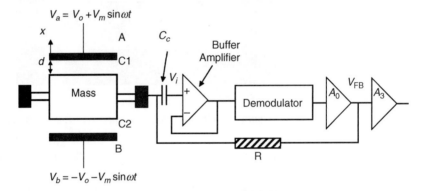

**Figure 8.32** Accelerometer with force-balanced sensing.

**Solution:** Note that when the voltage difference between the electrodes of a capacitor is $V$, the electrostatic force $F$ between two electrodes with spacing $d$ between them is obtained using the following relationship:

$$F = \frac{\text{Energy stored}}{d} = \frac{CV^2}{2}\frac{1}{d} \tag{8.28}$$

The capacitance of a parallel-plate capacitor $C$ having an area $A_m$ is given by the relation

$$C = \frac{\varepsilon_0 A_m}{d} \tag{8.29}$$

The permittivity of free space, $\varepsilon_0 = 8.854 \times 10^{-12}$ F/m. Substituting for $C$ from Eq. (8.29) in Eq. (8.28), the electrostatic force of attraction between the plates of the capacitor can written as

$$F = \frac{1}{2}\frac{\varepsilon_0 A_m V^2}{d^2} \tag{8.30}$$

Now, referring to the force-balance circuit shown above, we have the following situations. When $V_{\text{FB}} = 0$, the dc and ac voltage on the mass electrode is zero and the electrostatic force exerted by the electrode $A$ on the mass cancels the electrostatic force exerted by the electrode $B$ on the mass. When the dc feedback voltage $V_{\text{FB}}$ is applied to the mass as shown, the net dc voltage between the electrode $A$ and the mass changes from $Vo$ to $(Vo - V_{FB})$ and the magnitude of the dc voltage between the electrode $B$ and the mass changes from $Vo$ to $(Vo + V_{FB})$.

Therefore the electrostatic force of attraction between the electrode $A$ and the mass decreases, whereas it increases between the electrode $B$ and the mass, thus tending to restore the mass to its neutral position. As a result, the net electrostatic force $F_{es}$ on the mass when the feedback voltage $V_{FB}$ is applied to the mass is given by the relation (taking $A_m$ = area of the electrodes)

$$F_{es} = \frac{\varepsilon_0 A_m}{2} \left[ \frac{(V_0 - V_{FB})^2}{(d-z)^2} - \frac{(V_0 + V_{FB})^2}{(d+z)^2} \right]$$

$$= \frac{\varepsilon_0 A_m V_0^2}{2d^2} \left[ \frac{\left(1 - \dfrac{V_{FB}}{V_0}\right)^2}{\left(1 - \dfrac{z}{d}\right)^2} - \frac{\left(1 + \dfrac{V_{FB}}{V_0}\right)^2}{\left(1 + \dfrac{z}{d}\right)^2} \right] \tag{8.31}$$

Noting that $z/d \ll 1$, referring to Figure 8.32, denoting $A_0 = \frac{V_{FB}}{V_i}$, and observing from Eq. (8.22) that $V_{FB} = V_m A_0 \frac{z}{d}$, we can simplify Eq. (8.31) as:

$$F_{es} = -\frac{2A_m \varepsilon_0 V_0}{d^2} V_m A_0 \frac{z}{d}$$

When the mass is forced back to the equilibrium position, the inertial force $F = Ma$ is balanced by $F_{es}$. Hence we can write,

$$Ma = 2A_m \varepsilon \frac{V_0}{d^2} V_m A_0 \frac{z}{d}$$

This gives us

$$V_m A_0 = \frac{d^2 (Ma)}{2A_m \varepsilon_0 V_0 (z/d)} = \frac{4 \times 10^{-12} \times 6.8 \times 10^{-5}}{2 \times 10^{-6} \times 8.854 \times 10^{-12} \times 10 \times (0.1/2)} = 30.7$$

$$V_m = 1 \text{ volt}$$

Therefore, $A_0 = 30.7$

$$V_{FB} = V_m A_0 \frac{z}{d} = 30.7 \times 0.05 = 1.535 \text{ volts}$$

## ► 8.4 CASE STUDY OF A SMART STRUCTURE IN VIBRATION CONTROL

The control of noise generated by vibration of structural members is yet another application where smart materials can be used. This has great impact in increasing the integrity of the structure. Noise can be airborne or structure-borne. Airborne noises are generated due to the pressure difference caused by differential flow fields, as for example the aerodynamic noise generated by a moving aircraft or helicopter. The structure-borne noise is generated by vibrating structural members. These members act as a conduit for the vibration (disturbance) to propagate within the structure and are responsible for screeching or high-frequency noise. In such cases, reducing the vibration levels automatically reduces the structure-borne noise levels.

Vibration reduction can be achieved passively by identifying resonant conditions and modifying the geometry of the structure so that the natural frequency of the system is far away from the driving frequency of the system. Alternatively, one can do a detailed analysis to identify the regions in a structure having high vibration levels and design suitable damping mechanisms to remove these vibrations. If design constraints allow no modification to the existing structure, adaptive technology may be employed. One such methodology uses smart materials. These can generate the required additional force by mechanical coupling, which in turn can be used to create damping forces to remove vibrations in the structure. In the case study below, vibration levels in a beam modeled by a thin-walled box are controlled using sensors and actuators made of lead zirconate titanate, commonly known as PZT. This forms a class of smart materials that operate on the basis of the piezoelectric effect discussed in Chapter 2 (Section 2.9).

### 8.4.1 PZT Transducers

Piezoelectricity is attributed to an asymmetry in the unit cell and the resultant generation of electric polarization dipoles due to the mechanical distortion. Examples of piezoelectric materials include lead zirconate titanate (more popularly known by the acronym PZT), lead metaniobate, lead titanate, and their modifications. The changes in electrical charge are generated when mechanical stresses are applied across the face of a piezoelectric sheet or film. The converse effect is also observed in such materials.

PZT is arguably the most widely used component in smart systems. Its importance comes from its substantial piezoelectric properties. In these crystals, the force applied along one axis of the crystal leads to the appearance of positive and negative charges on opposite sides of the crystal along another axis. The strain induced by the force $F$ leads to a physical displacement of the charge $Q$ within the unit cell. This polarization of the crystal leads to an accumulation of charge

$$Q = \mathbf{d}\, F \tag{8.32}$$

In general, forces in the $x$-, $y$- and $z$-directions contribute to charges produced in any of the three directions. Hence, in Eq. (8.32) the piezoelectric coefficient $\mathbf{d}$ is a $3 \times 3$ matrix. Values of the piezoelectric coefficients of these materials are usually made available by the manufacturer. Typical values of the piezoelectric charge coefficients are 1–100 pC/N.

The reverse effect is used in piezoelectric actuators. Application of voltage across such materials results in dimensional changes of the crystal, a phenomenon discovered in 1880 by Jacques and Pierre Curie. The coefficients involved are exactly the same as in Eq. (5.118). The change in length $dL$ per unit applied voltage is given by

$$\frac{dL}{V} = \frac{FL/YA}{(d_{11}FL)/(\varepsilon_0\varepsilon_r A)} = \frac{\varepsilon_0\varepsilon_r}{Yd_{11}} \tag{8.33}$$

Here $A$ is the cross-sectional area, $L$ is the length of the beam, and $d_{11}$ is the piezoelectric coefficient. Strain in this expression depends only on the piezoelectric coefficient, the dielectric constant, and Young's modulus. Therefore, it may be inferred that objects of a given piezoelectric material, irrespective of its shape, undergo the same fractional change in length upon application of a given voltage.

The relationship between dipole moment and mechanical deformation is expressed as constitutive relations:

$$\sigma = cS - eE \tag{8.34}$$

$$D = \varepsilon_0 E + eS \tag{8.35}$$

where $\sigma$ is the mechanical stress, $S$ the strain, $E$ the electric field, $D$ the flux density, $c$ the elastic constant, $e$ the piezoelectric constant, and $\varepsilon_0$ the permittivity of free space. It may be noted that in the absence of piezoelectricity these relations reduce to Hooke's law and the constitutive relation for dielectric materials, respectively. A more generalized form of these relations is discussed in Section 5.7.

The effectiveness of a piezoelectric material is best expressed in terms of its electromechanical coupling coefficient $K^2$. By definition this is related to the other material parameters in the above constitutive equations by

$$K^2 = \frac{e^2}{c\varepsilon_0} \tag{8.36}$$

The direction of vibration of the piezoelectric depends on the dimensions of the slab. If $l \gg b$ and $h$, the slab will vibrate along the length direction. On the other hand, if $l$ and $b \gg h$ then the slab will vibrate in the thickness direction. Hence for a thin slab, the vibrations are in the thickness direction. The finite element model of such actuator is discussed in Section 5.8.

## 8.4.2 Vibrations in Beams

In this case study, vibration levels in a beam modeled by a thin-walled box are controlled using PZT actuators. The exact and finite element solutions of the vibrations in such a beam with attached piezoelectric patches are obtained in Section 5.8. Further, in Chapter 7, we discussed the possibility of selecting the coefficients for a controller for any process or system by following a few tuning steps. It was also shown that simple op-amp-based circuits can be designed to have the desired transfer function. Proportional-integral-derivative (PID) controllers can also be digitally implemented with microprocessors. In this section one application of controllers in microsystems and smart systems is discussed.

An experimental structure of a glass-epoxy composite box beam with two bimorph surface-mounted PZT patches is shown in Figure 8.33. Such thin-walled geometries are used extensively in aircraft structural components due to their inherent advantage of high stiffness and low weight. They are also used in many steel structures such as girders, bridges, scaffolds, etc. Analysis of these structures involves secondary effects such as warping, ovaling, etc. In addition, thin-walled structures in aircraft can produce coupled motions, such as bending, giving rise to torsional motion.

Here we want to suppress transverse bending modes. The PZT patches act as actuators through which both the exciting electrical load and the control signals are applied. The vibration in the box beam is induced through a single patch of PZT actuator using an oscillator. The generated vibration is sensed through an accelerometer that is then fed to the

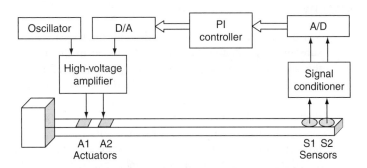

**Figure 8.33** Block diagram of the control system.

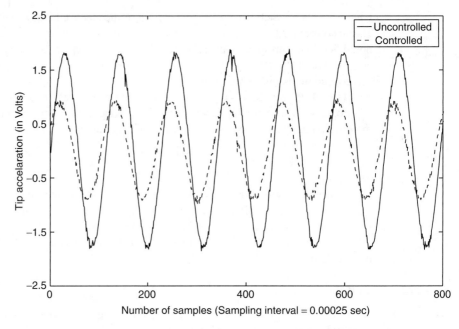

**Figure 8.34** Experimental results on the response of the sensor with and without the PI controller.

A/D converter after signal conditioning. The PI controller in this case can be implemented in a digital signal processing (DSP)-based control card. The sensed voltage is proportional to the acceleration; the control voltage from the controller is amplified by a high-voltage amplifier and then fed into the PZT actuators for bending actuation to suppress the transverse vibration.

First the closed-loop responses using PI controller are obtained using a SISO system and a single-output two-input system. The test specimen is excited in transverse bending mode by applying a single-frequency sinusoidal voltage through PZT actuators. Vibration suppression is performed by the PI controller. In this case, the tip acceleration is measured and the feedback is given as proportional gain $K_P$ times tip acceleration measured by the accelerometer sensor $S_1$. In addition, an integral gain $K_I$ times the velocity of the tip (obtained by integrating its acceleration) is also included in the feedback. A sinusoidal voltage is applied through actuator $A_1$ and the control signal is fed back through actuator $A_2$. $K_p$ and $K_i$ values used can be obtained by tuning using the Ziegler–Nichols rule described in Section 7.6.1. The experimental results in Figure 8.34 indicate significant suppression of vibrations in the beam.

## ▶ 8.5 SUMMARY

Various aspects of the integration of microsystems and microelectronics and their subsequent packaging have been discussed in this chapter. One of the key features in the success of microsystems is the possibility of integrating them with the required electronics. If the microsystem is indeed fabricated on silicon, integration can be achieved at the wafer level. Otherwise, it is achieved at the package level where the dies are interconnected before encapsulation. Various types of packages used in this context are briefly discussed and the processes involved in packaging are described. Unlike the processes discussed in Chapter 3, packaging steps are performed after dicing. It is often stated that packaging is the single most expensive step in the development of integrated circuits. As noted here, microsystems packaging, being even more intricate and specialized, becomes more critical for their success. Some reliability issues are discussed in this context.

Two example microsensors integrated with the required signal conditioning electronics circuitry are described in detail. The Wheatstone bridge and differential amplifier introduced in the previous chapter are used to improve the sensitivity of the piezoresistive pressure sensor. Various aspects of the fabrication and packaging of one such device are discussed in detail. A force-balanced feedback approach is used in the capacitive-sensed accelerometer, where electrostatic force is used to counter the force due to acceleration of the suspended seismic mass. Fabrication of these sensors using bulk micromachining and SOI-based approaches along with their extensive characterization were discussed in detail.

In yet another case study, the integration of a piezoelectric transducer onto a structure is discussed. This smart structure can reduce the structural vibrations by using suitable sensors and a microcontroller with a PI control loop. Control of structural noise and vibrations finds several applications in aerospace and civil engineering disciplines.

## ▶ REFERENCES

1. Yun, W., Howe, R. and Gray, P. (1992) Surface micromachined digitally force-balanced accelerometer with integrated CMOS detection circuitry, *Proceedings of the IEEE Solid State Sensor and Actuator Workshop*, 126–31, Hilton Head Island, SC, USA; published by IEEE, Piscataway, NJ, USA.

2. Gianchandani, Y. B., Kim, H., Shinn, M., Ha, B., Lee, B., Najafi, K., and Song, C. (2000) Fabrication process for integrating polysilicon microstructures with post-processed CMOS circuits, *Journal of Micromechanics and Microengineering*, **10**, 380–86.

3. Smith, J. H., Montague, S., Snieowski, J. J., Murray, J. R. and McWhorter, P. J. (1995) Embedded micromechanical devices for monolithic integration of MEMS with CMOS, *IEDM'95 Technical Digest*, 609–12.

4. Butler, J. T., Bright, V. M. and Comtois, J. H. (1998) Multichip module packaging of microelectromechanical systems, *Sensors and Actuators A: Physical*, **70**(1–2), 15–22.

5. Elwenspoek, M. and Wiegerink, R. (2001) *Mechanical Microsensors*, Springer-Verlag, Berlin, Germany.

6. Blackwell, G. R. (ed.) (2000) *The Electronic Packaging Handbook*, CRC Press, Boca Raton, FL, USA.

7. Varadan, V. K., Vinoy, K. J. and Gopalakrishnan, S. (2006) *Smart Material Systems and MEMS: Design and Development Methodologies*, John Wiley & Sons, Chichester, London, UK.

8. Kusamitsu, H., Morishita, Y., Marushashi, K., Ito, M. and Ohata, K. (1999) The filp-chip bump interconnection for millimeter wave GaAs MMIC, *IEEE Transactions on Electronics Packaging Manufacturing*, **22**, 23–28.

9. Malshe, A. P., O'Neal, C., Singh, S. and Brown, W. D. (2001) Packaging and integration of MEMS and related microsystems for system-on-a-package (SOP), *Proceedings of the SPIE Symposium on Smart Structures and Devices*, **4235**, 198–208.

10. Gerke, R. D. (1999) MEMS Packaging in *MEMS, Reliability Assurance Guidelines for Space Applications*, Brian Stark (ed.). Publication 99-1, Jet Propulsion Laboratory, Pasadena, CA, 166–91. (available at http://parts.jpl.nasa.gov/docs/JPL%20PUB% 2099-1H.pdf); accessed March 31, 2010.

11. Butler, J. T. and Bright, V. M. (2000) An embedded overlay concept for microsystems packaging, *IEEE Transactions on Advanced Packaging*, **23**, 617–22.

12. Reichal, H. and Grosser, V. (2001) Overview and development trends in the field of MEMS packaging, *Proceedings of the 14th IEEE International Conference on MEMS*, IEEE, 1–5, Piscataway, NJ, USA.

13. Cheng, Y. T., Lin, L., and Najafi, K. (2000) Localized silicon fusion and eutectic bonding for MEMS fabrication and packaging, *Journal of Microelectromechanical Systems*, **9**, 3–8.

14. Kumar, V. V., DasGupta, A. and Bhat, K. N. (2006) Process optimization for monolithic integration of piezoresistive pressure sensor and MOSFET amplifier with SOI approach, *Journal of Physics*, **34**, 210–15.

15. Bhat, K. N. (2007) Silicon micromachined pressure sensors, *Journal of Indian Institute of Science*, **87**(1), 115–31.

16. Bhat, K. N., et al. (2007) Wafer bonding—a powerful tool for MEMS, *Indian Journal of Pure and Applied Physics*, **45**, 321–16.

17. Matsumoto, Y., Iwakiri, M., Tanaka, H., Ishida, M., and Nakamura, T. (1996) A capacitive accelerometer using SDB-SOI structure, *Sensors and Actuators*, **A53**, 267–72.

18. Bhat, K. N. et al. (2003) Design optimization, fabrication and testing of a capacitive silicon accelerometer using an SOI approach, *International Journal of Computational Engineering Science*, **4**(3), 485–88.

19. Zubel, I. and Karamkowska, M. (2002) The effect of alcohol additives on etching characteristics in KOH solutions, *Sensors and Actuators*, **A101**, 255–61.

20. Bhat, K. N. (2009) Private communication based on laboratory work during January–May 2009 at the Center of Excellence in Nanoelectronics, IISc Bangalore, as part of the course *Micromachining for MEMS Technology*.

21. ADXL05 Single chip accelerometer with signal conditioning—datasheet, Analog Devices, Norwood, MA, USA, 1996.

22. Lemkin, M. and Boser, B. E. (1999) A three-axis micromachined accelerometer with a CMOS position-sense, *IEEE Journal of Solid-State Circuits*, **34**(4), 456–68.

## ▶ EXERCISES

8.1 A piezoresistive pressure sensor is realized by ion implantation of boron onto a n-type silicon diaphragm having dimensions $0.5\,\text{mm} \times 0.5\,\text{mm}$ and thickness $5\,\mu\text{m}$. Piezoresistors are arranged in the conventional manner on the membrane so that two of them experience maximum longitudinal tensile stress (connected in the opposite arms of a Wheatstone bridge) and other two experience maximum transverse tensile stress.

(a) Determine the burst pressure, ignoring stress crowding at corners.

(b) Determine the sensitivity of this pressure sensor for an input voltage of 1 V.

Assume that the Young's modulus of silicon is 170 GPa, the yield strength is 5 GPa, and Poisson's ratio is 0.2 (note that 1 bar $= 10^5$ Pa). Note that the gauge factor $G_L = -G_T = 100$ for boron-implanted single-crystal piezoresistors.

8.2 Polycrystalline p-type piezoresistors $R_1$, $R_2$, $R_3$, and $R_4$, each of them equal to $R = 1\,\text{k}\Omega$, are arranged as shown in Figure 8.35(a) on oxide grown on a single-crystal membrane having lateral dimensions $1\,\text{mm} \times 1\,\text{mm}$ and thickness $= 10\,\mu\text{m}$. The polysilicon resistor has

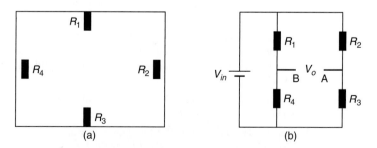

**Figure 8.35** Arrangement of resistors in Problem 8.2.

longitudinal gauge factor $G_L = 30$ and the transverse gauge factor $G_T$ is negligibly small ($G_T = 0$). These resistors are connected in the form of a Wheatstone bridge as shown in Figure 8.35 (b). Assuming $v = 0$ and $Y = 150\,$GPa for polysilicon, in this pressure sensor:

(a) Derive an expression for the output voltage $V_o$ in terms of $R$, $\Delta R$, and $V_{in}$ and mark the polarity of $V_o$.

(b) Determine the sensitivity for an input voltage $V_{in} = 10\,$V.

8.3 Repeat the above case when the four resistors are arranged as in Figure 8.36, keeping the bridge connection and all other parameter values the same as in Figure 8.35(b).

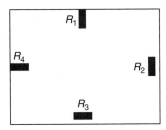

**Figure 8.36** Arrangement of resistors to Problem 8.3.

8.4 The sensitivity of a piezoresistive pressure sensor operating at 5 V power supply was found to be 100 mV/bar (1 bar $= 10^5\,$N/m$^2$). It gives a linear output up to 10 bar pressure.

(a) Design a differential amplifier circuit with differential gain $A_d = 5$ using an op-amp to amplify the signal output. Choose an appropriate power supply voltage for the op-amp.

(b) The Wheatstone bridge of the sensor is perfectly balanced with zero offset voltage. However, the amplified output gives an offset voltage when its input terminals are connected to the output of the pressure sensor. This is attributed to the finite CMRR of the amplifier whose differential gain is 5. Determine the above output offset voltage if the CMRR of the amplifier is 500.

8.5 The offset voltage of a piezoresistive pressure sensor is found to be 5 mV when the input voltage is 1 V. The sensitivity of the pressure sensor is 50 mV/bar for an input voltage of 5 V. Determine the output voltage when the pressure is 3 bar and the input is 5 V.

8.6 The offset voltage of a piezoresistive pressure sensor is found to be 10 mV when the input voltage is 1 V. The sensitivity of the pressure sensor is 50 mV/bar for an input voltage of 1 V. This pressure sensor is connected to an op-amp differential amplifier circuit operating from a dual 10 V power supply whose differential gain $A_d = 2$ and the CMMR is 200. Determine the output voltage of the amplifier when the pressure is **(a)** 5 bar and **(b)** 0 bar. The wheatstone bridge supply is 1 V.

8.7 Repeat Problem 8.6 for the same pressure sensor but with the differential amplifier having CMMR $= \infty$ and $A_d = 5$.

8.8 Repeat Problem 8.6 when the offset voltage is zero but the amplifier gain $A_d = 5$ and CMMR $= 200$.

# CHAPTER 9

# Scaling Effects in Microsystems

## LEARNING OBJECTIVES

After completing this chapter, you will be able to:

▶ Learn the scaling law and how to use it.

▶ Understand if inertial forces are significant in microsystems.

▶ Know why electrostatic force is favored at the microscale.

▶ Know how magnetic forces can be used at the microscale.

▶ Know about scaling in microfluidics.

▶ Know whether miniaturization is always preferred.

A question one should ask when developing or analyzing a microsystems device is whether miniaturization is really necessary if a macroscale counterpart already exists serving the same function. For example, pressure sensors, accelerometers, and video projection systems were available long before their microsystems counterparts came about. So, why are we now interested in miniature versions of them? Reductions in cost, weight, and power consumption might be some of the reasons for miniaturization, but this might not be the case always. Sometimes, a particular principle would not even work if the mircosystems device were made any larger: certain micro-opto-mechanical devices do not work at the macro level. In most other cases, miniaturization is preferred because *scaling* leads to several advantages, some of which we discuss in this chapter. Here we consider the scaling aspect in a variety of domains of microsystems devices to analyze and understand the effects of scaling. In this process, we also present the application of modeling techniques developed in earlier chapters.

At the outset, we should keep in mind a simple but profound scaling law that states:

> *. . . as the size is decreased, volume-related phenomena decrease much more rapidly than surface-related phenomena, which in turn decrease more rapidly than length-related phenomena.*

The simple scaling law is so important that we illustrate it pictorially in Figure 9.1 so as to etch it permanently in our minds.

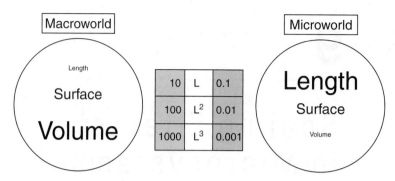

**Figure 9.1** Depiction of the scaling law between macro and microworlds.

Let us consider a box of size $1\,\mathrm{m} \times 1\,\mathrm{m} \times 10\,\mathrm{m}$. Its length—the largest of length, width and height—is $10\,\mathrm{m}$. Its total surface area and volume are given by $\{4 \times (10 \times 1) + 2 \times (1 \times 1)\} = 42\,\mathrm{m}^2$ and $10\,\mathrm{m}^3$, respectively. Now, if we reduce the three dimensions of this box by a factor of 10, that is, to size $0.1\,\mathrm{m} \times 0.1\,\mathrm{m} \times 1\,\mathrm{m}$, its length is reduced by the same factor. However, the surface area is reduced by a factor of 100 to $0.42\,\mathrm{m}^2$ and the volume by a factor of 1000 to $0.01\,\mathrm{m}^3$. We will see the implications of this reduction in various situations to understand the effects of scaling down to small sizes.

## ▶ 9.1 SCALING IN THE MECHANICAL DOMAIN

It is common knowledge that the self-weight of structures is important, especially in large civil engineering structures and large machines. A water tank designed without taking into account its own weight in addition to the weight of water it is to hold would most likely collapse. But is it important to consider the self-weight in microsystems components? Let us analyze this by considering the beam fixed at both the ends and deflecting under its own weight shown in Figure 9.2.

The weight of the beam is a distributed load equal to $q = g\rho_m A$ N/m, where $g$ is the acceleration due to gravity, $\rho_m$ the mass density of the material, and $A$ the cross-sectional area. Assume that the beam has a rectangular cross-section of dimensions $\alpha_1 l \times \alpha_2 l$, where $l$ is the length and $\alpha_1$ and $\alpha_2$ are constants of proportionality. According to the beam theory we learned in Section 4.2, the maximum deflection of the beam, which occurs at its midpoint, is given by[1]

$$\delta = \frac{ql^4}{384YI} \tag{9.1}$$

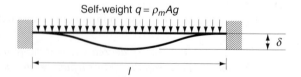

**Figure 9.2** A fixed–fixed beam deformed under its own weight.

---

[1] The reader is urged to derive this formula by solving for the entire beam deflection or just for the maximum deflection at the midpoint by using the Castigliano's first theorem discussed in Chapter 4, Section 4.3.

By noting that

$$I = \frac{(\alpha_1 l)(\alpha_2 l)^3}{12} = \frac{\alpha_1 \alpha_2^3 l^4}{12} \text{ and } A = \alpha_1 \alpha_2 l^2 \tag{9.2}$$

we can write the following:

$$\frac{\delta}{l} = \frac{12 g \rho_m (\alpha_1 \alpha_2 l^2) l^4}{384 Y (\alpha_1 \alpha_2^3 l^4) l} = \frac{8 \rho_m l}{32 Y \alpha_2^2} \propto l \tag{9.3}$$

That is, the relative deflection, which is a dimensionless quantity, is proportional to the size of the beam. So, if a 1 m long beam has a relative deflection of 0.01, a 1 μm long beam will have a relative deflection of $0.01 \times 10^{-6}$. Hence, we can conclude that the effect of self-weight for microsystems structures is negligibly small. This can be understood from the simple fact that self-weight per unit length decreases rapidly as it is proportional to the area, while stiffness decreases proportionally to length.

A question that immediately arises is: can we then say that inertial effects can be neglected in general because inertia is proportional to volume? Inertial effects are due to various accelerations experienced by a body. These include acceleration due to gravity, centrifugal and centripetal accelerations, Coriolis acceleration, etc. But it turns out that we cannot always neglect inertial effects. This is because things move faster at small scales and, hence, have significantly large inertial forces. We note that the natural frequency of a beam is inversely proportional to its size. So, as the size decreases, the natural frequency of free vibration goes up, indicating that small beams vibrate at a much greater rate than large beams of same proportions.

### Example 9.1

Show that the natural frequency of free transverse vibration of a slender beam is inversely proportional to its size. Comment on the effect of scaling on axial and torsional vibration frequencies.

**Solution:** Lumped models suffice for scaling analysis. As shown in the Appendix (Eq. A.3c), the natural frequency $\omega_n$ of the lumped one-degree-of-freedom model of an elastic structure is given by

$$\omega_n = \sqrt{\frac{k}{m}} \tag{9.4}$$

where $k$ and $m$ are the lumped stiffness and mass, respectively. The stiffness of a beam with a transverse load that causes it to bend is given by (see Eq. 4.29)

$$k_b = \alpha \frac{YI}{L^3} \tag{9.5}$$

where $\alpha$ is a number that depends on the boundary conditions of the beam ($\alpha = 3$ for a cantilever, $\alpha = 12$ for a fixed–guided beam, and so on). Let us denote the size by $l$. By noting that $I$ (the area moment of inertia of the beam's cross-section) is proportional to $l^4$ and $L$ (the beam's length) is proportional to $l$, it can be concluded that $k_b$ is linearly proportional to $l$. The mass is proportional to $l^3$. Hence, from Eq. (9.4), $\omega_n \propto l^{-1}$. Hence, as the size decreases, the transverse vibration frequency of a slender beam increases.

From Eq. (4.6), it can be seen that the axial stiffness $k_a$ is also linearly proportional to the size. So, the corresponding frequency is also proportional to the inverse of the size.

However, from Eq. (4.85), we can deduce the torsional stiffness, $k_t$, of a slender beam as

$$k_t = \frac{T}{\phi} = \frac{GJ}{L} \propto l^3 \tag{9.6}$$

and see that it is proportional to $l^3$. Since the rotational inertia ($mr^2$, where $r$ is the radius), which replaces $m$ in Eq. (9.4), is proportional to $l^5$, we conclude once again that the natural frequency is inversely proportional to size.

Next, let us consider an important observation by Galileo Galilei about the bones of large and small animals (Box 9.1). The gist of Galileo's observation was, "the smaller the body, the greater the *relative* strength." How do we understand this keen observation from our knowledge of mechanics today?

### Box 9.1: Galileo's Bones

*To illustrate briefly, I have sketched a bone whose natural length has been increased three times and whose thickness has been multiplied until, for a correspondingly large animal, it would perform the same function that the small bone performs for the small animal.*

*. . . whereas, if the size of a body be diminished, the strength of that body is not diminished in the same proportion; indeed, the smaller the body the greater its relative strength.* (From the *Dialogue Concerning Two New Sciences* by Galileo Galilei.)

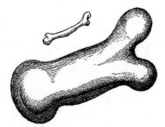

**Figure 9.3** Bones of a small animal and a large animal that have the same function but different proportions, as sketched by Galileo.

Let us look at the maximum stress in a cantilever beam due to its own weight. It is given by [refer to Eq. (4.18a) in Chapter 4]

$$\sigma_{\max} = \frac{Mc}{I} = \frac{M(t/2)}{I} = \frac{(ql^2)(p_t l)}{2p_l l^4} = \frac{(\rho g p_A l^2 l^2)(p_t l)}{2p_l l^4} = \frac{\rho g p_A p_t l}{2p_l} \propto l \tag{9.7}$$

where $p_A$, $p_t$, and $p_I$ are proportionality constants for the respective quantities indicated in the subscripts: area, thickness, and moment of inertia. That is, we express all geometric quantities with reference to the length of the beam by multiplying the length by a constant. Hence, length becomes an indicator of size. Here, $M = ql^2$ because it is the maximum bending moment for a fixed-fixed beam and $q = \rho_m gA$ is the same as in Eqs. (9.1) and (9.3). We see that the maximum stress due to the self-weight is proportional to the size. Since the maximum stress must lie below the permissible strength of the material, smaller beams (and hence the bones of smaller animals) are at an advantage. The bones of larger animals must be fat, whereas those of smaller animals can be slim. Here, we assume that the strength of the material does not change with size, a fair assumption as we move from the bones of elephants to the bones of rabbits. In fact, it may be noted that at the micron scale, the strength increases further as we move down to micron sizes.

**Problem 9.1**
We noted in Chapter 4 that a residual stress gradient makes released micromachined cantilever beams curl up/down. Does that happen in macromachined cantilever beams? Verify your answer by analyzing a microbeam and a macrobeam, assuming the same residual stress and stress gradient in the material.

## ▶ 9.2 SCALING IN THE ELECTROSTATIC DOMAIN

Electrostatic force is widely used in microsystems, but at the macroscale we hardly use it. A fair question to ask is: what makes electrostatic force so attractive at the microscale? The answer is the scaling effect. Let us look at the comb-drive actuator in Chapter 6 [see Figure 6.13(b) and Section 6.2]. With some algebra (left as an exercise to the reader), one can show that its relative deflection for a given voltage is given by:

$$\frac{\delta}{l} = \left(\frac{N\varepsilon_0 V^2}{4Y}\right)\frac{l^2}{gw^3} = \left(\frac{N\varepsilon_0 V^2}{4Y}\right)\frac{l^2}{(p_g l)(p_w l)^3} \propto l^{-2} \tag{9.8}$$

where $N$ is the number of interacting comb-finger pairs and $p_g$ and $p_w$ are proportionality constants for $g$ (the gap between adjacent comb-fingers) and $w$ (the comb height), that is, $g = p_g l$ and $w = p_w l$, respectively. This means that for a large comb-drive actuator of 1 m size and its microcounterpart of $100\,\mu$m size with the same proportions, the micro-comb-drive would have a relative deflection $10^8$ times larger than the large comb-drive actuator. Note that difference in size between 1 m and $100\,\mu$m is $10^{-4}$ and hence the relative deflection, according to Eq. (9.8), changes by a factor of $10^8$. We can thus see why miniaturization favors electrostatic actuation.

Another way of seeing the advantages of electrostatic force at the microscale is to write the expression for the voltage required for a certain relative deflection due to the electrostatic force. This is done by rearranging Eq. (9.8) as

$$V = \sqrt{\left(\frac{\delta}{l}\right)\left(\frac{4Y}{N\varepsilon_0}\right)\left(\frac{gw^3}{l^2}\right)} \propto l \tag{9.9}$$

Hence, we can see that if we need 1 V at the microscale ($100\,\mu$m size), we would need 10,000 V at the macroscale (1 m size). Who would use an actuator requiring such a large voltage! Thus, we can once again see that electrostatic actuation favors miniaturization. If we can figure out a practical way to tap the electrical power from lightning, we can use electrostatic actuators at the macroscale! Think about it.

▶ **9.3 SCALING IN THE MAGNETIC DOMAIN**

Electromagnetic actuation is quite common at the macrolevel: numerous motors used in a variety of applications clearly demonstrate this. However, magnetic actuation did not find favor among early researchers in microsystems in the late 1980s, and the conclusion at the time was that it could not be used at the microscale. Let us examine this using simple scaling arguments.

Imagine two current-carrying conductors. Note that it is not difficult to make them at the macroscale or the microscale. If two conductors of length $l$ are separated by a distance $d$ and carry currents $I_1$ and $I_2$, respectively, the magnetic force between them is

$$F = \frac{\mu_0}{2\pi} I_1 I_2 \frac{l}{d} \tag{9.10}$$

where $\mu_0$ is the permeability of the free space in which the conductors are placed. Now, let us see what happens as we change the size. In order to do this, we must keep something constant so that our comparison is fair. Let us say that we want to keep the current density ($J = I/A$, current per unit area) constant. Since area $A$ is proportional to the size of the conductor, we can write

$$I \propto JA = Jl^2 \tag{9.11}$$

Then, from Eq. (9.10) we can see that

$$F \propto l^4 \tag{9.12}$$

This is bad for miniaturization! If we get a 1 N force at the macroscale, we only get $10^{-24}$ N at the microscale, while we have the same current density in the macro and micropair of current-carrying conductors of the same proportions.

One might think that keeping $J$ constant in both micro- and macrosystems is unfair. So, we can keep something else constant. Since any current-carrying conductor heats up due to Joule heating, let us keep the temperature rise constant and see how the forces scale then.

We show that $k_t \Delta T \propto \rho_e J^2 l^2$, where $k_t$ is the thermal conductivity, $\Delta T$ the temperature rise, and $\rho_e$ the electrical resistivity. From *Fourier's law of heat conduction* equation, we have

$$k_t A \frac{dT}{dx} + \rho_e J^2 A = 0 \tag{9.13a}$$

which states that heat flux, $k_t(dT/dx)$, multiplied by the area of a 1D conductor such as a wire is equal to the Joule heat generated per unit length. On integrating Eq. (9.13a), we get

$$kT = \rho_e J^2 \frac{x^2}{2} + C_1 x + C_0 \tag{9.13b}$$

where $x$ is related to the size (the length of the wire). By noting that $C_0 = kT_0$, we can see that $k_t(T - T_0) = k_t \Delta T \propto \rho_e J^2 l^2$. For a constant temperature rise, we then have

$$J^2 \propto l^{-2} \Rightarrow J \propto l^{-1} \tag{9.13c}$$

Since $I = JA$, we can see that $I \propto l$. Therefore, as per Eq. (9.10), we can see that $F \propto l^2$. This is certainly better than what we had in Eq. (9.12), but it is still not favorable for small sizes.

It is interesting to consider an external magnetic field and one current-carrying conductor instead of two current-carrying conductors. The force in this case is given by invoking *Biot and Savart's law* for force on a conductor in a magnetic field

$$\mathbf{F} = I\mathbf{l} \times \mathbf{B} \tag{9.14}$$

where $I$ is the current, $\mathbf{l}$ the length vector of the conductor, and $\mathbf{B}$ the external magnetic field. Now, for constant current density we have

$$F \propto l^3 \tag{9.15}$$

and for constant temperature rise,

$$F \propto l^2 \tag{9.16}$$

This is somewhat better than for two current-carrying conductors. But the scaling is still not favorable. Does this imply that we cannot use magnetic actuation at the microscale? Well, we can use it provided that for reasonable values of current density and temperature rise, we get a suitably large force, relatively speaking, of course. Electrostatic force can give forces of tens of $\mu$N to possibly mN. If we can get the same with magnetic force, we can justify using magnetic force generated with current-carrying conductors. It is indeed possible because there are very inexpensive rare-earth magnets that can give very large (1 tesla or more) magnetic fields very close to the magnet. We can indeed get respectable magnetic forces at the microscale. This illustrates that we should use scaling arguments with caution.

## ► 9.4 SCALING IN THE THERMAL DOMAIN

Did you ever wonder why elephants have large ears and dinosaurs have fins on their backs? It has to do with scaling effects. Metabolic activity in living creatures produces heat, and warm-blooded animals must maintain a certain temperature. The heat produced is proportional to the volume (cube of the size) of the animal. This heat is generally lost through the skin, that is, the surface of the animal. Because of scaling, we can say that large animals produce much more heat than their surfaces can lose by convection as compared with small animals. This is because large animals have relatively more volume than surface area. In order not to overheat, large animals need special appendages for increasing surface area. Large ears on elephants and the fins on dinosaurs are such appendages.

This is again a matter of scaling, as we demonstrate through a simple calculation. If $p$ is the heat produced per unit time per unit volume, we can write:

$$hA\Delta T = pV$$

$$\Delta T = \frac{pV}{hA} \tag{9.17}$$

$$\Delta T \propto \frac{p}{h}l$$

where $\Delta T$ is the temperature difference between the animal's body and the ambient temperature, $V$ and $A$ are the volume and surface area of the animal, and $h$ is the heat transfer coefficient of convection. Thus, if we assume that $p$ and $h$ are the same for large and small animals, the temperature of the animal is proportional to its size. Hence, large animals must increase their surface area to keep them cool. No wonder that certain large animals like whales live in water: it is because $h$ is large under water.

What about small animals? Their surface is large relative to their volume, as per Eq. (9.17). This would imply that they will have difficulty keeping themselves sufficiently warm, since they lose more heat. Hence, small animals have increased metabolic activity to produce more heat and make $p$ larger. Nature has clearly found ways to fight the scaling laws.

The discussion so far on scaling in the thermal domain has been general; let us now consider a particular situation in the context of microsystems. We look at the electro-thermal microactuators discussed in Section 6.6 and analyze scaling effects in them.

Note from Eq. (6.131) that electrical resistance is inversely proportional to size. As size decreases, resistance increases. This means that resistances in microelectronics circuits are high and hence they draw small currents. Equation (6.136) shows that Joule heating is directly proportional to the square of the size. This implies that heat generation rapidly decreases with miniaturization. This is not surprising because heat generated is a volume-related phenome-non. However, microelectronic cooling is a significant problem in practice, since even a tiny amount of heat cannot be easily dissipated by conduction alone. Fans are often used to convect heat away from critical components. The thermal analysis in Section 6.6, Eq. (6.142) in particular, indicates that the temperature rise in an electrothermal actuator is independent of size. Hence, the thermally induced deformation is also independent of size. However, this is not strictly true as we did not consider convection and radiation, two other heat transmission modes. Scaling analysis of convection and radiation is not straightforward; it involves analysis of fluids and discussion of radiative heat transfer, respectively.

Analysis of transient heat flow is also important and complex. Consider the time constant for 1D heat transfer involving convection. By equating the heat lost to convection to the rate of internal thermal energy, we can write

$$-hA_c(T - T_\infty) = \rho_m c_v V \frac{dT}{dt} \tag{9.18}$$

where $h$ is the convective heat transfer coefficient, $A_c$ the area of the convecting surface, $T_\infty$ the ambient temperature, $\rho_m$ the mass density, $c_v$ the specific heat at constant volume, and $V$ the volume of the solid. For an initial temperature $T_0$ at $t = 0$, Eq. (9.18) has the following solution [1]:

$$\frac{(T - T_\infty)}{(T_0 - T_\infty)} = e^{-(hA_c/\rho_m c_v V)\, t} \tag{9.19}$$

The time constant, which is equal to $(\rho_m c_v V / h A_c)$, scales linearly with size if we assume that the material properties and the convective heat transfer coefficient are invariant with size. Hence, cooling times decrease with miniaturization. However, we note that the heat transfer coefficient cannot be assumed to be invariant with size. So, scaling analysis should only be used as a guideline; further analysis is always needed to verify the assumptions involved.

**Your Turn:**
Transient problems involving heat transfer in electrothermal microactuators also show interesting scaling effects. Use scaling analysis to argue how heating and cooling times vary with the size of electrothermal actuators. First model only conduction and then include convection and radiation.

## ▶ 9.5 SCALING IN DIFFUSION

Diffusion is the phenomenon of spreading of a species due to its concentration gradients without the help of an external force. Heat and electric currents flow because of diffusion. Diffusion happens when two different gases, liquids, or solids come into contact with each other. The rate of diffusion depends on the diffusion coefficient. A simple relationship can be given as follows (see Your Turn below):

$$x = \sqrt{2Dt} \tag{9.20}$$

where $x$ is the distance traveled by a diffusing species in time $t$ and $D$ is the diffusion coefficient given by

$$D = \frac{K_B T}{6\pi\eta r} \tag{9.21}$$

where $K_B$ is Boltzmann's constant, $T$ the temperature, $\eta$ the viscosity, and $r$ the hydro-dynamic radius.[2] We can see that $D$ is larger for smaller $r$. Hence, per Eq. (9.20), the diffusion distance is greater at the microscale. So, in a given time, species move faster over the required distance at microscales. Consequently, diffusion is sufficient for liquids to mix at small scales, whereas we need stirring and other active mixing modes at large scales. Since, as per Eqs. (9.21) and (9.20), small particles diffuse better than the large ones, effective separation of particles of various sizes can be achieved with minute samples introduced into microchannels. This is a useful feature in lab-on-a-chip applications, in which different particles in a complex fluid such as whole blood can be quickly and effectively separated by simply flowing them in microchannels.

Next, we see that nondiffusive mixing is a problem at small scales—another scaling effect.

---

**Your Turn:**

Diffusion is a fundamental phenomenon in the transmission of species. Fick's two laws of diffusion are the starting points for studying diffusion. Denoting $n$ as the concentration of the species, $t$ the time, $x$ the spatial variable, $D$ the diffusion coefficient, and $J_D = D\frac{\partial n}{\partial x}$ the diffusive flux, we can write Fick's laws as follows.

*Fick's first law*: species diffuse from high concentration to low concentration:

$$J_D = -D\frac{\partial n}{\partial x}$$

*Fick's second law*: concentration changes with time as

$$\frac{\partial n}{\partial t} = D\frac{\partial^2 n}{\partial x^2}$$

Study Fick's laws and their importance and think about where else you have seen or used these equations. Derive the following solution of the equation describing Fick's second law:

$$n(x,t) = \frac{C}{\sqrt{4\pi Dt}} e^{-\left(x^2/4Dt\right)}$$

where $C$ is a constant. Use this solution to verify Eq. (9.20).

---

[2] The hydrodynamic radius of a particle is the radius of the hard sphere that diffuses at the same rate as the particle.

## ▶ 9.6 SCALING IN FLUIDS

We discussed the mechanics of solids in detail in Chapter 4. Much of our discussion was based on the assumption that continuum theories remain applicable at the microscale. A continuum theory assumes that a large number of particles (atoms, molecules, grains, etc.) clumped together can be treated as a homogeneous medium with some average properties, rather than having to deal with properties of individual particles. This assumption is valid when there are sufficiently many particles in a particular size in the object we consider. This is true for solids at the microscale: a cube of micrometer size has millions of atoms of a particular kind. However, some of the continuum assumptions break down in fluids at the microscale.

As shown in Figure 9.4, there is a fundamental difference in the way a fluid flows in a micron-sized pipe. In traditional fluid mechanics for large pipes, we learn that the velocity profile across the cross-section of a fully developed flow is parabolic with zero velocity at the pipe walls. This is shown in Figure 9.4(a) and is called *nonslip* flow, meaning that the fluid particles do not slip along the pipe walls. This is not true for a micron-sized pipe flow: it has been observed that fluid particles slip along the walls with a finite velocity, as shown in Figure 9.4(b). This situation is usually characterized in fluid mechanics, by a nondimensional number. The one we use here is called the *Knudsen number (Kn)*, which is the ratio of the mean free path of a particle ($d_m$) in the flow to the characteristic dimension ($d$) of the flow:

$$Kn = \frac{d_m}{d} \qquad (9.22)$$

The mean free path, which is the average distance over which a particle in a fluid moves in between collisions with other particles, varies with the type of fluid and its temperature. At room temperature, it can be as small as 1 μm for air. This means that $Kn$ for microflows may be equal to 1 or may be even greater than that. For nonslip flow to occur, $Kn$ should be less than 0.1.

When $Kn$ is greater than 0.1, the flow slips on the walls of the pipe. One can still use continuum theories by modifying the viscosity, but when $Kn$ is above 10, this too does not work, and we must treat every particle of a fluid as a separate entity and calculate the behavior of the fluid. If this particle is a molecule, we call it molecular dynamic simulation. Thus, the Knudsen number introduced in Eq. (9.22) indicates in which regime a particular microfluidic device operates. Accordingly, the modeling relevant to that regime must be used.

Let us discuss another scaling effect in fluids. Another nondimensional number often used in fluid mechanics is the well-known *Reynolds (Re) number*, which is the ratio of inertial forces to viscous forces:

$$Re = \frac{\rho U L}{\eta} = \frac{\rho (\alpha L) L}{\eta} \propto l^2 \qquad (9.23)$$

where $U$ is the velocity of the fluid, $\rho$ the mass density, $L$ the characteristic length, and $\eta$ the viscosity. If we assume that $U$ is proportional to size, we see that $Re$ varies as the square of

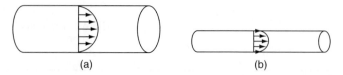

(a)                              (b)

**Figure 9.4** Velocity profiles in pipe flows: (a) nonslip flow in a macropipe; (b) slip flow in a micropipe.

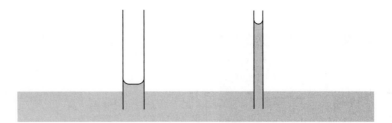

**Figure 9.5** Capillary effect in a large tube and a small tube.

the size. This means that *Re* is very small for microflows, which has serious implications because low-*Re* flows tend to be laminar, not turbulent. A nonturbulent flow does mix; however, all the mixing happens only by diffusion. If we put two microflows next to each other, mixing them takes a long time and happens over large distances. Hence, we have to resort to stirring to make them mix faster and better.

Let us consider another fluidic scaling effect that we are all familiar with in our daily lives. How do we pick a small *khas-khas* seed or a small bead? We wet our finger and touch it. There must be some force acting against gravity that does not let the seed or bead fall from our finger. This force is the surface tension that prevents the fall. We cannot pick up a heavy object by wetting our finger or our entire hand, as surface tension scales linearly with size. Because of the scaling effect noted at the beginning of this chapter (see Figure 9.1), surface tension is an enormous force, relatively speaking, at microscales.

Now, in Figure 9.5, which shows two capillary tubes and a liquid medium, we see that the liquid column rises more in the smaller capillary tube than in the larger one. Once again, this is due to the surface tension. To characterize this capillary effect, we use another nondimensional number, the *Bond number (Bo)*, which is the ratio of the gravitational force to the surface tension:

$$Bo = \frac{\rho g L^2}{\gamma} \tag{9.24}$$

where $\rho$ is the mass density, $g$ the acceleration due to gravity, $L$ the characteristic length, and $\gamma$ the surface tension. Clearly, surface tension dominates gravity at small sizes, making the value of *Bo* less than unity.

Since surface tension is such an enormous force at the microscale, it can be put to good use. Consider Figure 9.6a-b, which shows the interesting phenomenon *electrowetting*. Normally, a drop on a surface assumes the shape of an orb of a ball. Mercury on a glass surface looks like a complete ball, with a contact angle close to 180°. This is not true of a water drop on a glass surface. This is because the surface tension between glass and mercury is different from that between glass and water. This is a well-known phenomenon. But what is interesting is that this surface tension (and hence the contact angle) can be changed by various means. For example, if we coat the glass surface with a layer of material that is hydrophobic (water-repellent), even a water drop behaves like a drop of mercury with a large contact angle. There will be a rather permanent change in contact angle because we cannot remove the hydrophobic layer easily and quickly.

Surface tension can also be changed by changing temperature or applying an electrical field. Figure 9.6b shows an instance of the latter case. When the contact angle decreases, the drop spreads, pushing itself and the fluid surrounding it. Now, if we have a channel that has closely spaced electrodes with small gaps in between, we can make the drop move by

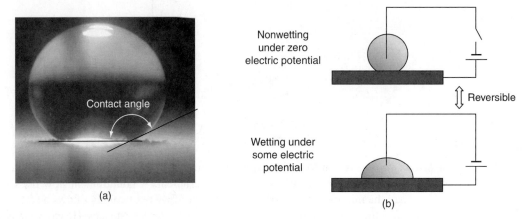

(a)

(b)

**Figure 9.6** (a) Surface tension and contact angle of a drop; (b) changing surface tension by applying an electric field. *Courtesy*: (a) M.S. Bobji.

sequentially changing the electric potential on the adjacent electrodes. Thus, we have a device that acts like a pump! This is a powerful way to "pump" small quantity of fluids because surface tension is the most favorable force at the microscale.

**Your Turn:**

In fluid mechanics, we often use nondimensional analysis, perhaps because experiments are needed that concern large systems such as flying bodies, currents in the ocean and atmosphere, and other large systems. Therefore, there are many nondimensional numbers in fluid mechanics. They are all named after people who studied fluid mechanics and found interesting and useful things. In this section, we have come across three such nondimensional numbers (Knudsen, Reynolds, and Bond number). In order to engage yourself with fluid mechanics, find out at least a dozen such numbers, get an intuitive understanding of them, and learn about the fundamental contributions of the eponymous fluids researchers.

## ▶ 9.7 SCALING EFFECTS IN THE OPTICAL DOMAIN

Optics and microsystems combine to give many benefits. This is because, as shown in Figure 9.7, the wavelength of the visible light, 400–700 nm, overlaps the range of motions possible with microsystems and the the range sizes that can be fabricated using micro-manufacturing techniques. We have already mentioned of MEMS elsewhere in this book. The

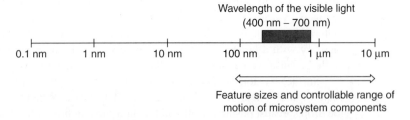

**Figure 9.7** A size chart showing the wavelength of the visible light (400–700 nm).

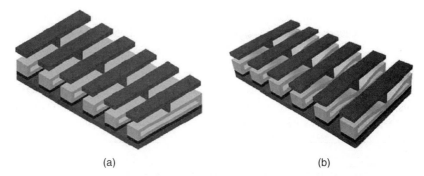

(a)                                    (b)

**Figure 9.8** An array of vertically movable microbeams helping generate a light beam of any wavelength of visible light: (a) unactuated and (b) actuated.

many systems that manipulate light would not be possible without the help of micromechanical devices. Reflecting tiny beams of light with micromirrors, creating interference patterns with moving beams, and making the entire optical workbench on a chip are some of the many possibilities. We present here just one example to show how scaling helps in micro-optics.

Figure 9.8(a) shows an array of beams, each of which can be moved down with electrostatic force. The widthwise spacing of these beams and the gaps under the beams are such that white light falling on them can be selectively converted to light of any color (i.e. wavelength). This can be done because of the interference between the light reflecting off the beams' top surface and the light reflecting off the bottom surface of the spaces between the beams. By holding some beams at specific heights, only certain colored light is made to have constructive interference; the rest would have destructive interference. This is used in a commercial device developed by Polychromix.[3] Note that this technique does not scale up to be any larger or any smaller. It must be precisely at the microscale of the wavelength of light.

In some optical phenomena, scaling down below a certain size is not helpful. This is an argument against miniaturization. Consider an absorption spectrometer that identifies a particular wavelength absorbed by a gas when light is passed through a chamber filled with that gas. Since all gases absorb certain wavelengths of light, we can identify a gas by the missing wavelength of the light exiting the chamber. For this to happen, we need to let the light (usually a laser beam) travel far enough inside the gas-filled chamber that the gas has enough time to absorb the specific color. It may appear that miniaturization is not useful here because miniature chambers obviously have small distances to travel. But one can always mount or etch small reflecting surfaces on the walls of the chamber so that the light travels a large distance before it comes out. However, doing this creates another problem. The spot size of the laser beam increases as it travels. The beam diameter $w(z)$ as a function of the distance $z$ it travels is given by

$$w^2(z) = w_0^2\left(1 + \frac{\lambda z}{\pi w_0^2}\right)$$
(9.25)

where $w_0$ is the initial diameter of the spot at the inlet into the absorption chamber and $\lambda$ the wavelength of the light. As the spot diameter increases, intensity decreases. So, the reflecting mirrors must be large enough to reflect all the light. Thus complete miniaturization creates a problem. If we take $\lambda = 4.5$ μm, we can see that 100 μm diameter spot will widen by 100%

---

[3] Visit www.polychromix.com to learn about commercial devices using this type of microbeam arrays.

within a distance of 16 mm, whereas a 300 μm diameter spot increases in size only by 3%. Clearly, here we do not want to make the device very small.

**Your Turn:**

Optics and miniaturization go well together for the reasons cited in this section. However, we have also given a counterexample. To make the negative case stronger, find two more counterexamples. Also, find four cases where miniaturized optics is advantageous.

## ▶ 9.8 SCALING IN BIOCHEMICAL PHENOMENA

Biological applications of microsystems are many. We have not covered biological applications and bio-MEMS extensively in this book as this would require some chemistry and much more detail about microfluidics. (For more information see references for this chapter.) However, we consider here some of the issues of scaling in such systems. We start by asking: why would one favor miniaturization when dealing with biochemical phenomena; in particular in labs-on-a-chip and microchemical reactors?

Miniaturization has the following advantages for biochemical labs-on-a-chip: small sample volumes; quick analysis; cost-effectiveness, because small amounts of reagents and power are required; and finally the possibility of combinatorial analysis. The last item perhaps needs more explanation. When screening a new medicine, one may want to try combinations of several chemicals or biological species to test effectiveness as well as side effects. Doing this with small volumes is definitely cost-effective. The selection of small sample volume in a lab-on-a-chip requires special caution, however, as we observe that the sample volume is given by

$$V_{sample} = \frac{1}{\eta N_A C} \tag{9.26}$$

where $\eta$ is the efficiency of the sensor, $N_A$ Avogadro's number, and $C$ the concentration of the species to be detected. As we decrease the sample volume, for a given concentration, the efficiency of the sensor must go up. Otherwise, we might miss the presence of a species and give a false negative result. Increasing the efficiency of the sensor might make the sensor more complicated and even more expensive. Therefore, we should be careful about making the sample too small.

Another minor problem arises in combinatorial analysis if the sample size is too small. Usually, a robot puts tiny droplets of a sample on a substrate or into a microwell. Then, chemicals are added. If the droplet is too small, it may simply evaporate! Note that smaller the drop, the larger its surface-to-volume ratio. And a larger surface area increases evaporation—an interesting scaling effect.

In considering chemical reactors, we should look at nature. Honeybees are perhaps the best example. They process and store honey in tiny near-hexagonal chambers rather than one big vessel. All living creatures are made of cells, tiny compartments that are full-fledged living things all by themselves. It is fascinating to ask: why did nature prefer this? There may be many reasons. And there are lessons to be learnt for chemical processes that take place in large reactors. In small reactors, it is easier to control the process parameters (pressure, temperature, concentration, etc.) perfectly so that the optimum chemical reaction can be achieved. However, the final conclusion is yet to be reached and this is an active research field. Detailed discussion of scaling effects in this requires knowing some fundamentals of chemical reaction kinetics; we leave this for interested readers to explore further.

## ▶ 9.9 SCALING IN DESIGN AND SIMULATION

Simple analysis, as in this chapter, reveals important scaling effects in various domains. Scaling of different quantities in many domains is summarized in Table 9.1. As noted in Table 9.1 in the context of fluidic quantities, simple scaling analysis may not always be entirely accurate, depending on the assumptions made. However, that does not reduce the usefulness of scaling analysis. We now discuss two uses of scaling analysis, one in design and another in simulation.

Imagine you are designing a micromechanical suspension such as those discussed in Chapter 4. A problem at hand requires you to have different stiffness values in different directions. As a designer, let us assume that you are free to increase or decrease the overall size of the suspension. Note that you must keep the changing stress in mind as you vary the stiffness. Table 9.1 aids in this context. If you halve the size, per Table 9.1, axial and bending stiffnesses decrease by a factor of two whereas torsional stiffness decreases by a factor of eight. Stress, on the other hand, increases by a factor of four in the axial direction and by a factor of eight in the bending and torsion modes, for the same loading. These scaling effects will indicate to the designer the extent to which the size can be changed. The altered device can then be simulated using FEA for a final check.

Scaling is often necessary in simulating microsystems devices and components. If a beam is $150 \, \mu m$ long, we may actually enter 150 [instead of $150 \times 10^{-6}$ m in the SI (*Système*

**Table 9.1 Approximate scaling effects in different domains**

| Domain | Characteristic | | Scaling |
|---|---|---|---|
| Mechanical: solids | Stiffness | Axial | $l$ |
| | | Bending | $l$ |
| | | Torsion | $l^3$ |
| | Stress | Axial | $l^{-2}$ |
| | | Bending | $l^{-3}$ |
| | | Torsion | $l^{-3}$ |
| | Inertial force | Translational | $l^3$ |
| | | Rotational | $l^5$ |
| | Natural frequency | | $l^{-1}$ |
| Mechanical: fluids* <br><br> *Continuum assumption breaks down at small scales. So, the scaling effects in fluids should be taken with a grain of salt; calculations based on simple analysis may sometimes be wrong. | Reynolds number | | $l^2$ |
| | Surface tension | | $l$ |
| | Bond number | | $l^2$ |
| | Coutte-flow-related damping coefficient | | $l$ |
| | Squeezed-film-effect-related damping coefficient | | $l$ |
| Electrical | Resistance | | $l^{-1}$ |
| | Joule heating | | $l^0$ |
| | Capacitance | | $l$ |
| | Pull-in voltage due to electrostatic force | | $l$ |
| | Relative deformation due to electrostatic force | | $l^{-2}$ |
| | Inductance | | $l^?$ |

**Table 9.1 (*Continued*)**

| Domain | Characteristic | | Scaling |
|---|---|---|---|
| Magnetic | Force between two current-carrying conductors | Constant temperature rise | $l^4$ |
| | | Constant current density | $l^2$ |
| | Force on a current-carrying conductor in an external magnetic field | Constant temperature rise | $l^3$ |
| | | Constant current density | $l^2$ |
| Thermal | Thermal resistance | | $l^{-1}$ |
| | Thermal capacitance | | $l^3$ |
| | Thermally induced deformation considering only conduction | | $l^0$ |
| | Thermally induced deformation considering also convection (and possibly radiation) | | $l^{-0.5}$ |
| | Thermal time constant (assuming that heat transfer coefficient is invariant with size) | | $l^{-1}$ |
| Miscellaneous | Diffusion coefficient | | $l^{-1}$ |

*I*nternational d'*U*nités, the International System of Units) system] in a software program because cubing the length would otherwise yield a very small number. This convenient scaling warrants attention with regard to the other quantities in the program. If this is a coupled electrostatic problem, we have to think of the value we would use for permittivity of free space. Recall that $\varepsilon_0 = 8.854 \times 10^{-12}$ F/m in SI units. Similarly, if it is an electrothermal problem, we must consider how to enter 8 V into the program. For $\varepsilon_0$, one may hasten to enter $8.854 \times 10^{-18}$ because a μm is entered as **m** for length. That is, there is a multiplicative factor of $10^6$ and $\varepsilon_0$ has units of F/m, suggesting that we should divide by $10^6$ before entering it. This will yield the correct result for capacitance calculation but not for the calculation of the electrostatic force. Observe the results and discussion in Example 9.2.

### Example 9.2

Consider a parallel-plate capacitor of area $A = 10^{-8}$ m$^2$ and gap $g = 10^{-6}$ m. Compute (i) its capacitance and (ii) the electrostatic force between the plates, first without scaling and then with scaling, wherein μm units are used. Let the potential difference between the plates be 5 V.

**Solution:**

(i) Without scaling,

$$\text{Capacitance} = \frac{\varepsilon_0 A}{g} = \frac{(8.854 \times 10^{-12})(10^{-8})}{10^{-6}} = 8.854 \times 10^{-14} \text{ F/m}$$

With scaling, if we use $\varepsilon_0 = 8.854 \times 10^{-18}$,

$$\text{Capacitance} = \frac{\varepsilon_0 A}{g} = \frac{(8.854 \times 10^{-18})(10^4)}{1} = 8.854 \times 10^{-14} \text{ F/m}$$

which agrees with the result obtained without scaling. Note that we were careful to use $A = 10^4$ μm$^2$.

(ii) Force, without scaling, is computed as

$$\text{Force between the plates} = \frac{\varepsilon_0 A V^2}{2g^2} = \frac{(8.854 \times 10^{-12})(10^{-8})(5)^2}{2(10^{-6})^2} = 1.1068 \times 10^{-6} \text{ N}$$

With scaling, if we keep 5 V as it is because there is no explicit m in its unit,

$$\text{Force between the plates} = \frac{\varepsilon_0 A V^2}{2g^2} = \frac{\left(8.854 \times 10^{-18}\right)\left(10^4\right)(5)^2}{2(1)^2} = 1.1068 \times 10^{-12} \text{ N}$$

How can we explain the discrepancy in the force calculation? The reason for the discrepancy is that we did not scale the voltage appropriately. For this, we should recall the dimensions of various quantities and consider the dimension $\underline{L}$ in them:

Dimensions of $\varepsilon_0 = M^{-1}L^{-3}T^4A^2$

Dimensions of electric potential, $\phi = ML^2T^{-3}A^{-1}$

Therefore, when we use $\mu$m scaling, we must multiply the SI unit values by $10^{-18}$ for $\varepsilon_0$ and $10^{12}$ for the electric potential (i.e. voltage). Then,

$$\text{Force between the plates} = \frac{\varepsilon_0 A V^2}{2g^2} = \frac{\left(8.854 \times 10^{-30}\right)\left(10^4\right)\left(5 \times 10^{12}\right)^2}{2(1)^2} = 1.1068(?\text{N})$$

The question arises whether this result is wrong. Notice that we have put (?N) for the unit. It is actually the result multiplied by $10^{-6}$ N, so that the force is $1.1068 \times 10^{-6}$ N; but why do we have to multiply by $10^{-6}$? Because the dimensions of force are $MLT^{-2}$. Hence, in reading the result of a calculation, we must multiply by $10^{-6}$. By the same token, we must scale the force up by $10^6$ on entering it into a calculation.

The problem in Example 9.2 and the scaling of the quantities in Problem 9.1 may seem confusing at first. Indeed, there is bound to be confusion when we scale only the lengths, especially when using commercial software or even our own programs involving multiple domains. To avoid confusion, a foolproof method is to observe the dimension of $L$ in the quantity in question. Let us say that a quantity has $L^x$ as the length dimension. Then, when we scale up m unit to $\mu$m, that quantity must be multiplied by $10^{6x}$ before entery into a calculation. Likewise, when we read that quantity back from the calculation, we must multiply by $10^{-6x}$ to get the result in SI units.

The next question is: how we do we know the length dimension (i.e. $L^x$) for just about any quantity, say, permittivity, permeability, resistance, specific heat, diffusivity, etc.? The procedure is simple: start with a known relationship in the dimensions of all the quantities are known except for the quantity in question. Let us take permittivity. The known relationship is Coulomb's law,

$$\text{Force} = \frac{\text{Charge}^2}{4\pi\varepsilon_0 \text{ gap}^2}$$

Writing its dimensions, we get

$$MLT^{-2} = \frac{A^2T^2}{\varepsilon_0 L^2}$$

from which it follows that permittivity's dimensions are $M^{-1}L^{-3}T^4A^2$. Similarly, to obtain the dimensions of electric potential, we can use the familiar relationship: power = electric potential × electric current. Therefore, the dimensions of electric potential are

**Table 9.2 Fundamental dimensions for some quantities; the factors for scaling lengths from m units to μm units**

| Quantity | Dimensions (only four are shown) | Multiplicative factor to enter into a calculation | Multiplicative factor to read from the result of a calculation |
|---|---|---|---|
| Density | $ML^{-3}T^0A^0$ | $10^{-18}$ | $10^{18}$ |
| Force | $MLT^{-2}A^0$ | $10^6$ | $10^{-6}$ |
| Young's modulus | $ML^{-1}T^{-2}A^0$ | $10^{-6}$ | $10^6$ |
| Residual stress | $ML^{-1}T^{-2}A^0$ | $10^{-6}$ | $10^6$ |
| Resistance | $ML^2T^{-3}A^{-1}$ | $10^{12}$ | $10^{-12}$ |
| Resistivity | $ML^3T^{-3}A^{-2}$ | $10^{18}$ | $10^{-18}$ |
| Permittivity | $M^{-1}L^{-3}T^4A^2$ | $10^{-18}$ | $10^{18}$ |
| Permeability | $ML^{-1}T^1A^{-1}$ | $10^{-6}$ | $10^6$ |
| Voltage | $ML^2T^{-3}A^{-1}$ | $10^{12}$ | $10^{-12}$ |
| Magnetic field | $ML^0T^{-2}A^{-1}$ | $10^0$ | $10^0$ |

$ML^2T^{-3}A^{-1}$. Here we used knowledge of the dimensions of some quantities in a familiar domain (mechanical and electrical). Sometimes, we may need to use two familiar relationships, especially when the domains involved are not so familiar to us. This is usually not a problem. Just note that there are only seven fundamental dimensions (and units): distance L (meter, m), mass M (kilogram, kg), time T (second, s), electric current A (ampere, A), intensity of light Cd (candela, Cd), quantity Q (mole, mol), and temperature K (Kelvin, K). This limited number—*only seven*—should comfort us that there may be many energy domains but fundamental quantities are few This means that what we really need to know or remember is limited. It is all in the details after that. This realization is particularly important in studying multidisciplinary fields such as micro and smart systems.

For convenience, Table 9.2 summarizes the MLTA dimensions of some frequently occurring quantities and the factors to multiply in entering it into a calculation and reading it off afterwards. The table is also useful when other quantities are scaled, say mass or time. Another use of this table and the dimensions-based method described above is in choosing the right length-scaling for a problem at hand. Once we know the factors to multiply for all the quantities, we can ensure that none of them become too large or too small.

**Problem 9.2**
Find the right scaling for an electrothermal-elastic simulation. Note that this involves lengths, electrical resistivity, thermal conductivity, specific heat, Young's modulus, voltage, temperature, and force. Try mm units, tens of mm, tenths of mm, hundredths of mm, etc., to judge where most quantities stay reasonable (not too small or too large).

## ▶ 9.10 SUMMARY

This chapter has two purposes. The first is the application of some of the modeling techniques discussed in Chapters 4 and 6; the second is to analyze and understand the effects of scaling as we move down to the microscale. We discussed scaling effects in a variety of domains, including their use in design and simulation. We used simple arguments and easy-to-follow analytical calculations, which are usually sufficient for scaling analysis. However, there is no substitute for detailed analysis, simulations, and experimentation. The main point of this chapter is that we must, when we see a micro-systems device, always ask if miniaturization is really warranted; and if so, how the device is benefitting. This kind of questioning is always rewarding and helps one find new ways to think about microsystems and also invent new concepts and designs.

## ▶ REFERENCE

1. Baglio, S., Castorina, S., Fortuna, L., and Savalli, N. (2002) Modeling and design of novel photo-thermo-mechanical microactuators, *Sensors Actuators A: Physical*, **101**, 185–193.

## ▶ FURTHER READING

1. Madou, M. (2001) Principles of microfabrication, *Scaling Laws*, CRC Press, Boca Raton, FL, USA.

2. Weigl, B. H. and Yager, P. (1999) Microfluidic diffusion-based separation and detection, *Science*, **283** (5400), 346–47.

3. Trimmer, W. (1989) Microrobots and micromechanical systems, *Sensors and Actuators A*, **19**(3), 267–87.

4. Chakraborty, S. (ed.) (2010) *Microfluidics and Microfabrication*, Springer, New York.

5. Baglio, S., Castorina, S., and Savalli, B. (2002) *Scaling Issues and Design of MEMS*, John Wiley & Sons, New York.

6. Incropera, F. P. and deWitt, D. P. (1996) *Fundamentals of Heat and Mass Transfer*, John Wiley & Sons, New York.

## ▶ EXERCISES

9.1 Consider a cantilever beam of length $l$, width $\alpha l$, and depth $\beta l$ made of a material with Young's modulus $Y$. Substitute these in the expression for the fundamental natural frequency of this beam. Choose some numerical values for $Y$, $\alpha$, and $\beta$ and plot how the natural frequency varies with $l$. What do you infer from this about the effect of size on the natural frequency?

9.2 Figure 9.9 shows a matchbox. Assume that each matchstick is 50 mm long and has cross-sectional area 4 mm². Assume also that each matchstick needs 1 mm² of surface coating for ignition. Consider a matchbox of size 50 mm × 30 mm × 4 mm. Is there enough ignitable area for all the matchsticks it can enclose? How large can the matchbox be made while keeping its proportions the same and having sufficient area to ignite all the matchsticks it can contain? Based on your analysis, can you argue for an optimum size for a matchbox?

9.3 A bar in a smart-material-based actuator must be stretched by about 1% of its length. If the failure stress is the same, does it matter if the size of the bar is increased or decreased? What happens if a cantilever is bent with a transverse load at its free tip to give a deflection of 1% of its length? Does size matter in this case?

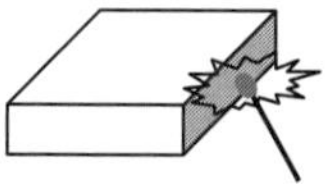

**Figure 9.9** A matchbox with a matchstick scratched over the ignitable area.

9.4 There is a helical compression spring in your ballpoint pen and a much bigger one in the suspension of a train. Measure the coil diameter, length, wire diameter, and the pitch of the spring in the ballpoint pen and gather or guess the same data for the spring in a train. Do they have the same proportions? Why or why not? Explain with a simple calculation.

9.5 A solenoid actuator consists of an electromagnetic coil and a spring-restrained core. How small can it be made? Take practical fabrication constraints into account.

9.6 Use scaling analysis to see if miniaturization favors the sensing of mass by monitoring changes in the resonance frequency of the transverse vibration of a beam. (*Hint*: Write the natural frequency formula in lumped form, using a lumped spring constant, beam mass, and mass to be detected.)

9.7 Obtain the fundamental dimensions of the following quantities: pressure, specific heat, diffusion coefficient, magnetic flux, heat transfer coefficient, emissivity, inductance, and capacitance.

9.8 Harvesting wind energy is proving to be a viable alternative today. Does miniaturization of wind turbines make sense? Use scaling analysis to support your answer. If you argue that it is indeed advantageous to miniaturize wind turbines, what applications do you foresee?

9.9 Micromachined gyroscopes have not succeeded as well as micromachined accelerometers in the commercial market or in cost-performance analysis. Are there any fundamental limitations for micromachined gyroscopes? Can scaling analysis help here?

# 10

# Simulation of Microsystems Using FEA Software

## LEARNING OBJECTIVES

After studying this chapter, you will be able to:

▶ Be familiar with a variety of commercial finite element software programs with which some micro and smart systems components can be simulated.

▶ Understand how to handle the simulation with realistic geometries of micromechanical components.

▶ Assimilate the modeling concepts and finite element analysis techniques learned in Chapters 4–6 and apply them in practice using commercial software.

## ▶ 10.1 BACKGROUND

The theory, analytical methods, and lumped modeling techniques discussed in Chapters 4 and 6 are useful in understanding the concepts and the principles underlying micro and smart systems devices. They are also useful in quickly computing the approximate behavior of a component, device, or system. They find use in the early stages of design as well. However, the simulation and design of a real system is much more complex and the simple geometries considered thus far are not adequate. Some simplifying assumptions (e.g. parallel-plate capacitor for electrostatic behavior) cannot be made if we want to accurately predict the behavior of a real system. Material properties may have to be varied across different layers of a device or within a layer. Nonlinearities of different kinds cannot be set aside and must be given due consideration. All this requires us to solve the governing partial differential equations (PDEs) and ordinary differential equations (ODEs) presented in earlier chapters. In Chapter 5, we discussed the finite element analysis (FEA) technique, keeping in mind the real devices and associated complications. Just as learning the concepts and the theory of physics governing the devices is not enough, so knowing the FEA theory is not sufficient.

Hence, in this chapter, we present the practical aspects of FEA. We do this by outlining the simulation of different types of micro and smart systems components using a variety of commercial FEA software programs. This, we hope, enables interested readers to try them on any software available to them.

While reading through this chapter, the reader should pay attention to the data presented in each example and observe the numerical values of the input and output quantities. This helps develop a quantitative intuition that is very valuable in practice. Another aspect that deserves attention has to do with the subtleties and nuances of FEA.

For example, which element should be used for what purpose? What kind of analysis is needed for simulating a component of interest? How does one extract lumped modeling parameters from FEA simulations?

A caveat is also pertinent at this stage: this chapter is by no means sufficient to learn enough about software programs. The help manuals for the software must be consulted when in doubt. One's experience, intuition, and creativity are also essential in using any software properly and well. The reader should also note that there are other software programs with capabilities similar to or different from the ones covered in this chapter.

In this chapter, we consider the following components analyzed for their responses using different software programs:

1. Nonlinear static elastic analysis of a silicon tapering helical spring using ABAQUS [1].

2. Modal analysis of a micromachined accelerometer using ANSYS [2].

3. Coupled electro-thermal-elastic analysis using COMSOL Multiphysics [3].

4. Nonlinear static elastic analysis of a comb-drive suspension using NISA [4].

5. Coupled analysis of a piezoelectric structure using a custom-made FEA program that interested readers may also develop on their own, using the material presented in Chapter 5.

6. An advanced example of computing the shift in the natural frequency of a tuning fork resonator, for an applied acceleration in a micromachined resonant accelerometer using ABAQUS.

7. Estimation of the electrostatic pull-in voltage of an RF-MEMS switch using IntelliSuite [5].

8. Determining the change in capacitance of a micromachined pressure sensor using Coventorware [6].

## ▶ 10.2 FORCE-DEFLECTION OF A TAPERING HELICAL SPRING USING ABAQUS

Consider a silicon wafer in which a spiral-shaped slot is etched though the thickness, leaving a central disk and a rim at the periphery. Let the radius of the central disk be sufficiently large to render it relatively rigid; see Figure 10.1. Also, let the width of the rim be sufficiently large to render it rigid relative to the "helical spring" that is cut out after etching along the spiral. We call this helical spring *a tapering helical spring*, since if we anchor the rim and lift the central disk up in the direction perpendicular to the wafer, we have a tapering helical spring. As we would learn from the simulation of this intriguing structure, it is an excellent linear spring over a large range of motion—much better than the cylindrical helical spring in a ball-point pen. Its relevance to microsystems is to how a flexible spring with a large range of motion can be obtained from a brittle material like single crystal silicon. It is all in the shape—which is the point of this exercise.[1]

Let the spiral be given by $r = a\theta$, where $a = 0.1910 \times 10^{-3}$ and $\theta$ ranges from $2\pi$ to $25\pi$. This means that the radius of the central disk is about 1.2 mm and the radius at the end of the spiral is about 15 mm. The spiral-cut wafer's schematic is shown in Figure 10.1. Let the width of the spiral be very small, say, 100 μm. Thus, the width of the spiral is a small fraction of the overall length of the spiral. Let the thickness of the wafer be 300 μm. The

---

[1] In fact, as heard in a U.S. MEMS conference in 1994, the early days of the field, an employee of a microelectronics company had made this spring by etching a silicon wafer in order to convince his boss that silicon can be flexible too and thereby give himself an opportunity to work on MEMS!

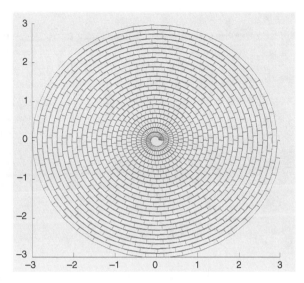

**Figure 10.1** Schematic of a spiral etched in a silicon wafer leaving an un-etched central disk at the center (the wide peripheral rim is not shown). The brick-like arrangement shows the finite elements used for analysis.

Young's modulus of silicon may be taken as 169 GPa and Poisson's ratio as 0.3. Let us ask this question: if we apply 1 N force to pull the central disk up in the direction perpendicular to the wafer while holding the rim (the outer peripheral edge of the spiral) fixed, how much does the central disk move up and what is the maximum stress experienced in the resulting tapering helical spring?

Let us use ABAQUS [1] finite element software to analyze this problem. First, we create the FE mesh for this structure using the C3D8 element in ABAQUS. The element code C3D8 in ABAQUS refers to the eight-noded displacement-based brick element with full integration. It uses linear interpolating functions. Each node has three degrees of freedom: the displacements in the $x$-, $y$-, and $z$-directions. The meshed model is shown in Figure 10.2. It has 6,606 nodes. The force at the center of the central disk can be seen in the

**Figure 10.2** The meshed model of the silicon wafer with spiral slot. The force boundary condition on the central spring and the displacement boundary condition at the outer edge of the spiral are shown in the spiral's undeformed configuration.

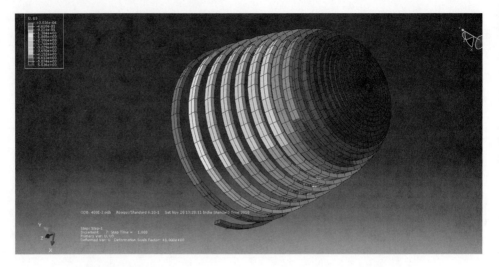

**Figure 10.3** The deformed model of a silicon slinky with von Mises stress.

figure, as can the displacement boundary conditions that fix the outer edge of the spiral. That is, we specify the three displacements of nodes on the outer edge of the spiral to be zero to simulate the anchored condition.

The nonlinear elastic deformation analysis in ABAQUS gives the $z$-direction displacement of the spiral as 13.4 mm and maximum von Mises stress as 5.12 kPa. Thus, this ''spring'' has a spring constant 74.63 N/m. This is a soft spring: the weight of a 1 kg mass will extend it by about 13 cm. Figure 10.3 shows the deformed structure in which we can see the nonlinearly tapering helical spring; the color in the figure shows the von Mises stress. It should be noticed that the stress is very small compared to the failure strength of silicon, which is about hundreds of MPa or 1 GPa. This means that this spring can be extended much more without fear of failure. One can see from Figure 10.3 that it resembles a toy known as a *slinky*. When the analysis was repeated for a range of force values, it was noticed that the force–deflection relationship remains linear over a wide range of deflection. Those interested may try making this *silicon slinky* and see for themselves. The analysis discussed in Example 4.6 may also be attempted to compare this numerical result with the analytical estimate.

## ▶ 10.3 NATURAL FREQUENCIES OF AN ACCELEROMETER IN ANSYS

Consider a micromechanical component in Figure 10.4. It is modeled in the commercial software ANSYS [2]. This is a $z$-axis accelerometer, that is, it can sense acceleration in the direction perpendicular to the central mass in the figure. In order to estimate the bandwidth of this sensor and determine its dynamic response, it is necessary to perform a modal analysis. The pertinent details and results are presented here. The session file for creating this model and performing the analysis is also given (see Box 10.1, page 434).

Each of the four suspension beams are 1.8 mm long and 90 μm wide. The square mass at the center has side 1.8 mm. The thickness of the beam and the central mass is 20 μm. The material properties are: $Y = 200$ GPa, $\rho_m = 2330$ kg/m$^3$, and $\nu = 0.3$. The model was discretized using a SHELL 63 finite element that has both bending and membrane behavior. Each element has three nodes and each node has six degrees of freedom: three translations and three rotations. Part of the meshed model is shown in Figure 10.5. The

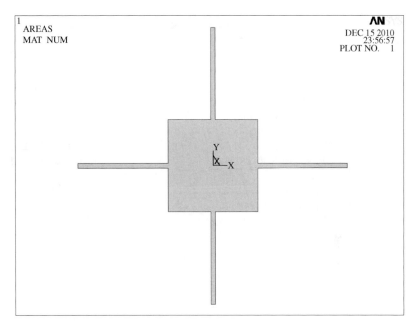

**Figure 10.4** A simple $z$-axis accelerometer modeled in ANSYS.

block Lanczos method of eigenvalue extraction is used. The four beams are anchored on the side that is not connected to the central mass.

In the modal analysis, the first three natural frequencies were found to be 3,616 Hz, 9,332 Hz, and 9,332 Hz. The corresponding mode shapes are shown in Figure 10.6. It should be noted that the first mode has the central mass moving up and down along with the beams. The second and third modes are degenerate, that is, they have the same frequency.

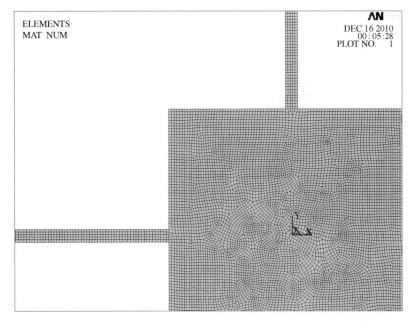

**Figure 10.5** Part of the meshed model of the $z$-axis accelerometer.

---

**Box 10.1:** Session file for modeling in ANSYS

---

```
FINISH
/CLEAR

/PREP7
K,1,0.9e-3,0.9e-3,,
K,2,0.9e-3,-0.9e-3,,
K,3,-0.9e-3,-0.9e-3,,
K,4,-0.9e-3,0.9e-3,,

K,5,45e-6,0.9e-3,,
K,6,-45e-6,0.9e-3,,
K,7,-0.9e-3,45e-6,,
K,8,-0.9e-3,-45e-6,,
K,9,-45e-6,-0.9e-3,,
K,10,45e-6,-0.9e-3,,
K,11,0.9e-3,-45e-6,,
K,12,0.9e-3,45e-6,,

K,13,45e-6,2.7e-3,,
K,14,-45e-6,2.7e-3,,
K,15,-2.7e-3,45e-6,,
K,16,-2.7e-3,-45e-6,,
K,17,-45e-6,-2.7e-3,,
K,18,45e-6,-2.7e-3,,
K,19,2.7e-3,-45e-6,,
K,20,2.7e-3,45e-6,,

FLST,2,20,3
FITEM,2,1
FITEM,2,5
FITEM,2,13
FITEM,2,14
FITEM,2,6
FITEM,2,4
FITEM,2,7
FITEM,2,15
FITEM,2,16
FITEM,2,8
FITEM,2,3
FITEM,2,9
FITEM,2,17
FITEM,2,18
FITEM,2,10
FITEM,2,2
FITEM,2,11
FITEM,2,19
FITEM,2,20
FITEM,2,12
A,P51X
```

```
MPTEMP,,,,,,,,
MPTEMP,1,0
MPDATA,DENS,1,,2330
MPTEMP,,,,,,,,
MPTEMP,1,0
MPDATA,EX,1,,200e9
MPDATA,PRXY,1,,0.3

ET,1,SHELL43
R,1,20e-6,,,,,
RMORE,,,

ESIZE,20e-6,0,
MSHKEY,0
CM,_Y,AREA
ASEL,,,,1
CM,_Y1,AREA
CHKMSH,'AREA'
CMSEL,S,_Y
AMESH,_Y1
CMDELE,_Y
CMDELE,_Y1
CMDELE,_Y2

FINISH
/SOL
ANTYPE,2

MODOPT,LANB,3
EQSLV,SPAR
MXPAND,3,,,0
LUMPM,0
PSTRES,0

MODOPT,LANB,3,0,0,,OFF
FLST,2,4,4,ORDE,4
FITEM,2,3
FITEM,2,8
FITEM,2,13
FITEM,2,18

/GO
DL,P51X,,ALL,

SOLVE
FINISH
/POST1
SET,LIST
```

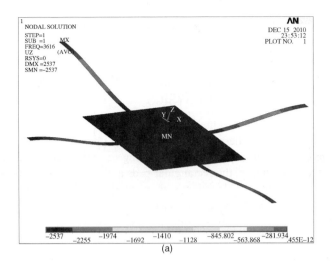

(a)

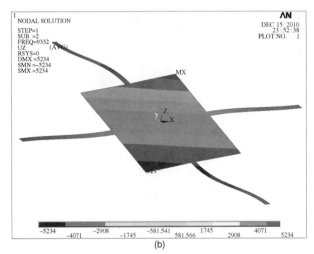

(b)

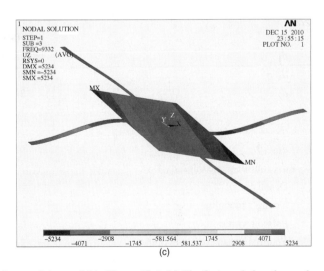

(c)

**Figure 10.6** Three mode shapes of the model in Figure 10.4. (a) The first mode has the up-down motion of the central mass; (b) and (c) are the tilting modes about the two orthogonal in-plane axes.

They are both tilting modes about the in-plane orthogonal axes. The degenerate eigen value is a consequence of the symmetry of the structure.

## ▶ 10.4 DEFLECTION OF AN ELECTRO-THERMAL-COMPLIANT (ETC) MICROACTUATOR IN COMSOL MULTIPHYSICS

In this section, we look at how to do simulations involving multiple domains. For this, we consider the electro-thermal-compliant (ETC) microactuator [7,8] shown in Figure 10.7. The geometric and material data pertaining to this problem are given in Table 10.1. As indicated in the figure, the rectangular pads at the either end of this folded beam structure are mechanically grounded, that is, they are anchored to the substrate. The temperature at the four edges of the pads is fixed at room temperature. The pad on the right side is electrically grounded while the one at the left is set at 3 V. The objective is to determine the

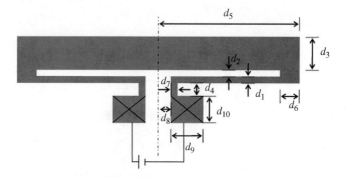

**Figure 10.7** An ETC microactuator.

**Table 10.1 Geometry and material data pertaining to the ETC microactuator in Figure 10.7**

| | |
|---|---|
| $d_1$ | 40 μm |
| $d_2$ | 20 μm |
| $d_3$ | 340 μm |
| $d_4$ | 100 μm |
| $d_5$ | 1800 μm |
| $d_6$ | 40 μm |
| $d_7$ | 300 μm |
| $d_8$ | 100 μm |
| $d_9$ | 1000 μm |
| $d_{10}$ | 900 μm |
| $d_{11}$ (out-of-plane thickness) | 20 μm |
| Density | 2329 kg/m$^3$ |
| Young's modulus | $179 \times 10^9$ Pa |
| Poisson's ratio | 0.28 |
| Thermal expansion coefficient | $2.6 \times 10^{-6}$ 1/K |
| Thermal conductivity | 130 W/m-K |
| Electrical resistivity | $1.72 \times 10^{-8}$ Ω-m |

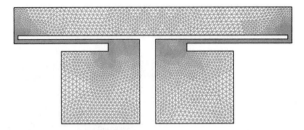

**Figure 10.8** Meshed model of the ETC microactuator in COMSOL MultiPhysics.

deformed shape of this actuator due to the application of a potential difference between the two pads. The equations for this problem were presented in Section 6.6. They involve sequentially coupled electrical, thermal, and elastic equations. We show the results of analysis of this problem using COMSOL MultiPhysics [3] software.

The meshed model of the ETC microactuator is shown in Figure 10.8. It is meshed with three-node planar elements in which quadratic interpolation functions are used. In electrical analysis, the electric potential is the degree of freedom at each node; temperature is the degree of freedom in thermal analysis, and displacements in the two planar directions are the degrees of freedom in the elastic deformation analysis. COMSOL MultiPhysics allows coupled analysis. The material properties and boundary conditions should be entered in each physics mode one by one. The coupling variables need to be set up. That is, electrical current-induced Joule heating term should be specified as the heat source term in the thermal analysis, and the temperature from the thermal analysis should be specified as the initial temperature in the elastic deformation analysis. When the coupled analysis is completed, we have the results pertaining to all three analyses.

Figure 10.9 shows the electric current density distribution in the structure. The maximum current density was observed to be $3 \times 10^9 \, \text{A/m}^2$. Figure 10.10 shows the deformed configuration with temperature distribution; it can be seen that the wide beam

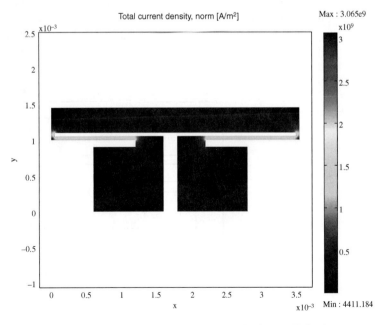

**Figure 10.9** Distribution of the electric current density in the ETC microactuator.

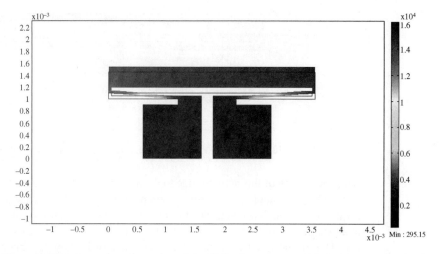

**Figure 10.10** Deformed configuration of the ETC microactuator with temperature distribution.

at the top is heated more than the narrow beams beneath it. A displacement of 2.6 μm was obtained.

Convection and radiation are not considered in this simulation. Interested readers may also attempt this with the data provided in [8].

## ▶ 10.5 LUMPED STIFFNESS CONSTANT OF A COMB-DRIVE SUSPENSION IN NISA

Much of Chapters 4 and 6 was devoted to lumped modeling of continuous systems. As we saw in those chapters, analytical solutions are possible only for simple geometries and boundary conditions. We now solve an example through which we can extract the lumped stiffness constant from the numerical solution of a comb-drive actuator using NISA [4] software.

The left symmetric half of the suspension of the comb-drive actuator is shown in Figure 10.11 (all dimensions indicated are micrometers). The side view at the bottom of Figure 10.11 shows that the bottom surface of the anchors of the folded beams are fixed. A force of 0.3 μN is applied on the central mass as shown. It was meshed using eight-noded

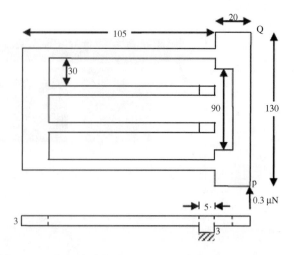

**Figure 10.11** Left symmetric half of the folded-beam suspension of the comb-drive actuator.

**Figure 10.12** Deformed model of left symmetric half of folded-beam suspension.

brick elements in NISA. Young's modulus of 169 GPa and Poisson ratio of 0.3 are used as material properties in this analysis.

Figures 10.12 and 10.13 show the results of the analysis. Figure 10.12 shows the deformed configuration of the suspension. The consequence of the fixed and vertically guided ends of the four beams can be seen clearly. The reader may recall Example 4.3 and the corresponding figure [Figure 10.17(a)] where this problem was solved analytically. The displacement of the central mass is 0.1185 μm and the maximum von Mises stress is 35.56 MPa.

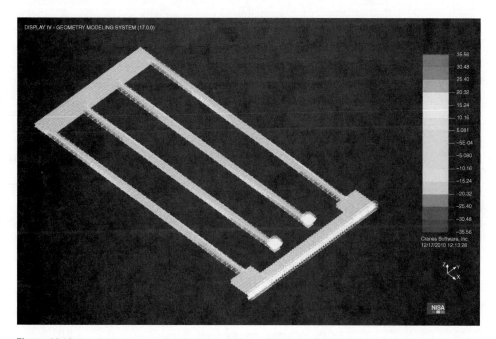

**Figure 10.13** Von Mises stress distribution in left symmetric half of folded-beam suspension.

## ▶ 10.6 PIEZOELECTRIC BIMORPH BEAM IN A CUSTOMIZED FEA PROGRAM

A piezoelectric isoparametric four-noded quadrilateral element was formulated in Chapter 5 (see Section 5.7). Here we use that element to study the behavior of the bi-morph beam under static voltage loading as well as combined voltage and mechanical loading. The latter case is important to see the effect of actuation due to the piezoelectric effect. This problem was studied in Section 5.8 using beam finite element and the solution was compared with an analytical solution. In this section, we study the same problem using the four-noded isoparametric piezoelectric finite element that was formulated in Section 5.7. The results obtained here are compared with both the analytical solution and those in the published literature.

The schematic of the beam and its dimensions and boundary conditions are shown in Figure 5.20. Chen et al. [9] presented a comparative study of bending of such a bimorph beam due to external applied voltage as part of verifying the accuracy of the piezoelectric FE solution, and this bimorph beam configuration was adopted from Woo-Hwang and Park [10]. The bimorph beam consists of two identical polyvinylidene fluoride (PVDF) beams laminated together with opposite polarities. The dimensions of the beam are taken as 100 mm × 5.0 mm × 0.5 mm. The material properties of the PVDF bimorph beam, taken as those in Chen et al. [9], are given in Table 10.2. The theoretical solution for the transverse displacement for the above problem was derived in Section 5.8.1, as

$$w(x) = 0.375 \frac{e_{31} V}{E} \left(\frac{x}{t}\right)^2 \tag{10.1}$$

where $V$ is the applied voltage and $t$ is the beam thickness. This beam is modeled using 200 formulated elements along the $x - z$ plane. When an external active voltage is applied across the thickness, the induced strain generates control forces that bend the beam. A unit voltage is applied across the thickness and the deflections at the nodes are computed. The deflection of the beam along the central longitudinal axis obtained from the 2D four-noded isoparametric FE formulation is compared with the theoretical value given by Eq. (10.1) and the work of Chen et al. [9], and Tzou and Tseng [11]. Figure 10.14 compares the deflection along the length of the beam for unit voltage applied across the thickness.

Next, the deflection of the beam is calculated for different applied voltages in the range 0–200 V. The calculated deflections from the present study are compared with those of Chen et al. [9], Tzou and Tseng [10], and Eq. (10.1). Chen et al. [9] used a first-order shear deformable plate finite element, while Tzou and Tseng [10] used hexahedral solid elements to

**Table 10.2 Material properties pertaining to the piezoelectric bimorph beam**

| Property of PVDF | Value |
| --- | --- |
| Young's modulus $E_{11}$ | $0.2 \times 10^{10}$ N/m$^2$ |
| Shear modulus $G_{12}$ | $0.775 \times 10^{10}$ N/m$^2$ |
| Poisson ratio $\nu_{12}$ | 0.29 |
| Poisson ratio $\nu_{21}$ | 0.28 |
| Piezoelectric constant $e_{31}$ | 0.046 C/m$^2$ |
| Piezoelectric constant $e_{32}$ | 0.046 C/m$^2$ |
| Piezoelectric constant $e_{33}$ | 0.0 |

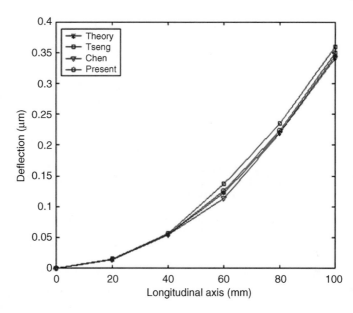

**Figure 10.14** Deflected profiles of the centerline (neutral axis) of the PVDF bimorph beam under an applied unit voltage using different methods.

model the same problem. These are plotted in Figure 10.15. Figure 10.16 shows the variation of the deflection along the central longitudinal axis for various values of applied voltages.

The problem solved here is associated with actuation with open-loop control. That is, for an electrical load caused by the voltage, strains are generated, which in turn cause deformations. In Figure 10.16, we can see that greater voltages produce larger deformations. The results presented here show close agreement between the theoretical and the formulated element and also with the established literature. The present study has thus

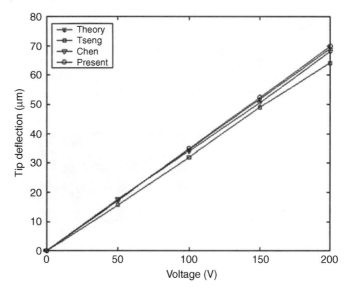

**Figure 10.15** Deflections of the free tip of the PVDF bimorph beam from FE program, beam-based analytical solution, and two results from the literature.

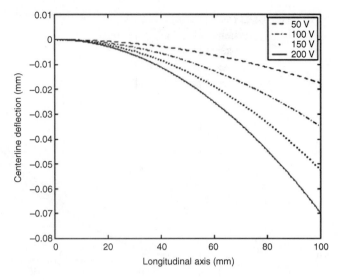

**Figure 10.16** Deflections of the PVDF bimorph beam for different voltages.

demonstrated the accuracy of the piezoelectric finite element, formulated in Chapter 5, in problems involving electromechanical coupling.

## ▶ 10.7 RESONANT MICRO-ACCELEROMETER IN ABAQUS

We now consider a slightly more advanced analysis that is necessary to compute the sensitivity of a resonant accelerometer. As shown in Figure 10.17, two pairs of beams forming a resonating tuning fork are attached to a central proof mass. The two tuning forks are set into vibration using electrostatic force (comb-fingers are not shown in the figure or in the close-up inset) and the resonance frequency is determined. When there is an acceleration, the inertial force experienced by the mass exerts axial force on the beams of the tuning forks. This changes the natural frequency of the tuning fork. This change in resonance frequency is correlated to the acceleration. Thus, it becomes an accelerometer.

The design shown in Figure 10.17 [12] has a special feature: a mechanical lever to amplify the axial force exerted on the beams. Our aim here is to compute the natural frequency of the resonating beams with and without the acceleration-induced force.

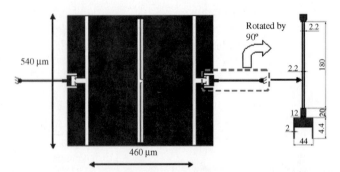

**Figure 10.17** A resonant accelerometer with (inset) detail of the resonating beam pair (drawn after [12]).

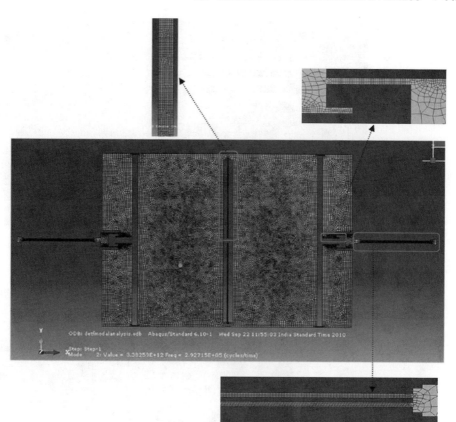

**Figure 10.18** A meshed model of the resonant accelerometer in ABAQUS.

The dimensions of the accelerometer are indicated in Figure 10.17. Material properties used are: Young's modulus = 150 GPa, mass density = 2330 kg/m$^3$, and Poisson's ratio = 0.3. The meshed model is shown in Figure 10.18. The CPS8R element in ABAQUS is chosen to mesh the model with mesh size 2.5–3 μm. There are 51,503 elements in this model. The block Lanczos method was chosen for eigenanalysis. The frequency around which to compute the natural frequencies was set at $10 \times 10^6$ Hz. The modal analysis was run first with no acceleration; then, static analysis was performed with NLGEOM (geometrically nonlinear analysis option) turned on and then modal analysis was conducted in the stressed state. Figure 10.19 shows the boundary conditions including the body force (i.e. acceleration-induced force).

The mode shape of interest, in which only the beams in the tuning fork vibrate, is not the fundamental (first) mode shape of this device. Hence, we need to compute several mode shapes for inspection. The mode shape of interest, shown in Figure 10.20, is found to have a natural frequency of 562,079 Hz. With 1g acceleration (9.81 m/s$^2$) applied, this frequency changes to 562,096. This means that there is a shift of 17 Hz per g. This is the sensitivity of this accelerometer. This change may seem very small, but the electronic circuitry monitoring the natural frequency of the tuning fork can resolve one part per billion. That is, the measurable change in frequency divided by the base frequency can be as small as $10^{-9}$. Here, we have $(17/562079) = 326 \times 10^{-6}$. Hence, this accelerometer, in principle, has the ability to detect $3 \times 10^{-6}$ g.

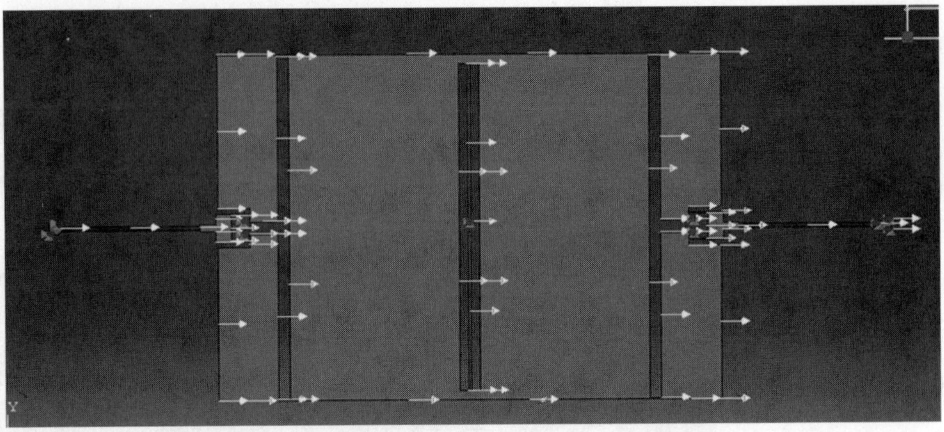

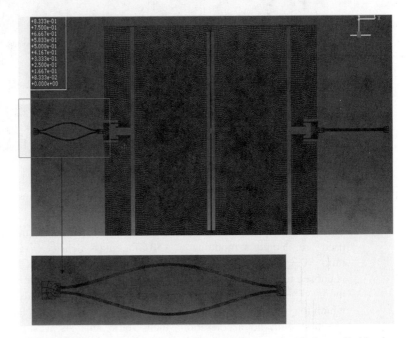

**Figure 10.19** Displacement boundary conditions and body force (acceleration-induced force) applied on the resonant accelerometer in ABAQUS.

**Figure 10.20** Mode shape of the resonant accelerometer with the applied load.

## ▶ 10.8 PULL-IN VOLTAGE OF AN RF-MEMS SWITCH IN INTELLISUITE

The RF-MEMS switch considered here is different from that described in Section 2.7.2; it is for high-power applications and is made of nonconventional material (Pt–Rh alloy) and microfabricated using nonconventional processes [printed circuit board (PCB)-technology]. It is also different in structure. It consists of a cantilever plate of varying cross-sectional area along its length. Its design, shown in Figure 10.21, is adapted from [13], where additional details of this switch can be found. Here we show how the pull-in voltage of this switch can be computed using IntelliSuite software [5]. The dimensions used in the IntelliSuite simulation are given in Figures 10.22 and 10.23.

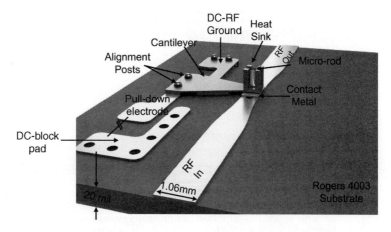

**Figure 10.21** An RF-MEMS switch manufactured on a PCB (adapted from [13]).

As can be seen in Figure 10.22, the electrode to actuate the switch does not cover the entire area beneath the plate. Therefore, the dimension $d$ in Figure 10.22 plays a crucial role in determining the pull-in voltage. Lumped modeling cannot easily take such features into account. During manufacturing [13], this dimension is kept between 400 and 500 μm. The reason for calling this structure a plate and not a slender beam is that the ratio of the transverse to longitudinal dimensions is greater than the range in which Euler–Bernoulli's beam theory holds true. Thus, one must use models from classical plate theory to obtain the equivalent

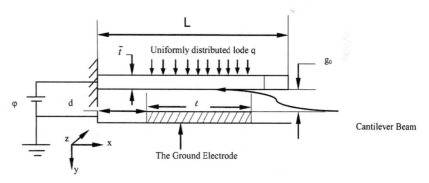

**Figure 10.22** Side view of the RF switch.

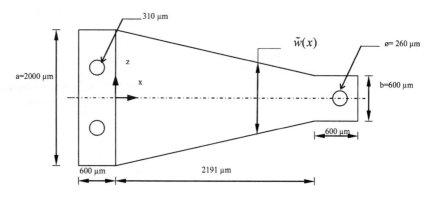

**Figure 10.23** Top view of the RF switch.

**Table 10.3 Details of the RF switch in Figure 10.23**

| Data | Value |
|---|---|
| $g_0$ | $10\,\mu m$ |
| $l$ | $350\,\mu m$ |
| $\tilde{t}$ | $50\,\mu m$ |
| $d$ | $400\,\mu m$ |
| Young's modulus (SS304) | $193\,GPa$ |
| Density | $7900\,kg/m^3$ |

lumped model for this configuration. Moreover, instead of calculating the lumped stiffness for a point-load assumption of uniformly distributed load, $q$ will give more accurate results. FEA-based results are needed take such details into account. The switch was simulated in IntelliSuite with the data in Table 10.3 to compute the pull-in voltage.

The 3D geometry of the RF switch was created using IntelliSuite's 3D Builder module. Note that a jagged edge was used in the model to simulate the tapering of the plate. The initial mesh sizes along the $x$-, $y$-, and $z$-directions (see Figure 10.24) were $48\,\mu m$, $48\,\mu m$, and $50\,\mu m$, respectively. This was later refined to mesh sizes $48\,\mu m$, $81.39\,\mu m$, and $100\,\mu m$, respectively (see Figure 10.25). Note that such refinement is mandatory in FEA to ensure that the mesh refinement is adequate for accurate results. The total number of nodes in this model was 11,989. Each node consists of three degrees of freedom as it is a 3D analysis. The geometry was then exported to the *Analysis* module for *ThermoElectro-Mechanical* FEA (note that meshing can also be done in the *Analysis* module if required; additional meshing for electrostatic analysis can also be done in this module for coupled simulation). The meshed structure is shown in Figure 10.24.

Determining pull-in voltage from IntelliSuite is not straightforward since it does not explicitly give the pull-in voltage. The user must obtain the pull-in condition by repeating the simulation several times, each time narrowing the bandwidth of the search. Whenever there is a sharp change in displacement for a small change in voltage (i.e. neutrally stable condition at pull-in, as discussed in Section 6.2.2), it can be concluded that pull-in has occurred. The pull-in voltage of the RF switch modeled here was found to be between 350 and 375 V.

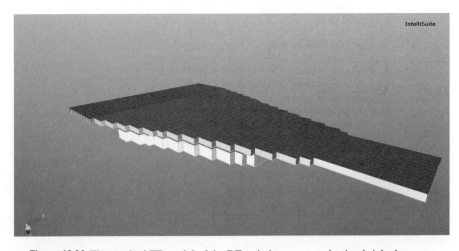

**Figure 10.24** The meshed FE model of the RF switch constructed using brick elements.

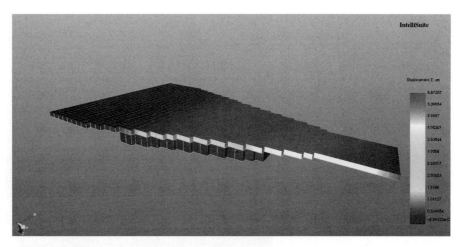

**Figure 10.25** The transverse displacement field at 375 V. The tip displacement is 5.57 μm.

## ► 10.9  A CAPACITIVE PRESSURE SENSOR IN COVENTORWARE

Finally, we consider another coupled simulation in Coventorware [6] using the example of a capacitive pressure-sensor. The pressure sensor we consider here has a circular diaphragm with a circular electrode underneath it separated by a gap. Such a pressure sensor can be micromachined using polysilicon surface-micromachining process.

Coventorware [6] allows the user to define a process flow (see Figure 10.27) and the lithography mask layouts (Figure 10.26). As shown in Figure 10.27, a substrate (taken as 10 μm thick for the purpose of modeling; its real thickness of hundreds of micrometers will make it difficult to see the other layers, which are much thinner in the software visualizations) is considered. A Poly0 layer of 1 μm thickness is deposited over it and then the Poly0 layer is etched by ''Straight Cut'' using *Poly0Mask*. An oxide layer of 2 μm thickness is then conformally deposited and etched by ''Straight Cut'' using *Oxide*

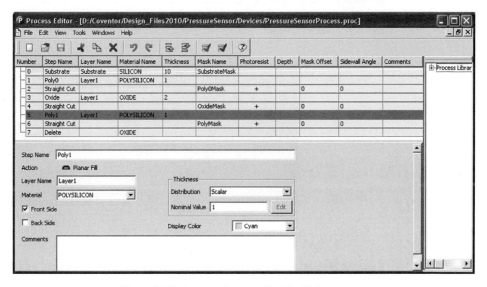

**Figure 10.26** Process flow specified in Coventorware.

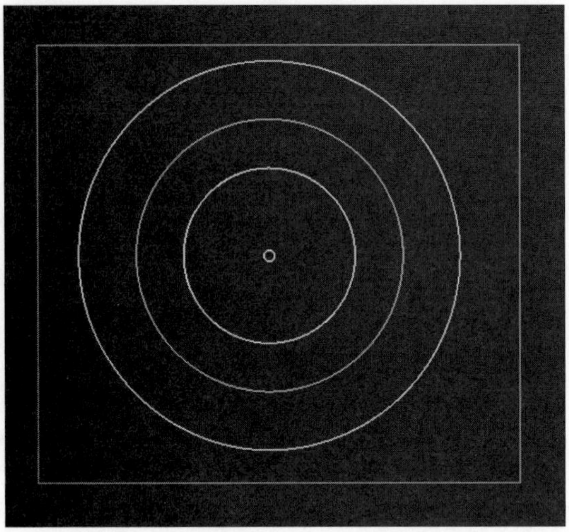

**Figure 10.27** Mask layouts defined in the Layout Editor of Coventorware.

*Mask.* This is followed by the conformal deposition of a Poly1 layer of 1 μm thickness and then etched using *PolyMask*. The oxide layer was deleted at the end to emulate the sacrificial step.

Note that the preceding process description is similar to a real process. Once we create the mask layouts of the masks named above (Poly0Mask, OxideMask, and PolyMask, all names chosen by the user), the 3D model that closely resembles the real structure is automatically created in the Designer module of Coventorware.

The mask layouts are shown in Figure 10.27. Three circles are shown inside a square. The square of 260 μm side corresponds to the *Substrate Mask*; this mask is artificial and is present merely to have the substrate in the model. The innermost circle of radius 45 μm is the size of the ground electrode and corresponds to the Poly0Mask. The second circle of 70 μm radius corresponds to the OxideMask. The outermost circle of 100 μm radius pertains to the PolyMask and defines the diaphragm of the pressure sensor.

The section view of the automatically built 3D model is shown in Figure 10.28 and the meshed model in Figure 10.29. A mesh size of 5 μm was used in creating this mesh.

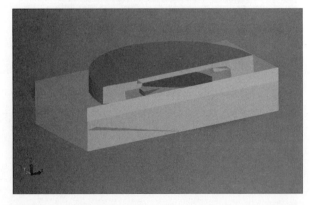

**Figure 10.28** Section view of the 3D model of the pressure sensor as built in Coventorware. The gap between the diaphragm and the bottom electrode is created after sacrificing (i.e. deleting) the oxide layer.

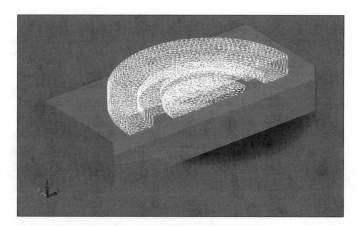

**Figure 10.29** Meshed FEA model of the pressure sensor. Note that the substrate need not be meshed.

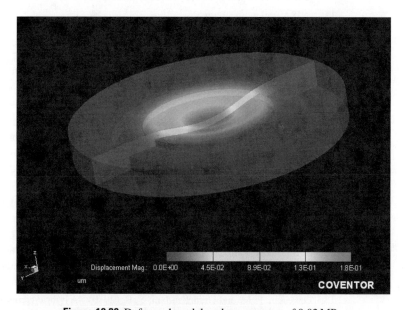

**Figure 10.30** Deformed model under a pressure of 0.02 MPa.

The surface patches for applying the pressure, anchoring the diaphragm, and specifying the voltage boundary conditions were defined for the model. After this, the Analyzer module of Coventorware was used to calculate the base capacitance as 0.0336 pF and the maximum deformation of the diaphragm under a pressure of 0.02 MPa as 0.18 μm. The deformed geometry with a slice view inside is shown in Figure 10.30.

## ▶ 10.10 SUMMARY

This chapter presented some simulation results using a variety of FEA software in which micro and smart systems components can be analyzed. Six commercial software programs (ABAQUS, ANSYS, COMSOL MultiPhysics, NISA, IntelliSuite, and Coventorware) and a custom-built non-commercial FEA program were used to simulate eight different components. The purpose of this chapter is to familiarize readers with different techniques and software available to simulate components of realistic geometry. The discussion in Chapters 4 and 6 was analytical modeling, which is of great use in the initial stages of design and analysis. Then, however, 3D geometries that are as close to reality as possible are needed for accurate prediction of the real performance. Hence, FEA software is an indispensable tool for micro and smart systems designers. The theory of FEA presented in Chapter 5 should be read in conjunction with the practical aspects discussed in this chapter. The reader is urged to try at least one of the six commercial programs, or any other software, to simulate a micromachined component. Writing one's own FEA code, of course, provides even more in-depth understanding of the subject.

## ▶ REFERENCES

1. ABAQUS, www.simulia.com

2. ANSYS, www.ansys.com

3. COMSOL MultiPhysics, www.comsol.com

4. NISA, www.nisa.com

5. IntelliSuite, www.intelleisuite.com

6. Coventorware, www.coventorware.com

7. Moulton, T. and Ananthasuresh, G. K. (2001) "Electro-thermal-compliant (ETC) microactuators" *Sensors and Actuators A: Physical*, **90**, pp. 38–48.

8. Mankame, N. and Ananthasuresh, G. K. (2001) "Comprehensive Analysis of an Electro-thermal Actuator," *Journal of Micromechanics and Microengineering*, **11**, pp. 452–462.

9. Chen, S. H. Wang, Z. D. and Liu, X. H. (1997) "Active vibration control and suppression for intelligent structures," *Journal of Sound and Vibration*, **200**(2), 167–177.

10. Hwang, W. S. and Park, H. C. (1993) "Finite element modeling of piezoelectric sensors and actuators," *AIAA Journal*, **31**(5), 930–37.

11. Tzou, H. S. and Tseng, C. I. (1990) "Distributed piezoelectric sensor/actuator design for dynamic measurement/control of distributed parameter systems: a piezoelectric finite element approach," *Journal of Sound and Vibration*, **138**(1), 17–34.

12. Seshia, A., Palaniapan, M., Roessig, T. A., Howe, R. T., Gooch, R. W., Schimert, T. R. and Montague, S. (2002) "A Vacuum-packaged Surface-micromachined Resonant Accelerometer," *Journal of Microelectromechanical Systems*, **11**, pp. 784–793.

13. Ozkeskin, F. M. and Gianchandani, Y. B. (2010) "A hybrid technology for Pt–Rh and SS316L high power micro-relays," *Solid State Sensors, Actuator and Microsystems Workshop*, Hilton Hill Island, SC, June 6-10, 2010.

# Mass-Spring-Damper or LCR Oscillator

By referring to Figure A1, we can write an equation to describe the dynamic behavior of the one degree-of-freedom mass-spring-damper system by summing all the forces acting on it:

$$m\ddot{u} + b\dot{u} + ku = F(t) \tag{A.1}$$

where $m\ddot{u}$ is the inertia force, $b\dot{u}$ the damping force, $ku$ the spring force, and $F(t)$ the applied force. The displacement and its first and second time-derivatives are indicated with $u$, $\dot{u}$, and $\ddot{u}$ respectively.

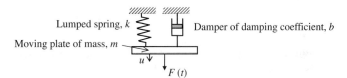

**Figure A.1** A mass-spring-damper model.

The dynamics of an electrical LCR (inductor-capacitor-resistor) circuit (see Figure A2) is also governed by the same equation but in terms of different variables.

$$L\frac{dI}{dt} + RI + \frac{\int I\,dt}{C} = V(t) \Rightarrow L\ddot{Q} + R\dot{Q} + \frac{Q}{C} = V(t) \tag{A.2}$$

where $Q$ is charge, $I$ is current, $L$ is inductance, $R$ is resistance, $C$ is capacitance, and $V(t)$ is the applied voltage.

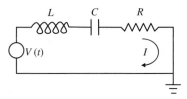

**Figure A.2** An electrical LCR (inductor-capacitor-resistor) circuit.

Equations (A.1) and (A.2) are ordinary differential equations (ODEs) of second order because they involve up to the second derivative of the variable to be computed, $u$ in the case of mass-spring-damper and $Q$ in the case of the LCR circuit. In the subsequent discussion, we consider only Eq. (A.1) but note that all of what is presented here applies to the LCR circuit as well.

## ▶ A.1. FREE VIBRATIONS

Let us first consider Eq. (A.1) with force (i.e., the right hand side term) and damping (i.e., $b$) equal to zero.

$$m\ddot{u} + ku = 0 \tag{A.3a}$$

$$\Rightarrow u(t) = A \sin(\omega_n t) + B \cos(\omega_n t) \tag{A.3b}$$

where

$$\omega_n = \sqrt{\frac{k}{m}} \qquad \text{(unit : rad/s; divide by } 2\pi \text{ to get in Hz or cycles/s)} \tag{A.3c}$$

It is easy to verify this solution by differentiating $u$ and substituting into the equation. The constants $A$ and $B$ can be found from the initial conditions: $u = 0$ and $\dot{u} = 0$ at $t = 0$.

$$B = u(0)$$
$$A = \frac{\dot{u}(0)}{\omega_n} \tag{A.4}$$

We can readily see that Eq. (A.3b) represents sinusoidal response with frequency, $\omega_n$. This is called the *natural frequency*. It is an important dynamic characteristic of an elastic system with inertia. This gives us the frequency at which the system would naturally oscillate if it were to be disturbed from its state of rest and then let to oscillate without any external influence.

## ▶ A.2 DAMPED FREE VIBRATIONS

As the first external influence, let us introduce the damping term (i.e., $b\dot{u}$) and write the new solution.

$$m\ddot{u} + b\dot{u} + ku = 0 \tag{A.5a}$$

$$\Rightarrow u(t) = e^{-(b/2m)t} \left\{ Ae^{\left(\sqrt{(b/2m)^2 - (k/m)}\right)t} + Be^{-\left(\sqrt{(b/2m)^2 - (k/m)}\right)t} \right\} \tag{A.5b}$$

where the constants $A$ and $B$ are to be determined from the initial conditions. The quantity under the square root in Eq. (A.5b) may be positive, negative, or zero depending on the values of $m$, $b$, and $k$. Let us keep $m$ and $k$ constant and vary $b$. The value of $b$ that makes the quantity under the square root sign in Eq. (A.5b) zero is called the *critical damping* and is denoted by $b_c$.

$$(b/2m)^2 - (k/m) = 0 \Rightarrow b_c = 2\sqrt{km} \tag{A.6}$$

We use this critical damping to define a non-dimensional quantity called the *damping ratio*, $\zeta$.

$$\zeta = \frac{b}{b_c} \tag{A.7}$$

With the help of Eqs. (A.3c, A.6) and (A.7), we can write Eq. (A.5b) as

$$u(t) = e^{-\zeta \omega_n t} \left\{ A e^{i\left(\sqrt{1-\zeta^2}\right)\omega_n t} + B e^{-i\left(\sqrt{1-\zeta^2}\right)\omega_n t} \right\} \tag{A.8}$$

When $\zeta < 1$, Eq. (A.8) gives *under-damped* oscillatory motion that can be re-written as

$$u(t) = e^{-\zeta \omega_n t} \left\{ C \sin\left(\omega_n t \sqrt{1-\zeta^2}\right) + D \cos\left(\omega_n t \sqrt{1-\zeta^2}\right) \right\} \tag{A.9}$$

so that there are two constants that are to be determined by initial conditions after Eq. (A.8) is expanded[1] and the imaginary terms are cancelled. The reader may verify that this solution satisfies the ODE, $m\ddot{u} + b\dot{u} + ku = 0$ for $\zeta < 1$. By examining Eqs. (A.3b) and (A.9), we can see that we can define *damped natural frequency*, $\omega_d$, as

$$\omega_d = \omega_n \sqrt{1 - \zeta^2} \tag{A.10}$$

Thus, the under-damped response gives an oscillatory motion of a slightly different frequency from the natural frequency. Furthermore, because of $e^{-\zeta \omega_n t}$ term in Eq. (A.9), the amplitude of this oscillation decreases exponentially with time. The solid-line curve in Figure A.3 shows this behavior schematically.

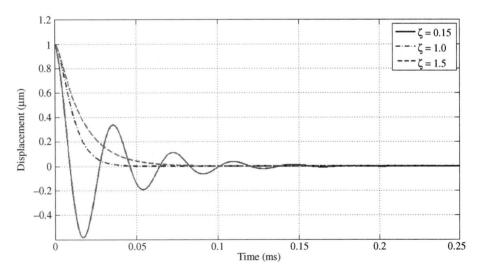

**Figure A.3** Under, critical, and over damped cases of a mass-spring-damper system.

---

[1] Use the identity: $e^{i\theta} = \cos\theta + i\sin\theta$.

When $\zeta = 1$ (called the *critically damped* case), Eq. (A.8) becomes $u(t) = e^{-\zeta \omega_n t}$ $(A + B)$. Since $(A + B)$ can be written together as one constant, we now fall short of one equation to satisfy two initial conditions: $u(0) = u_0$ and $\dot{u}(t) = \dot{u}_0$. So, the solution for $\zeta = 1$ should be re-written as

$$u(t) = e^{-\omega_n t}\{A + Bt\} \tag{A.11}$$

because it not only satisfies the ODE but also gives two constants to satisfy the initial conditions. The dash-dot-line curve in Figure A.3 shows this behavior schematically.

When $\zeta > 1$, Eq. (A.8) becomes directly applicable because $i = \sqrt{-1}$ in $e^{i\left(\sqrt{1-\zeta^2}\right)\omega_n t}$ disappears because $\sqrt{1 - \zeta^2}$ can be written as $i\sqrt{\zeta^2 - 1}$ yielding

$$u(t) = Ae^{\left(-\zeta - \sqrt{\zeta^2 - 1}\right)\omega_n t} + Be^{\left(-\zeta + \sqrt{\zeta^2 - 1}\right)\omega_n t} \tag{A.12}$$

The behavior of this *over-damped* case is illustrated by the dashed-line curve in Figure A.3.

It is important to get an intuitive understanding of the effect of damping on the dynamics of the system by studying Figure A.3 carefully. This is how a system would move from one state to another state when it is disturbed. The term 'state' here represents both the position and the velocity. High damping ($\zeta > 1$) means that it returns to original state quickly and low damping means that ($\zeta < 1$) it will oscillate around the original state for some time before settling. The time constant, $\tau$, which is the reciprocal of the damped natural frequency, indicates how fast the system responds to sudden changes in the state of the system or the external forces.

$$\tau = \frac{2\pi}{\omega_d} = \frac{2\pi}{\omega_n \sqrt{1 - \zeta^2}} \tag{A.13}$$

As can be seen in the curve corresponding to the under-damped case in Figures A.3, a few oscillations elapse before the displacement reaches the steady state. This means that the response time for the system is a multiple of $\tau$.

---

### Example A.1

If $k = 3.0$ N/m and $m = 0.1\text{E-9}$ kg, obtain the displacement, $u(t)$, for $\zeta = 0.15$, 1.0, and 1.5 with the initial conditions: $u(0) = 1\text{E-6}$ m, and $\dot{u}(0) = 0.0$ m/s.

**Solution:**

$$\omega_n = \sqrt{\frac{k}{m}} = \sqrt{\frac{3.0}{0.1E - 9}} = 0.1732E6 \text{ rad/s} = 27.5664 \text{ kHz}$$

$$b_c = 2\sqrt{km} = 2\sqrt{3 \times 0.1E - 9} = 34.6410\text{E-6 N/(m/s)}$$

*Case (i):* $\zeta = 0.15$ (under-damped case)

$$b = \zeta b_c = 0.15 \times 34.6410E - 6 = 5.1962\text{E-6 N/(m/s)}$$

From Eq. (A.9) and the given initial conditions, we get

$$u(t) = u(0) \; e^{-\zeta \omega_n t}\cos\left(\omega_n t \sqrt{1 - \zeta^2}\right) = 1E - 6\left(e^{-25.9087t}\right)\cos(0.1712E6 \; t)$$

This is plotted as a solid-line curve in Figure A.3.

*Case (ii)*: $\zeta = 1.0$ (critically-damped case)

$$b = \zeta b_c = 1.0 \times 34.6410E - 6 = 34.6410\text{E-6 N/(m/s)}$$

From Eq. (A.11) and the given initial conditions, we get

$$u(t) = e^{-\omega_n t}\{A + Bt\} = 1E - 6\left(e^{-0.1732E6\ t}\right)(1 + 0.1732E6\ t)$$

This is plotted as a dash-dot-line curve in Fig. A.3.

*Case (iii)*: $\zeta = 1.5$ (over-damped case)

$$b = \zeta b_c = 1.5 \times 34.6410E - 6 = 51.9615\,\text{E} - 6\,\text{N/(m/s)}$$

From Eq. (A.12) and the given initial conditions, we get

$$u(t) = Ae^{\left(-\zeta-\sqrt{\zeta^2-1}\right)\omega_n t} + B e^{\left(-\zeta+\sqrt{\zeta^2-1}\right)\omega_n t} = -0.1708E - 6\left(e^{-0.4534E6\ t}\right) + 1.1708E - 6\left(e^{-0.0662E6\ t}\right)$$

This is plotted as a dashed-line curve in Fig. A.3.

## ▶ A.3 DAMPED VIBRATIONS WITH A PERIODIC FORCE

For a periodic force, $F_0 \sin(\omega t)$, we have

$$m\ddot{u} + b\dot{u} + ku = F_0 \sin(\omega t) \tag{A.14}$$

whose solution is given by

$$u = U\sin(\omega t - \varphi) \tag{A.15a}$$

where

$$U = \frac{(F_0/k)}{\sqrt{\left(1 - (\omega^2/\omega_n^2)\right)^2 + (2\zeta(\omega/\omega_n))^2}} \tag{A.15b}$$

and

$$\tan\varphi = \frac{2\zeta(\omega/\omega_n)}{\left(1 - (\omega^2/\omega_n^2)\right)} \tag{A.15c}$$

It can be readily seen from Eqs. (A.15b-c) that when $\omega = \omega_n$, the amplitude $U$ equals $(F_0/2\zeta k)$ and the phase, $\varphi$, becomes 90°. This condition is referred to as *resonance*. If damping is very low, Eq. (A.15b) tells us that $U$ becomes very large. When there is no damping at all, it reaches infinity. That is, an elastic system vibrates with large amplitude when the frequency of the periodic force equals the natural frequency of the system. Figure A.4 shows the amplitude variation with applied frequency for the numerical data of Example A.1. Such a plot is called the *frequency response* of the system and it plays a very important role in characterizing the dynamics of a system. Figure A.4 shows $U$ in Eq. (A.15b) on a semi-log plot where the applied frequency (i.e., $\omega$) is plotted on a log scale. On the other hand, in Figure A.5, both axes are in the log scale.

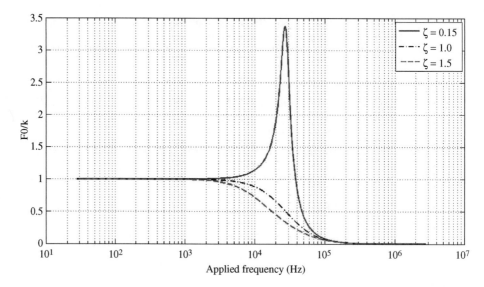

**Figure A.4** Frequency response on the semi-log scale for under, critical, and over damped cases.

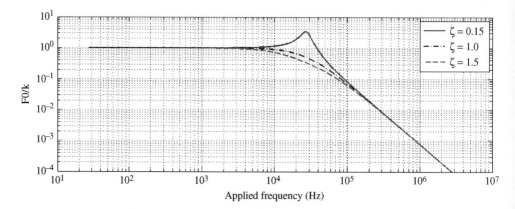

**Figure A.5** Frequency response on the semi-log scale for under, critical, and over damped cases.

It can be seen that the response, i.e., $(F_0/k)$, stays constant only for a certain band of frequency from zero to about one fifth to one third of $\omega_n$. This is called the *bandwidth* of the system wherein the output in the frequency response is independent of the applied frequency. Thus, the output of a sensor or an actuator is reliable only within the system's bandwidth. This is because, outside this range there will be ambiguity if the same $(F_0/k)$ arises due to different value s of $F_0$ and $\omega$. In other words, $\{F_{01}, \omega_1\}$ and $\{F_{02}, \omega_2\}$ may give the same $(F_0/k)$. To avoid such ambiguities, in practice, it is customary to take one third the natural frequency to be the bandwidth of the system.

One more feature worth noticing in Figures A.4-5 is how sharp the peak is at the resonance frequency. For many applications, we want this peak to very sharp so that the response is high only over a very narrow band of frequency around the resonance frequency. The sharpness of the resonance is quantified by what is known as the *quality factor, Q*. It is given by

$$Q = \frac{1}{2\zeta} \tag{A.16}$$

The range $2\zeta$ around the natural frequency has the property that $(F_0/k)$ is more than $(0.707/2\zeta)$. From Eqs. (A.6-7), we see that

$$Q = \frac{2\sqrt{km}}{b} \tag{A.17}$$

Therefore, high quality factor is obtained when damping is very low or mass and stiffness are very high.

# Glossary

**Amplifier bandwidth** The difference between the lower and upper half power points of an amplifier's transfer function. This is therefore also known as the −3 dB bandwidth.

**Amplifier efficiency** A measure of how much of the input power is usefully supplied to the amplifier's output. Class A amplifiers are very inefficient, in the range of 10–20% with maximum efficiency 25%. Class B amplifiers have a very high efficiency but are impractical because of high distortion levels. A tradeoff between these is the Class AB design. Modern Class AB amplifiers are commonly between 35% and 55% efficient with a theoretical maximum of 78.5%. Commercially available Class D switching amplifiers have reported efficiencies as high as 97%. The efficiency of the amplifier limits the amount of total power output that is usefully available.

**Anticlastic curvature** Creation of a saddle-like surface in a bent beam due to the Poisson effect.

**Bimorph effect** The deformation caused in a heated or cooled structure made of two materials with different thermal expansion coefficients.

**Buckling** An instability in a structural element; usually, compressive axial force in a slender structure results in transverse bending.

**Castigliano's theorems** Energy theorems that give force or deflection at a point upon differentiation of the strain energy or complementary strain energy; useful in lumped modeling of structures.

**Closed-loop control** This occurs when the output of the system is brought to the desired value by feeding the error signal to the controller under feedback control.

**Control variable** A quantity such as displacement, force, stress, or strain that is to be measured and/or controlled. These are necessarily the output variables. The control variable is normally controlled through an additional input to the system called the *manipulated variable*.

**Control** The sustained release of energy to limit or control the response of a desired (control) variable by inducing an additional input in the system in the form of a manipulated variable.

**Couette flow** One-dimensional flow between two plates.

**Coulomb's law** Fundamental law that gives the electrostatic force between two charges separated by a distance.

**Dopant diffusion** A process in which the dopant moves into a semiconductor substrate held at a high temperature.

**Dirichlet boundary conditions** Kinematic or essential conditions imposed on the boundaries of a structure; usually state variables.

**Disturbance** A signal that propagates through a system carrying a considerable amount of energy. In enforcing control of a system, one may require many disturbances, which can be internally generated (by using smart materials) or externally given as an input.

**Dopant** Impurity material introduced in semiconductor crystals to alter their properties. Doped regions are called n- or p-type depending on whether the impurity is donor or acceptor. Heavily doped n- and p- regions (concentration $> 10^{20}/cm^3$) are indicated by $n^+$ or $p^+$, respectively.

**Electric displacement vector** Electric field multiplied by permittivity of a material.

**Electrochromism** The ability of a material to change its optical properties (e.g. color) when a voltage is applied across it. Such materials are used as antistatic layers, electrochrome layers in liquid crystal displays (LCDs), and cathodes in lithium batteries.

**Electrostrictive materials** Materials that can change their dimensions significantly on application of an electric field. The effect is reciprocal as well. Although the changes thus obtained are not linear in either direction, these materials have found widespread application in medical and engineering fields.

**Epitaxial layer** A structured thin film layer that follows the crystal structure beneath (Greek *epi* = on, *taxis* = arrangement).

**Error signal** The difference between the output signal and the feedback signal. In many cases, the feedback signal may be a function of the output signal or its derivatives. In structural application, the output signal is normally displacement or strain and its derivatives, namely velocities and accelerations.

**Etching** A chemical process (wet or dry) of removing materials to define shapes or thinning.

**Euler–Bernoulli theorem** Bending moment in a beam is proportional to curvature.

**Feedback control** If due to a disturbance, the difference between the output of the system and some reference input is reduced, and if this reduction was obtained based on this difference, then we call the operation feedback control.

**Free-body diagram** A diagram showing all forces, both external and reaction forces, acting on a structure or a portion of a structure.

**Gain** In amplifiers the ratio of output to input power or amplitude. This is usually measured in decibels: $G[dB] = 10 \log (P_{out}/P_{in})$.

**Gauss quadrature** An approximate method for numerical integration over a domain; it is based on the rigorous theory of polynomial approximation.

**Gauss's law** A fundamental law in electrostatics that the charge enclosed in a volume is equal to the surface integral of the dot product of the electric displacement vector and the normal to the boundary surface; it can be derived from Coulomb's law.

**Gyroscope** An instrument that measures the angular rate of a body on which it is mounted.

**Implantation** Introduction of dopants by ionizing the source and transporting the ions to the substrate; these ions are neutralized by electrons entering from the back of the substrate.

**Isoparametric element** A finite element in which the coordinates and the variable are interpolated with functions of the same order.

**Lift-off** An indirect approach for pattern transfer in which the unwanted parts of the patterned film are lifted off by depositing the film above an easily removable film that is already patterned.

**Linear system** The principle of superposition holds for such a linear system: that is, the response to several inputs can be obtained by treating one input at a time and adding the result.

**Linearity of an amplifier** Although an ideal amplifier would be totally linear, real circuits are linear only within certain practical limits. When the signal drive to the amplifier is increased, the output also increases until some part of the amplifier becomes saturated, resulting in distortion. Several linearization techniques can be used to avoid the undesired effects of these nonlinearities.

**Lithography** A process for transferring patterns onto a substrate or thin film for patterning.

**Lumped modeling** Simplified but energetically accurate modeling of a system in terms of a few variables.

**Magnetostrictive materials** Quite similar to electrostrictive materials, except for the fact that they respond to magnetic fields. The most widely used magnetostrictive material is TERFENOL-D, made from rarest of the rare-earth materials called teribium. This material is highly nonlinear and can produce large strains that in turn can produce large block forces. These materials are also used in similar applications.

**Neumann boundary conditions** Conditions imposed on the derivatives of the state variables on the boundary force or moment boundary conditions imposed on a structure; also called natural boundary conditions.

**Neutral axis** The axis along the length of a slender beam where the normal stress due to bending is zero; it coincides with the centroidal axis for a straight beam.

**Noise** An undesirable but inevitable product of electronics devices and components; measured as the peak output voltage produced by the amplifier when no signal is applied.

**Ohmic contact** An interface between dissimilar materials that allows current flow in either direction with negligible voltage drop across the interface. A typical rectifier is a nonohmic contact, allowing current flow only in one direction.

**Open-loop control** A system in which the outputs play no role in the control action of inducing the response. The controller may provide the actuating signal required for a specific task based on a set of previously known parameters. In other words, the error signal is not needed in the control action. Fixed operating conditions exist for all input, so that, the accuracy depends on the calibration and cannot be assured.

**Output dynamic range**   The range between the smallest and the largest useful output levels. The lowest useful level is limited by output noise, while the largest is limited by distortion. The ratio of these two is the dynamic range of the amplifier.

**Overshoot**   The amount by which the output exceeds its final, steady-state value in response to a step input.

**Passivation**   Treatment of the surface to make it less reactive.

**Permittivity**   A material property that indicates the ability to permit and retain electric charge.

**Piezoelectric materials**   Ceramics or polymers that can produce a linear change of shape in response to an applied electric field: application of the field causes the material to expand or contract almost instantly. These materials have found several uses in actuators in diverse fields of science and technology. The converse effect has also been observed and has led to their use as sensors.

**Plant**   The physical object that requires control, such as a mechanical device (e.g. a cantilever beam),or a helicopter, aircraft, spacecraft, etc.

**Principle of virtual work**   External virtual work is equal to internal virtual work.

**Proof mass**   The inertial mass in an accelerometer; also called seismic mass.

**Pull-in phenomenon**   A snap-through-like instability occurring in an electrostatically actuated structure at a critical voltage called the pull-in voltage.

**Residual stress**   Unrelieved stress retained in a structure after fabrication.

**Rheological materials**   Materials in the liquid phase that can change state instantly on application of an electric or magnetic charge. These fluids find applications in breaks, shock absorbers, and dampers for vehicle seats.

**Sensor**   A device that responds to a physical stimulus (heat, light, sound, pressure, magnetism, or a particular motion) and transmits a resulting impulse (as for measurement or operating a control).

**Shape function**   An interpolating function for the variables in finite element analysis.

**Shape-memory alloy**   Alloy materials such as Au–Cd and Ni–Ti that retain their original shape through a specified temperature range.

**Slew rate**   The maximum rate of change of an output variable, usually quoted in volts per microsecond. Many amplifier circuits have limited slew rate because the impedance of a drive current must overcome capacitive effects in the circuit; this may limit the full power bandwidth to frequencies well below the amplifier's frequency response when dealing with small signals.

**Squeezed film effects**   The spring and damping effects due to squeezing a thin film of a fluid between two structural elements.

**Transducer**   A device actuated by power from one system that supplies power to a second system, usually in another form.

# Index

# About the Authors

**Prof. G.K. Ananthasuresh** (B. Tech. IIT-Madras, 1989; PhD, Michigan, 1994) is Professor of Mechanical Engineering at the Indian Institute of Science, Bangalore, India. His previous positions include Associate Professor at the University of Pennyvania and visiting professorships at Cambridge University, UK, and Katholike Universiteit, Leuven, Belgium, and post-doctoral fellowship at Massachusetts Institute of Technology, Cambrudge, USA. His current research interests include compliant mechanisms, kinematics, multidisciplinary design optimization, microsystems technology, micro- and meso-scale manufacturing, protein design, and cellular biomechanics. He serves on the editorial boards of six journals and is co-author of more than 180 journal and conference papers as well as two edited books and 11 book chapters. He is a recipient of the NSF Career Award in the U.S. and the Swarnajayanthi Fellowship and Shanti Swarup Bhatnagar Award in India as well as six best-paper awards at international and national conferences. He is a Fellow of the Indian National Academy of Engineering. He has been working in the area of microsystems since 1991 with emphasis on modeling and design.

**Prof. K. J. Vinoy** is Associate Professor in the Department of Electrical Communication Engineering at the Indian Institute of Science (IISc) Bangalore. He received his bachelor's degree (1990) from the University of Kerala, India, his master's (1993) from Cochin University of Science and Technology, India, and his PhD (2002) from Pennsylvania State University, USA. From 1994 to 1998 he worked at National Aerospace Laboratories, Bangalore, India, and he did post-doctoral research at Pennsylvania State University's Center for the Engineering of Electronic and Acoustic Materials and Devices (CEEAMD) from 2002 to 2003. His research interests include RF MEMS, smart materials, and several aspects of microwave engineering. He has published over 100 papers in technical journals and conferences. He is on the editorial board of the *Journal of the Indian Institute of Science* and is a referee for several international and national journals. He is on the Technical Advisory Board of the International Symposium on Antennas and Propagation and is a coordinator of the Indian Nanoelectronics Users Program at IISc. He is a Fellow of the Indian National Academy of Engineering. He is currently Chair of the Bangalore Chapter of IEEE MTT/AP Societies. This is his third book in the area of microsystems.

**Prof. S. Gopalakrishnan** received his master's degree in engineering mechanics from the Indian Institute of Technology, Madras at Chennai and his PhD from Purdue University's School of Aeronautics and Astronautics. Subsequently he was a post-doctoral fellow in the School of Mechanical Engineering at Georgia Tech. He is currently Professor in the Department of Aerospace Engineering at Indian Institute of Science, Bangalore, India and a Resource Executive in the Society for Innovation and Development at the Indian Institute of Science. He works in the areas of wave propagation, structural health monitoring, smart structures, and modeling of microsystems and nanosystems. He has contributed significantly to the understanding of wave propagation in anisotropic and inhomogeneous media and has been key in popularizing the spectral finite-element method as an analysis tool. He is a Fellow of the Indian National Academy of Engineering and received the Satish Dhawan Young Scientist Award from the Government of Karnataka, India. He is Associate

Editor of the journal *Smart Materials and Structures and Structural Health Monitoring* and is on the editorial board of five other journals. He is Chair of the Aerospace Applications and Structural Health Monitoring group of the National Programme on Micro and Smart Systems, Government of India. He has written three books, eight book chapters, 113 refereed papers in international journals, and 80 conference publications.

**Prof. K. N. Bhat** received his PhD degree from IIT Madras in 1978 and was Professor in the EE department there. He spent two years at Rensselaer Polytechnic Institute, Troy, NY as a post-doctoral research associate, and while on sabbatical leave from IIT Madras served as a Visiting Professor at the University of Washington, Seattle. After retirement from IITM in 2006, he has been a Visiting Professor in the ECE Department at the IISc Bangalore. He has authored one book and over 200 technical papers in journals and conferences, has guided several MS and PhD scholars, and has contributed immensely to the growth of VLSI and MEMS technology, education, and manpower development in India. Prof. Bhat is a Fellow of the Indian National Academy of Engineering.

**Prof. V. K. Aatre** received his BE (Mysore, 1961), ME (IISc, 1963) and PhD (Waterloo, 1967), all in electrical engineering. He was Professor of Electrical Engineering (1968–1980) at the Technical University of Nova Scotia, Halifax, Canada. He worked in India's Ministry of Defence in the Department of Defence Research and Development as a Principal Scientific Officer (1980–1984), Director of Naval Physical Oceanographic Laboratory (1984–1991), Chief Controller (1991–1999), and Director General of DRDO and Scientific Advisor Defence Minister (1999–2004). During this period he designed and developed sonar suites for surface ships, submarines and the air arm of the Indian Navy. He was also instrumental in the development of integrated electronic warfare systems for the Indian Army, Navy, and Air Force, and established GaAs MMIC fabrication and VLSI design centers for Ministry of Defence. Dr. Aatre is the founding president of the Institute of Smart Structures and Systems and runs national programs on smart materials and MEMS. He has published over 60 papers in the fields of active filters, digital signal processing, and defense electronics, and his book *Network Theory and Filter Design* was published by John Wiley & Sons. He is a Fellow of the IEEE (USA) and the National Academy of Engineering (India), a Distinguished Fellow of IETE (India) and several other societies. Dr. Aatre is the recipient of the prestigious Padma Bhushan Award of the Government of India.